CW00547255

Ministero BB.CC.AA.
Soprintendenza per i Beni Ambientali e
Architettonici di Napoli e Provincia
Palazzo Reale - Napoli

Centro Materiali Compositi
AMME-ASMECCANICA

Università degli Studi di Napoli
"Federico II"

The
Institute of
Materials

EUROPEAN ASSOCIATION
FOR COMPOSITE MATERIALS

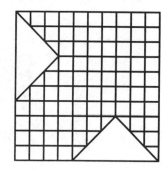

ECCM - 8

EUROPEAN CONFERENCE
ON
COMPOSITE MATERIALS

SCIENCE, TECHNOLOGIES and APPLICATIONS

3 - 6 JUNE 1998   NAPLES - ITALY

VOLUME 4

EDITOR: I. CRIVELLI VISCONTI

WOODHEAD PUBLISHING LIMITED

Published by Woodhead Publishing Limited,
Abington Hall, Abington, Cambridge CB1 6AH, England

First published 1998, Woodhead Publishing Limited

© 1998, Woodhead Publishing Ltd

Conditions of sale
All rights reserved. No part of this publication may be reproduced or transmitted in any form or by any means, electronic or mechanical, including photocopy, recording, or any information storage and retrieval system, without permission in writing from the publisher.

While a great deal of carc has been taken to provide accurate and current information, neither the authors nor the publisher, nor anyone else associated with this publication, shall be liable for any loss, damage or liability directly or indirectly caused or alleged to be caused by this book.

British Library Cataloguing in Publication Data
A catalogue record for this book is available from the British Library.

ISBN 1 85573 410 9 (Vol 4)
ISBN 1 85573 377 3 (Four volume set)

ECCM-8
EUROPEAN CONFERENCE ON COMPOSITE MATERIALS
SCIENCE, TECHNOLOGIES AND APPLICATIONS

EXECUTIVE COMMITTEE

I. Crivelli Visconti	ITALY (Chairman)
M.G. Bader	GREAT BRITAIN
H. Benedic	FRANCE
A.R. Bunsell	FRANCE
G. Caprino	ITALY
K. Friedrich	GERMANY
J. Hawkins	GREAT BRITAIN
A. Kelly	GREAT BRITAIN
A. Langella	ITALY
H. Lilholt	DENMARK
L. Longoni	ITALY
A. Massiah	FRANCE
R. Naslain	FRANCE
W. Nicodemi	ITALY
K. Schulte	GERMANY
R. Teti	ITALY
I. Verpoest	BELGIUM

ECCM-8
EUROPEAN CONFERENCE ON COMPOSITE MATERIALS
SCIENCE, TECHNOLOGIES AND APPLICATIONS

CONTENTS

SYMPOSIUM 6

SYMPOSIUM 6

Progress in the Development of SiC/SiC Composites for Advanced Energy Systems: *CREST-ACE* Program

A. Kohyama*, Y. Katoh*, T. Hinoki*, W. Zhang* and M. Kotani

*Institute of Advanced Energy, Kyoto University
Gokasho, Uji, Kyoto 611, Japan
&
Crest, Japan Science and Technology Corporation
4-1-8 Honmachi, Kawaguchi, Saitama 332 Japan

Abstract

Under the title of "R & D of Environment Conscious Multi-Functional Structural Materials for Advanced Energy Systems", a new R & D activity to establish high efficiency and environmental conscious energy conversion systems, as one of the programs of Core Research for Evolutional Science and Technology (CREST), has been initiated from October 1997 to September 2002.

This program cares for R & D of high performance materials and materials systems for severe environments and production of model components for energy conversion systems is carried out. The emphasis is on R & D of SiC/SiC, W/W with their system studies to establish sound material life cycles. The program outline and preliminary results on SiC/SiC are provided.

1. Introduction

It has been well recognized that we have to achieve balance among our increasing need for energy at reasonable price, our commitment to a safer, healthier environment and to reduce dependence on potentially unreliable energy suppliers[1,2]. It is also important to have better flexibility and efficiency in the way energy is transformed and used.

As a key technology to establish high efficiency and environmental conscious (low impact systems on environment) energy conversion systems, multi-functional (structural) materials R & D is emphasized in this program. The nickname of this program; CREST-ACE stands for CREST- Advanced Material Systems for Conversion of Energy. This program cares for R & D of high performance materials and materials systems for severe environments. For this purpose, starting from materials design, process developments, applications of those materials to advanced energy systems towards the end of their material cycles are systematically carried out. The final goal is to produce model components for high efficiency and environmental conscious energy conversion systems.

As important energy options for the future, nuclear fission energy and nuclear fusion energy cannot be ignored. In these materials systems, nuclear reactions and transformations by high-energy beams and particles such as neutrons and γ rays have strong impacts on environment through the production of radioactive elements and emissions of electromagnetic waves. Therefore, low activation materials R & D have been major efforts in fusion and fission energy research.[3-6].

To meet the program goal, high temperature ceramics composites, such as silicon carbide (SiC) fiber reinforced SiC matrix composite materials (SiC/SiC) and high temperature metal composites, such as tungsten (W) alloy fiber reinforced W composite materials (W/W) have been selected as the base material systems.

This objective of this paper is to provide outline of the program and some recent results on SiC/SiC.

2. Material Systems
2.1. SiC/SiC Composite Materials

There is a strong demand to make high performance ceramic matrix composites (CMC) for advanced energy systems, such as nuclear fusion reactors, advanced gas turbine engines, SiC/SiC composites are considered to be the most potential candidates for them because of its advantages;

(1) high specific strength, (2) high temperature strength, (3) fracture toughness compared with ceramic materials, (4) insulating material (prevent energy loss by conduction), (5) controllable to improve conductivity, (6) low induced radioactivity under nuclear environments, etc.

This is beneficial to achieve high plant heat efficiency with higher reliability on safety.

R & D of SiC/SiC under CREST-ACE program can be divided into three tasks; (1) Process development of material production into composite material, (2) Evaluation and prediction of materials performance and (3) Design and fabrication of multi-functional components for energy conversion systems. The first task consists of three sub-tasks; (1) improvement and innovation of SiC fibers, (2) process development of composite material production including matrix materials R & D, (3) design and control of interfacial microstructure to optimize material performance. The second task is on (1) mechanical properties, (2) thermal and electrical properties, (3) establishment of evaluation test methodology for SiC/SiC composite materials and fibers. In this task, studies on irradiation effects and on severe environmental effects are emphasized. For the third task, elements of energy conversion components for fusion reactor and for high temperature gas reactor will be designed and be fabricated as the goal of this program which should verify the specifications of the elements. Low activation characteristics are the most important technological challenges and selection of low activation elements and elimination of harmful elements (to produce high purity SiC or SiC(x), where X is element(s) to improve thermal and electrical properties) are to be extensively carried out. In this task, Chemical Vapor Infiltration (CVI) method, Polymer Impregnation and Pyrolysis (PIP) method, and Reaction Sintering (RS) Process have been studied. SiC fiber R & D are improvement of Poly-Carbo-Silane (PCS) type, innovation of PCS+(M) type, new polymer type and their combinations. Interfacial microstructure control is also studied by CVI, PIP and RS methods.

2.2. W/W Composite Materials

potential materials, in many cases refractory metals represent the best choice despite their high density compared with ceramics. The refractory metal alloys are based on V, Nb, Mo, Ta and W, where the V-alloys may have too low a low temperature capability for the use. Although materials considered for use in space reactors range from Nb alloys to W alloys [7,8], these alloys have an issue of low fracture toughness and a concern about embrittlement under irradiation environment and some other severe environments. In general, metal fiber reinforced metal matrix composite materials (MMC or FRM) are designed to use high strength but not highly ductile fibers as reinforcement and ductile matrix as stress transport and catastrophic fracture inhibitor, such as W/Cu, Ti-Al/Ti. Whereas, W/W composites are utilizing high strength and improved ductility by fiber and low strength W with higher fracture toughness provides optimized mechanical properties. The utilization of ultra-high strength W alloy fibers is also beneficial to eliminate difficulty in large monolithic block fabrication. There have been some R & D efforts in space reactor application nearly twenty years ago but the activities were interrupted in mid 80's. This task is to renew the R & D activity of W-alloy/W or W-alloy/Nb-alloy composite materials for high efficiency and environmental conscious conversion systems. The task includes (1) improvement of ultra-high strength W fibers, (2) development of composite material fabrication process by CVI, (3) Evaluation and prediction of materials performance and (4) Design and fabrication of multi-functional components for energy conversion systems. In this task, low activation characteristic and high resistance to high heat flux exposure are important requirements for the applications to nuclear fission and fusion reactors.

2.3. Component for Energy Conversion

Design and fabrication of multi-functional components are included as a part of composite materials R & D, but in this task, based on the currently existing conceptual design studies, such as DREAM, SSTR, CREAT, more realistic component design will be done and the demonstration components will be fabricated and verification of their performance to meet the design values will be done. Figure 1 is an preliminary conceptual plan of the demonstration component, where energy transportation medium contacting with Cu is supposed to be water and other media are He gas and liquid metals or molten salts.

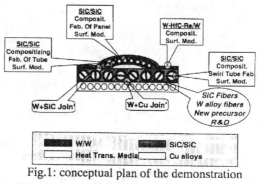

Fig.1: conceptual plan of the demonstration component

2.4. Technological Issues to Establish Reliable Material Systems

The important object of
this program is a
contribution to make a
social system, which has
low impact on
environment and has a
sufficient public
acceptance. For this
purpose, this program
tries to cover major
issues in materials life
cycle as shown in Fig.2.
As can be seen in the
figure, reduction of
radioactive wastes and
establishment of waste
management scheme is

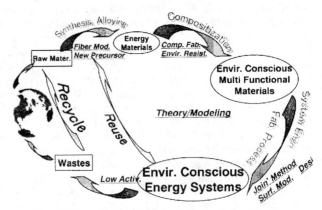

Fig. 2: Materials Life Cycle and Technological Issues

important and recycling and reuse of composite materials are big challenges.
Technological issues are also indicated in Fig.1 where the importance of joint
technology and surface coating or modification technologies is emphasized.

3. Organization

The CREST-ACE
program activity is
supported by JSTC
and is operated as a
JSTC activity with
the participation of
scientists/engineers
and professors from
university, institutes
and industries to
make a CREST-
ACE team. The
major participants
are shown in Fig.3
together with main
facilities used in this

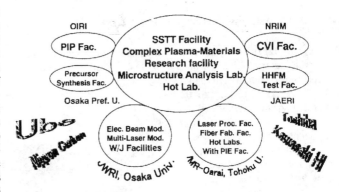

Fig. 3: Research Structure and Main facilities

program. Institute of Advanced Energy (IAE), Kyoto University is the central
organization of this program and Institute for Materials Research, Oarai Branch (IMR-
Oarai), Tohoku University, Joining and Welding Research Institute (JWRI), Osaka
University, Osaka Prefecture University are the participants from universities. National
Research Institute of Metals (NRIM), Japan Atomic Energy Research Institute (JAERI)
and Osaka Industrial Research Institute (OIRI) are from national institutes and Ube,
Nippon Carbon, Toshiba and Kawasaki Heavy Industries are participating from
industries.

4. Preliminary Results on SiC/SiC R & D
4.1. PIP Process R & D by PVS and Other New Plymers[9]

As one of the most promising fabrication processes, there have been many efforts on PIP process R & D where improvements of performance and reliability of SiC/SiC composites for structural application are stressed.

In this study, to reduce porosity in composites and to control microstructures of matrix and matrix/fiber interface are emphasized. The objective is to develop a fabrication scheme and process of SiC/SiC composites by PIP process with an improved matrix integrity under variations of polymer-precursors and their blends. As is well known, the microstructure of ceramics derived from pre-ceramic polymer is very complicated and influenced by the polymer and fabrication process, in general. Thus, poly-vinyl-silane (PVS) is selected as an important candidate for the PIP process development.

PVS is a pre-ceramic polymer with low viscosity at an ambient temperature. The polymer-to-ceramic conversion chemistry of PVS was studied by means of thermo-gravimetric analysis (TGA), differential thermal analysis (DTA), infrared spectroscopy, gas chromatography and X-ray powder diffraction.

Figure 4 is the preliminary summary of the pyrolysis process of PVS.

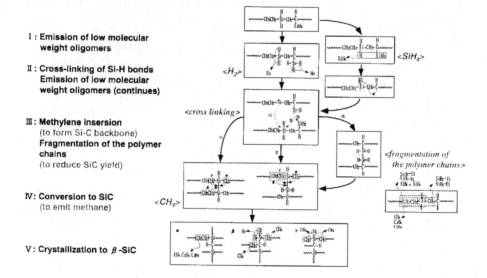

I : Emission of low molecular
weight oligomers

II : Cross-linking of Si-H bonds
Emission of low molecular
weight oligomers (continues)

III : Methylene insersion
(to form Si-C backbone)
Fragmentation of the polymer
chains
(to reduce SiC yield)

IV : Conversion to SiC
(to emit methane)

V : Crystallization to β-SiC

Figure 4: preliminary summary of the pyrolysis process of PVS

The fabrication procedure of composites is as follows:

(1) infiltration PVS into a 2D woven sheet of Hi-Nicalon at an ambient temperature in vacuum,

(2) thermosetting of PVS (e.g., 1K/minute up to 573K and then holding for 10 minutes) , and

(3) hot-press for densification and curing.

The composites fabricated were then subjected to densitometry for fundamental study of process parameter optimization.

Ceramic yield from PVS depended on the ceramization process. Decomposition was a five staged reaction; as follow: (I) loss of low molecular weight oligomers (II) cross-linking (III) chain scissions (IV) growth of Si-C backbone (V) crystallization. Based on this analysis, process optimization methodology is to be established.

4.2. Effect of Fiber Coating on properties of SiC/SiC Composites [10-12]

In order to evaluate mechanical properties of fibers quantitatively, matrices and their interfaces in fiber reinforced SiC/SiC composites, nano-indentation tests have been carried out. Using the same technique, fiber push-out test was performed. Also, in order to see relationship between interfacial coating thickness of carbon and bending strength, three point bending tests were carried out with various carbon coating thickness SiC/SiC composites. Due to insufficient uniformity in carbon coating thickness, the results on bending strength have a large scatter. Fig. 5 presents data obtained by bending tests. From the figure, carbon coating on the fibers improved the fracture toughness of SiC/SiC under bending test. In this result, SiC/SiC composites with around 1 μm carbon coating thickness had a peak in bending strength. This tendency was similar with that of interfacial shear strength, although carbon-coating thickness at the peak seemed different. These results suggest that it is possible to control mechanical property by controlling interfacial shear strength. This result is consistent with the result by Snead et al. [6], that adequate carbon coating improves mechanical performance of SiC/SiC composites. Further work on improved materials will bring us a clearer insight about effects of interface on performance of SiC/SiC composites. The specimens were analyzed by means of scanning electron microscopy (SEM), before and after

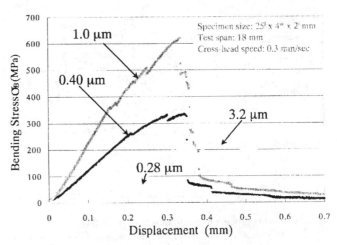

Fig.5: Effect of Carbon Coating Thickness on Three Point Bending Strength

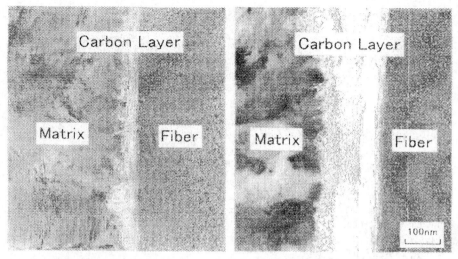

Undebonded Interface   Debonded Interface
(prior to Test)   (after Test)

Figure 6: Debonded Interface After Bending Test

indentation tests. From the indentation load vs. displacement relations, the fiber pushed out process has been discussed and the initiation loads for interfacial debonding and those for interfacial sliding were defined and were discussed in comparison with the C/C composites. The load at push-in was likely to increase in proportion to the fiber circumference. The initiation load at push-out increased with increment of contact area between fiber and matrix. From this relation interfacial shear strength was defined. The effects of fiber coatings on interfacial shear strength were discussed with the results obtained from TEM observation. Figure 6 is an example and shows the microstructures of the fiber-matrix interfaces of the specimen with 50 nm-thick fiber coating prior to and after a push-out test. Site at which the debonding occurred during the push-out test was identified as the carbon coating adjacent to the fiber. Preliminary investigation suggests that carbon coating thickness effects microstructure of carbon coating adjacent to fibers.

5. Conclusion

This paper introduces the new R & D activity, "R & D of Environment Conscious Multi-Functional Structural Materials for Advanced Energy Systems". The present status of the program is provided. The kick-off meeting of this CREST-ACE program was held on December 1997 and every research activities have been started towards each task goals.

Preliminary results on SiC/SiC by CVI method and PIP method are promising and those efforts are currently most extensive and concentrated.

Acknowledgement

The author expresses his appreciation to the members of CREST-ACE Program for their efforts to run this program and their excellent accomplishments. The sincere appreciation is due Japan Science and Technology Corporation (JSTC) for supporting this activity.

References

[1] National Energy Strategy, US DOE, DOE/S-0082P, 1991
[2] Y. Kaya, Proceedings of 10th Pacific Basin Nuclear Conference, p15, 1996.
[3] A. Kohyama, H.Tezuka, N.Igata, Y.Imai, J. Nucl. Mater., 141-143, p.513, 1986.
[4] A. Kohyama,, H. Matsui and A. Hishinuma, Proceedings of 10th Pacific Basin Nuclear Conference, p883, 1996.
[5] A. Kohyama, A. Hishinuma, D.S.Gelles, R.L.Klueh, W.Dietz and K.Ehrlich, J. Nucl. Mater., 233-237, p.138, 1996.
[6] L.L.Snead, R.H. Jones, P.Fenici and A. Kohyama, J. Nucl. Mater., 233-237, p.26, 1996
[7] R.H.Titran, J. R. Stephens and D. W. Petrasek, in "Refractory Metals: State of the Art", ed. P. Kumar and R. L. Ammon, TMS, p1, 1989
[8] F. W. Wiffen, CONF-8308130, US-DOE, p.252,1983
[9] M.Kotani , A.Kohyama , K.Okamura and T.Inoue, Proc. Of 2nd IEA International Symposium on SiC/SiC, ed. A.Kohyama, R.H.Jones and P.Fenici, 1998
[10] A.Kohyama, H.Hinoki, H.Serizawa and S.Sato, Proc.11th International Conference on Composite Materials, 1997
[11] T.Hinoki, A.Kohyama, S.Sato and K.Noda, Proc. ICFRM-8(1997), in print for J. Nucl. Mater.
[12] W. Zhang, T. Hinoki and A. kohyama, Proc. ICFRM-8(1997), in print for J. Nucl. Mater.

C/C-SiC COMPONENTS
FOR HIGH PERFORMANCE APPLICATIONS

W. Krenkel, R. Renz
DLR (German Aerospace Center)
Institute of Structures and Design, Pfaffenwaldring 38-40, 70569 Stuttgart, Germany

ABSTRACT

To date, the use of ceramic matrix composites (CMC) for high temperature components is mainly limited to aerospace applications. Due to the wide variability in material properties and the low cost manufacturing route of the in-house developed Liquid Silicon Infiltration (LSI) process, other markets are now accessible. Carbon fibre reinforced carbon silicon carbide materials (C/C-SiC) offer a high thermal and corrosive resistance and can be fabricated in a near-net shape and extremely lightweight design. This paper describes the actual status of the material's development and shows some new applications where C/C-SiC materials are attractive alternatives to conventional materials.

KEYWORDS: Ceramic Matrix Composites, C/C-SiC, Liquid Silicon Infiltration, Lightweight Structures

1. INTRODUCTION

Originally, the DLR developed C/C-SiC materials for structural aerospace applications. Heat shields for re-entry capsules and intake ramps for hypersonic aircraft have been realized successfully as demonstrators [1]. By varying the constituents e.g. matrix precursor, fibre reinforcement and the processing parameters, different qualities of C/C-SiC materials can be achieved [2]. They are all characterized by a low density and low amounts of porosity, low thermal expansion, high thermomechanical properties and an extreme thermoshock resistance. Typically, the use of coatings on the carbon fibres has been avoided to cut down the costs of the process. Also, all infiltration steps of matrix materials are carried out only once to reduce the manufacturing time.

Within the first manufacturing step of the LSI process a green body is formed by common CFRP techniques like RTM or filament winding. Commercially available carbon fibre preforms (e.g. woven fabrics) with standard sizings and one-part thermoset systems as precursors can be used to fabricate laminates with mainly 2D-reinforcements. To avoid costly machining steps in the subsequent ceramic phase, these green bodies can be produced in near net and simple shapes like rings, disks, flat or thick panels or tubes (Figure 1).

After curing, the CFRP green bodies are pyrolized under nitrogen atmosphere. During pyrolysis the matrix adopts a crack pattern as a result of the precursor's shrinkage and induced forces by the stiffer fibres. After this thermal treatment at 900°C, the composites show a system of translaminar channels resulting in an open porosity of 15-30%. The shrinkage of the C/C preform depends on the type of precursor, the fibre/matrix-bonding forces and the fibre content, and usually lies in the range of 0-10%, depending on the fibre orientation.

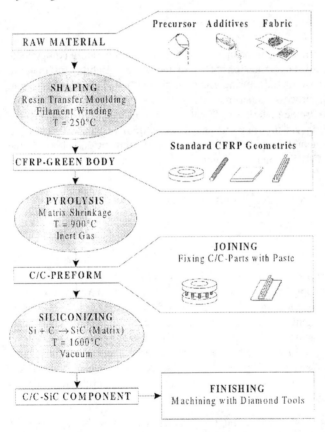

Figure 1: LSI process for the manufacturing of modular C/C-SiC components

In the third step, the silicon carbide matrix is formed by a controlled diffusion process under vacuum. Liquid silicon is infiltrated into the porous C/C preform at approximately 1600°C. Due to the high purity and the low viscosity, the molten silicon is driven mainly by capillary forces into the porous microstructure. The infiltration is accompanied by a simultaneous chemical reaction between silicon and carbon to form silicon carbide, whereby the silicon preferably reacts with the amorphous structure of the carbon matrix. To produce complex structures, an in-situ joining method has been developed [3] which allows the manufacture in a modular design. Different C/C parts are preformed and fixed together with a paste consisting of a special binding resin and a solid filler. The subsequent siliconizing step forms a mechanically as well as thermally stable joint between the siliconized C/C-SiC parts.

C/C-SiC materials can be fabricated with quite different properties depending on their microstructure. Table 1 summarizes the typical mechanical properties of so-called XB and XT qualities which have been used as structural materials for the first prototype components.

Quality		XB	XT
Flexural Strength	[MPa]	160	300
Tensile Strength	[MPa]	80	190
Strain	[%]	0,15	0,35
Young's Modulus	[GPa]	60	60
Density	[g/cm³]	1,9	1,9
Open Porosity	[%]	3,5	3,5

Table1: Room temperature properties of different C/C-SiC materials

2. BRAKE DISKS AND CLUTCHES

The feasibility of C/C-SiC materials for high performance brakes and clutches is the objective of different tribological projects. Friction between rotating disks and stationary pads causes them to heat up to over 1000°C, so adequate thermal shock resistance and a stable high coefficient of friction with low wear rates are required. Different test results show a high ablative and oxidative stability of C/C-SiC under friction and wear conditions. As known from other braking materials, the coefficient of friction increases with decreasing velocity, reaching the highest frictional values just before braking ends.

Due to the anisotropic thermal behaviour of the standard C/C-SiC, the friction surface can be overheated and the coefficient of friction decreases to an unacceptable level of about 0.2. Nevertheless, C/C-SiC can be adapted for high performance brakes, which show much higher and more stable friction levels. High wear resistant coatings and variations in fibre orientation and conductivity increase and stabilize the COF considerably (Figure 2).

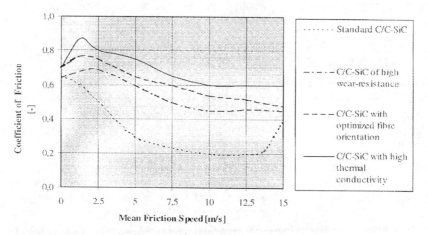

Figure 2: Friction behaviour of different C/C-SiC materials used in a multiple disk brake

As the density of C/C-SiC is in the range of 2 g/cm³, vehicles like automobiles, trucks and high speed trains achieve essential weight savings in the unsprung mass, when eqiupped with these materials. The DLR is also investigating new braking designs for these vehicles and fabricates internally ventilated as well as massive brake disks (Figure 3).

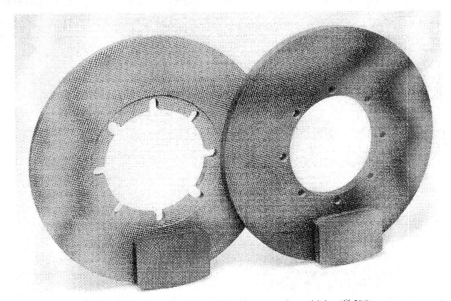

Figure 3: C/C-SiC brake disks and pads for automotive vehicles (Ø 280 mm)

Similar tribological requirements exist for the linings of clutches. The extreme thermal stability of C/C-SiC makes them an attractive candidate to substitute organic or sintered materials in special cases.

3. BEARINGS FOR CHEMICAL PUMPS

Monolithic ceramics bearings (e.g. SSiC) are commonly used in water or chemical pumps but their low thermal shock resistance and low damage tolerance lead to a non-acceptable fracture behaviour. Fibre reinforced ceramics with their higher fracture toughness, their low thermal expansion and their applicability in large dimension structures may overcome these restrictions [4].

Corrosion experiments have been conducted with C/C-SiC samples exposed for 360 hours in sea water at 40°C (pH=7,6) and sodium hydroxide at 180°C (pH=11). In sodium hydroxide an average mass loss of about 1% could be observed, whereas in sea water no corrosion occurred. Accompanying strength tests showed that the materials are only slightly influenced by the corrosive agents and the resulting flexural strength values are within the normal scatter of the material's properties (Figure 4).

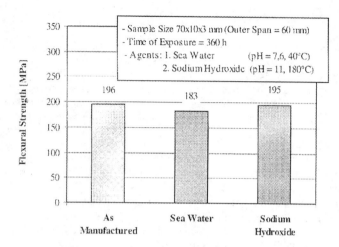

- Sample Size 70x10x3 mm (Outer Span = 60 mm)
- Time of Exposure = 360 h
- Agents: 1. Sea Water   (pH = 7,6, 40°C)
             2. Sodium Hydroxide (pH = 11, 180°C)

Figure 4: Flexural strength values of C/C-SiC after exposure in corrosive agents

First tribological tests of standard C/C-SiC bearings in water-lubricated sliding contact mated with a monolithic SSiC bearing (both surfaces were lapped) showed at a sliding speed of v = 0,01 m/s a COF of 0.16. Increasing the velocity to v = 15 m/s the COF decreased to a low friction level of about 0.06. Depending on the constituents of C/C-SiC and the microstructure a lubricating film separates the friction surfaces in case of a high velocity resulting in a low COF. At high friction levels the surface temperature increases and the lubricant partially evaporates. As a consequence, the amount of solid state friction increases and undesirable wear occurs.

To obtain a lower COF at high speed, bearings with a fibre orientation perpendicular to the surface were fabricated (Figure 5) which reduces the surface temperature because of the higher thermal conductivity in the fibre direction. Due to the inherent low porosity

of the standard C/C-SiC (normally lower than 5%), the absorption of fluids into the microstructure is limited. Therefore, other improvements covered the tailoring of the microstructure in order to increase the open porosity to 10% with adapted pore distribution on the friction surface.

Figure 5: Bearings of C/C-SiC with fibre orientation parallel (left) and perpendicular (right) to the surface

4. GEAR WHEELS

The demand for lighter and smaller structural components in power transmissions require lighter gear wheels. Carbon fibre reinforced plastics (CFRP) have recently become the subject of attention in the automobile and aerospace industries [5]. The lower wear rates of ceramic matrix composites promises a great potential for wheels in gear boxes with non-oil lubricant. Additionally, the low density and the high thermal stability of C/C-SiC are unique properties for future lightweight gear wheels, for example in helicopters.

Generally, the substitution of ductile metals by fibre reinforced composites in gear wheels requires a new design philosophy. Referring to the active loading, the teeth are subjected to compressive, bending and shear forces during operation. Due to high material strength values in the fibre direction, every tooth was designed with a fibre orientation parallel to the active stresses. Figure 6 shows a prototype of a gear wheel, where all eight teeth are machined from a flat panel with a 2-D reinforcement. To transmit the torque, two base disks and eight bolts are used to assemble the final component. All parts are joined together with an special paste which leads, after siliconizing, to an homogeneous and thermally stable component. The final contour of the involute gear wheel was machined by erosion technique.

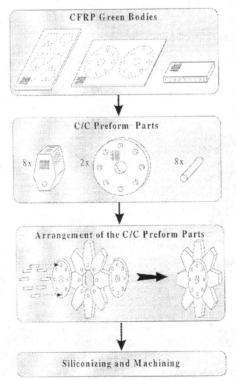

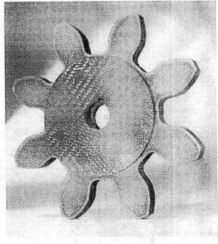

Figure 6: Right: Prototype of a gear wheel made of C/C-SiC, each tooth with the same fibre orientation (crown line diameter: 160 mm); Left: Modular design of the gear wheel

5. COMPONENTS FOR TURBO ENGINES

Due to the low density and the resulting low moment of inertia, C/C-SiC composites are also promising materials for components in turbo engines. Generally, C/C-SiC can be used at high temperatures (above 500°C) and in oxidative atmosphere only for short life times. For long term applications in gaseous and liquid mediums, protective coatings are necessary to reduce oxidation as well as erosion. Suitable ceramic coatings have already

been developed, for example CVD-SiC, which increase the stability of CMC considerably.

As an example, a first design study of a radial pump wheel has been realized as shown in Figure 7. The design consists of circular plates with inserted blades, which allows the adaptation of each blade to the mechanical and fluid mechanical requirements. Further tests have to be done to prove the concept.

Figure 7: Radial pump wheel made of C/C-SiC

6. CONCLUSIONS

C/C-SiC materials have been developed using the Liquid Silicon Infiltration process. Taking advantage of the simplicity of the LSI process, the short manufacturing times and the comparatively low raw material costs, C/C-SiC composites offer a great potential for thermally stable, highly integrated structures in addition to the traditional space applications where economic aspects are more restrictive. Complex and highly integrated structures have been realized in a modular design by joining the parts via an in-situ joining step during siliconizing. First successful tests show, that C/C-SiC materials are a new material class which is also feasible in special fields of mechanical engineering.

REFERENCES

[1] Kochendörfer, R.; Krenkel, W.: *CMC Intake Ramp for Hypersonic Propulsion Systems*, HT-CMC 2, Santa Barbara, California, August 21 - 24, 1995

[2] Krenkel, W.; Fabig, J.: *Tailoring of Microstructure in C/C-SiC Composites*, ICCM-10, Whistler, B.C., Canada, August 14 - 18, 1995

[3] Krenkel, W.; Henke, T.; Mason, N.: *In-Situ Joined CMC Components*, CMMC-9, San Sebastian, Spain, September 9 - 12, 1996

[4] Leuchs, M.; Prechtl, W.: *Anwendungsspezifische Entwicklung von faserverstärkter Keramik für rohrförmige Komponenten des Maschinenbaus*, Werkstoffwoche, Stuttgart, Germany, 1996

[5] NG, E.-T.; Lu, C.-H.; Lin, H. H.: *Finite Element Modelling of Composite Spur Gears*, ICCE-3, New Orleans, USA, 1996

Application of Aluminium-Based Composite Materials

Prof. V.R.Ryabov, A.Ya.Ishchenko*, Dr. M.M.Monnin**

* E.O.Paton Electric Welding Institute, National Academy of Sciences of Ukraine

11 Bozhenko str., 252650, Kyiv, Ukraine

** Universite Montpellier. Centre National de.la Recherche Scientifique, France

Abstract

Described is the application of pressure welding for manufacture of the pipes and bars of composite materials. The conditions of magnetic-pulse welding of the pipes of composite material of aluminium-boron system with aluminium shells are calculated. Analyzed are the peculiarities of the motion of aluminium shell up to the moment it collides with the pipe of composite material, shown is its extreme character. Calculating dependencies of the speed, with which the shell collides with the pipe, on the technical parameters of the welding are plotted, which allows the optimal gap, corresponding to the maximum rate of collision, to be selected.

Introduction

At present the industry tends to expand the application of metallic composite materials (MCM) of aluminium-boron, aluminium-carbon, aluminium-silicon carbide and aluminium-steel systems. The main volume of their application is in aerospace systems. Here, the typical assemblies are tubular load-carrying structures, with provide reduction of weight by 25-30%.

Magnetic-pulse welding is one of the methods of producing such joints and structures.

Magnetic-discharge facilitates the welding of shaped fittings, tubular adapters and solid tips of up to 60 mm diameter.

The quality of welded joints of aluminium alloys when magnetic-discharge welding, a form of pressure welding, - is determined by the character of the plastic flow of materials in the contact zone. In this case, the temperature on the external contact surfaces of specimens to be joined does not exceed 300-350°C due to heating by eddy currents. It should be noted that, due to the short-time duration of the current (less than 10 microseconds) crystallization processes are hardly possible. In investigations of the effect of the impulse-force action on the process of substance transfer an abnormally rapid migration of atoms was discovered which exceeds by an order of magnitude the rate of migration in the diffusion in gases. Evidently, this transfer is determined by the high rates of plastic deformation proceeding in the metal during impulse loading. With a growth in the rate of deformation, the content of dissolved elements in the crystal lattice of welded metals and alloys increases.

The schematic circuit diagram of the installation intended for magnetic-discharge welding (MDW) is shown in Fig. 1. The voltage of 220/380 V at a commercial frequency is fed to the primary winding of a high-voltage transformer (1), the secondary winding of which is connected to a rectifier converting the high-voltage alternating current into direct current. High-voltage capacitors (2) are charged to a voltage, determined by the production-process conditions. The discharge-thyratron (3) is intended for 'firing' by instantaneous connection of the circuit and the capacitor begins to discharge to inductor (4), inside which, a pulsed magnetic field is induced, accelerating the aluminium tube (shell) 5 to high velocities.

With all other conditions equal, the rate of plastic deformation by magnetic-discharge welding depends on the speed of collision of the thrown shell with the static matrix.

In paper [1] the characteristics of the motion of the thin-walled shell were considered up to its impact upon the static cylindrical matrix.

In the formulation of a mathematical model of the motion of the shell up to the time of impact, we proceed from the system of equations of an electromagnetic field and from the formulae of motion under the action of a ponderomotive force. The main assumptions of the model are as follows:

- the induced magnetic field is considered to be homogeneous and axially symmetric;
- end effects, determined by the final length of the inductor, can be neglected.

Then from the system of Maxwell's equations and from the formulae of incompressibility of a shell the expression follows for a tangential component of the magnetic field intensity H:

$$\frac{1}{r}\frac{\partial}{\partial r}\left(r\frac{\partial H}{\partial r}\right) - \mu_0 \gamma v \frac{\partial H}{\partial r} = \mu_0 \gamma \frac{\partial H}{\partial r},$$

$$R_1(t) < r < R_2(t), t > 0,$$

where: r - radial coordinate; $v(r_1 t)$ - instantaneous speed of motion of point on the shell with coordinate τ at a time instant t; $R_1(t)$; $R_2(t)$ - inner and auter radius of the shell; μ_0 - magnetic constant.

Let us introduce the designations: R_3 and R_4 - inner and outer radii of the inductor; b - its length (in axial direction); r_c, L_c, C - circuit parameters (resistance, inductance and capacitance respectively); U_0 - capacitor charge.

Let $\alpha(t)$ and $\beta(t)$ be a trajectory of motion of two derivative points of the shell. Then we shall obtain an integral identity:

$$\mu_0 \gamma \int_{t_k}^{t_{k+1}} dt \int_{\alpha(t)}^{\beta(t)} \frac{\partial H}{\partial T} r\, dr = \int_{t_k}^{t_{k+1}} dt \int_{\alpha(t)}^{\beta(t)} \frac{\partial}{\partial r}\left(r\frac{\partial H}{\partial r}\right) dr - \mu_0 \gamma \int_{t_k}^{t_{k+1}} dt \int_{\alpha(t)}^{\beta(t)} v\frac{\partial H}{\partial r} r\, dr. \quad (2)$$

By approximating the integrals included in the expression (2), we shall obtain the analogue difference of the equation (1) on the Lagrangian grid

$$\omega^{(k)} = \left\{ r_i^{(k)}, i = \overline{O,N} \right\},$$

where: $r_i^{(k)} = r_i(t_k)$ - Lagrangian grid nodes which at the time instant $t = t_k$ are the coordinates of the trajectory of the points on the shell, originating from the points of the uniform grid

$$\omega^{(0)} = \{r_i = R_1 + ih; \; i = \overline{0,N} \},$$

where: $h = (R_2 - R_1)/N$ - grid spacing at the initial time instant.

The auxiliary problem of the calculation of the intensity of electric fields $E_s^{(1)}(t), E_s^{(2)}(t)$ is also solved by the grid method according to the impact implicit difference scheme.

From the method of the calculation experiment we analyzed the processes of motion of the shells and penetration of a magnetic field into them for workpieces 2.0 mm thick under the following parameters of the installation and electromagnetic circuit: $b = 15$ mm, $R_3 = 15$ mm, $R_4 = 65$ mm, $r_c = 0.010$ m, $L_c = 30$ nH, $C = 120$ mF. The capacitor charge U_0 and clearance between a matrix and a shell were varied.

Shown in Fig. 2 are the characteristics of the motion of a 2.0 mm thick shell for a case in which the matrix inside the shell was omitted. The dependence of the speed of the median point of the shell on the time has an extremal character determined by a change of the sign of the magnetic field intensity H_{R2} on the shell external boundary. When the amplitude of magnetic field intensity H_{R2} is equal to zero, the magnetic force vanishes and the shell moves solely because of the inertial force and the force of deformation resistance resulting in its deceleration. Then with an increase of the amplitude, the magnetic force increases proportionally by H^2_{R2}.

The experimental character of the dependence of the velocity of collision on the gap $\Delta = R_1 - R_0$ (Fig. 3) at three different charges of the capacitor (for a shall 2.0 mm thick) is determined by the dynamic characteristics of the shell motion. In practice, this means that at a minimum value of capacitor charge it is possible to join shells of greater thickness by attaining maximum velocities of collision. Therefore the production-process opportunities of the existing magnetic-discharge welding units become considerably expanded.

Thus, the design dependences of the shell collision velocity on the production-process parameters have been discussed that allows the optimum value of a technological gap to be chosen according to the maximum collision velocity corresponding to it.

The assembling-welding technology for joining tubular CM members of aluminium-boron to aluminium alloys has also been discussed and enables the high-strength quality welds to be obtained.

Using the dependences obtained, magnetic-discharge welding was carried out of aluminium-boron composite materials to aluminium alloy AMg6 fittings using the N-170 machine. A diagram illustrating the assembly of the workpieces and their dimensions is shown in Fig. 4. The welding conditions are as follows: capacitor capacitance $C = 120$ mF, charge value $U_0 = 1.48 \times 10^4$ J, collision velocity $v = 248$ m/s.

The pulsed magnetic field is formed by a massive cylindrical single-turn inductor connected to the electromagnetic circuit.

When welding the composite material pipes with a aluminium or titanium alloy tip steel sleeve is inserted into the pipe to avoid deformation during welding.

The tentative conditions for the magnetic-discharge welding (MDW) of pipes are given in Table.

Pipe diameter, (mm)	Charge voliage, (kV)	Gap value, (mm)	Lap value, (mm)	Quantity of weldings,
6	10.2	0.5	15	2
20	16.2	0.75	25	3

As an example, Fig. 5 shows the structure of a connecting link truss in which the rods and pipes are made of aluminium-boron composite material.

Metallographic investigations of the welds confirmed the absence of pores, cracks and poor fusion in the joint.

There is no failure of the upper of reinforcing fibres and their order of arrangement remains undisturbed. The breaking load in static breaking tests was 55-60 % of the breaking load of the composite material pipe in the direction of reinforcement.

Thus, the magnetic-pulse welding process can be used for joining solid and tubular cylindrical parts from composites.

References

1. Khrenov K.K., Chudakov V.S. Production of welded joints in magnetic-pulse welding of cylindrical parts. - Svarochnoye proizvodstvo, 1978, ¹ 9, p. 13-14.

2. Demchenko V.F., Ryabov V.R., Bocharnikov I.V. et al. (1992). Design estimates of the velocity of collision of aluminium shells in magnetic-discharge welding. Avtomaticheskaya Svarka, ¹ 6, p. 23-25.

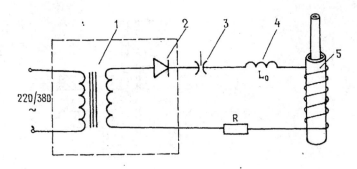

Figure 1. Schematic circuit diagram of the installation intended for magnetic-discharge welding.

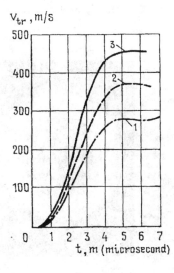

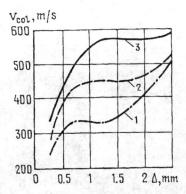

Figure 3. Dependence of the collision velocity on a gap between the matrix and the shell at different voltages of the charge of capacitors (1-3 - ref. to Fig. 2).

Figure 2. Dependence of the travelling speed of the median point of the shell on time at different charges of the capacitor: 1 - $U_0 = 13$ kV; 2 - 15 kV; 3 - 16 kV.

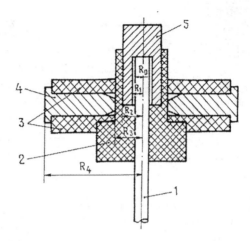

Figure 4. Diagram of assembling the workpieces for magnetic-discharge welding:
1 - rod made of CM; 2 - fluoroplastic sleeve; 3 - textolite straps; 4 - inductor;
5 - tip.

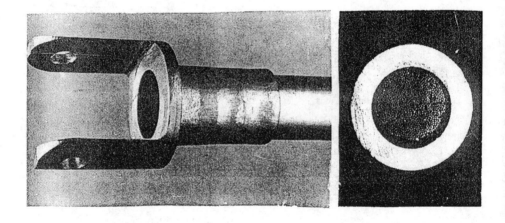

Figure 5. Appearance (a) and macrostructure (b) of a combined welding joint.

Development of the combustor liner composed of ceramic matrix composites (CMCs)

Kenichiroh Igashira*, Kozi Nishio*, Takeshi Suemitsu**

*   Research Institute of Advanced Material Gas-generator - 4-2-6 Kohinata, Bunkyo-ku, Tokyo, 112-0006 Japan
**   Material Research Department of Akashi Technical Institute, Kawasaki Heavy Industries, ltd. - 1-1 Kawasaki-cho, Akashi, 673-8666 Japan

ABSTRACT

The Research Institute of Advanced Materials Gas-Generator (AMG), which is a joint effort by the Japan Key Technology Center and 14 firms in Japan, has, since fiscal year 1992, been conducting technological studies on an innovative gas generator that will use 20% less fuel, weigh 50% less, and emit 70% less NOx than the conventional gas generator through the use of advanced materials.

Within this project, there is an R&D program for applying ceramic matrix composite (CMC) liners to the combustor shown in Figure 1, which is a major component of the gas generator.

In the course of R&D, continuous Si-Ti-C-O fiber-reinforced SiC composite (Si-Ti-C-O^F/SiC) was selected as the most suitable CMC for the combustor liner because of its thermal stability and formability.

An evaluation of the applicability of the Si-Ti-C-O^F/SiC composite to the combustor liner on the basis of an evaluation of thermal stability of a Si-Ti-C-O^F/SiC composite was carried out.

Figure 1 A traial Si-Ti-C-O^F/SiC combustor liner for AMG. (max. ϕ500mm x 150mm^h)

1. Introduction

Si-Ti-C-O^F/SiC composite materials have greater heat resistance, corrosion resistance, and specific strength than conventional heat-resistant super-alloys and higher damage tolerance than monolithic SiC, which makes them highly suitable for use in the high-temperature components of gas turbine engines[1,2,3,4]. The most important subject of research for improving the characteristics of Si-Ti-C-O^F/SiC composites is how to control the fiber/matrix interface characteristics, because the strength of the fiber/matrix interface must not be too high in order to allow the continuous fiber-reinforced ceramic matrix composite to exhibit superior toughness by undergoing a series of complex fracture processes, starting with the debonding of the fiber/matrix interface[2,3,4,5]. To accomplish this, much attention is being focused on forming a boundary layer (interphase) featuring a lubricative function in the fiber/matrix interface. Various research organizations have reported that both strength and toughness are remarkably improved by the formation of carbon or boron nitride at the interface[6,7]. However, when such an interphase is exposed to a high-temperature oxidation environment and is oxidized, it either disappears or forms oxides, resulting in a decrease in strength[8,9,10]. For this reason, the durability of Si-Ti-C-O^F/SiC composites is judged to be greatly dependent on not only the durability of the reinforcing fiber, which plays a major part in giving them their strength, but also the durability of the interphase. However, few systematic studies have been made on the durability of Si-Ti-C-O^F/SiC composites, the modality of the fiber and the matrix, and their relationship to the interface.

In our study, Si-Ti-C-O^F/SiC composite materials were oxidized at high temperatures under various conditions and the strength of the material and the mechanism of the decline in toughness were estimated based on the results of observation of the microstructure by SEM and TEM, aiming at finding a method for improving durability, while also seeking a method for estimating service life.

2. Test method

2.1 Preparation of test samples

The process for manufacturing the Si-Ti-C-O^F/SiC composites used as test samples is shown in Figure 2. An orthogonally three-dimensionally woven SiC fabric was used as

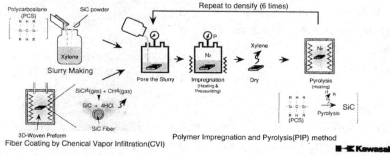

Figure 2   Schematic illustration of the fabricating procedure of Si-Ti-C-O^F/SiC composites.

the reinforcement fiber preform. The SiC fiber was Tyranno LoxM from Ube Industries Ltd. (Because this SiC fiber is an amorphous fiber containing a certain amount of titanium and oxygen, we shall refer to it as Si-Ti-C-O fiber and as Si-Ti-C-O^F when in combination with SiC.) In the first step, the surface of each Si-Ti-C-O fiber in the preform was coated with SiC to form a layer about 1μm thick using the CVI method, which we shall refer to as the CVIed-SiC layer. The CVIed-SiC layer is formed on the fiber surface to prevent direct contact between the Si-Ti-C-O fiber and the matrix, which contains impurities after being formed by the PIP method, and because its anti-oxidation characteristics are superior to those of carbon and boron nitride. The pre-ceramic polymer used in the PIP process was Polycarbosilane from Nippon Carbide Co., Ltd. In the impregnation process, the pre-ceramic polymer is dissolved in xylene and a slurry in which fine SiC particles are suspended is used. The SiC-coated preform is impregnated with the slurry, then undergoes heat-pressurized impregnation in a pressure vessel that allows heating. Next, the preform is dried to evaporate the xylene, heated in a nitrogen atmosphere, then undergoes pyrolization. By repeating the impregnation and pyrolization treatment for six cycles, an Si-Ti-C-O^F/SiC composite with a pore residual volume fraction of 10 vol% and a bulk density of about 2.5 g/cm^3 was obtained for use as the test sample. The component elements and their characteristics are listed in Table 1.

Table 1 Components of Si-Ti-C-O^F/SiC composite

	Filament	Preform
Reinforcement	Si-Ti-C-O fiber (Tyranno LoxM; Ube Co., Ltd.) :d~11μm, σ$_f$~3.5GPa, E$_f$~188GPa	V$_F$ ~ 40% X : Y : Z ~ 1 : 1 : 0.1
Interface layer	CVI-SiC Thickness: 1 - 2μm	
Matrix	SiC(derived from PCS*) (Nippon Carbon Co., Ltd.)	
	SiC particle (Betarundum: Ibiden Co., Ltd)	

2.2 Durability evaluation test

Durability was evaluated by heating the test sample in an ambient atmosphere furnace or in a vacuum of about 10^{-2} Pa for a certain period of time, then measuring its weight and conducting a bending test at room temperature. The test piece used was a small slab 55 mm long by 12 mm wide by 4 mm thick. The bending test conditions were four-point bending, 50 mm outer span and 16 mm inner span, with a cross-head speed of 0.5 mm/min. The measuring instrument was an Instron universal tester and load and cross-head displacement were measured. The stress-strain relationship was obtained by using the beam bending equation. The maximum stress value was set as the residual stress. To estimate fracture toughness, a test piece in which a chevron notch had been made was heated for a certain period of time, as mentioned above, followed by a four-point bending test at room temperature.

2.3 Observation of the microstructure

The microstructure before and after the durability test was observed using SEM and TEM.

3. Results and Discussion

The cross-sectional microstructure of the as-processed material observed by SEM is shown in Figure 3. The round part measuring about 10 μm are the Si-Ti-C-O fibers and the white part about 1μm around the fibers are the CVIed SiC layer. The remaining part is the matrix formed by the PIP method.

Figure 4 is a TEM image of the fiber/CVIed-SiC layer interface. The part lacking contrast is the amorphous Si-Ti-C-O fiber. The part where much streak-like contrast can be seen is the crystalline CVIed-SiC layer. In the interface between the two, there is a interphase layer several dozen nanometers thick, which was found to contain a relatively large amount of carbon through TEM-EDX analysis.

Figure 5 plots the residual bending strength of the test sample at room temperature after having been exposed to the ambient atmosphere for 500 hours at various temperatures ranging from 1173K to 1473K. There was almost no drop in strength up to the vicinity of 1273K but there was a sudden decline after passing 1273K and the strength of the material exposed for 500 hours at 1473K dropped to nearly 1/3 that of the as-processed material. Figure 6 shows residual bending strength and changes in weight when the test sample was exposed to air at 1473K. Strength declined regularly in relation to the logarithmic time but the weight increased regularly, suggesting the formation of oxides.

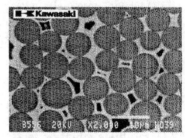

Figure 3 Cross section of as-processed
Si-Ti-C-O^F/SiC composite.

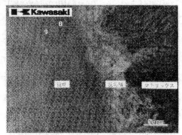

Figure 4   TEM image of fiber/CVIed-SiC interface
of as-processed Si-Ti-C-O^F/SiC
composite.

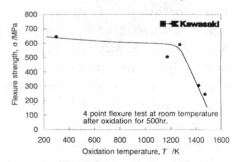

Figure 5   Strength degradation of Si-Ti-C-O^F/SiC
composite after oxidation for 500hr

We attempted to collate the durability of the Si-Ti-C-O^F/SiC composite by applying the Larson-Miller parameter [LMP = T x (C+log t_{ex})/1000; T: exposure temperature (K), t_{ex}: exposure time (hr), C: constant], which is used for evaluating the creep characteristics of metals etc., aiming at integrally dealing with the thermal load, which is provided by the heating time and the exposure time. Figure 7 shows residual bending strength after heating in the ambient atmosphere and in a vacuum collated with the LMP. The value of the constant in the LMP is believed to vary depending on the material: it is normally in the range of 16 to 22 for heat-resistant alloys but for the Si-Ti-C-O^F/SiC composite, the linearity of data was obtained by setting the constant at 10. The strength of the test sample was found to decline greatly under a thermal load of LMP = 17 or above in the oxidation environment and in the vacuum. However, the decline in strength was remarkably greater in the vacuum than in the oxidation environment. Figure 8 shows the residual tensile strength of the Tyranno LoxM fiber, which is the reinforcement fiber for the Si-Ti-C-O^F/SiC composite, collated with the LMP after

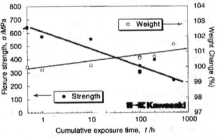

being exposed to the oxidation environment under various conditions. As in the results for the composite shown in Figure 7, there was a large drop in strength at LMP = 17 or above. This would indicate that the residual strength of the Si-Ti-C-O^F/SiC composite after exposure to a high-temperature environment is greatly affected by the residual strength of the reinforcement fiber. The decline in strength of the reinforcement

Figure 6   Strength degradation and weight change of Si-Ti-C-O^F/SiC composite after oxidation at 1473K.

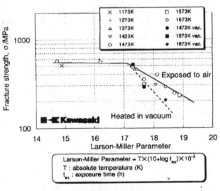

Figure 7   The relationship between remaining flexure strengths of Si-Ti-C-O^F/SiC composite and Larson-Miller parameter.

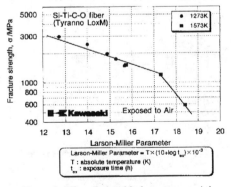

Figure 8   The relationship between remaining tensile strengths of Si-Ti-C-O fiber(Tyranno LoxM) and Larson-Miller parameter.

fiber by exposure to high temperatures is due to the fact that the Tyranno LoxM fiber in the amorphous state is crystallized while undergoing thermal decomposition under a high temperature in a reaction which can be expressed by the following formula.

$$Si - Ti_{0.02} - C_{1.33} - O_{0.44} \rightarrow SiC + TiC + 0.13O + 0.31CO (g) \text{ — (1)}$$

Figure 9 shows changes in the weight of the Si-Ti-C-O^F/SiC composite collated with the LMP. In the vacuum, weight loss clearly occurred under a thermal load of LMP = 17 or above but in the oxidation environment the weight rose regularly, although at a very low rate. The weight loss in the vacuum is clearly due to the decomposition of fiber. SEM observation reveals that the shrinkage of the fiber, due to decomposition, in the Si-Ti-C-O^F/SiC composite in the oxidation environment, even the weight rose regularly, was also occurred.

Figure 10 shows the fracture toughness values after oxidation collated with the LMP. Because the number of measuring points was small, it is not clear whether there was a sudden decrease in the vicinity of LMP = 17 as was seen in residual strength but it was confirmed that the fracture toughness value, which was 22 for the as-processed material, declined as the LMP increased.

In Figure 9 again, the differing fracture styles of the Si-Ti-C-O^F/SiC composite according to exposure environment and thermal load are expressed using symbols. The fracture modality in which the fracture surface presents much fiber pull-out after the bending test is represented as "pull-out", that with a smooth fracture surface like that of monolithic ceramics is represented as "catastrophic", and that

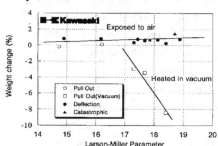

Figure 9   The relationship between weight changes of Si-Ti-C-O^F/SiC composite and Larson-Miller parameter.

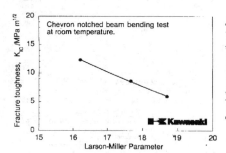

Figure 10   The relationship between fracture toughness of Si-Ti-C-O^F/SiC composite and Larson-Miller parameter.osite.

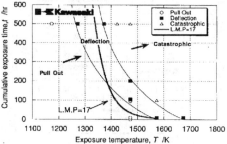

Figure 11   Fracture mode map of Si-Ti-C-O^F/SiC composite.

which is halfway between the two, presenting almost no pull-out and an irregular fracture surface due to crack deflection is represented as "deflection". It was confirmed that the material exposed to a high-temperature oxidation environment fractured pseudo-plastically accompanied by the pull-out phenomenon when the LMP was low and changed to a brittle fracture modality accompanied by the pull-out phenomenon as the LMP rose. However, the material heated in a vacuum fractured accompanied by the pull-out phenomenon even under thermal loads with high LMP values. Figure 11 is a fracture modality map of the Si-Ti-C-O^F/SiC composite after exposure to a high-temperature oxidation environment, showing that the material moved towards the brittle fracture modality as the thermal load increased. The thermal load of LMP = 17 is found in the vicinity of the area defined as deflection. Figure 12 is a TEM image of the vicinity of the fiber/CVIed-SiC layer interface

of a sample of the material exposed to air for 500 hours at 1473K. The many spots seen in the fiber are the SiC crystal that became coarser. A layer that did not exist in the as-processed material has formed between the fiber and the CVIed-SiC layer, and it was identified as silicate through TEM-EDX analysis. It has been reported that silicate is formed on the fiber/matrix interface by oxygen that has penetrated cracks in the matrix and this silicate locally increases the debonding strength in the fiber/matrix interface, causing stress concentration, which makes it harder for fiber pull-out to occur[6,12]. We believe that the embrittlement of this material due to oxidation was caused by the formation of a silicate layer in the fiber/CVIed-SiC layer.

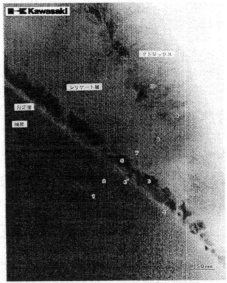

Figure 12   TEM image of fiber/CVIed-SiC interface of Si-Ti-C-O^F/SiC composite after oxidation at 1473K for 500hr(L.M.P=18.7).

4. Conclusions

The decline in strength of Si-Ti-C-O^F/SiC composite is caused mainly by the thermal decomposition of the fiber but embrittlement due to the formation of a silicate layer in the fiber/CVIed-SiC interface also has some effect on the decrease of strength.

Regarding these two major factors that contribute to the decline in strength of Si-Ti-C-O^F/SiC composite under high temperatures, thermal decomposition of the fiber and formation of a silicate layer, the diffusion of elements inside the fiber or silicate is believed to be the rate control process. For this reason, we attempted to apply the LMP.

In order to improve the durability of Si-Ti-C-O^F/SiC composite materials, improvement of the heat-resistance of the reinforcement fiber is of primary importance. However, this problem may eventually be solved by applying high-heat-resistant fiber (Tyranno LoxE or Hi-Nicalon) offered by certain textile manufacturers. If oxidation in the fiber/CVIed-SiC interface is not prevented, the embrittlement of the material will progress and will not be possible to achieve reliability for prolonged use in a high-temperature environment. This problem may be dealt with by applying a crack sealant covering the vicinity of the fiber with a material that remains ductile and stable under high temperatures.

Acknowledgmens

This study is being carried out under the Advanced Materials Gas-Generator R&D program with a joint investment from the Japan Key Technology Center. The authors wish to express their gratitude to the Japan Key Technology Center for making this study possible and permitting this paper to be published.

Reference

[1] P. J. Lamicq, G. A. Bernhart, M. M. Dauchier and J. G. Mace: ibid., 65(1986), 336.

[2] Y. Kagawa, H. Hatta : Tailoring Cemamic Composite,(1990);ISBN4-900508-15-2

[3] R. Tanaka : J. Japan Soc. Heat Treatment, 30 (1990) , 134

[4] K.Okamura : J. Japan Soc. Composite Material, 20 (1994) , 34

[5] H. C. Cao, E. Bishoff, O. Sbaizero, M. Ruhle, A. G. Evans, D. B. Marshall and J. J. Brennan: J. Am. Ceram. Soc., 73(1990), 1691.

[6] A. J. Caputo, D. P. Stinton, R. A. Lowden and T. M. Besmann: Am. Ceram. Soc. Bull, 66(1987), 368.

[7] R. Naslain, O. Dugne, A. Guette, J. Sevely, C. Robin-Brosse, J. P. Rocher and J. Cotteret: J. Am. Ceram. Soc., 74(1991), 2482.

[8] L. Filipuzzi, G. Camus, and R. Naslain: J. Am. Ceram. Soc., 77(1994), 459.

[9] L. Filipuzzi and R. Naslain: J. Am. Ceram. Soc., 77(1994), 467.

[10] S. Baskaran and J. W. Halloran: J. Am. Ceram. Soc., 77(1994), 1249.

[11] K. Kakimoto, T. Shimoo and K. Okamura: J. Ceram. Soc. Japan, 103(1995), 557.

[12] R. F. Cooper and K. Chyung: J. Mater. Sci., 22(1987), 3148.

Fibre Composites for exceptional Service Conditions, Composed of Metallic Fibres and Electrodeposited Multi-Layer Matrices

H.Weiß*, W.M.Semrau**

* Labor für Oberflächentechnik, Universität-GH-Siegen - Germany
** IWS Ingenieurbureau Wolfgang Semrau für werkstoffbasierte
   Systementwicklung, Korrosion und Oberflächentechnik, Bremen - Germany

Abstract

Metallic Matrix Fibre Reinforced Compounds are usually developed to serve as light-weight high-strength materials. Mostly they consist of ceramic fibres in a ductile matrix and are manufactured by casting, squeeze casting or extrusion. The fibres are intended to provide the high strength required. However, as the volume fraction is generally limited to about 40%, their contribution to the mechanical strength is not so pronounced.

Combining high strength metal wires with a multi layer matrix material, comprising a gradient of mechanical properties within the matrix in different areas relative to the fibre, wire/matrix-composites were obtained that exhibit high tensile strength, an excellent behaviour during dynamic bending and a retarded failure mode showing cracks stopping at interfaces as is known from wooden structures. Fibre contents vary from 88 to 96 vol.-%.

Furthermore, taking into account the needs for economical production, a new approach was made, creating fibres by the forging process itself.

1. Introduction

In order to increase the performance of materials in technical applications, high strength fibres are used to improve the behaviour of low strength materials. In the field of polymer compounds this method is well established. Glass fibre- and, in high strength applications, carbon fibre reinforced polymer compounds are well known, not only to the technical expert.
To transfer these benefits into the large field of metallic materials, high strength fibres are used to reinforce metallic materials and their alloys, [1]. They are usually developed to serve as light-weigth high-strength materials. Mostly they consist of ceramic fibres in a ductile matrix and are manufactured by casting, squeeze casting or extrusion. The fibres are intended to provide the high strength required. However, as the volume fraction is generally limited to about 40%, their contribution to the mechanical strength is not so pronounced.

Moreover, they are very difficult to handle and, e.g. in case of Al-matrices, do not exceed the performance of high strength Al-alloys as much as its costs do, their applications remained rather limited. In the field of turbine blades, fibre reinforced nickel based compounds, [2], are known to comprise a rather low fibre content.

2. Metallic Fibre Composites with a high Fibre Content

In this investigation, another approach is described. The intention was to take advantage of the high strength level of the fibres without the disadvantages connected with high strength materials such as brittleness and poor behaviour under dynamic loading. Thus, the fibre content should be raised and the matrix properties should be adaptable to a wide range of characteristics. Also a retarded failure behaviour should be achieved.

In order to meet these objectives, a model material was designed and manufactured in the laboratory consisting of cold worked high strength metal wires with tensile strength levels about 2000 MPa and a metal matrix material, which was added by coating the wires by PVD, [3], or by electroplating,[4]. Then, the wires were bundled up and compacted by a thermomechanical treatment, Fig.1.

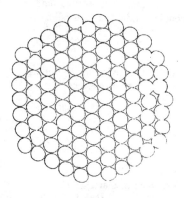

10 : 1   non etched

Figure 1: Fibre reinforced composite model material after compaction
          by thermomechanical treatment

This treatment consisted of a forging operation, which bonded the coatings by friction bonding and resulted in a dense microstructure, without pores or holes, Fig.2. As the highest temperature applied was below the melting point and just above the beginning of recrystallization of the matrix metal, no liquid phase occurred, Fig.3.

The fibre content was varied from 88 to 96 vol.-% without affecting the compaction process in a detrimental way. The tensile strength of the fibres could be stabilized at the high level of the starting material.

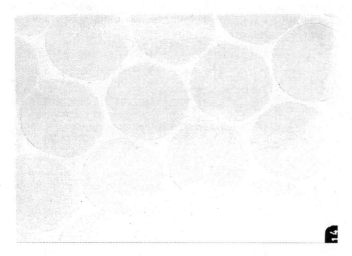

40 : 1   non etched

Figure 2: Material after compaction with dense microstructure

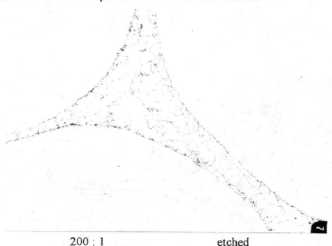

200 : 1   etched

Figure 3: Recrystallized nickel-matrix

Characterization of the compound by means of static and dynamic testing revealed the matrix to be the weakest link of the system. Ideally, the matrix material should have different properties in different areas relative to the fibre. Therefore, a matrix material comprising a gradient of mechanical properties should be desirable.

3. Multi-Layer structured Matrix

The need for a matrix material with different mechanical properties in different areas relative to the fibre led to an electrochemically deposited coating composed of different metals, [5], alternately deposited on the wire, Fig.4.

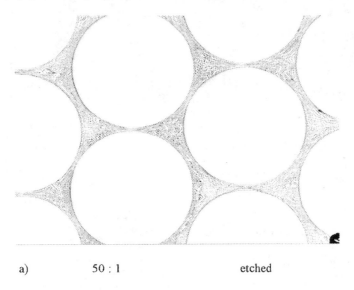

a)   50 : 1   etched

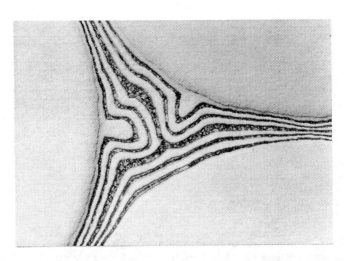

b)   250 : 1   etched
Figure 4: Fibre composite model material with multi-layer-matrix after compaction

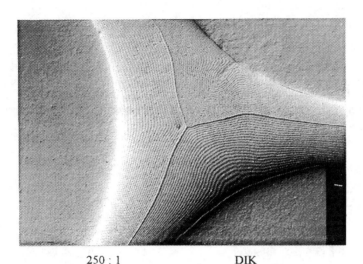

250 : 1   DIK

Figure 5: Multi-layer matrix

Partial alloying by the diffusion process combined with thermomechanical compaction led to smoothing down abrupt changes of prerequisites within the transition zones of the directly neighboured layers. This is one of the important conditions in such a material to avoid high peaks of internal stresses which could lead to early failure. By varying thickness and succession of two or three different types of coating metals and by varying temperature and time of application, a widespread field of different characteristics of the matrix material could be produced, [6].

As a result, wire / matrix-composites were obtained that exhibit high tensile strength, an excellent behaviour during dynamic bending and a retarded failure mode, showing cracks stoppimg at interfaces as it is known from wooden structures, Fig.6. Thus, the matrix layers serve as very effective crack stoppers.

Figure 6: Fracture of wire/matrix composite resembling fracture of wooden structures

4. Approach to simplified Production Conditions

Furthermore, a new approach was made creating fibres by the forging process itself. Coarse grained metallic powders, used as a starting material instead of metallic wires, were covered by electroplating with a metallic layer serving as matrix and compacted by the same forging process described for the wire/matrix-material The grains deform into fibres during compaction, Fig. 7.

17 : 1   etched

Figure 7: Fibre/matrix composite with electroplated coarse powder as starting material

This technology resulted in a first model fibre composite with fibre diameters of about 50 to 80µm and a fibre length of about 30 to 60mm. Using other dimensions, fibre composite materials with fibre diameters down to about 10µm with fibre lengths of several hundred millimeters seem to be within reach.

5. Powder Metallurgical Aspects

In addition, the described technology provides an innovation potential for powder metallurgy. The lack of both, a liquid phase and high temperatures with their risks for grain structures and grain boundaries, leads to easy handling systems from the viewpoint of the metallurgist.

As the matrix material is ideally distributed no boundary section exists not covered by matrix material, Fig. 8. This is a good basis for a uniform product quality.

500 : 1   non etched
Figure 8: Grain boundaries of the former powder grains all covered with matrix metal

References

[1] *E. ElMagd, E.Mokhtar,* Mechanical Behaviour of Fibre-Reinforced Metals
Current Advances in Mechanical Design and Production Pergamon Press, 1980
[2] *R. Warren, G.S. Upadhaya,* Mechanical Properties of Fibre Reinforced
Superalloy Composites -Proc. 3rd Intern. School of Sintered Materials
Amsterdam 1984
[3] *H. Stuke,* Fibre reinforced Compounds with metallic Matrix
IFAM - Techn. Report, Bremen 1984
[4] *W.M. Semrau,* Tungsten-Fibre Metallic Compound with electroplated
Matrix, - IFAM - Techn. Report, Bremen 1987
[5] *W.M. Semrau,* Composition of electrodeposited Multi-Layer-Structures
comprising a gradient of properties for Surfaces in Technical Service
IFAM - Techn. Report 1989
[6] *W.M. Semrau,* CuNiFer-Layers for Seawater Application, composed of
galvanically deposited Multi-Layer-Coating
12th Scandinavian Corrosion Congress & Eurocorr '92, Helsinki, June 1992

Heat-resistant oxide-fibre composites

S.T.Mileiko
Solid State Physics Institute of the Russian Academy of Sciences, Chernogolovka
Moscow district, 142432 RUSSIA

Recent advance in the author's laboratory in the field of heat-resistant composites is reviewed. Since previous review paper[1] was published an essential progress has been made mainly due to three attainments:

1. Usage of the fibres produced by the internal crystallization method (ICM)[2] becomes possible in a variety of the matrices including those of high corrosion resistance. This attainment is mainly based, first, on the understanding of high strength of oxide fibres produced by ICM,[2,3] and secondly, on a finding of the method to transfer the fibres from the molybdenum matrix into a permanent matrix, e.g. that of nickel aluminide alloy. Calculation of the stress rupture time of such composites shows that a creep strength of 150 MPa at 1200°C on an appropriate time base can be reached.

2. Development of so called MIGL fibres produced by infiltrating a moving molybdenum micro-rope with an oxide melt and subsequent crystallizing it.[4] This allows crystallizing oxide-based fibres at very high rates, up to 85 mm/s, which can definitely yield an essential decrease in the cost of the fibres.

3. Development of a CVD-procedure for coating internal surface of the channels inside an oxide matrix[5] to approach to optimization of the fibre/matrix interface in oxide/oxide fibrous composites produced by the internal crystallization method.[6] Hence, a class of the oxide-fibre/oxide-matrix composites produced by ICM is now look more prospective as it promises obtaining sufficiently tough composites made of inherently strong components with high gas corrosion resistance.

The work was performed under financial support of International Science and Technology Center, Project # 507-97, and INTAS-RFBR, Project # 95-0599.

APPLICATION OF CERAMIC MATRIX COMPOSITES TO ROTATING COMPONENTS FOR ADVANCED GAS-GENERATOR

N. Suzumura[*], T. Araki[*], T. Natsumura[*], S. Masaki[*], M. Onozuka[*], H. Ohnabe[*], T. Yamamura[*], K. Yasuhira[*]

[*]Research Institute of Advanced Material Gas-Generator, 4-2-6 Kohinata, Bunkyo-ku, Tokyo, 112-0006, Japan

Abstract

Ceramic matrix composites (CMCs) were examined in spin tests to assess applicability to rotating components (bladed disks) for advanced gas-generators. The alignment of SiC fiber yarns was studied and matrix forming was developed. CMC circular disks, consisting of three-dimensional (3-D) woven fabrics and a SiC matrix formed by CVI/PIP were prepared and tested. Stress distribution of rotating disks was calculated assuming the orthotropy of 3-D CMC materials. Results showed that the burst strength can be estimated sufficiently accurately based on the "maximum hoop stress" criterion. The machined bladed disk was also tested at room temperature. The maximum blade tip speed reached to 644 m/s without apparent damage on the blade root surface. Results showed that the 3-D CMCs we developed were potentially applicable to rotating components (bladed disks).

1. Introduction

Research Institute of Advanced Material Gas-Generator (AMG) was established in 1993 to develop basic key technologies for next-generation high-performance gas-generators using advanced material systems. The research program of AMG is now in progress for three major targets, i.e., improvement of 20% specific fuel consumption, 50% reduction of weight, and 70 to 80% reduction of NOx emission. Advanced material systems under development at AMG include CMCs, metal matrix composites (MMCs), and heat-resistant polymer matrix composites [1].

CMCs are generally considered key materials for advanced gas-generators and studied for application to turbine components[2,3,4]. Many activities focus on applications to gas turbine stator components. In the research program of AMG, we studied CMC turbine rotors. The following technical indexes are set for high-performance turbine rotors.

- Material temperature: maximum 1400°C
- Blade tip speed: 570 m/s

We discuss the CMC bladed disk strength at room temperature based on the results of component tests and stress analysis. We demonstrated the potential applicability of 3-D CMCs to bladed disks.

2. Material Design and CMC Disk Manufacturing

Three dimensional (3-D) woven fabrics using Si-Ti-C-O fiber (Tyranno[TM] Lox-M) were selected to prevent delamination. Chemical Vapor Infiltration (CVI) and Polymer Impregnation and Pyrolysis (PIP) were applied to obtain sufficient mechanical properties for rotating components as matrix forming of 3-D CMCs (Fig. 1). In typical stress-strain curves measured at room temperature (Fig. 2) [5,6], the tensile strength of 3-D CMCs was about 500 MPa, equal to the strength calculated from fiber volume fractions. Yarns of CMC disks were arranged following the cylindrical coordinate system (r- θ -z) and yarn alignment was designed to meet the strength requirement of stress distribution. The strength in hoop and radial directions required from the stress distribution of rotation. The optimization of fiber volume fractions was carefully studied for the disk. The obtained disk added yarns of the hoop direction in the inner region and in the radial direction in the outer region. Using the developed manufacturing process, circular CMC disks were fabricated. After forming of the matrix, all components such as disks and blades were machined.

3. Component Test and Stress Analysis

In applying 3-D CMCs to rotating components such as bladed disks, technical issues for bladed disks are as follows:
- Burst strength of 3-D CMC disks
- Stress analysis approach to evaluate disk strength
- Effect of stress concentration at blade root

The burst strength of rotating disks composed of ductile metallic materials is generally described based on average hoop stress criteria such that the burst occurs when the average hoop stress reaches ultimate material strength. 2-D CMC disks with a CVI matrix reportedly burst in the same manner[7].
The burst strength of the 3-D CMC disk was assessed in spin tests at room temperature (Fig. 3). The test disk was attached to a drive shaft driven by an air turbine (Fig. 4). Disks are 250 mm outer diameter (o.d), 50 mm inner diameter (i.d), and 7 mm thick (t).
In tests, 3-D CMC disk burst to 4 pieces in hoop mode at about 36,200 rpm (Fig. 5). Many fiber pullouts were observed on the fracture surface of the disk, indicating that the 3-D CMC disk did not brake brittly.

Stress distribution was analyzed by FEM, in which material constants were calculated by the unit cell model corresponding to the structure of 3-D woven fabrics[8]. Material constants were calculated at each portion of the radial direction, because the 3-D CMC disk has a distribution of fiber volume fractions in the radial direction. The accuracy of this analysis approach was confirmed by the measuring of strain distribution of another disk. Rotating burst speeds were estimated from maximum hoop stress and average hoop stress criterion. The maximum hoop stress criterion was defined such that the disk burst occurs when the maximum hoop stress (= bore stress) reaches the maximum material hoop strength (= bore hoop strength). The average hoop stress criterion was defined such that the disk burst occurs when the average hoop stress reaches the average ultimate hoop strength which average the ultimate hoop strength distributing in the radial direction. Ultimate hoop strengths of each region were estimated from fiber volume fractions and plate sample tests (Table 1). We found that the disk burst occurred when the maximum hoop stress reached the strength in the hoop direction. Other 3-D disks reported yielded the same results[9]. Thus, the burst strength of 3-D CMC disks can be predicted based on the maximum hoop stress criterion, but more study is necessary to clarify the difference of behavior between 2-D and 3-D CMC disks.

The fracture mechanism of the 3-D CMC structure with stress concentration is not clear. Monolithic ceramic components fracture when the peak stress such as the blade root reaches the material strength. If CMC fracture occurs at the same manner, CMC bladed disk fracture will depend on the peak stress at the blade root.
Blade strength was assessed in tensile tests by model specimens (Fig. 6). Specimens were machined from the portion corresponding to blades in a disk. The stress concentration factor at fillet R was about 1.6 for minimum section of specimens. Specimens were set to the testing machine (Fig. 6) and tested at room temperature. All specimens broke in the parallel section and no root fracture was observed. The average tensile strength of all specimens was nearly equal to the strength calculated from fiber volume fractions at the minimum section. This shows that the stress concentration of the blade root does not affect the strength of the blade. The allowable stress of blades is obtained from the strength at the minimum section of the blade root.

Test results of the disk and the model blade indicates the capacity of the 3-D CMC bladed disk.

4. Bladed Disk Spin Test

The bladed disk (Fig. 7) was designed based on provisional design criteria set by the above test results and tested in spin tests at room temperature. Sizes of the bladed disk are 246 mm (o.d) - 50 mm (i.d) - 22 mm (t) and with blades 23 mm long. The bladed disk was attached to the drive shaft and tested under the same condition as the circular disk.
Testing was completed with the 3-D CMC bladed disk being accelerated to 50,000 rpm. The blade tip speed reached to 644 m/s. It surpassed the AMG target (570 m/s). No apparent damage was seen on the disk surface or at the blade root.
The stress distribution of the 3-D CMC bladed disk was analyzed (Fig. 8). We found that the 3-D CMC blade disk sustained higher hoop stress in the bore than that of the simple circular disk. The difference in strength lay in the dispersion (strength, density, etc.) of 3-D CMC disks, but more study is needed to clarify the difference between the simple circular disk and the bladed disk.
Peak stress calculated from the above model sample results and average stress of blade root, reached to the blade strength at maximum rotating speed. Thus, this test ensured that the stress concentration of the blade root does not affect strength of blades.

5. Discussion

The burst mechanism of 3-D CMC disks is considered to operate as follows:
3-D CMCs consisting of yarns of radial, hoop (θ), and thickness directions are nonuniform (Fig. 1). The strength of each direction depends on the amount of fiber in each direction. The 3-D CMC disk has a distribution of hoop strength in the radial direction and the maximum hoop strength is set for the disk bore. When the 3-D CMC disk is rotating, the hoop stress distribution is macroscopic stress distribution across yarns of θ-direction and maximum hoop stress appeared at the disk bore. The stress level of adjacent yarns is close, so yarn fracture is continuous. Thus, the fracture of 3-D CMC disks is caused in the fracture of the disk bore at the same time.

The blade fracture mechanism is considered to operate as follows:
As the blade is machined from the disk, fillet R is put in yarn. The stress distribution at fillet R is microscopic stress distribution in yarn, which is damaged by peak stress, and peak stress is released. As the peak stress does not affect other yarns, the fracture of blades occurs when the average stress of the net section reaches the strength of the radial direction.

The fracture mechanism of disks and blades was studied in component tests. Test results led to provisional design criteria of the 3-D CMC bladed disk, and the 3-D CMC bladed disk was designed and demonstrated in spin tests.
The 3-D CMCs we developed are potentially applicable to rotating components.

6. Conclusion

The 3-D CMC disks we developed were fabricated and evaluated. We obtained the following conclusions:
(1) The stress distribution was analyzed by FEM, in which material constants were calculated by the unit cell model corresponding to the structure of 3-D woven fabrics. Analysis and spin test results indicated that the burst strength of 3-D CMC disks can be predicted based on the maximum hoop stress criterion.
(2) The stress concentration of the blade root does not affect blade strength.
(3) The bladed disk was designed based on provisional design criteria set by reflecting the above test results and tested at room temperature. The blade tip speed reached 644 m/s, exceeding the AMG target of 570 m/s.
These tests showed that the 3-D CMCs we developed are potentially applicable to rotating components.
High-temperature (maximum 1400℃) possibilities of 3-D CMCs will be assessed and improved at AMG.

References

[1] *M. Hiromatsu, et al.*, **RESEARCH AND DEVELOPMENT STATUS OF ADVANCED MATERIALS GAS-GENERATOR (AMG) PROJECT** - ASME, 1995, 95-GT-287

[2] *R.A. Lowden, D.P. Stinton, T.M. Besmann*, **CERAMIC MATRIX COMPOSITE FABRICATION AND PROCESSING : CHEMICAL VAPOR INFILTRATION** - Handbook on Continuous Fiber-Reinforced Ceramic Matrix Composites, Published by CIAC, CINDAS/Purdue Univ. and ACerS, 1995

[3] *C.P. Beesley, M. JL Percival*, **OPPORTUNITIES AND CHALLENGES FOR CMCS IN GAS TURBINES** - 7th European Conference on Composite Materials (ECCM-7) -London-UK, 14-16 May, 1996, Vol.1

[4] *H. Ohnabe, et al.*, **POTENTIAL APPLICATION OF CERAMIC MATRIX COMPOSITES TO AERO-ENGINE COMPOSITES** - Proceeding of International CMC Symposium Part A : applied science and manufacturing, Elsevier Science publication, -31 October, 1997 (to be published)

[5] *S. Masaki, et al.*, **DEVELOPMENT OF Si-Ti-C-O FIBER REINFORCED SiC COMPOSITES BY CHEMICAL VAPOR INFILTRATION AND POLYMER IMPREGNATION & PYROLYSIS** - HT-CMC-2-Santa Barbara-USA, 1995

[6] *T. Araki, et al.*, **MANUFACTURING OF CERAMIC MATRIX COMPOSITE ROTOR FOR ADVANCED GAS-GENERATOR** - 22nd Annual Cocoa Beach Conference and Exposition, 20-24 January, 1998 (to be published)

[7] *H. Ohnabe, et al.*, **BURST STRENGTH AND NON-LINER STRAIN ANALYSIS OF ROTATING SiC/SiC DISCS** - Proceeding of the Annual Meeting of JSME/MMD, 13-14 October, 1994, Vol.B

[8] *H. Ohya, et al.*, **STRENGTH EVALUATION AND STRUCTURAL DESIGN OF FIBER REINFORCED CERAMIC MATRIX COMPOSITES** - Proceeding of The 10th International Symposium on Ultra-High Temperature Materials, 1996

[9] *N. Suzumura, et al.*, **DEVELOPMENT OF DISK FOR ADVANCED GAS-GENERATOR (COLD SPIN TEST OF SIC/SIC DISK)** - Proceeding of the 1996 Annual Meeting of JSME/MMD, 3-4 October, 1996, Vol.B (in Japanese)

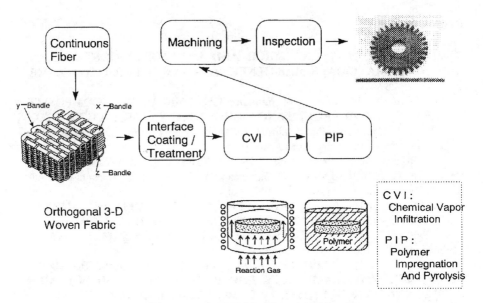

Figure.1.   Developed 3-D CMCs Manufacturing Process

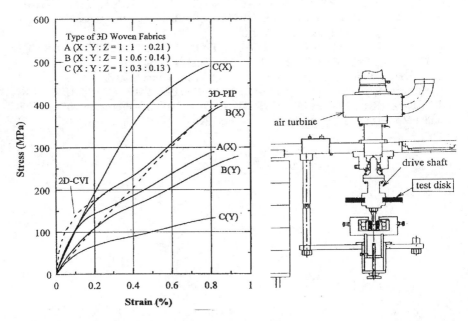

Figure.2.   Stress-strain curves of 3-D CMC[5]   Figure.3.   Spin Test Apparatus

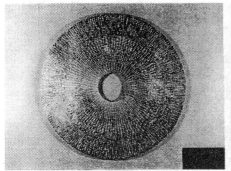
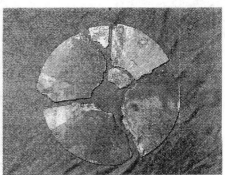

Figure.4. 3-D CMC Disk   Figure.5. Fracture Mode of 3-D CMC Disk

Table 1.   Burst Speed Value Comparison

		Estimated results (Na)	
	Experimental (Nt)	Maximum Hoop Stress = Bore Hoop Strength	Average Hoop Stress = Average Hoop Strength
rpm	36200	35600	43700
% (Na/Nt)	--	98	121

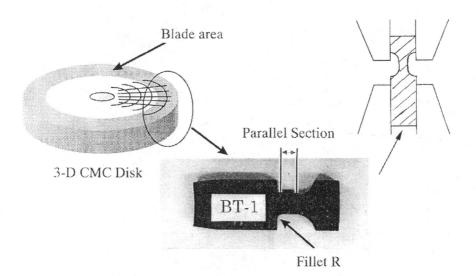

Figure.6. Model Specimens of Blade Root and Set to Testing Machine

Figuer.7. 3-D CMC Bladed Disk

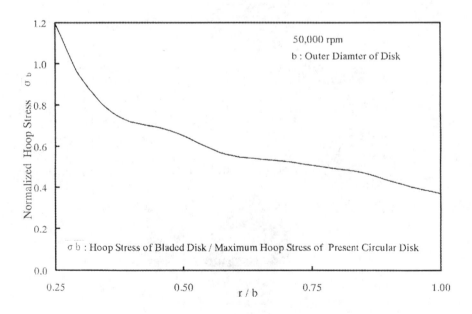

Figure.8. Hoop stress distribution of 3-D CMC Bladed Disk

Blade area

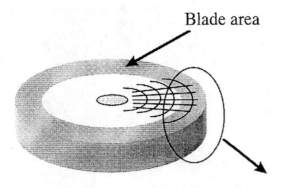

Figure.6. The model Specimens of The Blade Root

Investigation About The Interface Nature Of The Ni₃Al Intermetallic Reinforced Particles With Al Alloy Matrix Composite Obtained By Powder Metallurgy.

Carlos Ferrer* , Vicente Amigó*, MªDolores Salvador*, David Busquets*, José Manuel Torralba**
*Departamento de Ingeniería Mecánica y Materiales. Escuela Técnica Superior de Ingenieros Industriales. Universidad Politécnica de Valencia. Camino de Vera s/n. 46022 Valencia. España
**Departamento de Ingeniería. Universidad Carlos III de Madrid. C/ Buterque, 15. Leganés 28911 Madrid. España

Abstract

Ni₃Al particulate reinforced AMC's (aluminium matrix composites) are emerging as very interesting materials for structural use. The correlation of the increased mechanical behaviour is based on the interface continuity of the matrix with Ni₃Al particles and their toughness characteristics. This investigation analyses heat treatment parameters and the reaction products of the interface. Specimens were prepared by powder metallurgy process comprising mixing of powders, compaction, heating and finally extrusion of the composite. Different heat treatments covering a whole range of temperatures and times were carried out. Several layers of reaction products appeared in the interface, depending upon temperature and time. Several analysis were realised including SEM, EDX microhardness, and fractography of specimens fractured by impact under cryogenic temperatures.

1. Introduction

In recent years, Aluminium Matrix Composites (AMC's) have gained importance broadly in applications where good mechanical properties and weigh savings are required. Reinforcements based on particles, whiskers and fibres of Al₂O₃ and SiC have been well investigated and understood [1-5].

Nowadays, intermetallics are beginning to be used as reinforcements in aluminium alloys, being aluminium nitride Ni₃Al one of the most interesting. Some subjects of research have been focused on studying the improvements made in mechanical and friction characteristics of some aluminium alloys[6, 7].

On the other hand, aluminium alloys of interest are those with increased strength by means of precipitation hardening. These processes involve the use of relatively high temperatures, mainly in the solution treatment, and time is controlled in order to assure the whole solution of alloying elements. However, these heat treatments have an important influence over the matrix-reinforcement interface. Since the correlation of the increased mechanical behaviour is based on the interface continuity of matrix with Ni₃Al particles and their toughness, we think that the study of the intermetallic-matrix interface evolution would be of interest regarding different treatment conditions.

2. Experimental

The aluminium alloy used was obtained by mechanical alloying of the elements in the theoretical composition in weight pct. as follows :Cu: 4.5, Si:0.7, Mg:0.5, Al: bal. [8]. Ni_3Al intermetallic, obtained by rapid solidification atomising process (RST), was supplied by CENIM-CSIC (Spain). An evaluation of the production process and morphology of the powder obtained was reported by Pérez et Al [9]. Both powders were alloyed in a lab mixer, with a total amount of intermetallic compound of 5% in weight, compacted, heated (500°C, 30 min.) and extruded with a reduction ratio of 25:1 to form a bar shape of about 5 mm. in diameter. This process is described by da Costa et. al. [6], and has proven to give good results.

The heat treatment parameters where chosen in order to cover a whole range of possibilities. They could be summarised in three main groups:

• as extruded and ageing at different temperatures and times:

Temperature, °C	Times
160° C	2, 18, 48 h
177° C	2, 8, 24 h

These treatments were carried out in order to simulate a process of ageing, by considering that the solution treatment was developed by the heating prior to the extrusion (500°C, 30 min.), and supposing a quenching in calmed air.

• as extruded and solution treatment 30 minutes at different temperatures:

Temperatures(°C)	Time
250, 350, 400, 450, 475, 500	30 min.

These treatments were developed to see the evolution of the interface matrix-reinforcement for a standard solution time and different temperatures.

• as extruded and solution treatment at 500°C, and different times:

Temperature	Times
500° C	2, 4, 6 h

These last treatments were done to emphasise the diffusion processes in order to study the development of the interfaces beyond the usual procedures for solution treatment of aluminium alloys.

Different techniques were used to study the evolution of the Ni₃Al-matrix interface: Optical Microscopy, Scanning Electron Microscopy (SEM) by means of a JEOL JSM 6300, and Energy Dispersive X-Ray analysis (EDX) with a Link Isis system. Also, microhardness tests (Matsuzawa MHT 2) were conducted to ascertain the relative values of hardness between the different phases that appeared.

3. Results and Discussion

For the two first groups of heat treatments it was found that there was not an appreciably evolution of the interface, and only some tiny particles of secondary phases appeared attached to the Ni₃Al intermetallic particles. The microstructures of these specimens where closely similar to the one developed after extrusion. This appears to be in consonance with the literature regarding the stability of these compounds for temperatures below 300° C [10]. Even for the specimens of the second group, with some heat treatments above this temperature, and because the relatively short time for solid diffusion used, analogue results were obtained (Fig. 1a).

However, in the third group, involving higher temperatures and times where diffusion mechanisms are enhanced, several changes were observed. For 500°C and 2 hours, a higher amount of reaction phase developed around de Ni₃Al intermetallic particles, in comparison to the samples in the as extruded condition or the ones from the first two groups (Fig. 1b).

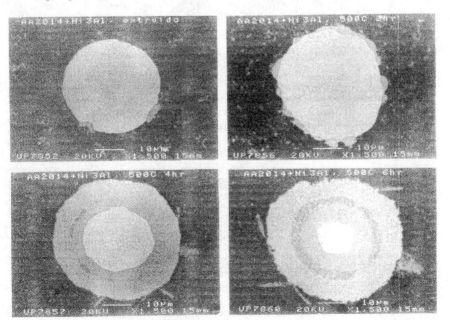

Figure 1. Interface development for different heat treatments: a) as extruded, b) 500°C, 2 hours, c) 500°C, 4 hours, d) 500°C, 6 hours.

Again, for the same temperature and 4 hours of treatment, it was clearly seen the development of two new phases along with the observed before. In this case, the Ni_3Al particles appeared surrounded for two concentric and continuous rings or layers and another discontinuous between the previous layers (Fig. 1c). The outer one (layer 1), had a similar composition that the intermetallic phases attached to the reinforcements initially obtained after extrusion (see microanalysis below).

At last, for 6 hours at the same temperature, the same three different reaction phases appeared but being wider than in the sample before. Also, the intermediate layer (layer 2) appeared now as an almost continuous ring.

Qualitative and quantitative line analysis where performed to see the evolution of the main elements involved in the diffusion process. The first of these comprised of different line scan and mapping of zones of interest. In these figure (Fig. 2) is clear the development and increase in width of the reaction layers and evolution of elements with respect to the time of heat treatment.

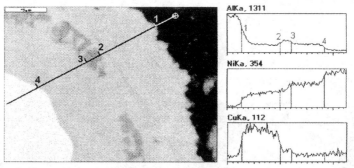

Figure 2. X-ray linescan of intermetallic particle in specimen heat treated at 500°C for 6 hours. a) SEM microgreaph showing the line of analysis. b) X ray counts.

A qualitative difference in composition of the three layers is shown, yielding some interesting results. Layer 1, the first to form, has a relatively high amount of copper. On the other hand, the layers that form in later stages and are inside the first one (layer 2 and 3) have a very low content of copper and in several cases negligible. This result may also be seen in the mapping of these layers (Fig. 3).

By comparison with the SE image, it is clearly shown the difference between layer 1 and 3:

1. the intensity of X-ray detection for Ni is higher in layer 3 than in layer 1, as could have been expected, because of closeness of layer 3 to the initial Ni_3Al.
2. in the case of copper, its presence is important in layer 1, and one remarkable thing is that this content is higher than in the matrix, as an indication of some degree of migration of atoms and concentration in this layer. In the other layers this content is very low.
3. for aluminium, the intensity of X-ray detection is homogeneous in both layers.

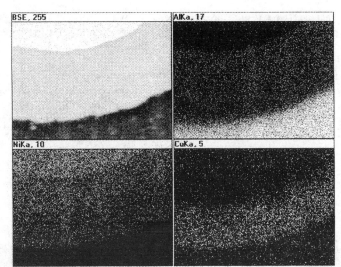

Figure 3. Map of elements of a intermetallic particle and reaction layers in specimen heat treated at 500°C for 4 hours

Quantitative analysis consisted of data collection in several points of X-ray analysis aligned in a row with a constant separation, where a quantification of elements present were performed. Figure 4 shows the points of analysis and the composition of each layer graphically. From these data, new interesting results may be inferred, considering composition and phases present:

1. The outer layer, number 1, can be assimilated to an $Al_3(Ni, Cu)_2$ phase, analogue to the Al_3Ni_2 phase shown to develop in diffusion pairs in literature [11-14]. This is in agreement with the total solution relation between Ni and Cu, that could explain the substitution of some atoms of Ni by others of Cu.
2. In layer 2, there is an steep decrease of Cu content and also an increase in aluminium, being this phase stoichiometrically closer to Al_3Ni.
3. Finally, composition of layer 3 falls in the zone of Al_3Ni_2.

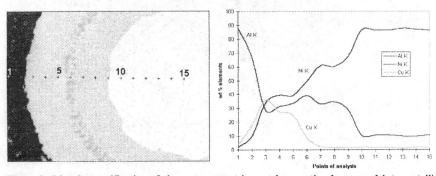

Figure 4. Lineal quantification of elements present in matrix, reaction layers and intermetallic particle of the specimen heat treated at 500°C, 6 hours.

This last two results are in agreement with Tsao et al. [14] who found an appreciable absence of Cu in the Al_3Ni and Al_3Ni_2 intermetallics developed in diffusion pairs between aluminium copper alloys and nickel.

For determining the relative strength of the phases developed, microhardness test were conducted. Figure 5 shows how the reaction layers formed have higher hardness even than the Ni_3Al intermetallic, being layer 3 the one with the highest. Due to the smallness of the Vickers hardness indentation marks done, their values were obtained from SEM images.

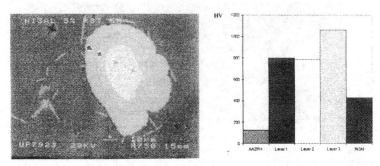

Figure 5. SEM image of vickers microhardness test indentations and hardness correlation.

Finally, to evaluate the fracture behaviour of the materials under study, some cryofracture test were carried out. This technique was used to try to observe the fragile behaviour in a more accentuated manner. Temperature used was that of liquid nitrogen (77°K). Fracture was produced by an impact tool manually driven. This was located 'in situ' in a special chamber of the SEM apparatus for immediate observation, thus avoiding any possible contamination.

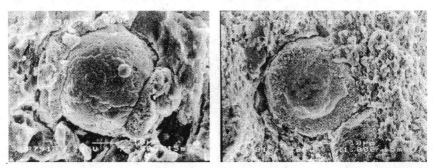

Figure 6. a) Fracture of specimen heat treated 2 hours at 500°C. b) Fracture of specimen heat treated 4 hours at 350°C.

These tests showed the different fracture mechanisms involved with the heat treatments carried out. As shown in figure 1, for low temperatures and times, where the reaction layer 1 is not completely developed around the intermetallic particles (below 2

h. at 500°C), the fracture progressed encircling the interface of these. In the other case, for higher times, although it has to be considered that together with the increased hardness there is a loss in toughness, the fracture opened out through the reaction layers and the remaining Ni$_3$Al particle. Nevertheless, this was indicative that for the latter case the interface continuity was assured. Therefore, it could be inferred from these results that in the former case the composites have a weak interface whereas in the latter this bond is much stronger. This result could be important considering the enhancement in strength and elastic modulus promoted by a strong interface between matrix and reinforcement in metal-matrix composites

4. Conclusions

When dealing with the manufacture of aluminium matrix composites reinforced with intermetallics, the development of reaction layers due to heat treatments is a very interesting point to take in consideration because the changes in the interface nature that come together this processes.

This work has demonstrated that the matrix-reinforcement continuity of these materials could be achieved trough the adequate degree of reaction between the intermetallic and the aluminium alloy. In this case, this was accomplished at least after a heat treatment more than 4 hours at 500°C,· when a complete reaction layer was developed.

Three different reaction layers were observed to develop for these specimens for high temperatures and times, with an important concentration of copper in the outer layer.

5. Acknowledgements

This research was supported by project MAT96-0722-C02-02 "Plan Nacional de I+D", CICYT, Spain.

6. References

[1]   T. S. Srivatsan, I. A. Ibrahim, F. A. Mohamed and E. J. Lavernia. Journal of Materials Science, Vol. 26, 5965-5978, (1991).
[2]   Aluminium-Matrix Composites.   Aluminium and Aluminium Alloys, ASM International 160-179, (1993).
[3]   D. J. Lloyd. International Materials Reviews, Vol 39, 1-23, (1994).
[4]   B. C. Pai, G. Ramani, R. M. Pillai and K.G. Satyanarayana. Journal of Materials Science, Vol.30, 1903-1911, (1995).
[5]   A. S. Chen, R.S. Bushby,· M. G. Phillips and V. D. Scott. Proc. R. Soc. Lond. A, 450, 537-552, (1995)

[6]   C.E. da Costa, W. Zapata, J.M. Torralba, J.M. Ruiz-Prieto and V. Amigó. "P/M MMC's base aluminium reinforced with Ni_3Al intermetallic made by mechanical alloying route". Materials Science Forums. Vols. 217-222, pages 1859-1864, (1996).

[7]   C. Díaz, J.L. González-Carrasco, G. Caruana and M. Lieblich. "Ni_3Al intermetallic particles as wear-resistant reinforcement for Al-base composites processed by powder metallurgy". Metallurgical and Materials Transactions A, Volume 27^a, 3259-3266, (1996).

[8]   C.E. da Costa.Thesis for PhD degree. "Obtención de materiales compuestos de matriz de aluminio reforzados con intermetálicos vía pulvimetalúrgica. Estudio y optimización de la aleación base y los intermetálicos obtenidos por aleación mecánica", Thesis for PhD degree, UPM, Madrid (1998).

[9]   P. Pérez, J.L. González-Carrasco, G. Caruana, M. Lieblich and P. Adeva. Mater. Charact., vol. 33, 349-356, (1994).

[10]   J.L. González-Carrasco, F. García-Cano, G. Caruana and M. Lieblich. Materials Science and Engineering, vol. A183, L5-L8, (1994)

[11]   A. J. Hickl and R. W. Heckel. Metallurgical Transactions A, Vol. 6^a, 431-440, (1975).

[12]   S. B. Jung, Y. Minamino, T. Yamane and S. Saji. Journal of Materials Science Letters,Vol. 12, 1684-1686, (1993).

[13]   D. C. Dunand. Journal of Materials Science, Vol. 29, 4056-4060, (1994).

[14]   C. Tsao, S. Chen. Journal of Materials Science, Volume 30, 5215-5222, (1995).

Interfacial reaction and applicability of Al$_{18}$B$_4$O$_{33}$/Al composites

H.Fukunaga, J.Pan, G.Sasaki and L.J.Yao

Department of Mechanical Engineering, Hiroshima University, Kagamiyama 1-4-1,
Higashi-Hiroshima, 739- Japan

Abstract

In the present work, Al$_{18}$B$_4$O$_{33}$ whisker reinforced pure Al, Al-9%Cu, 6061Al and
AC8A composites have been fabricate by squeeze casting process. Pure aluminum did
not react with whisker during casting, either by XRD for extracted whisker from
composite, or by TEM observation of the interface. Tensile strength and bending
strength of Al$_{18}$B$_4$O$_{33}$/pure Al were 248MPa and 436MPa. After heated at 530°C for 1h
they were not down(kept 267MPa and 459MPa respectively). But if Al$_{18}$B$_4$O$_{33}$/pure Al
was heated at a temperature higher than 726°C, it was easy to detect interface product
Al$_2$O$_3$. Furthermore, a whisker was found to have some particle-like substances sticking
on it, after heating at 730°C for 10min and to be run out after heated at 800°C for
10min. Composite bending strength decreased along with heating temperature. In Al-
9%Cu matrix composite, interfacial reaction feature was similar to pure aluminum.
After T6 treatment composite strength was enhanced from 560MPa to 670MPa. 6061Al
and AC8A matrix reacted with whisker during casting. A spinel phase MgAl$_2$O$_4$ was
observed at interface and grew drastically after composite was re-heated at about 520°C.
The applicability of Al$_{18}$B$_4$O$_{33}$/Al composites have been discussed from the view of
interfacial compatibility.

1. Introduction

Many kinds of whiskers have been developed as reinforcements of metal matrix
composites. As the representative ones, SiC and Si$_3$N$_4$ whisker are often mentioned for
their chemical stability and excellent mechanical properties in aluminum matrix
composites. But their cost may become a fatal factor to limit their application. In recent
years, some inexpensive whiskers have been of great interest, such as K$_2$O·6TiO$_2$,
Al$_{18}$B$_4$O$_{33}$, MgO, TiO$_2$, for reinforced aluminum composites [1-5]. Among these whiskers,
aluminum borate(Al$_{18}$B$_4$O$_{33}$) whisker is considered as a desirable one because of with a
low price and a higher strengthening ability [6-8]. If a whisker is to become a competitive
reinforcement candidate, besides its property and cost, it is hopeful to use a simple
process in fabrication of composites, and to keep a fairly stable compatibility between
whisker and aluminum matrix in composites preparation and application environment.
In the present work, a squeeze cast process was used to fabricate Al$_{18}$B$_4$O$_{33}$ whisker
reinforced aluminum alloy composites. Chemical reaction of whisker with different

aluminum alloy matrices was investigated by means of SEM, XRD, HRTEM and so on. The purpose is to evaluate this whisker's compatibility with aluminum alloy and present a reference for the application of $Al_{18}B_4O_{33}$/Al composites.

2. Experiment Methods

Aluminum borate($Al_{18}B_4O_{33}$) whisker, supplied by Shikoku Chemicals Industry in Japan, was used as reinforcement. The whiskers are octagonal prisms generally with four wide {120} side surfaces and four narrow {100} side surfaces. The whisker growth direction belongs to [001] [7]. The following aluminum matrices were employed respectively: pure aluminum(Al>99.99%), 6061Al(Si:0.4-0.8%, Mg:0.8-1.2%, Cu:0.15-0.4%, Al:bal.), AC8A(Si:12%, Mg:0.96%, Cu:0.91%, Al:bal.) and Al-9%Cu. The composites were prepared by a squeeze casting method at the aluminum melt temperature of 720℃ and 800℃, and the preform preheating temperature of 500℃ and 700℃, respectively. After extracted from the composites in different states by 15%-HCl aqueous solution, the whisker and composites were analyzed by means of XRD(JEOL/JRX-12VC type X-ray diffractometer), SEM(HITACHI/S800 type scanning electron microscope) and TEM(PHILLIPS/M12 type and JEOL/2000EX-II type transmission electron microscope), respectively.

3. Results and Discussion

3.1 Pure Al matrix composites
$Al_{18}B_4O_{33}$/pure Al composites were fabricated at different melt pouring temperatures from 720℃ to 840℃. Flexural test result shows that their bending strengths were in the same level. Tensile strength was 248MPa and 267MPa in as-cast state and after heated at 530℃ for 1h, respectively. Furthermore, composites were heated for 10min at different temperatures and their whiskers were extracted from the heated composites. Fig.1 shows SEM appearance of the extracted whiskers. Whisker surfaces in Fig.1(a) and Fig.1(b) are clear and smooth, no reaction product can be seen. But in the case of the whisker extracted from composite after heated at 750℃ some particle-like products have been found sticking on whisker surface as shown in Fig.1(c). From the observation of the extracted whisker from the composite after heated at 800℃, whiskers have almost become fragments as shown in Fig.1(d). Fig.2 corresponds to the XRD results for the cases above. For the composite after heated at 660℃ and 700℃, no other diffraction peaks appeared except whisker itself(Fig.2a and Fig.2b). In the whisker from the composite heated at 750℃ there existed peaks of γ-Al_2O_3 and δ-Al_2O_3 (Fig.2c). For the whisker from composite heated at 800℃, the peaks of δ-Al_2O_3 and γ-Al_2O_3 increased in a large scale and ones of whisker went down as shown in Fig.2(d). This indicates that by the reaction between whisker and aluminum, whisker will be run out finally.

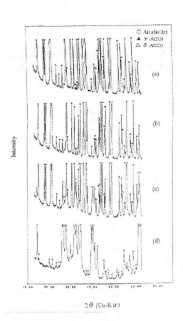

Fig.1 SEM photographs of the whiskers extracted from $Al_{18}B_4O_{33}$/pure Al composites after heated for 10min at different temperatures: (a) 660℃, (b) 700℃, (c) 750℃ and (d) 800℃.

Fig.2 X-ray diffraction patterns of the whiskers extracted from $Al_{18}B_2O_{35}$/pure Al composites after heated for 10min at different temperatures: (a) 660℃, (b) 700℃, (c) 750℃ and (d) 800℃.

DTA (differential thermal analysis) result shows that there was an obvious exothermic peak from about 726℃ during heating $Al_{18}B_4O_{33}$/pure Al composites at a heating rate of 5℃/min [9]. By thermo-dynamical calculation, Gibbs free energy of the reaction between $Al_{18}B_4O_{33}$ and aluminum at 726℃ is –689kJ/mol. There-fore, the possible reaction during heating higher than 726℃ is considered as:

$$Al_{18}B_4O_{33} + 4Al \rightarrow 11\gamma\text{-}Al_2O_3 + 4B \quad (1)$$

By means of a high resolution TEM, nuclea-tion of reaction product at the beginning of interface reaction was identified. Fig.3 shows such an interface structure of $Al_{18}B_4O_{33}$/pure Al composite after heated at 730℃ for 5min.

Fig.3 HRTEM image of an extracted whisker from $Al_{18}B_4O_{33}$/pure Al composite after heated at 730℃ for 5min.

γ - Al_2O_3 phase can be seen nucleating on the whisker.

Flexural test results show that bending strengths of the composite after heated at 750°C and 770°C were 143MPa and 120MPa, respectively, whereas one in as-cast state was 235MPa. This means that the chemical reaction is detrimental for the composite property.

3.2 Al-9%Cu matrix composites

In the case of $Al_{18}B_4O_{33}$/Al-9%Cu composites, it has been noticed that the bending strength enhanced from 560MPa(as-cast) to 670MPa(T6 state). This indicates that copper existence does not aggravate interfacial reaction. Similar to pure aluminum matrix composites, γ-Al_2O_3 and δ-Al_2O_3 began to be found in the extracted whisker from composite after heated at 750°C for 10min in its XRD pattern. Particle-like products can be observed on whisker surface. Moreover, if Al-9%Cu matrix composites and pure Al matrix composites were heated in the same condition of 800°C for 72h, respectively, their extracted whisker had a similar phase constituent: dominant phase Al_2O_3 and less phase $Al_{18}B_4O_{33}$. No any product phase containing copper was found. Therefore, copper in matrix does not participate in the interfacial reaction between whisker and aluminum.

3.3 6061Al and AC8A matrix composites

In these two aluminum alloys, magnesium amount is about 1wt%. The difference is that 6061Al alloy only has 0.4-0.8wt% silicon and AC8A alloy has 12wt% silicon. Bending strength of $Al_{18}B_4O_{33}$/6061Al composite(as-cast) was 649MPa, did not increase after T6 treatment(bending strength was 636MPa in T6 state). From Saito et al's results, it can be known that $Al_{18}B_4O_{33}$/6061Al exhibited 350MPa and 393MPa in tensile strength for as-cast and T6 state, respectively, whereas the matrix strength was increased one times after T6 treatment[10]. Such experiment result indicates that there should be a chemical reaction happened during T6 treating. In fact, from the as-cast composite, the reaction product as shown in Fig.4 was found. It can be seen that whisker has lost its smooth surface and became uneven. By EDP(electron diffraction pattern) and EDS(energy dispersive spectrometer) the product sticking on the whisker has been identified as a spinel phase $MgAl_2O_4$. After T6 treatment, there were more and larger $MgAl_2O_4$ in composite interface [11]. The formation of $MgAl_2O_4$ nibbles whisker and hence damages whisker. As for AC8A alloy matrix composites, the interface reaction occurred during heating at about 520°C. Fig.5 gives TEM image comparisons of pure Al and AC8A matrices composite after heated at 520°C for 3h. Under the same treating conditions, there was not a reaction between whisker and pure aluminum, but so many particle-like reaction products were seen around the whisker in the case of AC8A composites. From HRTEM observation, this reaction product was confirmed as $MgAl_2O_4$ (Fig.6). DTA result shows that there was an exothermic peak at about 517°C, which also verified the reaction between whisker and AC8A alloy near this temperature.   Because of interface

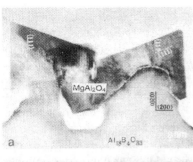

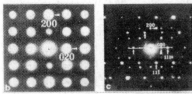

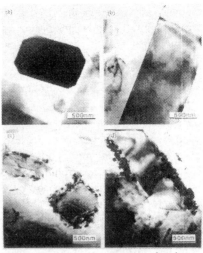

Fig.4 HRTEM image of an interfacial reaction product in as-cast $Al_{18}B_4O_{33}$/6061Al composite (**b** and **c** are EDP of whisker and the combined EDP of product and whisker).

Fig.5 TEM photographs showing interface states of $Al_{18}B_4O_{33}$/pure Al (a,b) and $Al_{18}B_4O_{33}$/AC8A (c,d) composites after heated at $520^\circ C$ for 3h.

reaction, the bending strength of $Al_{18}B_4O_{33}$/AC8A composites reduced from 660MPa(as-cast state) to 590Mpa (T6 state). Moreover, the bending strength of this composite went down drastically along with increase of heating time at $520^\circ C$. It is considered that $Al_{18}B_4O_{33}$ whisker begins to react with magnesium in aluminum alloy from about $520^\circ C$ as:

$$4Al_{18}B_4O_{33} + 33Mg \rightarrow 33MgAl_2O_4 + 16B + 6Al \qquad (2)$$

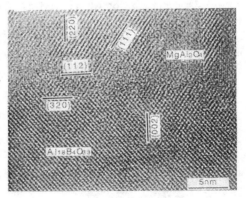

Fig. 6 HRTEM image of interface structure between $Al_{18}B_4O_{33}$ whisker and $MgAl_2O_4$ phase.

If heating temperature is higher than about $726^\circ C$, meanwhile reaction between $Al_{18}B_4O_{33}$ and aluminum (eq.1) will occur.

3.4 Discussion on interfacial reaction and applicability of $Al_{18}B_4O_{33}$/Al composites

If a whisker reinforced aluminum alloy composite become available in practice, at least two aspects should be considered. One is that their properties must be superior to the conventional materials. Another is that whisker-matrix must be stable chemically. Although $Al_{18}B_4O_{33}$ has a good compatibility with pure aluminum matrix, the mechanical property of its composite cannot be competitive with aluminum alloy. $Al_{18}B_4O_{33}$ whisker reinforced Al-Cu alloy may become an applicable composite, because there is not any observable interfacial reaction during squeeze casting, and composite property can be improved obviously by matrix age hardening(T6 treatment). Regarding to Mg-containing aluminum alloy(e.g. 6061Al , AC8A, ADC12 etc.), whisker/matrix interface is fairly unstable, a spinel phase $MgAl_2O_4$ can be found easily whether during squeeze casting or re-heating at higher than 520℃, which has been confirmed by the present work or others [6,7,12,13]. Therefore, some routes have been tried to prevent $Al_{18}B_4O_{33}$ whisker from reacting with these aluminum alloys. Norio et al [14] developed an Al_2O_3 coating process. By this coating treatment $Al_{18}B_4O_{33}/AC4C$ composite strength could increase about 50MPa compared with that of non-coated composite. But this Al_2O_3 coating did not improve the strength of composite after T6 treatment. Kitamura et al [15] reported an oxide coating method. $MgAl_2O_4$ coating can be created by heat treatment after mixing small particles of $Mg(OH)_2$ and Al_2O_3 uniformly with whisker. In surface treated $Al_{18}B_4O_{33}/ADC12$ composite, no reaction product was observed because of existence of $MgAl_2O_4$ layer as a reaction barrier. Compared with non-surface treated composite, the tensile strength at 350 ℃ for surface treated composite increased about 30%. Shintari et al [16] introduced a low-cost surface modified method for $Al_{18}B_4O_{33}$ whisker. By a NH_3 gas-nitriding process, AlN, θ-Al_2O_3 and α-Al_2O_3 could grow on whisker surface. TEM observation of the surface modified whisker/AC8A composite showed that θ-Al_2O_3 and α-Al_2O_3 work as reaction barrier. This composite has a high tensile strength(160MPa) and fatigue strength(90MPa) at 623K, so has been used to a piston and a combustion chamber edge of diesel engine.

4. Conclusions

1. Interfacial reaction between $Al_{18}B_4O_{33}$ whisker and pure aluminum matrix was not found both in as-cast composites and in heated composites below 660℃. After heating at 726℃ for 10min, reaction product phase Al_2O_3 was discovered in composite interface. Whisker would be exhausted completely by interfacial reaction if raising temperature or extending time for heating.

2. Copper in matrix did not participate in the reaction of whisker with aluminum. After T6 treatment, $Al_{18}B_4O_{33}/Al$-9%Cu composite strength could be improved by matrix age hardening.

3. In the interface of 6061Al and AC8A matrix composites, both in as-cast state and in heated state, it was easy to observe the product of $MgAl_2O_4$. Because of such reaction, T6 treatment could not enhanced composite strength.

References

[1] H.Fukunaga, J.Pan and D.M.Yang, Fabrication and properties of new whisker reinforced aluminum alloy composites, C-MRS International'90 Beijing, China, 1990.6, Advanced Structural Materials, Ed.by Y.Han, Elsevier Science Publishers B.V.(1991)45-50.

[2] K.Suganuma, T.Fujita, N.Suzuki and K.Niihara, Aluminum composites reinforced with a new aluminum borate whisker, J. Mater. Sci. Lett., 9(1990)663-635.

[3] H.Harada, Y.Kudoh, Y.Inoue and I.Tsuchitori, Preparation and mechanical properties of AC8A aluminum alloy composite reinforced with potassium titanate whisker, J. Japan Inst. Metals, Vol.58, No.1(1994)69-77.

[4] J.Pan, X.G.Ning, J.H.Li, H.Q.Ye, H.Fukunaga, Z.K.Yao and D.M.Yang, Properties and interfacial structures of $K_2O \cdot 6TiO_2$ whisker reinforced aluminum composites, Composite Interfaces, Vol.4, No.2(1996)95-109.

[5] J.Pan, J.H.Li, H.Fukunaga, X.G.Ning, H.Q.Ye, Z.K.Yao and D.M.Yang, Microstructural study of interface reaction between titania whisker and aluminum, Composites Science and Technology, 57(1997)319-325.

[6] K.Nagatomo and K.Suganuma, Fabrication of aluminum borate whisker preform without binder and characteristics of 6061 aluminum alloy matrix composites, J. Japan Inst. Metals, Vol. 58, No.1(1994)78-84.

[7] X.G.Ning, J.Pan, K.Y.Hu and H.Q.Ye, Transmission electron microscopy studies of the interface microstructures in a $Al_{18}B_4O_{33}w/6061Al$ composite, Materials Letters, 13(1992)377-381.

[8] L.J.Yao, G.Sasaki and H.Fukunaga, Reactivity of aluminum borate whisker reinforced aluminum alloys, Materials and Science Engineering, A224(1997)59-68.

[9] L.J.Yao, G.Sasaki, M.Yoshida and H.Fukunaga, Interfacial reaction and interface structure of $Al_{18}B_{33}w/Pure$ Al composites, Proc. 5th Japan International SAMPE Symposium, Tokyo, Oct.28-31, 1997, pp.423-428.

[10] N.Saito, M.Nakanish and Y.Nishida, Effect of heat treatment on the mechanical properties of aluminum-borate whisker reinforced 6061 aluminum alloy, J. Japan Inst. Light Metals, Vol.44, No.2(1994)86-90.

[11] X.G.Ning, J.Pan, J.H.Li, K.Y.Hu, H.Q.Ye and H.Fukunaga, The whisker-matrix interfacial reactions in SiC, Si_3N_4 and $Al_{18}B_4O_{33}$ whisker reinforced aluminum matrix composites, J. Mater. Sci. Lett. 12(1993)1644-1647.

[12] K.Suganuma, G.Sasaki, T.Fujita and N.Suzuki, Interfacial reaction between aluminum borate whisker and AC8A and 6061 aluminum alloys, J. Japan Inst. Light Metals, Vol.41, No.5(1991)297-303.

[13] N.Nishino and S.Towata, Effects of Si and Mg contents on mechanical properties of aluminum-borate whisker reinforced aluminum alloys, Proc. 5th Japan International SAMPE Symposium, Tokyo, Oct.28-31, 1997, pp.429-434.

[14] N.Norio, I.Tuchitori, M.Iwasaki and H. Hata, Mechanical properties and heat treatment of Al_2O_3 coated aluminum borate whisker/Al alloy composites, Report of Western Hiroshima Prefecture Industrial Research Institute, No.36(1993)48-51.

[15] T.Kitamura, K.Sakane, H.Wada, H.Hata and Y.Shintari, Development of composite materials of aluminum borate whisker, Proc. International Workshop on Advanced Materials for Functional Manifestation of Frontier and Environmental Consciousness, Tokyo, Sept.17-18, 1997, pp. 11-18.

[16] Y.Shintari, Y.Okochi and M.Sugiyama, Aluminum base metal matrix composite reinforced with low-cost surface-modified $9Al_2O_3 \cdot 2Al_2O_3$ whisker, Proc. 5th Japan International SAMPE Symposium, Tokyo, Oct.28-31, 1997, pp.399-404.

Fibre-Matrix Interactions in an Intermetallic Matrix Composite, Base Ni₃Al, Reinforced With SiC Fibre

C.Testani*, M. Di Stefano°, M.Marchetti°, A. Ferraiuolo**.

* Centro Sviluppo Materiali, S.p.A, CSM Via di Castel Romano 100, 00128-Rome- Italy.

° Aerospatial dept. , University of Rome "La Sapienza", Rome Italy.

** CSM Consultant.

Abstract

Fibre-matrix interactions in an intermetallic matrix composite have been investigated. A Ni₃Al-based intermetallic alloy, under the form of thin foils, was first characterised to assess the material properties. The foil matrix was then assembled with long SiC fibres tape, as fibre-foil lay-up. The composite obtained after compaction and densification by H.I.P.-D.B. (Hot Isostatic Pressing-Diffusion Bonding) technology, was subjected to microstructural and mechanical characterisation. A complete compaction of the matrix occurred by means of a fibre-matrix interface reaction. The chemical analysis highligted a deep Ni diffusion in the fibre core. Definite discontinuities in the atomic composition were detected which highlighted the boundaries of different interaction zones. The tensile tests carried out on the composite specimens confirmed the deleterious influence of the fiber matrix interaction. It is suggested that the fiber matrix interaction produced formation of Ni₅Si₂ and carbon precipitation.

The further advancement of SiC fibre reinforced Ni₃Al matrix composite requires the development of a chemically inert, compact and homogeneous coating of the SiC fibers, prior their insertion between the Ni₃Al foils and the subsequent HIP compaction.

Introduction

The performance of turbo-engines is conditioned by the maximum operative temperature of some critical components, such us turbine blades and discs, combustion cases, etc.., hence by their constituent materials.

Superalloys are successfully employed, but the gain obtained through their successive generations is becoming progressively lower [1].

Significant improvements of the performance of the future turboengines can be expected by considering new families of materials for high temperature applications, [1].

Intermetallics, and moreover, Intermetallic Matrix Composites (IMC) show potential applications, although there are still problems to be solved before their production and utilisation, [2, 3].

In particular, the assessment of the fibre-matrix interactions is a problem actually object of several research works [4, 5, 6, 7]. These interactions may affect the structure and performance of the material in terms of temperature and mechanical stress during service at long time exposure to extreme conditions.

The present work concerns an investigation of the interactions occurring during the compaction and densification by HIP-DB (Hot Isostatic Pressing Diffusion Bonding) of a Ni₃Al based matrix, reinforced by SiC long fibres and assembled as fibre-foil lay-up.

EXPERIMENTAL

The adopted matrix has been intermetallic alloy foils, Ni₃Al-based and its the chemical composition is given in Tab. I. SiC long fibres (Commercial Avco SCS-6) having 140 μm of diameter were selected as reinforcement. The EDS-SEM fibre section analysis has put in evidence an inner core of carbon 33 μm of diameter, surrounded by a SiC layer. Moreover the fibre is coated with a 4μm Carbon film enriched, at the outer surface.

The matrix foils (100 μm of thickness) were subjected to mechanical testing and microstructural analysis by the optical and the Scanning Electron Microscope (SEM).

Al	Zr	B	Ni
19,5	0,6	450ppm	Bal

Tab.I: Ni₃Al matrix Chemical composition (%at)

The mechanical tests were performed according to the standard UNI-EN 10002. The tensile tests performed at room temperature on the Ni₃Al foil show results in good agreement with the literature (Tab.II). After the composite compaction the mechanical tests, on three specimens, according with the ASTM-1012, showed a clear brittle behaviour with a definite cleavage fracture at 260 MPa.

The matrix microstructure shown in Fig. 1, (the specimens, has been etched with "Marble solution", 10g. CuSO₄ + 50 cc. HCl (37%) + 50 cc. H₂O -ASTM E-407- for 20 seconds) is equiaxial with a mean grain size of 5,5 μm.

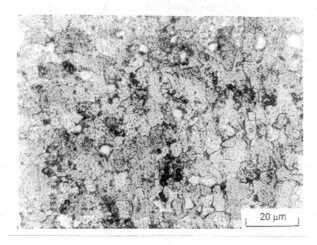

Fig. 1 Ni3Al matrix microstructure

	YS$_{0.2\%}$ (Mpa)	U.T.S. (MPa)	A(%)	E(GPa)
Mean value on three specimens	521±16	1189±11.5	23.1±3.5	183±5.5

Tab.II: Room Temperature tensile tests (mean value on three specimens) on thin foils of Ni$_3$Al.

Nine thin foils of Ni$_3$Al based matrix were assembled with eight SiC fibres-layers and densified by HIP-DB technology. The hold-conditions and the parameters of the densification cycle were chosen in agreement with the literature [8,9,10,11] and taking into account the temperature range for matrix plasticity because the intermetallic foils had to accommodate its shape around the fibres.

After curing at 1150°C and 1200bar for 3 hours, two flat-plates of composite (70 x 140 x 1,28 mm) were obtained without shape distortions with an optimum outer appearance (fig.2).

Mechanical [and metallographic] tests were carried out on specimens cut from two composite plates. SEM with Energy Dispersive Microprobe (EDS) was used to understand the interactions at fibre-matrix interface.

Fig. 2: Flat plates composites as HIPped.

Optical microscopic analysis revealed that the composite was fully and completely densified, and that the interactions deeply involved both the fibres and the matrix.

In particular, it was found that the fibres were subjected to an apparent increase in diameter, from the initial 140 μm up to about 200 μm, (Fig. 3). The 4 μm Carbon fibre coating was broken into fragments which were radially displaced outward. Several distinctive zones were found, with marked differences in appearance, while the carbon core of the fibres was unchanged.

A semi-quantitative SEM analysis revealed that the difference in appearance of the various zones is associated with differences in composition due to Carbon precipitation. Fig.3 reports the SEM micrography of the composite with zones named A, α, β, α$_1$, α$_2$ of different appareance.

This micrography shows that the atomic ratios between Ni and Al inside zones "B", "α$_1$" and "α$_2$" were different from the original-ones, thus it appears that these zones were involved in diffusive phenomena.

The "β" zones located far from the fibres interfaces were left unchanged and resulted of the same composition of the original matrix.

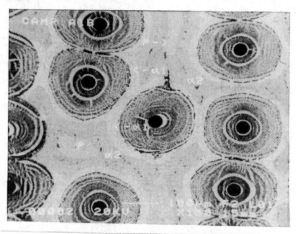

Fig. 3: SEM micrography of the composite

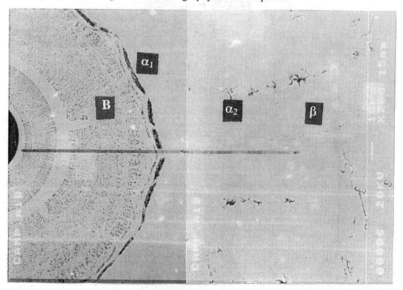

Fig. 4: SEM-EDS Profile location at the fibre-matrix interface

A semi-quantitative profile analisys were performed by SEM-EDS, proceeding radially from the carbon core of the fibre across the matrix, untill within the "β" zones.

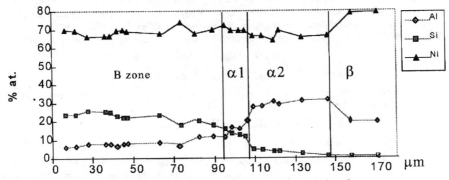

Fig. 5 SEM-EDS composition profile at the fibre-matrix interface

Definite discontinuities in the atomic composition were detected, which highlighted the borders of the different interaction zones.

From the EDS analysis, (profile is shown in Fig.4), it is possible to observe that:

- The "B" zone was characterised by an atomic ratio Ni/Si = 2,8 and Ni/Al = 9, with the presence of carbon precipitates.

- In the "α_1" zone, after a discontinuity in the Al atomic percentage at the boundary B-α_1 ($\Delta Al_{at.\%}$=5), the atomic percentage of this element rises again up to about 13%, while the percentage of Si decrease down to 10% .

- In the "α_2" zone, a Aluminium content of $30_{at.}$%, was detected while Si is practically absent.

- The "β" zone appears as the only one left untouched by the interactions and its composition is actually like that of the original matrix material.

DISCUSSION

The 4 μm Carbon coating applied by the producers on the SiC fibres has revealed to be completely ineffective as a barrier against such a reaction, probably because it was not intended for the purpose in concern. SEM-EDS quantitative analyses in the various zones of the composite shows that Ni is deeply diffused from the matrix into the fibres where it is completely absent in the as-received product.

The occurred reactions should be characterised by a significant thermodynamic driving force (ΔG) and by fast kinetics, because the interactions involved practically all the composite volume in the three hours, (relative short time), of curing.

The presence of the Carbon precipitates and the atomic ratios Ni/Al, Ni/Si detected at the different zones imply that the main interface reaction involve Ni diffusion, from Ni_3Al matrix, and Si from SiC fibres. If the possible binary compounds (no thermodynamic data are available for ternary phases) are considered, the reaction due to the Ni migration, should result in a Ni_xSi_y compound and Carbon precipitates, as products.

From a thermodinamic point of view all the involved phases are relatively pure at any temperature and show an activity value which can be taken as unity, [12], and hence even a moderate free energy difference between products and reactants is in principle capable of pushing the reaction down to a complete disappearance of the reactants.

The literature [4, 5, 6, 7], suggests a reaction between Ni_3Al in the matrix, and the SiC of the fibres, which produces Ni_2Si with $\Delta G_{1150 \,°C}= -15600$ J*mol^{-1} of Ni_2Si. However, a critical examination of the available thermodynamical data regarding binary compounds involving Ni and Si (Tab.3), shows that this product of reaction is liable and further reacts with Ni_3Al to produce Ni_5Si_2, according to reaction:

$$Ni_3Al + 4Ni_2Si => 2Ni_5Si_2 + NiAl$$

reaction ΔG is: $\Delta G_{1150 \,°C} = - 11800$ J*mol^{-1} of Ni_2Si (1)

The above considerations indicate the reaction (2) as the most probable among the ones giving out binary phases in agreement with the found atomic ratios:

$$Ni_3Al + 4/5\ SiC => NiAl + 2/5\ Ni_5Si_2 + 4/5\ C$$

reaction ΔG is: ($\Delta G_{1150 \,°C} = -55000$ J*mol^{-1} of Ni_5Si_2 (2)

Besides, on the basis of the data found in literature for the density of the reaction products (which are noticeably lower than those of the reactants), a simple calculation shows that reaction (2) is able to explain both the actual increase of the diameter of the fibres and the atomic composition of the "α_1" zone.

However, the deep Nickel migration inside the fibres, thus impoverishing the "α_1" and also the "α_2" zone of the matrix, accounts for the excess of Aluminium occurring in these zones. This situation and the relevant presence of the brittle Ni_5Si_2 phase inside the fibres justify the definite brittle behaviour of the entire composite, as revealed from mechanical tests which impairs the benefical effects of the Boron addition to Ni_3Al, [13, 14, 15].

Compound	ΔH (j*mol^{-1})	ΔS (j*mol^{-1} K^{-1})	ΔG 1150°C (j*mol^{-1})	Reference
$Ni_3Al^{(13)}$	-140000	-32,2	-103747	10
$SiC^{(17)}$	-111740	-8,4	-102260	16
$Ni_2Si^{(15)}$	-128770	5,28	-134710	7
$NiAl^{(13)}$	-118500	-28,04	-86955	4
$Ni_5Si_2^{(15)}$	-304430	-2,74	-301348	7

Tab.3: Literature data of free energy of formation for some Ni-Si compound.

Because of the relatively high free energy accompanying reaction (2) an effective kinetic barrier (coating) should be developed in order to separate the SiC of the fibre from the Ni_3Al of the matrix.

The coating should be inert, homogeneous and compact for the entire length of the fibre; in fact, even a point-defects, like localised porosities, randomly distributed among the fibres, are liable to allow reaction (2). A compact and homogeneous coating made of Alumina, or directly the use of Alumina fibres, might satisfy all the requirements, but in both cases the presence of Zr inside some Ni_3Al base alloy, should be avoided. This element in fact, is reactive with respect of Alumina and can again lead to the Aluminium enrichment of the Ni_3Al base matrix around the fibres, and therefore to the formation of brittle phases at least along the surface of fibres.

CONCLUSIONS

The problem of the reactivity between SiC-fibre and Ni_3Al-matrix has been studied in the present work. In particular, it was shown that during the composite production process the Ni_3Al matrix extensively reacts with SiC to form Ni_5Si_2 and Carbon precipitates in the fibres. This fact led to a significant increase of the Al/Ni atomic ratio in the matrix area. That impairs the beneficial effects of the Boron addition into Ni_3Al resulting in a complete enbrittlement of the matrix.. As a conclusion only inert fibres such as Al_2O_3 fibres might satisfy all the requirements for obtaining a composite based on Ni_3Al matrix showing good mechanical properties.

REFERENCES.

1)   J. R. Stephens: "Intermetallic and Ceramic Matrix Composite for 815 to 1370 °C (1500 to 2500 °F) gas turbine engine applications", in proceedings of TMS annual meeting in Anaheim, CA, U.S.A., "Metal and Ceramic Matrix Composites", 1990

2)   S. Froes: "The intermetallics: what they are and what they do offer?", Material Bullettin Monthly, pag. 34-39, 1991

3)   P. Brozzo: "Struttura e proprietà dei materiali metallici", vol. II parte I , E.C.I.G.,1979

4)   T.C. Chou: Scripta Metall. et Mat., vol.24, pag. 409-414, 1990.

5)   X. Liang, H. K. Kim, J. C. Earthman, E. J. Lavernia: Mat. Sci. Eng., A153, pag. 646-653, 1992.

6)   J. M. Yang, W. H. Kao, C. T. Liu: Mat. Sci. Eng., A107, pag. 81-91, 1989.

7)   M. Lindholm, B. Sundman: Metall. and Mat. Trans. A, vol.27A, pag. 2897-2903, 1996.

8)   M.F.Ashby et al.: Metallurgical trans., vol.14a, pag.211-221, 1983.

9)   Vinod K. Sikka: Mat. Res. Soc. Symp. Proc., vol.81, pag.487-492, 1987.

10)   A. Bose, M. Moore: High Temperature/High Performance, MRS vol. 120, 51-56, Pittsburg, PA,1988.

11)   W. B. Li, M. F. Ashby, K. E. Eastering: Acta Metallurgica,vol.35, n°12c, pag. 2831-2842, 1987E. M. Schulson and D.R. Barker, Scr.Metall, vol. 17, 519, 1983

12)   P. Silvestroni:"Fondamenti di Chimica" 7ª ed., Veschi ed. Roma, 1987.

13)   C. T. Liu, C. L. White, et al.: "High Temperature Materials Chemistry", II L.A.Munir & D Kubicciotti, electrochemical Society, vol 83, n°7, pag.32-41, Pennigton, N.J., 1983

14)   K. Aoki & O. Izumi: Nippon Kizoku Takkaishi, 43-1190, 1979

15)   J. M. Yang, W. H. Kao, C. T. Liu: Mat. Sci. Eng., A107, pag. 81-91, 1989.
       16)Handbook of Chemistry and Physics, 62nd ed., CRC press inc, 1982.

Ultralight MgLi matrices reinforced with discontinuous oxidic fibres and continuous carbon fibres

S. Kúdela
Institute of Materials and Machine Mechanics SAS, Bratislava, Slovakia

Abstract

Composite materials based on binary Mg-8%wt.Li and Mg-12%wt.Li matrices reinforced with either discontinuous δ-Al_2O_3 fibres or continuous T300 carbon fibres were manufactured by the pressure infiltration process. The chemistry of the reinforcements damage resulting from the reactive diffusion of both Li and Mg elements into fibre bulk is briefly depicted. Deformation behaviour and fracture of composite samples have been investigated regarding the phase composition of MgLi matrices and the extent of the fibre/matrix interaction in particular.

1. Introduction

Very low density (1.3-1.6 gcm^{-3}) and excellent ductility of Mg-Li based alloys make them an attractive candidate for aircraft and aerospatial applications in particular. Nevertheless, poor stiffness and low creep resistance of MgLi alloys are inferior to another competing light-metal materials and several works were carried out on polycomponent Mg-Li-X alloys (typically X=Al,Ag,Zn) with the aim to improve their strength and deformation characteristics. However, usual strengthening mechanisms (solid solutions, age hardening, work hardening) are either of low efficiency or the strengthening effect is instable due to high mobility of Li atoms and high vacancy concentration which encourage overageing phenomena and reduce activation energy for creep [1].

The composite approach offers the possible way how to overcome these drawbacks, in particular concerning the creep behaviour, as demonstrated by creep characteristics of MgLi alloys reinforced with continuous steel wires [2] and discontinuous SiC whiskers [3]. From technological and economical points of view the pressure infiltration is the favourable manufacturing method of these composites. In order to hold the advantage of low density of MgLi matrix the high-strength carbon fibres (1.8 gcm^{-3}) can be considered the most suitable reinforcement and also δ-Al_2O_3 fibres (3.2 gcm^{-3}) enhance the density of the composites only sligtly in case of small fibre volume fraction (< 20vol.%). However, the lithium presence increases radically the reactivity of MgLi melts against these reinforcements during infiltration process and the attempts to manufacture these composites have only limited success so far [4]. On the other hand, lithium is known as an alloying element improving the wetting of fibres [5].

Present paper is intended to characterize the structure, tensile deformation and fracture of composite materials based on Mg-8%wt.Li and Mg-12%wt.Li matrices reinforced with either discontinuous δ-Al_2O_3 fibres or continuous T300 carbon fibres which have been prepared by the pressure infiltration technique.

2. δAl₂O₃ /MgLi composites

Composite samples based on Mg, Mg8Li and Mg12Li matrices were manufactured by the pressure infiltration at 690°C/30sec of the preheated fibrous preforms (~600°C) in an autoclave under argon pressure (up to 6 MPa). The fibrous preform consisting of ~10%vol. of planar randomly arranged δ-Al₂O₃ fibres (Saffil RF) was prepared by the suction-assisted sedimentation of discontinuous fibres dispersed in the liquid medium and subsequent drying. The mean fibre diameter and length were ~3-5 μm and ~100 μm respectively. The preform was inserted into perforated steel holder without using any binder medium. As known, Mg8Li matrix displays duplex *hcp/bcc* structure while those Mg and Mg12Li are homogeneous *hcp* and *bcc* alloys.

Fig.1 shows δAl₂O₃ fibre population occuring in Mg8Li matrix in the cut area parallel to the fibre array planes where *hcp* islands surrounded with *bcc* phase are seen in the matrix region without any significant relation between occurrence of *hcp/bcc* localities and the fibre distribution. The structure of Mg and Mg12Li based composites is trivial.

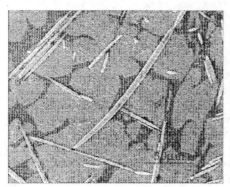
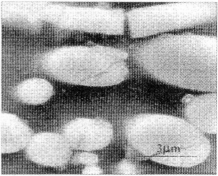

Fig 1. Distribution of δ-Al₂O₃ fibres in the dual structure of Mg8Li matrix, SEM image. (*hcp* - gray, *bcc* - dark)

Fig.2. Detail SEM image of δ-Al₂O₃ fibres in Mg8Li matrix demonstrating the fibre embrittlement.

During infiltration of δ-Al₂O₃ fibres with molten MgLi alloys a reactive diffusion of both Mg and Li elements into fibre volume occurs attaining Li/Al≤0.3 and Mg/Al≤0.06 concentration levels (in terms of atomic fraction) when simple Li₂O and MgO oxides are formed in the fibre bulk by the displacement redox reactions and Li⁺ ions (from Li₂O) are topotactically incorporated into δ-spinel lattice in the next step [6]. Divalent Mg²⁺ ions do not enter the δAl₂O₃ lattice so that discrete MgO phase remains accumulated within fibre bulk causing the fibre embrittlement together with Li₂O and other reactions products (elemental Al and Si) [7]. Significantly higher concentration of Mg at the fibre periphery was found compared to the fibre inside but no remarkable fibre/matrix reaction zone was recognizable there [8]. δAl₂O₃ fibres are prone to the cracking during metallographical sample preparation (grinding, polishing) as shown in Fig.2. The infiltration time reduction (≤ 30 s) minimizes such fibre damage wherein a favourable effect of Li on the wetting of fibres is still keeped. No bulk affection has been observed in δ-Al₂O₃ fibres infiltrated with pure Mg.

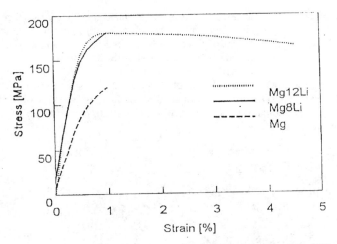

Fig. 3. Stress - strain diagrams of tensile loaded Mg, Mg8Li and Mg12Li
based composites reinforced with short δ-Al$_2$O$_3$ fibres (~10%vol)

Series of composite samples with Mg, Mg8Li and Mg12Li matrices were tensile tested
at room temperature. Typical stress-strain curves obtained are shown in Fig.3 and
determined strength- and deformation characteristics are summarized in Tab.1. There
are apparent sufficiently higher values of U.T.S. and Young modulus of MgLi
composites comparing to those Mg based as well as fairly good tensile ductility of
Mg12Li composites and a rather brittle behaviour of those Mg and Mg8Li based.

Quantity	Mg		Mg8Li		Mg12Li	
	matrix	composite	matrix	composite	matrix	composite
U.T.S.[MPa]	135	124	112	176	99	181
Elongation [%]	3	0.8	16	0.9	28	4.2
E-modulus [GPa]	45	38	46	56	46	58
Composite/matrix U.T.S. ratio	–	0.91	–	1.57	–	1.83

Table. 1   Strength- and deformation characteristics of Mg, Mg8Li and Mg12Li alloys
and related composites

SEM observation of the fracture surfaces of tensile failed composite samples provided
the following results: (i) brittle fracture mode was characteristic of matrix regions in
Mg and Mg8Li matrix composites compared to those Mg12Li based displaying fairly
high matrix ductility and, (ii) extent of the fibre "pull-out" clearly decreased with the
increase in Li alloying. The latter phenomenon confirms the favourable effect of Li on
the strength of the created interfacial bond. No traces of the chemical affection were
found on the fibre jacket surfaces.

In Fig.4 is shown SEM structure of the zone adjacent to the fracture surface in tensile failed Mg8Li matrix composites suggesting that fibre fracture and/or debonding occur predominantly at the intersection of fibres with *hcp/bcc* regions boundaries when the failure of composite body appears to be innitiated with the fibre fracture. This can be considered the reason for the premature fracture and poor ductility of Mg8Li based composites. On the other hand, the multiple fibre breakage was the prominent feature of the reinforcement fracture in Mg12Li matrix composites whereas fibre debonding occured only rarely indicating the effective fibre/matrix strength transfer took place during composite sample loading (Fig.5).

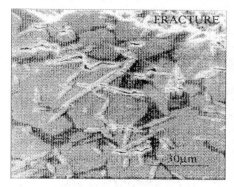 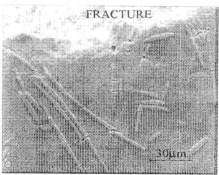

Fig.4. SEM micrograph of the zone ocuring below   Fig.5. SEM image ilustrating the fracture mode
the fracture surface in tensile failed Mg8Li   of δ-Al$_2$O$_3$ fibres in Mg12Li composites
based composites   after being tensile failed

The results presented above demonstrate that Li alloying of MgLi matrix can influence deformation behaviour and fracture of δAl$_2$O$_3$/MgLi composites through following factors: (i) fibre embrittlement, (ii) fibre/matrix bond increase, (iii) matrix plasticity increase and, (iv) change in matrix phase composition. Although these effects are rather versatilely displayed on mechanical properties of related composites the reinforcing efficiency expressed as the composite/matrix U.T.S. ratio clearly increases with increased Li alloying (Tab.1).

3. C(pyC)/MgLi composites

These composites were fabricated by the similar pressure infiltration process as described above using the same Mg, Mg8Li and Mg12Li matrix alloys. The preform of continuous carbon fibres (approx. 50%vol.) consisted of commercial PAN based carbon fibres T300 (Torayca) covered with the CVD pyrolytic carbon layer (pyC). The CVD procedure was carried out at M&T Verbundtechnologie (Mittweida, Germany) using of hydrocarbon precursor under thermal gradient conditions. Two different thickness of pyC coatings were obtained in this way : 314 nm and 600 nm, respectively The preform was made by the winding of continuous T300(pyC) fibres on the steel core. The distribution of T300(pyC) fibres in Mg12Li matrix is shown in Fig. 6.
The essence of the attack of carbon fibres from molten MgLi alloys lies in the rapid diffusion of Li and subsequent formation of ionic Li$_2$C$_2$ carbide in the fibre bulk

wherein no discernible zone of the reaction product (Li$_2$C$_2$) at the fibre/matrix interface was found. This process is accompanied with a very large volume change ($\Delta V/V \approx 1$) and subsequent strong internal stresses that cause the fibre destruction in case of large reaction extent. Although Mg was also detected in the carbon fibres, no Mg-related reaction product was found within the fibre bulk because the formation of MgC$_2$ and Mg$_2$C$_3$ carbides by direct reaction is impossible [9]. After loosing its electron, the diffusion of Li into carbon structure is strongly interplanar which is connected with exceedingly high energy barrier for the cross-plane passage of Li$^+$ ions ($\sim$13 eV) compared with its in-plane movement (0.7 eV) [10]. As shown elsewhere [11], the pyC layer inhibites effectively the Li passage into fibre bulk which can be attributed to the pyC coating texture where relatively well-organized large blocks of sp^2 hexagones are oriented mainly parallel with the fibre axis containing reduced number of hexagone edges with uncompensated high-energy electron pairs that increase critically the reactivity of carbon bodies.

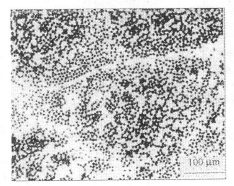 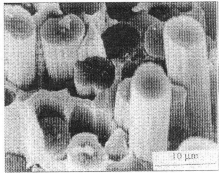

Fig. 6. Distribution of T300(pyC) carbon fibres in Mg12Li matrix, light micrograph.

Fig. 7. Fracture surface of T300(pyC)/Mg12Li composite sample after the tensile test (infiltration at 690°C/10sec), SEM.

The protective function of pyC coating against MgLi melts cannot be explained by thermodynamical reasons. This is associated only with retarded penetration of Li into fibre bulk, so that the fibre damage progress cannot be entirely prevented as shown by tensile tests of composite samples (longitudinal loading) where U.T.S. values up to ~ 1000 MPa are attained in case of very short infiltration time ($\leq$10s) whereas rapid U.T.S. decrease can be observed in both Mg8Li and Mg12Li based composites if infiltration time is prolonged (Fig. 8). It is seen that no significant effect of Li content in MgLi matrices is displayed on U.T.S. values of these composites.

Effect of the thickness of pyC coating on U.T.S. value of Mg-, and Mg12Li-based composites (infiltrated at 690°C/10 s) is demonstrated by Fig. 9, providing following observations: (i) both 314nm pyC-coated and uncoated T300 carbon fibres display in Mg matrix approximately the same reinforcing effect, (ii) reinforcing efficiency of 314nm pyC-coated fibres in Mg12Li matrix is comparable with that attained in the innert Mg matrix, (iii) very thick pyC layer (600 nm) is ineffective in terms of achieved strengthening effect in Mg12Li composites. The latter observation can be explained by the occurrence of craks in thick pyC coatings which cause the drop in tensile strength of monofilaments [12]. The cracks in the coating promote the penetration of the Li in the fibre bulk during composite fabrication so that the pyC layer loos its protective function.

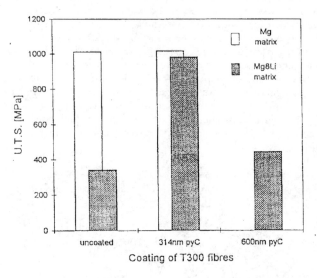

Fig. 8. Effect of the thickness of pyC coating on the reinforcing efficiency of T300(pyC) fibres in Mg and Mg8Li matrices. Infiltration was carried out at 690°C/10s.

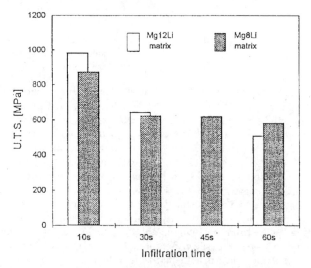

Fig. 9. U.T.S. values of Mg8Li and Mg12Li composites reinforced with T300(314 nm pyC) carbon fibres demonstrating the effect of the infiltration time prolongation.

Tensile stress-strain diagrams of longitudinally loaded MgLi/T300(pyC) composites display the usual course when the deformation of tensile samples occurs up to failure in the elastic region with the strain values of $\leq 1\%$. SEM micrographs confirmed that the composite fracture occurred during tensile deformation predominantly with the fibre "pull out" when the fibre/matrix debonding is characteristic of these composites, although many of the fibres were broken flush with the fracture surface. This indicates that the wetting of T300(pyC) fibres with MgLi melts during composite manufacturing is relatively poor which prevents the formation of the strong interfacial bond, although the tensile strength of these composites is remarkable high. The possible solution of these problems lies in the additional alloying of MgLi matrix with elements displaying the chemical afinity against carbon.

4. Concluding remarks

Structure, tensile behaviour and fracture of Mg, Mg8Li and Mg12Li matrix composites reinforced with discontinuous δAl_2O_3 fibres and continuous T300 carbon fibres were briefly characterized, with regard to the fibre/matrix compatibility in particular. Due to the reactive diffusion of Mg and Li elements into fibre bulk, taking place during composite manufacturing, the damage of fibres occurs decreasing critically their reinforcing efficiency. However, in case the kinetics of these processess is effectively retarded (minimal contact time, pyC protective layer) a considerable strengthening effect can be attained.

Acknowledgements
This work was supported by the Grant Agency VEGA of the S.R. (Project 2/2048/98).

References

[1] J. H. JACKSON, P. D. FROST, A. C. LOONAM, L. W. EASTWOOD and
     C. H. LORING, *Trans. Am. Inst. Met. Eng.* 185 (1949) 149
[2] B. A. WILCOX and A.H. CLAUER, *Trans. Met. Soc. AIME* 245 (1969) 935
[3] T. W. CLYNE and P. J. WHITHERS : *An Introduction to Metal Matrix
     Composites*, Cambridge University Press 1995, p.509
[4] J. F. MASON, C. M. WARWICK, P. J. SMITH, J. A. CHARLES,
     and T. W. CLYNE, *J. Mat. Sci.* 24 (1994) 5071
[5] F. DELANNAY, L. FROYEN and A. DERUYTERE, *J. Mat. Sci.* 22 (1987) 1
[6] S. KÚDELA, V. GERGELY, L. SMRČOK, S. OSWALD, S. BAUNACK
     and K. WETZIG, *J. Mat. Sci.* 31 (1996) 1595
[7] S. KÚDELA, V. GERGELY, S. BAUNACK, A. JOHN, S. OSWALD
     and K. WETZIG, *J. Mat. Sci.* 32 (1997) 2155
[8] S. KÚDELA, R. RENNEKAMP, S. BAUNACK, V. GERGELY, S. OSWALD
     and K. WETZIG, *Mikrochimica Acta* 127 (1997) 243
[9] J. VIALA, P. FORTIER, G. CLAVEYROLAS, H. VINCENT and J. BOUIX,
     J. Mat. Sci. 26 (1991) 4977
[10] R. C. BOEHM and A. BANERJEE, *J. Chem. Phys.* 96 (1992) 1150
[11] S. KÚDELA, V. GERGELY, E. JÄNSCH, A. HOFMANN, S. BAUNACK,
     S. OSWALD and K. WETZIG, *J. Mat. Sci.* 29 (1994) 5576
[12] T. HELMER, H. PETERLIK and A. KROMP, *J. Am. Ceram. Soc.* 78 (1995)133

Processing and Characterisation of a Ti-15-3/SiC Fibre Composite via Tape Casting

C. M. Lobley & Z. X. Guo

Department of Materials, Queen Mary and Westfield College, University of London,
London, E1 4NS, UK.

Abstract

Potential applications of Ti/SiC fibre metal matrix composites have been largely hindered by high cost. This paper investigates the use of a low-cost tape casting technique for the manufacturing of a Ti-15-3 based composite reinforced by Sigma SiC monofilaments. The powder tape thickness was characterised for the control of composite fibre distribution. The as-consolidated Ti-15-3/SiC composite shows very uniform fibre distribution. Tensile testing of the composites was carried out using parallel-sided specimens. The results show an average failure strength of 1342MPa, strain 0.91% and Young's modulus 168GPa, which are comparable to those obtained from other manufacturing processes.

1. Introduction

SiC fibre-reinforced titanium is currently being investigated for use in gas turbine engines. The composite material exhibits excellent specific strength and specific stiffness, with a service temperature up to 200°C higher than the matrix alloy. Components which have been specifically targeted include compressor discs[1] and nozzle activator pistons[2]. Other aeroengine applications such as fan blades[3] have also received attention. Various Ti alloys have been considered as a viable matrix for the composites, Ti-15V-3Cr-3Al-3Sn, (wt%) is of interest due to its good ductility and strength.

The widely-known manufacturing methods for these composites are Foil-Fibre-Foil (FFF) consolidation, Physical Vapour Deposition (PVD) and Vacuum Plasma Spraying (VPS). However, the FFF route relies upon expensive matrix foils, while PVD and VPS require high pre-processing costs. Furthermore, control of the fibre distribution can be difficult with FFF, PVD is a relatively slow process and VPS results in fibre damage from impact by the molten metal droplets[4]. What is clearly required is a cost-effective method of producing high-quality composite materials.

Tape Casting is a recently developed fabrication process for Ti/SiC fibre composites[5], as schematically illustrated in Figure 1. The process uses relatively large inexpensive matrix alloy powders mixed with an organic binder to form a slurry, which is then cast to form a thin uniform powder tape. Once dried, the tape is laid-up with filament-wound SiC fibres and consolidated in a vacuum hot press, with a burnout dwell at moderate temperatures to ensure removal of the organic components.

The objective of this investigation is to study the feasibility of manufacturing high-quality Ti-15-3 based SiC fibre composites using the tape casting method. This was carried out firstly by the examination of the powder tape uniformity, then the microstructure and tensile properties of the as-consolidated composites and the monolithic matrix. The results were discussed and compared with previous findings from the conventional manufacturing processes.

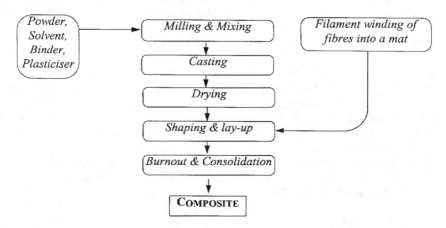

Figure 1: Schematic illustration of the tape casting process

2. Materials and Methods

A slurry was prepared using Ti-15V-3Cr-3Al-3Sn wt% alloy powder (<150μm diameter), poly(vinyl butyral), and benzyl butyl phthalate, dispersed in an azeotropic mixture of ethanol and methyl ethyl ketone. This was milled and mixed in a planetary ball mill for 1 hour and then batch cast using a speed of 0.25m/s and a doctor-blade gap height of 0.35mm. Once dried, the tape was peeled and its thickness was measured using a micrometer along each side of the 80mm wide tape, 10mm from the edge. One measurement was recorded every centimetre along the tape. The thickness at a given length of the tape was calculated from the measured data of the two sides.

DERA Sigma 1140+ SiC fibres were filament wound with an inter-fibre spacing of approximately 150μm and held in place using a binder. Sections of the tape were alternately laid up with the fibre mats, and consolidated in a Hot Isostatic Press using a proprietary undisclosed pressing schedule by DERA Sigma. For comparison, a matrix-only sample was produced by the consolidation of powder-tapes under identical conditions to the composites.

Parallel-sided longitudinally-reinforced specimens were cut from the composite panel and polished to remove surface damage arising from the machining. GFRP end-tabs were applied using a high-strength two-part structural adhesive; relevant dimensions are shown in Figure 2. The specimens were tested in an Instron 6025 using wedge-action friction grips, with a cross-head speed of 0.25 mm/min and an extensometer for longitudinal strain measurement.

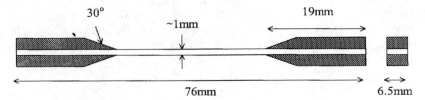

Figure 2: Dimensions of the composite test specimens

3. Results and Discussion

The thickness and uniformity of the cast powder tape largely determines the uniformity of the fibre distribution in the final composites, hence study and control of tape thickness are of great importance. The results of the thickness measurements are shown in Figure 3, which show that there is a variation along the length of the tape, especially in the first 10cm. The variation in this portion of the tape is over 100μm, which is typical of the batch tape casting operation. For the rest of the tape the variation is within 12%, with the average thickness for this section around 175μm, variation ±20μm. The thickness appears to reduce slightly towards the end of the tape.

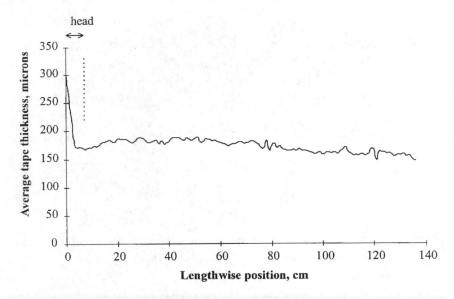

Figure 3: The variation in tape thickness along the length

The initial thick 'head' of the tape occurs mainly because an excessive amount of slurry was released and accumulated over a short length before the casting chamber reached its full speed. The 'head' should be longer and more gradual if the casting chamber reaches full speed less rapidly. This relatively short 'head' is due to the fast-reacting pneumatic operation of the batch caster; an electric motor would otherwise lead to a longer 'head'. The small fluctuation of the thickness may be attributed to the relatively large matrix powder size involved, particularly when the tape thickness is very close to the largest powder size (~150µm), as in this case.

The slight reduction in tape thickness towards the end of the tape may be due to a reduction in the height of slurry in the chamber. This is an expected and unavoidable occurrence in batch casting as the pressure head due to the slurry height reduces as the chamber is emptied. The method of slurry introduction to the chamber is an important one affecting the pressure head. Currently, slurry is introduced manually as the chamber moves, which understandably does not maintain a constant level of slurry. Further modification of the process will be carried out to ensure a more consistent slurry level and uniform tape thickness. Nevertheless, over the majority of the length, the tape thickness is relatively uniform.

As a result, the consolidated composite specimen, Figure 4, exhibits good fibre distribution. Moreover, powder particles may also occupy the inter-fibre spaces during consolidation and resist fibre movement, avoiding the occurrence of fibre swimming and impingement. Also visible here is a low level of matrix porosity which suggests incomplete consolidation by the HIPing process.

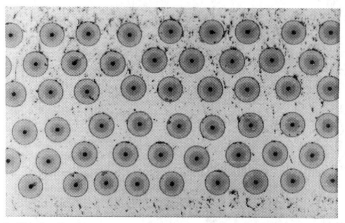

Figure 4: Typical microstructure of the tape cast composite material; the fibres are 100µm in diameter

These composites were used for subsequent tensile testing, data from the tests are shown in Figure 5. Four composites and one monolithic specimen were tested, with data shown in Table I. The tensile strength of the tape cast monolithic material was found to be 7% lower than one study[6], and 4% higher than another[7], the latter of which was made using foil consolidation. The strain to failure however was much lower than either, although the Young's modulus was higher than both previous cases. The low ductility of the specimens may be due to incomplete removal of the organic slurry components, or incomplete consolidation of the specimens. Further investigation is required to clarify this point, and to see how the ductility may be improved.

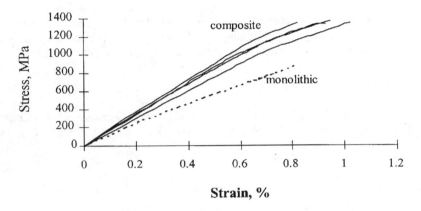

Strain, %

Figure 5: Longitudinal stress-strain behaviour for Ti-15-3-3-3/σ1140+ composites produced by tape casting

Table I: Comparison of tensile test data, this study in *italic*

	Failure Stress	Strain to Fail	Young's Modulus	Ref.
Monolithic	*883MPa*	*0.81%*	*119.9GPa*	
	950MPa	3%	115GPa	6
	800MPa	2.9%	90.3GPa	7
Composite	*1342MPa*	*0.91%*	*167.8GPa*	
	1450	0.95%	179.5GPa	7
	1770	1.25%	179.5GPa	7

Average results from the tape cast composite specimens compare well to one data set (which was heat treated), but less well to the other. Both comparative data sets used specimens reinforced by SCS-6 fibres with a volume fraction of 34%, whereas the tape cast specimens used Sigma 1140+ fibres with a V_f of 25%. This study also used small parallel-sided specimens, which will make the effect of any physical defects on the specimen more significant, and may lead to an under-estimation of the properties. Moreover, the existence of residual porosity in these specimens may also adversely influence the mechanical properties of the composites.

4. Conclusions

Tape casting has been successfully applied here to the manufacture of Ti-15-3/SiC fibre composites.

Apart from an initial 'head', the powder tape possesses a thickness variation of less than 12% throughout its length, such a variation may be due to the relatively large powder size and to the variation in the height of slurry in the casting chamber.

Even with low-level matrix porosity, the composite exhibits an average tensile strength of 1342MPa, a strain of 0.91% and Young's modulus of 168GPa, which are comparable to values obtained on similar composites manufactured by an alternative process.

5. Acknowledgements

This work is supported by the EPSRC and the Structural Materials Centre at DERA Farnborough (on contract SMCU/4/875). The authors are grateful to the staff at the SMC, for helpful discussions and assistance.

References

[1] *G.F. Harrison, and M.R. Winstone*, AEROENGINE APPLICATIONS OF ADVANCED HIGH TEMPERATURE MATERIALS - 'Mechanical Behaviour of Materials at High Temperature', (ed. C. Moura Branco *et al*), 309-325; 1996, Kluwer Academic

[2] *C.M. Ward-Close, F.H. Froes, and S.S. Cho*, SYNTHESIS AND PROCESSING OF LIGHTWEIGHT METALLIC MATERIALS- AN OVERVIEW - 'Synthesis/Processing of Lightweight Metallic Materials II' (ed. C. M. Ward-Close *et al*), 1-15; 1997, Warrendale, PA, TMS.

[3] *R. Leucht, K. Weber, H.J. Dudek, and W.A. Kaysser*, PROCESSING OF SiC-FIBRE REINFORCED TITANIUM PARTS - Proceedings of 7th European Conference on composite Materials London, UK, 1996.

[4] *Z.X. Guo*, submitted to Materials Science and Technology, 1998.

[5] *C.M. Lobley and Z.X. Guo*, submitted to Materials Science and Technology, 1998.

[6] *S. Jansson and K. Kedward*, DESIGN CONSIDERATIONS AND MECHANICAL PROPERTIES OF SCS-6/Ti 15-3 METAL MATRIX COMPOSITE AFTER DEBOND - Composite Structures, Vol. 39, no. 1-2, 11-19, 1997.

[7] *B. S. Majumdar and G. M. Newaz*, INELASTIC DEFORMATION OF METAL MATRIX COMPOSITES: PLASTICITY AND DAMAGE MECHANISMS - Philosophical Magazine A, Vol. 66, No. 2, 187-212, 1992.

Processing of Complex Shaped MMC Products through Sintering by Infiltration of Loose Powder Mixtures

S. Domsa*, R. Orban*

*Department of Materials Science and Technology - Technical University of Cluj-Napoca – Blv. Muncii 103-105, 3400 Cluj-Napoca - Romania

Abstract

Sintering by infiltration of loose powders (SILP) combines the advantages of PM (large variety of materials) with those of casting (complex shaped and/or large dimensions) in the MMC parts production. Studying the SILP phenomena, the authors have established that infiltration process occurs both by flow of molten alloy and by capillarity. Thus, the "infiltrability" appears as a complex property of the system, which involve both an optimum porosity of loose powder, determined by the particle size distribution of the mixtures, and a high fluidity of infiltrating alloy, as well as its good wettability in respect with solid components.

The authors have studied some systems constituted from diamond – [fusion] tungsten carbide (both as reinforcement components) – W-Ni loose powder mixtures infiltrated with several special bronzes and brasses.

Good results in production of wear resistant composite materials are obtained, successfully used, e.g. in diamond drilling tools fabrication.

Introduction

Sintering by infiltration of loose powder (SILP) consist, from the practical point of view [1], in a certain mould cavity (usually complex shaped) filling with a selected mixture of different loose powders, followed by its gravitational or hydrostatic infiltration with an appropriate molten infiltrating alloy (Fig.1).

The last one should be capable to act as a liquid phase and sinter together the powder particles, forming, in this way, a new material with certain imposed properties. Usually, this new material should be also able to incorporate, by mechanical encapsulation or by diffusion liaisons, some non-metallic (particulate or fibrous) reinforcement components. In this way, after cooling, results various metal matrix composites (MMC). Even a

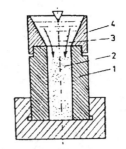

Fig.1. Principle of the SILP process

new reinforcing phase ca by synthesized by a reactive infiltration.

Thus, SILP processing method combines the advantages of PM (large variety of materials) with those of casting (complex shaped and/or large dimensions) in the MMC parts production. Although this method have been applied for some composites/parts fabrication (e.g. diamond drilling tools [2]) for several years, its mechanism have not yet been extensively investigated, probably being initially accepted as identical with those of traditional melt infiltration of powder compacts. In fact, the peculiarity of SILP is notable.

2. Brief Study of the SILP phenomena

Like the classical infiltration of pre-sintered skeletons, SILP has two stages: infiltration of loose powder with an infiltrating alloy and its liquid phase sintering. Since the last one does not differ essentially from those of classical systems [3], the present paper will be especially oriented toward the first, infiltration, stage of SILP.

Unlike of classical infiltration, in SILP, the loose powder particles don't form initially a rigid skeleton. They may have a different behavior than this one at the contact with the flowing infiltrant. Moreover, in the SILP practice the last one is introduced in a granular state, i.e. in a solid instead of a liquid phase. So, during the SILP, the solid infiltrant and loose powder from mold cavity are heated in the some time.

From this reality may arise two main peculiarities, as follows.

a) Sintering of loose powder particles prior to the infiltrant melting. Indeed, in a W-Ni system considerable shrinkage (Fig.2) occurs even at temperatures of a ~ 1100°C, proving a solid state sintering. So, if an infiltrant with an operating temperature of a ~ 1200°C, e.g. a Cu-Sn-Ni alloy is used, at a reasonable processing time a quite rigid skeleton may be formed by the loose powder prior to the infiltrant melting.

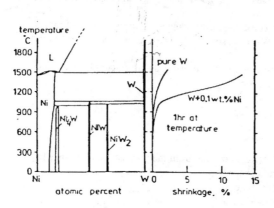

Fig.2. W-Ni phase diagram and sintering shrinkage of a W and of W+0,1 [w/o] Ni vs. temperature

b. Predominance of the capillary flow mechanism of infiltration. It was found [3] that, if the granular infiltrant melts enough slowly, even if any solid state sintering prior to the infiltrant melting takes place, the infiltration process of loose powders occurs almost exclusively by capillary phenomena. On the contrary, at higher melting rates, the granular infiltrant suddenly becomes liquid and, by quick flowing, determines the apparition of dynamic forces that can move powder particles from one another. This phenomenon was put into evidence at infiltration of a loose mixture of fusion WC-W-Ni powders with the some Cu-Sn-Ni alloys as previously, at a quite high heating rate (~ 100°C/min) – Tab.1 [4]. Both for higher initial porosity and proportions of the fine fraction (- 40 μm) into the powder, the determined volumetric proportion of the infiltrant in the obtained composite was higher than the initial porosity of powder. At lower porosity and coarser powder, a shrinkage occurred instead. In the common SILP practice, the solid grains of infiltrant, being of different average dimensions (~ 1 to 10 ÷ 15 mm) are getting melt at different moments. Consequently, it is to be expected that usually the infiltrating rates are low enough to lead to the predominance of capillary flow.

As a proof, the measured hard ness along a sample processed from the same materials as above, is quite constant up to the proximity of the top (Fig.3), where, the excess of infiltrant determined its decreasing.

The assurance of the optimum conditions for capillary flow imposes:

Proportion of the fine fraction (-40 μm) in the loose powder	Porosity of the loose powder	Proportion of infiltrant in the SILP processed sample	Variation of virtual porosity during the SILP process
[%]	[%]	[%]	[%]
100	64.79	81.21	+ 16.42
45	55.97	57.81	+ 3.28
26.5	52.80	52.31	- 0.93
8.4	52.18	49.40	- 2.78
5	54.41	45.57	- 8.84

Tab.1. Variation of virtual porosity of loose powder during the SILP process

i) an optimum value of the capillary radius, i.e. an optimum porosity of the loose powder in order to realize both a maximum height and a reasonable rate of infiltration. However, since loose powders typically have a much higher porosity than a pre-sintered skeleton [5], the capillary radius could be of a too high value to assure proper conditions for capillary flow. So, usually an as low as possible porosity of the loose powder is practically required. Consequently, at the powder mixture preparation, a proper particle size distribution, able to assure such a porosity, should be realized. A compactization by vibration of the loose powder was established to be favorable [5];

ii) an as good as possible wetting of the solid particles by the molten infiltrant. In SILP, wetting is determined by the nature of the solid particles and liquid infiltrant, as well as by the surface and environmental conditions [6]. Indeed, a successful infiltration requires a quite complete wetting. As a general rule, wetting is favored if the solid and liquid are chemically compatible, i.e. the liquid is capable of alloying with solid substrates or some compounds are formed at the interface.

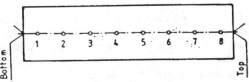

1 - o Coorse powder
2 - ▲ Medium powder
3 - x Fine powder

Fig.3. Hardness variation along a sample processed by SILP

In the case of carbide-metal systems only few published data for the surface energy of solid carbides are available – Tab.2 [7]. Owing to the large range of their variation, it is not possible to establish a general rule concerning carbides wetting by the molten metals. However, all the considered carbides are complete wetted by molten Ni. Carbides with lower values of surface energy (ZrC) are wetted by Cu only over 1200°C. Carbides with higher surface energy (TiC, TaC, VC, WC) are wetted by almost all metals.

Carbide	Surface energy, γ_{SV}, at 1100^0C [J/m^2]
ZrC	0.800 ± 250
UC	1.000 ± 300
TiC	1.190
TaC	1.290 ± 390
VC	1.675 ± 500
WC	1675

Tab.2. Surface energy of some important carbides

3. Experimental application of SILP and results

In this paper we will show only the main results of the authors in the SILP application. They were focused on the obtaining of some particulate reinforced MMC of a high abrasion wear resistance for diamond drilling tools.

Loose powder mixtures, %						
Components	A$_0$	A$_1$	A$_2$	A$_3$	A$_4$	Grain size, µm
Fusion WC	30	55	72	-	-	-150+37
	-	-		83	92	-400+37
W	65	40	23	-	-	-37+0
Ni	5	5	5	-	-	-37+0
	-	-		17	8	-400+100
Infiltrants, %						
Symb.	Cu	Sn	Ni	Zn	Si	Fe-Mo+Mn
B$_1$*	67	-	-	30	3	-
B$_2$*	70	1	-	24	3	2
B$_3$**	79	14	-	-	1	6
B$_4$**	75	6	10	-	1	7
						1 Cu-Li

For SILP of tools with: * synthetic diamond
** natural diamond

Tab. 3. Typical powder mixtures and infiltrants for the processed MMC

Typical compositions of fusion WC (FCT)-W-Ni mixture of loose powders of various granulation and Cu-Sn(-Ni) and Cu-Zn-Si infiltrants established as adequate on the basis of the above principles, are given in Table 3 [2, 3, 4]. Nickel and tungsten powders necessary to form the metal-matrix alloy-actually a heavy alloy – of as high as possible strength and toughness were introduced into the loose mixture of powders, together with WC. In the as formed matrix alloy, the W particles were diffusionaly bonded by the Cu-Ni-Sn and Cu-Ni-Zn solution in the same way as WC ones.

In some infiltrating pre-alloys were supplementary added some amounts of Ni in order to determine both WC grain diffusional liaison and a heavy alloy matrix formation. The required good wetting characteristics of both pre-alloy and in situ formed matrix alloy were obtained by adding small amounts of reactive elements, like Si, up to 3-4 wt.%. These are able to induce chemical reactions at the interface and weaken any present oxide, to reduce the surface tension of the melt and the solid/liquid interfacial energy. Other alloying elements – like Sn, Fe, Mo and Mn in quite little proportions improve the mechanical properties and the corrosion resistance as well.

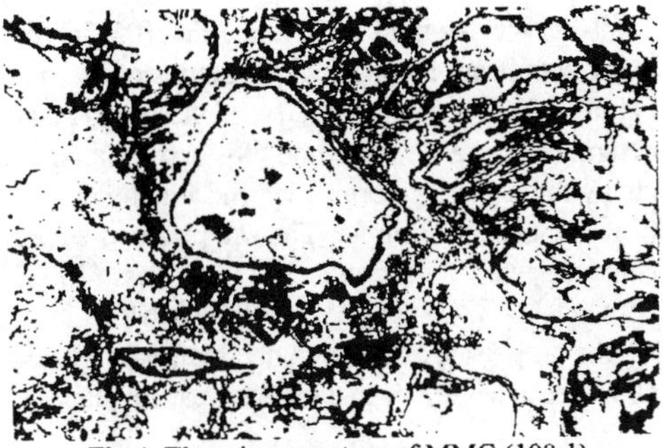

Fig.4. The microstructure of MMC (100:1)

The microstructure of such composites (with out diamond) obtained by SILP are given in Figure 4. As can be seen, the quite large particles of WC are uniform distributed and well embedded, by diffusional liaisons, in the metal matrix of a W-Ni-Cu-Zn (or-Sn) fine grain alloy.

The SILP was carried out by gravitational infiltrating technique, in an electrical heated room furnace, using dry argon as protective atmosphere (d.p.= -30°C). The process has been performed in graphite molds for 15 minutes at temperatures with about 150°C over the infiltrant liquidus, i.e. at

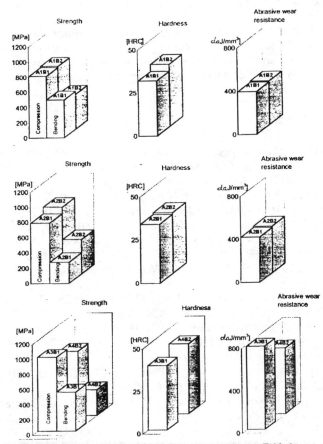

Fig.5. Mechanical properties of some elaborated MMCs (see Table 3)

1150-1215°C for the Cu-Sn and 1000-1050°C for the Cu-Zn infiltrants, respectively.
Some significant mechanical properties of the obtained by this route MMCs are presented in
Figure 5. The main property of these composite materials – abrasive wear resistance was
determined by a new method [8], based on an energetically evaluation of the wear process,
i.e. by the energy amount required for the displacement through wear of the unit of volume
from the material sample. The samples was subjected to friction against a black SiC grinding
stone, selected on account of its hardness (26 GPa), higher than those of the reinforcement
component –WC (24 GPa) – of the MMC.
All the properties of the obtained materials are comparable with those of common hardened
alloy steels, having even a higher wear resistance.

4. Conclusion

SILP is a very promising method for fabrication of composite products of complex shape and
also large dimensions.
The properties of these composite materials can be designed in a large range.
The SILP predominant mechanism, in usual practical conditions, is the capillary flow. For its
occurring, an optimum porosity of loose powder and good wetting of powder particles by the
liquid infiltrant should be assured by proper components and processing conditions.

References

[1] *V. Constantinescu, R. Orban, S. Domsa*, PROPERTIES OF SOME WEAR
RESISTANT SPECIAL SINTERED MATERIALS "Horizons of Powder Metallurgy" -
Proceedings of PM'86 International Conference, Dusseldorf, 1986, Part I, p.575.
[2] *R. Orban, S.Domsa, V. Constantinescu*, NEW COMPOSITE MATRICES FOR
TOOLS AND WEAR PARTS WITH INCREASED WEAR RESISTANCE, Proceedings of
the 13-th International Plansee Seminar, Reutle, 1993, Vol.2, p.307.
[3] *R. Orban, V. Constantinescu, S. Domsa, T. Dobra*, ON THE COMPLEX SHAPED
AND LARGE STRUCTURAL PARTS PRODUCTION BY SINTERING BY
INFILTRATION OF LOOSE POWDERS - Proceedings of EUROPM'96, Munchen, 1997,
p.130.
[4] *V. Constantinescu, R. Orban*, FUSION TUNGSTEN CARBIDE PARTICULATE
REINFORCED METAL MATRIX COMPOSITES FOR ABRASION AND WEAR
APPLICATIONS - Proceedings of the PM'94 WORLD Congress, Paris, 1994, Vol.1, p.487.
[5] *W. Schatt*, PULVERMETALLURGIE SINTER - UND VERBUNDWERKSTOFFE, A.
Huthig Verlag, Heidelberg, 1986.
[6] *K.M. Raals et. al.*, INTRODUCTION TO MATERIALS SCIENCE AND
ENGINEERING, John Wiley&Sons, 1976.
[7] *O. Kubaschewski et.al.*, MATERIALS THERMOCHEMISTRY, Pergamon Press,
New York , 1993.
[8] *S. Domsa*, TESTING METHOD AND APPARATUS FOR WEAR/ABRASION
RESISTANT MMCs, Proceedings of ECCM 7, London, 1996, Vol.1, p.443.

MICROSTRUCTURE AND PROPERTIES OF PM STEEL REINFORCED BY DISPERSED OXIDES AND CARBIDES PHASES

J. LECOMTE-BECKERS*, A.MAGNEE*

*University of Liege, Dept. of Metallurgy and Material Science, Rue A. Stevart, 2, Bât. C 1, B - 4000 - LIEGE (Belgium)

Abstract

Two ferrous matrix, a high speed steel and a maraging steel has been reinforced with alumina, titanium and vanadium carbides. The composite was realised by powder metallurgy. The particles were mechanically mixed with the prealloyed powder before compaction. To optimize the composite properties, the processing parameters (i.e. prealloyed powders, size, mixing conditions, temperature of sintering, ...) and the effect of reinforcement (size and composition of particles, volume, fraction, ...) were studied. It is necessary to obtain a full density with the minimum porosities and to control both the size, the distribution and the dispersion of the reinforced particles. The optimal sintering temperature depends on the reinforcement particles used and is of great importance for the properties. Processing parameters (mixing of particules) can greatly affect the final properties. It has been showed that the use of Al_2O_3 increases the wear resistance in the two matrix.

1. Introduction

For long and for many applications, a material with properties between hard metal and high speed steel is necessary. This material should have the high resistance to wear of the carbides together with the toughness of high speed steel matrix. An attractive solution consists to combine in a composite the properties of a tough ferrous matrix with these of hard reinforced particles. In varying the nature and the proportions of reinforcement and matrix, the properties of the composite can cover many applications. The fabrication route choosen is PM metallurgy. In fact, PM high speed steels have been a commercial reality for nearly 20 years and is an easy way to produce composite.

2. Choice of matrix and reinforcement - Fabrication method

Two standard matrix have been studied :

- a high speed steel : 1,5 % C, 4,5 % Cr, 12,5 % W, 5 % V, 5 % Co, Fe
- a maraging steel : 18 % Ni, 8 % Co, 5 % Mo, 0,4 % Ti, Fe.

The high speed steel powder is obtained by water or gaz atomization. The maraging steel is obtained only by gaz atomization due to the presence of Ti and Al.

The fabrication route involves consolidation to full density by cold pressing and sintering. The particles are mechanically mixed with the prealloyed powder before compaction. This operation can be done directly on water atomized powder. But, due to the stape of the gaz atomized powder a specific preparation has to be done to obtain a powder that could be mechanically mixed.

The reinforced particles have to present some important characteristics :

- a high hardness (min 1700 HV)
- a good interaction with the matrix
- a low solubility in the steel.

Some carbides such as TiC or NbC as well as TiN on Al_2O_3 could be used. The interest of VC is lower because the eutectic Fe-VC appears at temperature near the sintering temperature.

3. Experimental results

Most of experimental work have been done with the high speed steel matrix to determine the best compacting and sintering conditions with the different particles. These results have been used for the maraging steel.

a. High speed steel matrix + alumina

Different types of alumina have been used. In fact particles vary due to the different powder - manufacturing methods used. Particles characteristics such as shape, density and purity have a significant influence.

Two types of alumina have been particularly studied : calcinated alumina and electro-furnace alumina. Table I and II gives some properties of these alumina particles.

The high speed steel powder used is obtained by water atomization and his granulometry is ≤ 150 µ.

The ferrous powder have been mechanically mixed with the particules (10 % in volume) during 24 hours. The powder have been mechanically compacted without lubricant. These samples have been sintered under vacuum at temperature from 1000 to 1400°C to determine the optimum conditions. The best sintering conditions seems to be 1270°C during one hour. For these conditions, the hardness is maximum (662 HV) and the residual porosity rather low (< 0.04 %) ; moreover the grain size is not too high (25 µm). If the sintering temperature is increased the grain size and the particles size increase also.

The obtained microstructure is made of the ferrous matrix together with round carbides (V, W) C and alumina particles (fig. 1). The alumina has no reaction with the matrix so the carbides distribution is the same as in the matrix alone.

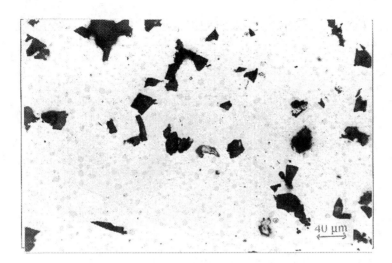

Fig. 1 - Microstructure of high speed steel marix + 10 % alumina.

b. High speed steel matrix + carbides

Particles of VC (10 % in volume) have been mixed to the high speed steel powder under the same conditions. The best sintering temperature is again about 1270°C. At this temperature, the hardness is high (680 HV) and the porosity low (0.43 %). An increase in the sintering temperature leads to bigger carbides and grain size.

The obtained microstructure is made of (V, W) C carbides dispersed in the ferrous matrix (fig. 2).

Particles of TiC (10 % in volume) have been mixed to the high speed steel powder under the same conditions. The best sintering temperature is higher (1320°C). The hardness is high (657 HV) and the porosity low (0,60 %). The microstructure is made of an acicular TiC particles with the (V, W) C carbides (fig. 3). Because of the high temperature, the size and dispersion of (V, W) C carbides is rather heterogeneous.

c. Properties

Mechanical properties have been determined on the best composite obtained after compaction with alumina particles. The properties used to classify the composite is the hardness and the porosity level. For the alumina particles, the electro-furnace alumina with low titanium content gives the best results (table III). The size of the particles must not be to small.

The composites studied were evaluated in tension and wear tests.

Table I - Characteristics of Electro-furnace alumina

N°	size ≤ μm	Al_2O_3	SiO_2	Fe_2O_3	TiO_2
1	50	99.37	0.16		
2	120	95.93	0.68	0.22	2.6
3	80	83.10	-	0.63	14.21
4	120	49.82	0.07	0.02	49.93
5	60	57.21	-	0.33	42.33
6	45	94	2	1	2.5

Fig. 2 - Microstructure of high speed steel + 10 % VC.

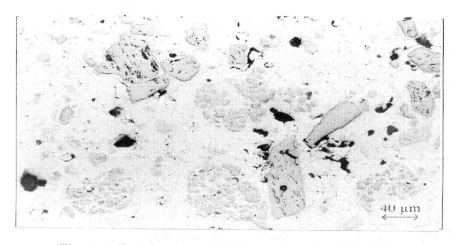

Fig. 3 - Microstructure of high speed steel + 10 % TiC.

Table II - Characteristics of Calcinated alumina

N°	Size	Al_2O_3	SiO_2	Fe_2O_3	TiO_2
	$\leq \mu m$		max	max	
7	120	99	0.03	0.03	0.005
8	45	99	0.03	0.03	0.005
9	10	99.8	0.03	0.03	0.005
10	BET-80-100(m2/g)	98	0.04	0.03	
11	BET-8-15(m²/g)	98	0.04	0.03	
12	BET-0.5-1(m²/g)	98	0.04	0.03	

Table III - Hardness and porosity level in composite (high speed steel powder + 10 % Al_2O_3)

Particles	Hardness	Porosity level
	(HV 20°	
Electrofurnace alumina		
1	491 + 25	9.4 + 7.2
2	593 + 1	3.9 + 4.6
3	571 + 31	3.3 + 2.7
4	282 + 23	4.3 + 5.9
5	295 + 12	9.5 + 7.9
6	535 + 23	6.5 + 8.8
Calcinated alumina		
7	473 + 1	5.4 + 4.8
8	412 + 1	8.1 + 2.8
9	441 + 1	11.5 + 6.3
10	287 + 22	4.1 + 2
11	386 + 1	6.3 + 3.2
12	326 + 6.8	9.8 + 6.6

Table IV - Tensile Tests at room temperature

Alloy	Rupture strength
High Speed Steel	1268 MPa
High Speed Steel + 10 % Al2O3 N° 6	825 MPa
Maraging Steel	2248 MPa
Maraging Steel + 10 % Al2O3 N° 6	1715 MPa

The results of the tensile tests are illustrated in Table IV. Fig. 4 gives the results of the abrasion tests. Fig. 5 shows the results of the wear tests (metal - metal).

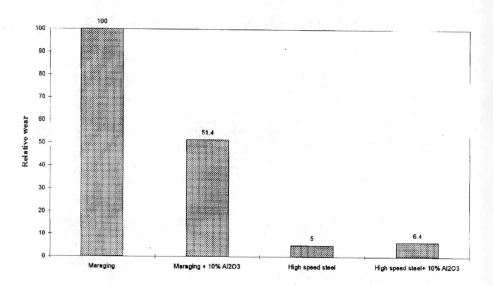

Fig. 4 - Results of abrasion tests.

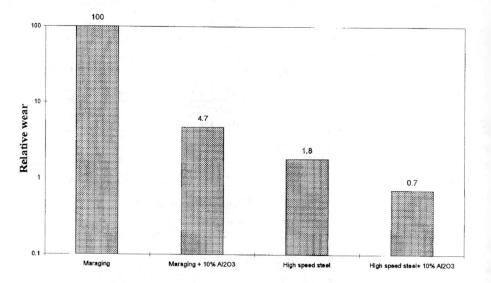

Fig. 5 - Results of wear tests (metal-metal).

Conclusions

The use of ceramic reinforcement in these matrix lower the rupture strength of the composite in comparaison with the results obtained in the matrix alone but the tensile resistance of the obtained composite remains high (825 Mpa - 1715 Mpa).

On the other hand, the use of alumina particles increases the wear strength of the composite. In rubber wheel abrasion tests (V = 3 m/s, abrasif = quartz) (1), the composite maraging steel + alumina possesses a good behaviour : the relative wear is nearly twice better as compared to the maraging steel. The high speed steel is caracterized by a better wear resistance compared to the maraging steel. The carbides present in the high speed steel matrix are really very hard and the use of alumina does not change the wear resistance. This one is similar in matrix with and without Al_2O_3.
In metal-metal wear tests the use of alumina seems interessant in the two matrix. The use of Al_2O_3 is particularly efficient in the maraging matrix : the relative wear is reaching level as good as in a high speed steel.

References

1. *A. Magnée* - Wear 162 - 164 - 1993, p. 848-855.
2. *K.M. Kulkarni* - MPR - Sept. 1998 - p. 629 - 633.
3. *E. Köhler - C. Gutsfeld - F. Thümmler* - pmi, vol. 22, N° 3/990 p 11.14
4. *R.E. Lawn - F.G. Wilson - C.D. Desforges* - Powder Metallurgy, 1976, N° 4 - p. 196-201.
5. *K.S. Kumar - A. Lawley - M.J. Koczak* - Metall. Trans. A., Vol. 22A, 1991, Nov. - p. 2733-2759.
6. *C. Zhouw - J.R. Moon* - Powder Metallurgy, 1991, Vol. 34, n° 3, p. 205-211.

The Reactive Liquid Processing of Metal Matrix Composites

Varuan M. Kevorkijan[*]

[*]Independent researcher, Lackova 139, 2341 Limbuš, Slovenia

Abstract

In this work a new concept in studying reactive wetting in $Al_{(l)}$-$SiC_{(s)}$ and $Al_{(l)}$-$AlN_{(s)}$ systems is demonstrated. The basic idea is to determine the extent and nature of reactive wetting by using some easily observable parameters.

The experimental technique is based on monitoring the volume fraction of SiC and AlN particles successfully incorporated into the matrix after controlled solidification of selected species from the mold. In this way, the effectiveness of reactive wetting of ceramic particles by liquid metal is **quantitatively** expressed by the volume fraction of ceramic particles successfully incorporated into the matrix.

1. Introduction

Fig. 1 represents schematically an isolated ceramic phase with volume V(R) and surface area S(R) which is completely immersed into a melt. Simultaneously with the process of immersion, a fast interfacial chemical reaction also occurs and, as a result, a thick layer of new ceramic phase with volume V(P) is grown at the interface. The reaction front is localized in a thick reaction layer with volume $V(P_i)$. Recall that the chemical reaction at the interface was selected to be very rapid and that proceeds faster than mass flow through the interfacial boundary layer. Accordingly, one can consider such a process of reactive wetting under chemical equilibrium conditions.

Generally, the immersion of ceramic phase is thermodynamically possible only if the energy released by the interfacial chemical reaction is sufficient for the formation of the new surface between immersed particles and the melt. Referring to this, an exergonic interfacial reaction should be selected to promote wetting in the system. For the process of reactive wetting performed in real time, it is also important that the surface layer growing to the critical thickness is fast enough to be finished during the time of immersion (i.e. before the melt rejects the particles). Based on this, the kinetics of the selected interfacial reaction also becomes very critical.

Let $\Delta G_r^\#$ (J/mol) represents the free energy per one mol of products of interfacial chemical reaction released during the complete chemical conversion, $V(P)$ the volume of ceramic phase formed at the interface (see Fig. 1) , $S(P)$ the surface of immersed ceramic phase, T_0 the temperature of the melt before immersion, T^* the temperature of the melt after immersion, and T_i the local temperature at the interface. Based on this, one can write:

$$\Delta G_r^\# V(P)\ \rho(P)/\ M(P) = \gamma_{sl} S\ (R) + V(P_i)\ \rho(P_i)\ c(P_i)\ (T_i\text{-}T_0) + V(M)\ \rho(M)c(M)\ (T^*\text{-}T_0) + V(P)\ \rho(P)\ c(P)\ (T^*\text{-}T_0) + V(R)\ \rho(R)\ c(R\ (T^*\text{-}T_0) \tag{1}$$

where ρ is density, M molar mass, c specific heat capacity and symbols P, P_i, R and M are used to denote products, products at the interface, reactants and melt respectively. Note that the term products refers to the new ceramic phase formed during interfacial reaction, products at the interface means the interfacial layer of the new ceramic phase, reactants represent the raw ceramic powder introduced into the melt and the term melt denotes the molten metal without ceramic particles.

Thermodynamically, Eq. 1 represents the sum of energetically different terms:

$\Delta G_r^\# V(P)\ \rho(P)/\ M(P)$ is the energy released by the interfacial chemical reaction, and $V(P)\ \rho(P)/\ M(P)$ represents the number of moles of product formed at the surface of the reactant. Note that V(P)-the volume of product layer-is time dependent and is diffusion controlled.

$\gamma_{sl} S\ (R)$ is the energy for the formation of the new surface in liquid metal. One can assume that the surface of the reactant covered by the thick product layer, $S(R+S)$, is approximately equal to the surface of uncovered reactant, $S(R)$. Finally, the surface of uncovered reactants can be expressed as follows: $S(R) = S_0(R)\ \rho(R)\ V(R)$.

$V(P_i)\ \rho(P_i)\ c(P_i)\ (T_i\text{-}T_0)$ is the energy for local heating of the interface to temperature T_i which is usually higher than the temperature inside the product layer or inside the melt.

The last three terms in Eq. 1 also represent energies necessary for heating the different parts of the system (melt, product layer and reactants, respectively) to the final temperature of the melt.

Based on this, it is evident that only a part of the free energy of the system released by the interfacial chemical reaction affects the interfacial energy of the ceramic phase, γ_{sl}.

This portion of energy can be expressed by $\eta\Delta G_r^\#\ V(P)\ \rho(P)/\ M(P)$ where η represents the efficiency. In this way, one can write:

$$\eta\Delta G_r^\#\ V(P)\ \rho(P)/\ M(P) = \gamma_{sl} S(R) \tag{2}$$

Note that $S(R) = \chi\ V(R)$ where χ is the geometrical parameter (e.g. for a spherical particle, $\chi = R/3$ where R is the radius of the particle).

Based on this, one can write: $V(P)/V(R\ ) = [\gamma_{sl}\ M(P)]/[\eta\Delta G_r^\#\rho(P)\ \chi]$.

Taking into account that $V(P) = S(R+P) x = S(R) x$ where x is the thickness of the product layer , the following relation can be obtained:

$$x = [\gamma_{sl} M(P)]/[\eta \Delta G_r^{\#} \rho(P)] \tag{3}$$

According to Eq. 3, it is evident that the interfacial energy of the newly formed ceramic layer is influenced by the thickness of the product layer, x, and by the free energy of the interfacial chemical reaction, $\Delta G_r^{\#}$. Here x represents the critical thickness of product layer which enables the successful incorporation of ceramic particles into the melt. In this way, x and γ_{sl} are time dependent parameters. In contrast, $\Delta G_r^{\#}$ is constant. Of course, this assumption is valid only if the rate of the interfacial chemical reaction is faster than the rate of diffusion through the product layer. Based on this, the conversion of reactants at the reaction front is complete. However, the movement of the reaction front through the bulk of non-reacted reactants is diffusion controlled and can be described by the unreacted core model. In this way, the basic assumption that the interfacial chemical reaction is very fast does not mean that the process of reactive wetting is not time dependent. In this particular case, the rate of reactive wetting is determined by the mass flow through the reaction layer which is the rate controlling step. Theoretically, the rate of reactive wetting w_w is the permanent mass or volume flow of reactants which can be introduced into the melt without rejection:

$$w_w = m(R)/ [t\, m(M)] = V(R)\, \rho(R)/ [t\, V(M)\, \rho(M)] = n(R)\, M(R)/ [t\, n(M)\, M(M)] \tag{4}$$

where m(R) represents the mass of reactants introduced into the melt and m(M) the mass of molten metal.

Assuming that the growth of the product layer is a diffusion-controlled process [1] one can write:

$$x = k\, t^{1/2} \tag{5}$$

where t represents the time of immersion.

The growth of the interfacial layer is a thermally activated process, and the growth rate (k) is strongly temperature dependent. This temperature dependence can be described by the following formula:

$$k = k_0 \exp(- E_a/2RT) \tag{6}$$

k_0 is a frequency factor and E_a is the activation energy for layer growth.

Combining Eqs. 3 and 5 one can calculate the time necessary for the immersion of n moles of products:

$$t^{1/2} = [\gamma_{sl}\, M(P)]/[\eta \Delta G_r^{\#}\, k\, \rho(P)] \tag{7}$$

Introducing $w_w = n(R)M(R)/[t\, m(M)]$ and taking into account that $n(R)/n(P) = \eta \Delta G_r^{\#} \zeta\, \rho(R)/[\gamma_{sl}M(R)\,)$, the rate of reactive wetting can be expressed as follows:

$$w_w = = n(R)\, M(R)/ [t\, m(M)] = [n(P)/t]\, \eta\, [\Delta G_r^{\#} \zeta\, \rho(P)/[\gamma_{sl}\, m(M)] \tag{8}$$

Note that $n(P)/ [t\, m(M)]$ corresponds to the rate of chemical conversion at the interface, and η is the efficiency with which the free energy $\Delta G_r^{\#}$ released by the chemical reaction is used for the formation of new surface in the melt. Assuming that the number of moles of product n(P) formed at the interface corresponds to the thickness of the product layer x, and is diffusion controlled, one can obtain: $n(P)/t \propto x/(k^2 x^2) \propto k^2/x$ where k is defined by Eq. 6 . According to Eq. 3, one can finally derive: $n(P)/ [t\, m(M)] \propto k^2 \eta\, \Delta G_r^{\#}\, \rho(P) /[\gamma_{sl}\, M(P)]$.

In this way, Eq. 8 can be rewritten as follows:

$$w_w = V(R)\, \rho(R)/[t\, m(M)] \propto e^{-Ea/RT}\, \eta^2\, [\Delta G_r^{\#}/\gamma_{sl}\,]^2 \tag{9}$$

This theoretical prediction indicates that a constant processing temperature should result in a constant wetting rate. Moreover, the volume fraction of ceramic reinforcement dispersed in the matrix should increase linearly with the time of immersion and exponentially with the increase in the temperature of melt.

These predictions can be easily investigated in practice , as will be discussed later in this work.

Unfortunately, the thermodynamic formalism applied can not predict the maximal concentration of ceramic particles successfully incorporated into a melt. By applying the model, one can only predict the maximal wetting rate at which rejection will not occur.

For practical usage of the model, the efficiency η, the specific surface energy γ_{sl} of the product layer, and the activation energy E_a for the selected interfacial reaction should all be known. The efficiency η can be estimated for those systems for which experimental values of γ_{sl} and $\Delta G_r^{\#}$ are available By TEM and SEM inspection of the product layer formed on the surface of the ceramic particles during immersion, one can also determine the average value of x. Based on this, one can now calculate the efficiency using Eq. 1. As will be discussed in detail later, for the interfacial chemical reactions considered in this work, the value of η lies between 10^{-4} and 10^{-3}. It seems that only a small amount (0.01-0.1%) of the energy released by the interfacial reaction may be transformed to the energy of the new surface created during the immersion.

The measurement of γ_{sl} is difficult and it is more convenient to estimate this value on the basis of thermodynamic considerations [2, 3]. For example, in the simple thermodynamic model suggested by Warren [2], the value of γ_{sl} for pseudo-binary systems in contact with liquid metal is correlated with the melting point of the ceramic species T_m, the molar volume of the compound V_M (cm^3) and the number of atoms in the molecule, by the following relation:$\gamma_{sl}^{\#} = kT_m/b(V_M/b)^{2/3}$ where k is an empirical constant between 5×10^{-4} and 8×10^{-4}.

2. Experimental procedure

The following as received ceramic powders with a fraction of particles less than 10 μm were used in this study: SiC (HSC 1200 Microgits, Superior Graphite), Si_3N_4 (H. C. Starck), AlN (Advanced Refractory Technology, Inc.), Mg_3N_2 (Alpha, Inc.), fused silica (BPI, Inc.) and TiO_2- rutile (Cometals, Inc.).

In some experiments, SiC and AlN particles were covered with a TiO_2+C layer using phenolic resin and fine TiO_2 powder (99.9% pure, rutile, average particle size 1 μm, Cometals, Inc.).

SiC particles were also covered by a SiO_2 layer by heating in air.

Standard 356-T6 aluminum alloy with a nominal composition of 7 wt. % of Si and 0.3 wt.% of Mg were used in all casting experiments.

SiC and AlN reinforcements used in this study were coated with a carbon layer doped with TiO_2. The SiO_2 layer on SiC was thickened by heating in air at 1000 K for 1h. In a separate set of experiments, the wetting tendency of the as- received SiC and AlN reinforcements were also evaluated. Fused silica, TiO_2 (rutile), Si_3N_4

and Mg_3N_2 powders were   directly introduced into a melt without surface modification.

3. Results and discussion

In order to verify the proposed model of reactive wetting, the increase of the volume fraction of ceramic particles in the matrix was monitored as a function of time and the temperature of immersion.

The volume fraction of ceramic particles was inspected in rapidly solidified samples (25 cm^3) taken from the top of the crucible every 5 minutes during 1 hour of continuous isothermal mixing (at about 800 rpm).The experiments were performed at different processing temperatures:1100, 1200, 1300 and 1300 K and using different ceramic species. The samples were solidified in small stainless steel crucibles. These specimens were later cut, polished, etched, and analyzed.

The chemical reactions selected in this work to promote the reactive immersion of ceramic species into the melt are listed in Table 1:

Table I: The list of chemical reactions used in this work

No	Reaction	ΔG (kJ/mol of product)	E_a(kJ/mol)
I	$Si_3N_4+4Al=4AlN+3Si$	-363	not available
II	$Mg_3N_2+2Al=2AlN+3Mg$	-293	not available
III	$3TiO_2+4Al+3C=2Al_2O_3+3TiC$	-931*	294$\pm$31
IV	$3SiO_2+4Al=2Al_2O_3+3Si$	-325	160$\pm$20

*per 3mols of TiC and 2 mols of Al_2O_3

The change of Gibbs free energy was calculated using thermodynamic data from the third edition of the JANAF tables [4] and assuming unit reactivity of the constituents.

Displacement reactions (I) and (II) proceed between highly reactive nitrides and the aluminum melt to yield thermodynamically stable AlN. Unfortunately, in the literature there is no available kinetic data for these two reactions. However, some novel observations performed by Schiroky et al. [5] indicate that reaction (II) proceeds rapidly at temperatures slightly higher than the melting point of Al .

The aluminothermic reaction (III) has been investigated recently by Choi et al. [6] Based on measurement of the real-time temperature profile, the authors [6] reported that the reaction starts at a temperature approximately 210 K higher than the melting point of Al and is controlled by carbon diffusion through solid TiC. The calculated activation energy is $\sim$230 $\pm$35 kJ/mol.

Reduction of molten aluminum by SiO_2 (Reaction IV) has been also studied by several authors [7, 8].   At 1200K the reaction is thermodynamically favorable. According to [8], the activation energy is $\sim$160 $\pm$ 30 kJ/mol.

Fig. 2 shows the experimentally measured volume fraction of ceramic particles in the matrix at different processing times. The reactive immersion of oxide and non-oxide ceramic species was performed   using different interfacial reactions. As evident, the experimentally collected results fit well the predicted linear increase in the volume fraction of ceramic particles in the matrix. These results also suggest that the reduction of molten aluminum by SiO_2 results in the most rapid reactive immersion. Other reactions, especially the displacement reactions between nitrides and the aluminum melt are less effective.

However, the rate of reactive immersion does not necessarily correspond to the rate of interfacial chemical reaction. It is reasonable, for example, to expect that the efficiency of an interfacial reaction, η, decreases with increasing reaction rate. In this way, a very rapid interfacial reaction could result in only a modest rate of reactive immersion.

Without kinetic data and a knowledge of the efficiency of the interfacial reactions, it is not possible to use the model in some particular case. However, the experimentally monitored temperature dependence of the rate of reactive immersion can be used to estimate the activation energy of the interfacial reaction, which in some cases is a known parameter. In this way it is possible to investigate the validity of the model. The procedure is demonstrated in Fig. 3. The slope of the line ln W_w - 1/T corresponds to the E_a/R. Generally, the results obtained for E_a are consistent with previously reported values of the activation energy for reactions III and IV. It was also found that the activation energies for displacement reactions I and II are 328±50 kJ/mol and 374±50 kJ/mol respectively. This indicates that the rate of displacement reactions I and II is significantly lower than the rate of the alumothermic reaction III, and the rate of the reduction of molten aluminum by SiO_2 (reaction IV).

Comparing the volume fraction of ceramic reinforcement successfully introduced into the matrix with the values of the activation energy for the corresponding interfacial reaction, it becomes evident that the volume fraction of ceramic reinforcement increases with the lowering of the activation energy of the interfacial chemical reaction. In this way, use of a interfacial reaction with a low activation energy (reaction IV) resulted in a composite with more that 20 vol% of ceramic reinforcement in the matrix.

4. Conclusions

In conclusion, it should be emphasized that the linear time dependence and the exponential temperature dependence of the volume fraction of ceramic reinforcement dispersed in the matrix has been addressed theoretically and has been verified experimentally.

It was demonstrated that the rate of reactive wetting W_w depends on the exergonic nature of the interfacial reaction, the kinetics of the interfacial chemistry involved and the efficiency of system in using the energy released by the interfacial chemical reaction for the formation of new surface in the melt.

Consequently, the rate of reactive wetting does not necessarily correspond to the rate and the exergonic nature of the interfacial chemical reaction. It is reasonable, for example, to expect that the efficiency of an interfacial reaction would decrease with increase in the reaction rate and its exergonic nature . In this way, a very rapid and exergonic interfacial reaction could result in only a modest rate of reactive wetting.

Hence, further understanding of the ability of a system to convert the energy released by the interfacial reaction into energy of the new surface created in the melt becomes necessary.

Experiments and theoretical analysis, such as those described here, are in progress for purposes of achieving this understanding.

References

1. Fan, Z., Guo Z. X. and Cantor, B., THE KINETICS AND MECHANISM OF INTERFACIAL REACTION IN SIGMA FIBRE-REINFORCED Ti MMCS- *Composites Part A,* 1997, 28A, 131-140.
2. Warren, R. SOLID-LIQUID INTERFACIAL EMNERGIES IN BINARY AND PSEUDO-BINARY SYSTEMS-*J. Mater. Sci.,*1980, 15, 2489-2496.
3. Eustathopoulos, N, ENERGETICS OF SOLID7/LIQUID INTERFACES OF METALS AND ALLOYS-*Inter. Met .Rev.,* 1983, 28/4), 189-210.
4. JANAF THERMODYNAMICAL TABLES, 3rd edition (pub. ACS and AIP, 1985).
5. Schiroky, G. H., Miller, D. V., Aghajanian, M. K. and Fareed, A. S., FABRICATION OF CMCs AND MMCs USING NOVEL PROCESSES- *Key Engineering Materials,* 1996, 127, 141-152.
6. Choi, Y. and Rhee, S. W., REACTION OF TiO$_2$-Al-C IN THE COMBUSTION SYNTHESIS OF TiC-Al$_2$O$_3$ COMPOSITE- *J. Am. Ceram. Soc.,* 1995, 78(4), 986-992.
7. Hida, G. T., Lin, I. J. and Nadiv, S., KINETICS AND MECHANISM OF THE REACTION BETWEEN SILICON DIOXIDE AND ALUMINIUM - *AiChE Symp. Ser.,* 1988, 84(263), 69-72.
8. Prabriputaloong, K and Piggott, M. R., REDUCTION OF SiO$_2$ BY MOLTEN Al. *J. Am. Ceram. Soc.,* 1973, 56(4), 184-185.

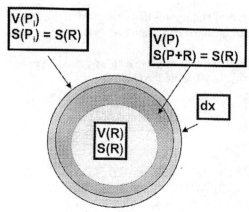

Fig. 1: Ceramic reinforcement with volume V(R) and surface area S(R).

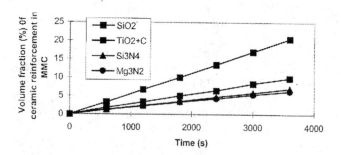

Fig. 2: Linear time dependence of the volume fraction of different ceramic reinforcements immersed into the matrix.

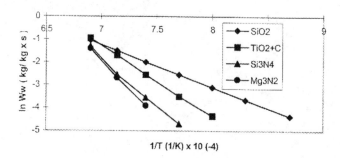

Fig. 3: Logarithms of the wetting rate for different ceramic reinforcements in MMC versus inverse temperature of the melt.

Fabrication of glass and ceramic matrix composites containing 2-dimensional metallic reinforcement using electrophoretic deposition

A. R. Boccaccini[1] , P. A. Trusty[2], W. Winkler[1], H. Kern[1]

[1]FG Werkstofftechnik, Technische Universität Ilmenau, D-98684 Ilmenau, Germany

[2] IRC in Materials for High Performance Applications, The University of Birmingham, Birmingham B15 2TT, England

Abstract

The infiltration of commercially available metallic fibre mats by silica or boehmite sols using the electrophoretic deposition (EPD) technique was investigated. The nanosized silica and boehmite particles were negatively and positively charged in colloidal suspensions at pH 9 and 4, respectively. Upon application of an electric field, the particles migrated to the metallic fabric acting as the deposition electrode in the EPD cell. Three different 316L stainless steel fabrics were considered and it was found that the quality of the infiltration depended mainly on the fibre architecture used. The EPD parameters, i.e. applied voltage and deposition time, were optimised for obtaining a high solids loading in-between the fibre tows and a firm adherent deposit. The infiltrated fibre mats, being of high quality, i.e. low macroporosity with no significant microcracking, serve as prepregs for the manufacture of silica or alumina matrix composites reinforced with a 2-dimensional metallic phase. The densification of the matrix at temperatures sufficiently low to avoid fibre degradation, is identified as being the critical step to be optimised in subsequent studies.

1. Introduction

The development of fibre reinforced ceramic and glass matrix composites is a promising means of achieving lightweight, structural materials combining high-temperature strength with improved fracture toughness and damage tolerance [1]. Considerable research effort is being expended on the optimisation of ceramic composite systems, with particular emphasis being placed on the establishment of reliable and cost-effective fabrication procedures. In this context, whilst the initial efforts were in fabrication of unidirectional composites, they are increasingly shifting towards the more isotropic, 2-dimensional (2-D) reinforced materials using woven-fibre mats as reinforcing elements [2]. The majority of the research undertaken so far on the 2-D reinforcement of ceramics has been conducted using ceramic fabrics, including SiC based (e.g. Nicalon), alumina and mullite woven fibre mats [2-5]. Metallic fabrics are commercially available also and are made from a variety of metals including stainless steel and especial alloys (e.g. Hastelloy X). These fabrics may provide interesting reinforcing elements for the fabrication of ductile phase reinforced brittle matrix composites, including glass matrix composites.

In general, the reinforcement of ceramic matrices by continuous ductile elements has not been as much investigated as their ceramic-ceramic counterparts, despite the advantages they may offer. These include an increased resistance to damage during composite processing due to the intrinsic ductility of metallic fibres and the possibility of exploiting their plastic deformation for composite toughness enhancement [6]. Moreover, it has been suggested [7], that if an optimised matrix/reinforcement interface bonding strength is achieved, both plastic deformation of the reinforcement and frictional sliding between the components can be utilised in the composite, leading to a significant fracture toughness improvement. The main limitation of these composites, however, is the relative low thermal capability and poor chemical resistance of the metallic reinforcement, which restricts the application temperature and environment. The research interest concerning ceramic/metal composites has focused traditionally on discontinuous particulate reinforcement [8,9] and, more recently, has shifted to composites with three-dimensional interpenetrating microstructures prepared by chemical reaction processes, the metal infiltration of porous ceramic preforms, or sol-gel techniques [10-12]. The fabrication technologies in these cases, however, are complex. Besides a few reports on continuous ductile reinforcement of glass matrices [6,13], including a biocompatible glass [14], no significant research has been devoted in recent years to metallic continuous fibre reinforcement of ceramics and certainly even less work has been performed using 2-D metallic fibre mats to reinforce brittle matrices [15].

Ceramic composites incorporating 2 or 3-D fibre reinforcements are particularly prone to exhibiting uncontrolled microstructures and residual porosity. This is because it is extremely difficult to achieve complete infiltration of the matrix material into the fibre tows (where the intra-tow openings may be down to the order of ≤ 100 nm). A novel, simple and inexpensive method for achieving complete infiltration of tightly woven fibre preforms has been developed recently [4]. It is based on the electrophoretic deposition (EPD) of colloidal ceramic sols into the fibre preforms. Using nanoscale ceramic particles in a stable non-agglomerated form and exploiting their net surface electrostatic charge characteristics whilst in colloidal suspension provides an appropriate means of effectively infiltrating the densely packed fibre bundles. A schematic diagram of the basic EPD cell is shown in Figure 1. If the deposition electrode is replaced by a conducting fibre preform, the suspended particles will be attracted into and deposited within it. The movement of ceramic sol particles in an aqueous suspension within an electric field is governed by the field strength, and the pH and ionic strength of the solution [4]. In a recent review article by Sarkar and Nicholson [16], a complete description of the EPD technique and its applications in ceramic technology is presented. The feasibility of the process to infiltrate *ceramic* woven fibre preforms has been demonstrated for single and mixed component sols in earlier studies [4,17-19]. Mainly SiC-based [4,17,18] and alumina [5, 19] woven fibre mats have been employed. In this contribution, the suitability of the EPD technique to prepare other composite systems is considered, using 2-D *metallic* fabrics as the reinforcement.

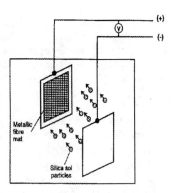

Figure 1: Schematic diagram of the electrophoretic deposition cell for obtaining ceramic deposits on metallic fibre mats. The metallic fibre mat acts as the positive electrode when the particles in the sol are negatively charged (e.g. silica particles at pH=9).

2. Experimental

Boehmite (γ-AlOOH) and silica sols are considered as precursors for the production of alumina and glass matrix composites, respectively. In this investigation, a commercially available boehmite sol (pH=4, Remal A20, Remet Corporation, USA) and a silica sol (Nyacol 2040 NH_4, pH=9, Akzo-PQ Silica, Amersfoort, The Netherlands) were used for the electrophoretic infiltration of the metallic fibre mats. The particle morphology and the relative spatial arrangement of the particles within the sol have been investigated by transmission electron microscopy elsewhere [20,21]. The boehmite particles have a fibrilar morphology and a mean size of 50 nm. The silica particles are spherical with an average diameter of 20 nm.

A broad variety of metal fabrics are commercially available (Bekaert SA, Zwevegem, Belgium) [22]. The following fibre architectures were chosen for the present study: i) a woven satin fabric (Bekitherm FA), ii) a three dimensional web of loose fibres in a non-woven labyrinth structure (Bekipor WB) and iii) a sintered solid filter felt (Bekipor-ST). The individual fibres have an average diameter of 12 μm. All fabrics are made of 100% standard 316L stainless steel. There are other fibre mats available also, made from alloys such as Inconel 601, Hastelloy X and Fecralloy, which have better high temperature properties than 316L, but for this study, the more cost-effective 316L stainless steel was considered adequate to demonstrate the feasibility of this particular processing technology.

For the EPD experiments, 15 mm x 15 mm squares were cut from the as-received fibre preforms. They were placed in the colloidal suspension, and then vacuum degassed before being infiltrated using EPD. A standard d.c. power supply was used for providing the electric field. A d.c. voltage of 4V was applied to the electrodes, which were 3 cm apart. The 4V voltage was chosen on the basis of previous EPD studies of silica and mullite sols onto SiC Nicalon fibre mats [4,17], where it was shown that increasing the voltage over 4V resulted in significant formation of bubbles between the deposited particles. This is a consequence of oxygen and hydrogen evolution at the anode and the cathode, respectively, due to the electrolytic decomposition of the aqueous medium [4]. At the working pHs of 4 and 9, the boehmite and silica particles in the sol are positively and negatively charged, respectively [23]. Accordingly, the metallic fabric to be infiltrated was placed as the cathode (for the boehmite experiments) or as the anode (for the silica experiments) (see Figure 1). The deposition time was varied to find the optimum time for the complete infiltration of the sol into the intra-tow regions, and for obtaining the desired thickness of the deposited surface layer on the mat. The infiltrated fibre fabrics were dried slowly in a humid atmosphere (~80% humidity). The dried infiltrated fibre mats were subsequently impregnated with resin and polished to a 1 μm finish for scanning electron microscopy (SEM) examination and SEM energy dispersive X-ray analysis (EDX). Some dried boehmite impregnated fibre mats were heat-treated for 1 hour at temperatures between 900 and 1200 °C in order to investigate the matrix densification. The sintered mats were prepared for SEM examination as described above.

3. Results and Discussion

Figures 2a-d are SEM micrographs showing the different metallic fibre fabrics after infiltration with silica (Fig. 2a) or boehmite (Figs. 2(b-d)) particles using EPD with an applied voltage of 4V. In general, the fibre fabrics investigated could be infiltrated successfully with the sols and a firm matrix deposit which adhered to the fibres was produced. Some differences were noted, however, between the infiltration quality of the different fabrics. It can be observed that EPD is capable of producing a high level of particle infiltration into the electrically conducting fibre tows for the Bekitherm and Bekipor WB fabrics using both sols and a low deposition time (1 min) (Figs. 2(a-c)). Owing to the small particle size of the ceramic (nano)particles, they could infiltrate the spaces within the fibre tows efficiently. It has been suggested in previous reports, that the infiltration process is enhanced in conductive fabrics by opening up the spaces in the fibre tows through the mutual repulsion of the charged fibres [17]. The looser architecture of the Bekipor WB mats seems to allow for the highest particle packing and solids-loading. On the contrary, the most difficult mat to infiltrate was the Bekipor ST fabric (Fig. 2d), which comprises of a very tight and rigid structure. As the SEM image in Figure 2d shows, infiltration of the boehmite sol was only partially achieved for this fabric, even after 3 minutes of deposition time. Similar results were achieved with the silica sol. Indeed, since this metallic web is designed to be used as a filter medium with high holding capacity and gel retention capability [22], the poor infiltration achieved by the ceramic sol is not surprising. The presence of ceramic material in the intra and inter-tow spaces was confirmed qualitatively by EDX spot analyses by detecting the peak for aluminium or silicon. The results of these analyses are presented in separate reports for the silica [21] and boehmite [24] sols. The optimum deposition time in each case was determined by a systematic trial-and-error approach. For the Bekitherm and the Bekipor WB fabrics, a deposition time of 1 minute was found to yield good infiltration for both sols. Due to this short deposition time, the deposited layer thickness was low enough to prevent significant cracking upon drying in air. Thin films (i.e. < 1 μm thick) can be dried in air without significant cracking because the tensile stresses generated on shrinkage are too low to cause the growth of cracks from microscopic flaws [25]. The infiltrated fibre fabrics in this study were dried slowly in a humid atmosphere (80% humidity); as a consequence, crack formation in the ceramic matrix between the fibres was minimised.

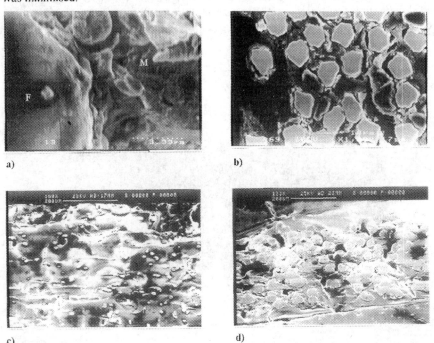

a)   b)

c)   d)

Figure 2: SEM micrographs of EPD-infiltrated fibre mats: Bekitherm infiltrated with (a) silica and (b) boehmite sol, Bekipor WB infiltrated with boehmite sol (c), and Bekipor ST infiltrated with boehmite sol (d). The voltage used was 4 V and the deposition time was 1 min (a-c) and 3 min (d).

Results on the densification behaviour of the silica infiltrated mats and on their incorporation into glass matrices to form composites have been presented elsewhere [26,27]. In this study, the quality of the boehmite infiltrated mats after heat-treatments at temperatures between 900 and 1200 °C was investigated. It is expected that in this temperature range the boehmite material will densify and transform to α-alumina [28]. However, the heat-treated samples exhibited extensive microcracking of the matrix upon cooling and this is shown, for example, for a Bekipor WB fabric, heat-treated for 1 hour at 960 °C, in Figure 3. This result indicates that the heat-treatment conditions used (e.g. heating rate, holding time at temperature, cooling rate, etc.) were not optimal and that the presence of the rigid fibres had a detrimental influence on the densification behaviour of the ceramic matrix, leading to serious defects in the microstructure. The further study of the densification and crystallisation behaviour of the boehmite matrix precursor at temperatures between 900 and 1200 °C is the focus of current research. It is envisaged that the choice of a suitable nanosized boehmite sol precursor as matrix material will allow densification temperatures < 1200 °C, which is a mandatory requirement due to the moderate temperature capability of the metallic fibres. In order to enhance the densification behaviour of the boehmite gel, the original sol will need eventually to be seeded with fine (~0.1 μm) α-Al_2O_3 particles in small concentration (<2.0 wt%), according to a method proposed by Messing and co-workers [28,29].

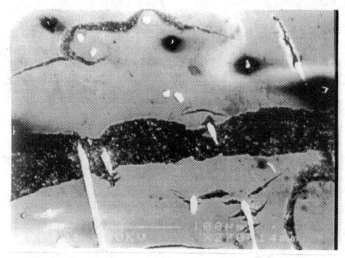

Figure 3: SEM micrograph showing the development of extensive microcracking in a Bekipor WB fibre mat infiltrated with boehmite sol, after heat-treatment (960°C, 1 h).

4. Conclusions

It has been demonstrated that the EPD sol infiltration technique can be used to successfully infiltrate silica and boehmite sols into different commercially available 316L stainless steel fabrics. The quality of the infiltration depended on the architecture of fibre mat employed. The parameters of the EPD infiltration process, i.e. voltage applied and deposition time, can be optimised to obtain a high solids loading in the intra-tow regions and a firm, adherent, ceramic deposit. The infiltrated metallic fabrics of Bekitherm and Bekipor WB type are of sufficient quality (high infiltration, no macroporosity and minimal microcrack development) to be used as preforms for the fabrication of alumina or silica (glass) matrix composites. The heat-treatment conditions necessary to obtain a dense, defect-free matrix need to be investigated further, particularly for the boehmite material. The subsequent steps would involve a slurry dipping procedure of the infiltrated fibre mats to achieve the desired ceramic volume fraction and to even out the non-uniform nature of the EPD deposit; the formation of the composite green body by stacking the prepregs; and the high-temperature consolidation by pressureless sintering or hot-pressing.

Acknowledgements
Dr. B. de Bruyne (Bekaert NV, Belgium) is acknowledged for providing the metallic fabrics. The assistance of Mrs. Sabine Awabi with the preparation of the manuscript is gratefully acknowledged.

References

[1] *R. W. Davidge*, FIBRE REINFORCED CERAMICS- Composites, 18, 1987, 92-98.

[2] *P. K. Liaw*, FIBER REINFORCED CMCs:PROCESSING; MECHANICAL BEHAVIOR; AND MODELLING - JOM 47, 1995, 38-44.

[3] *D. L. Davidson*, CERAMIC MATRIX COMPOSITES FATIGUE AND FRACTURE – JOM 47, 1995, 46-52.

[4] *T. J. Illston, C. B. Ponton, P.M. Marquis, E. G. Butler*, THE MANUFACTURE OF WOVEN FIBRE CERAMIC MATRIX COMPOSITES USING ELECTRO-PHORETIC DEPOSITION- *Third Euroceramics* Vol. 1, Edited by P. Duran and J. F. Fernandez, Faenza Editrice Iberica, Madrid, 1993, 419-424.

[5] *C. Kaya, C., P. A. Trusty, C. B. Ponton*, MANUFACTURE OF ALUMINA FIBRE/MULLITE MULTILAYER NANOCERAMIC MATRIX COMPOSITES USING ELECTROPHORETIC FILTRATION DEPOSITION (EFD) - Proc. 9th International Metallurgy & Materials Congress, Istanbul, Turkey, 1997, 657-662.

[6] *I. W. Donald, B. L. Metcalfe*, THE PREPARATION; PROPERTIES AND APPLIATIONS OF SOME GLASS-COATED METAL FILAMENTS PREPARED BY THE TAYLOR-WIRE PROCESS - J. Mat. Sci. 31, 1996, 1139-1149.

[7] *T. K. Lee, K. N. Subramanian*, OPTIMISATION OF INTERFACIAL BONDING TO ENHANCE FRACTURE TOUGHNESS OF CERAMIC MATRIX REINFORCED WITH METALLIC RIBBON- J. Mat. Sci. 30, 1995, 2401-2405.

[8] *H. J. Edress, A. Hendry*, METAL REINFORCED CERAMIC MATRIX COMPOSITES - Silicates Industriels 7/8, 1990, 217-222.

[9] *P. A. Trusty, J. A. Yeomans*, THE TOUGHENING OF ALUMINA WITH IRON – EFFECTS OF IRON DISTRIBUTION ON FRACTURE TOUGHNESS - J. Europ. Ceram. Soc., 17, 1997, 495-504.

[10] *E. D. Rodeghiero, O. K. Tse, E. P. Giannelis*, INTERCONNECTED METAL-CERAMIC COMPOSITES BY CHEMICAL MEANS - JOM 47, 1995, 26-28.

[11] *W. Liu, U. Köster*, MICROSTRUCTURES AND PROPERTIES OF INTERPENETRATING ALUMINA/ALUMINIUM COMPOSITES MADE BY REACTION OF SiO_2 GLASS PREFORMS WITH MOLTEN ALUMINIUM - Mat. Sci. Eng. A210, 1996, 1-7.

[12] *M. Sternitze, M. Knechtel, M. Hoffman, E. Broszeit, J. Rödel.*, WEAR PROPERTIES OF ALUMINA/ALUMINIUM COMPOSITES WITH INTER-PENETRATING NETWORKS - J. Am. Ceram. Soc. 79, 1996, 121-   28.

[13] *B. Wielage, M. Penno*, BEITRAG ZUR HERSTELLUNG UND CHARAKTERISIERUNG VON FASERVERSTÄRKTEM GLAS - VDI Berichte 1151, 1995, 579-583.

[14] *J. Wilson*, COMPOSITES AS BIOMATERIALS – Glass ... Current Issues. Edited by: A. F. Wright and J. Dupuy, Martinus Nijhoff Publishers, Dordrecht, 1985, 574-579.

[15]   A. E. Rutkovskij. P. D.   Sarkisov, A. A. Ivashin, V. V. Budov, GLASS CERAMIC-BASED COMPOSITES - Ceramic- and Carbon-matrix Composites. Edited by V. I. Trefilov, Chapman and Hall, London, 1994, 255-285.

[16] P. Sarkar,   P. S.   Nicholson,   ELECTROPHORETIC   DEPOSITION   (EPD): MECHANISMS; KINETICS AND APPLICATION TO CERAMICS - J. Am. Ceram. Soc. 79, 1996, 1987-2002.

[17]   A. R.   Boccaccini,   C. B.   Ponton, PROCESSING CERAMIC-MATRIX COMPOSITES USING ELECTROPHORETIC DEPOSITION - JOM 47, 1995, 34-37.

[18]   P. A. Trusty, A. R. Boccaccini, E. G. Butler, C. B.   Ponton, NOVEL TECHNIQUES   FOR   MANUFACTURING   WOVEN   FIBER   REINFORCED CERAMIC MATRIX COMPOSITES. I. PREFORM FABRICATION - Mat. and Manuf. Processes 10, 1995, 1215-1226.

[19]   P. A. Trusty, A. R. Boccaccini, E. G. Butler, C. B.   Ponton, THE DEVELOPMENT OF MULLITE MATRIX CERAMIC FIBRE COMPOSITES USING ELECTROPHORETIC DEPOSITION - Advanced Synthesis and Processing of Composites and Advanced Ceramics 2. Edited by K. V. Logan, The American Ceramic Society, Ohio, 1996, 63-70.

[20] A. R. Boccaccini, I. MacLaren, M. H. Lewis, C. B. Ponton, ELECTROPHORETIC DEPOSITION INFILRATION OF 2-D WOVEN SiC FIBRE MATS WITH MIXED SOLS OF MULLITE COMPOSITION – J. Europ. Ceram. Soc. 17, 1997, 1545-1550.

[21] A. Nowack, STUDIENARBEIT – RWTH - Aachen University of Technology, Aachen, Germany, 1996.

[22] Bekitherm, Heat Resistant Separation Materials, N.V. Bekaert S.A., Zwevegem, Belgium, Produkt Information.

[23] J. C. Huling and G. L. Messing, SURFACE CHEMISTRY EFFECTS ON HOMOGENEITY CRYSTALLISATION OF COLLOIDAL MULLITE SOL-GELS – Ceram. Trans. 6, 1990, 220-228.

[24]   A. R.   Boccaccini,   P. A.   Trusty,   ELECTROPHOERTIC   DEPOSITION INFILTRATION OF METALLIC FABRICS WITH A BOEHMITE SOL FOR THE PREPARATION OF DUCTILE-TOUGHENED CERAMIC COMPOSITES – J. Mat. Sci. 33, 1998, 933-938.

[25] C. J. Brinker, G. W. Scherer, SOL-GEL SCIENCE, Academic Press, New York, 1990.

[26] A. R. Boccaccini, J.   Ovenstone, P. A. Trusty, FABRICATION OF WOVEN METAL REINFORCED GLASS MATRIX COMPOSITES - Appl. Comp. Mat. 4, 1997,145-155.

[27] P. A. Trusty, A. R. Boccaccini, ALTERNATIVE USES OF WASTE GLASSES: FABRICATION   OF   METAL   FIBRE   REINFORCED   GLASS   MATRIX COMPOSITES - Appl.Comp. Mat., in press.

[28] G. L. Messing, M. Kumagai, LOW-TEMPERATURE SINTERING OF α-ALUMINA-SEEDED BOEHMITE GELS - Ceram. Bull. 73, 1994, 88-91.

[29] M. Kumagai, G. L. Messing, ENHANCED DENSIFICATION OF BOEHMITE SOL-GELS BY α-ALUMINA SEEDING - J. Am. Ceram. Soc. 67, 1984, C-230 - C-231.

STABILIZATION OF COMPOSITE CERAMICS STRUCTURE AT HIGH TEMPERATURES VIA NANOPOLYMETALLOCARBOSILANES

A.M.Tsirlin[*], V.G.Gerlivanov[*], N.A.Popova[*], S.P.Gubin[**]
E.K.Florina[*], B.I.Shemaev[*], E.B. Reutskaya[*].

* State Scientific Center of Russian Federation State Scientific-Research Institute of Chemistry and Technology for Organoelement Compounds, 38, Shosse Entusiastov, 111123 Moscow, Russia.
** Institute General and Inorganic Chemistry of Russian Academy of Science, 31, Leninsky Prospect, 117071 Moscow, Russia.

Abstract

New method of ceramic phase structure stabilization using cluster (or nano-) metal particles in the polycarbosilane chemistry is devised. Specimens of nanopolytitano- and nanopolyzirconocarbosilanes having direct «metal-silicon» links have been prepared and investigated. Polymers are strictly monophase. They have improved structure, no additional oxygen content, narrow molecular mass distribution (MMD) and good spinning ability in the melt state. Nano-particles were found to be very active in reactions of regrouping and polycondensation during polysilane to polycarbosilane transformation. Optimal stage of synthesis which is appropriate to introduce metal particles in the reactive mixture by fast thermodestruction of special organometallic compounds was established. A tenfold decrease in time of synthesis can be achieved.

1. Introduction

The working out of high oxidation resistant and reasonable cost material for a long-term work at 1500 - 1700 ^{0}C in civil aviation and automobile engines is a subject of priority nowadays [1]. Transition from working temperature 1100-1200 (now maximum) to 1500-1700 ^{0}C makes it possible to decrease fuel consumption and increase flight distance by 25%. Cooling systems in some cases become unnecessary, the adiabatic cycle of Diesel engine can be realized [2,3]. One of perspective materials is CMC of SiC/SiC type reinforced with coreless SiC fiber. Unfortunately all variants of this material now known can not withstand work for a long time at temperatures more than 1100 ^{0}C and are too expensive. There are many difficult problems to overcome in order to satisfy the high new demands. The most acute one is the stabilization of fine ceramic structure of fiber and matrix at high temperature application.

The ceramic structure in the polymer technology is tightly connected with a composition of ceramic polymer precursors. That is why a lot of important new works appeared last several years on the chemistry of such polymers [4, 5]. In this paper some positive and partly new results in chemistry and technology of the problem are also represented.

Nowadays the metallic clusters (nano-particles, < 10nm) attract serious interest because of their high chemical reactivity and unique physical properties. Some versions
have been proposed to introduce such metallic particles into matrices of linear organic
polymers. Methods of introducing of appreciable quantities (up to 15 mass.%) of
metallic nano-particles were developed. The composition and structure of the clusters
and the changes in a polymeric matrix were investigated in detail. It was shown that
such materials have a monophase nature. Metallic particles are very active not only in
relation to oxygen, but to nitrogen also, including the activation of strong covalent
bonds (C-H, C-F, C-C etc.). As we can be aware, an introduction of metallic clusters
(nano-particles) into polycarbosilane class of polymers is made for the first time. We
hope that this work can allow to improve the properties of SiC-ceramic polymer
precursors and the extent of ceramic stabilization.

2. Nano (cluster) - polymetallocarbosilanes

Such metals as Ti, Zr, Hf, B and some other are widely used in ceramics as effective
stabilizers of a fine ceramic structure. But the problem of development of additional
new methods to introduce metals into ceramic is still very acute. A very interesting
way is the introducing metals via ceramic polymer precursors. These methods ought
to provide the homogeneous distribution of metal atoms and no oxygen use. As can be
seen, the traditional methods using, for example, such compounds as polyborosilox-
anes or tetraalkoxititaniums (No. 2 and 3 on Fig.1 and in Table 1) debase polymers.

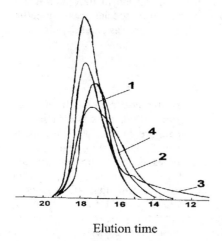

Elution time

Figure 1. MMD of synthesized PCS.
1-high pressure, 2-low pressure, initiated by boron, 3- low pressure, titanium
initiated, 4- NPMCS

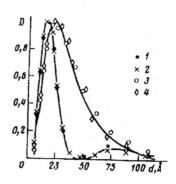

Figure 2. Iron cluster particles size
distribution in high pressure polyethylene.
  Content of Fe, mass. %: 1-0,7; 2-1,0; 3-
5,0;4-7,0.

Our work is directed to development of oxygen-free nanopolymetallocarbosilanes (NPMCS). We go from our experience on preparing of organic linear electrocon-ducting monophase metallopolymers. In this way the metals are introduced in a cluster form into vacancies which naturally exist in every polymer. These voids arise during the packing of polymer molecules. Nano-particles of metals are formed *in situ* as a result of high velocity thermodestruction of organometallic compounds [6]. Carbonyls, acetates, formiates and some other classes of metal compounds are applicable. Small angle X-ray scattering (Fig. 2) gives a notion of the method possibility. In the most interesting for ceramics stabilization interval of iron concentrations (1 to 3%) the particle sizes in polyethylene are in a narrow interval 5 to 50 Å [7]. A separate metal phase is absent at these concentrations.

Table 1. Polymer precursors characteristics.

Characteristics	$[HSiC_3]/$ $[SiC_4]$	Content, % mass.			MMD		
		O_2	H at Si	Metal	M_n	M_w	M_z
PCS autoclave	0.9-0.95	0.3 - 0.8	0.65 - 0.75	–	830	1310	2290
PBCS	0.8-0.85	1.0 - 3.0	0.55 - 0.65	0.5 - 1.0	1050	3010	9100
PTiCS	0.2-0.3	4.0 - 6.0	0.30 - 0.60	≤ 5.0	890	7300	25000
NPMCS	1.1-1.3	< 0.5	0.80 - 1.00	0.5 - 3.0	1000	1800	4200

PCS are different from linear polymers in structure and thermal properties. It is necessary to use specific metals and other initial organometallic compounds to stabilize them. Regimes of thermodestruction are also different for each of them. Besides new chemical phenomena of polymer forming process due to nano-particles are very important
The differences between syntheses of PMCS and of NPMCS in the main are as follows. Tetrabutoxytitanium, for example, reacts with polymer molecule saving all its oxygen atoms. It forms so called «oxygen bridges» between Si and Ti atoms. In the second case bis(organyl) dichlorides of metals, for example, decompose fast and completely *in situ* to metal and volatile ligands. Metal atoms link Si atoms directly:

$$nR_2MCl_2 \xrightarrow{t^0C} 2nR + n(-\overset{|}{\underset{|}{M}}Cl_m)$$

$$(-\overset{Me}{\underset{Me}{\overset{|}{\underset{|}{Si}}}}-)_n \xrightarrow{t^0C} (-\overset{Me}{\underset{Me}{\overset{|}{\underset{|}{Si}}}}-\overset{Me}{\underset{Me}{\overset{|}{\underset{|}{Si}}}}-)_x(-\overset{H}{\underset{Me}{\overset{|}{\underset{|}{Si}}}}-CH_2-)_y \xrightarrow{n(-\overset{|}{\underset{|}{M}}Cl_m),-mnHCl}$$

$$(-\overset{H}{\underset{Me}{\overset{|}{\underset{|}{Si}}}}-CH_2-)_k-\overset{|}{\underset{|}{M}}-(-\overset{CH_2-}{\underset{Me}{\overset{|}{\underset{|}{Si}}}}-\overset{|}{\underset{|}{C}}H-)_l \xrightarrow{t^0C} crosslinkal \cdot nano-PMCS,$$

I
-M- = *metal nano-particles, m = 0,1,2; x=2-10, y=5-8, n=30, k/l=1.1-1.3*
I

Table 1 shows that the new (cluster) method provides nearly the same metal content, no additional oxygen content, more regular polymer structure and even increasing of Si-H - groups. The MMD is as good as the best PCS (autoclave method) has: a narrow unimodal molecular mass distribution with the polydispersity index (M_w/M_n) about 2. The average molecular mass is in the range from 1000 to 1200. High molecular mass components (tails) are absent. According to the rheological characteristic, NPMCS can be used as a spinning melt at lower temperatures in comparison with unmodified PCS. Viscosity of about 1000 P is achieved at about 130-150 ^{0}C instead of 250-280 ^{0}C. Preliminary data showed that contents of metal up to 3% do not prevent stable spinning of fibers.

The role of metal clusters in the polycarbosilane synthesis is represented by the results of a detail investigation of the process of polysilane to polycarbosilane conversion. Intermediate and end products were analyzed by NMR-, UV-, IR-spectroscopy, chromatomass-spectrometry, gas and high pressure liquid chromatography.

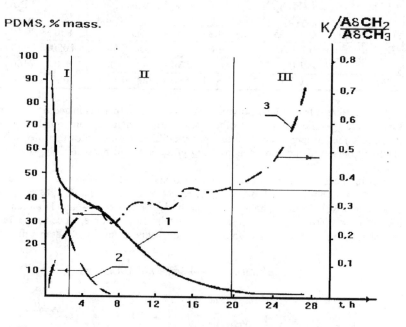

Figure 3. Conversion of polydimethylsilane to polycarbosilane.
1- without catalytic additives; 1- in presence of metalloparticles; 3- criteria K

Without any catalytic additives this process at the temperature about 350^{0}C proceeds about 28 hours (Fig. 3) and more. This period can be divided into three stages: the first - during first 3 hour, the second - to the 20th hour and the third - to the 28th hour.

It was established that decomposition of more than half of the initial polydimethylsilane takes place during the first stage and about forty compounds are formed of both polysilane and polycarbosilane structures with 1 to 8 silicon atoms (Fig.4). Linear, cyclic and bicyclic components are identified. Cyclic products such as decamethylcyclopentasilanes and dodecamethylcyclohexasilanes are the most stable in this mixture (Fig.5, curve 1). A lot of linear methylpolysilanes with up to 20 atoms of Si were found by UV- spectroscopy with wave length 294 nm as well.

At the second stage the IR - spectroscopy criteria "K", that is the ratio of intensity of methylene (σ 1360 cm^{-1}) and methyl (σ 1405 cm^{-1}) peaks, was used for investigation. Its theoretical value was calculated using standard linear polymer (CH_2-CH_2-$SiMe_2$). For ideal units of PCS (CH_2SiHMe), having the same quantities of CH_2 and CH_3 groups, this ratio is equal 0.37. Towards the end of the second stage K reaches this value. That indicates the polysilane to polycarbosilane transformation process is practically completed. But stable cyclic silanes disappear from the reactive mass only at the very end of the synthesis. During the third stage relative quantity of (CH_2) groups doubly rises as a result of condensation processes of hydride-silyl and methyl-silyl groups.

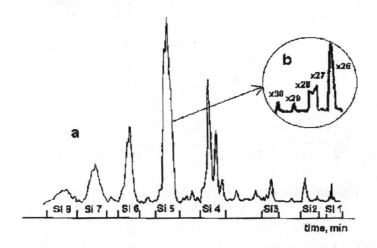

Figure 4: Products of PDMS pyrolysis.
a) gas-liquid chromatography; b) chromatomass spectrometry.

If the reaction is carried out in the presence of thermodestruction products of Ti or Zr - compounds, the time of the reaction may be reduced to 3-5 hours (even at lower temperature). After metals introduction the process speeds up but without any negative by-effect. Si-Si-fragments fully disappear, no insoluble or high fusible products are formed. High vacuum topping is only necessary after the synthesis to remove volatile products.

We suppose that this phenomenon is connected with catalytic effect of active metallic nano-particles. They can act as centers of spatial branching of macromolecules without diminishing of Si-H groups consentration. This assumption is supported by enlarging of the quantity of reactive Si-H groups, responsible for embranchment in unmodified PCS.

The introduction of metals at the intermediate stage of syntheses turned to be the most effective. The study showed that in this case nano-particles took part in chemical processes of rearrangement and polycondensation. It is very important to establish the optimal period of contact of nano-particles with reaction mixture.

The choosing of organometallic compounds is a special problem in this field. For example, halogen substituted compounds can react with hydride atoms of PCS during thermolysis and give off undesirable hydrogen halides.

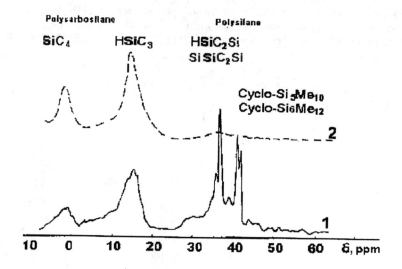

Figure 5: NMR (^{29}Si)-spectrum of intermediate (curve 1) and final (curve 2) products of nano-PMCS synthesis.

Observed more active interaction of nano-particles and carbosilane polymers as compared with linear organic polymers gives very useful results. The presence in the polymer of additional Si-H groups gives possibilities to graft new reactive functional groups and alloying compounds. The reason of greater activity is probably connected with less energy, more length of organosilicon bonds in comparison with organic bonds (Table 2) and differences in electronegativities of Si and C atoms (Si=1.8, C=2.5, H=2.1).

To examine the metal particles status in polymers a polyferrocarbosilane was synthesized using Fe(CO)$_5$ as a dopant of iron atoms. It is a good model for examination of the object with NGR-spectroscopy. According to NGR - spectrum (Fig.6) more than 80% of particles have a size of no more then 5 nm, 10% of Fe atoms are in the form of superparamagnetic clusters with a size near 1 nm. About 40% are in a form of low spin complexes of Fe, having valence 2 or 3. In this case we can't exclude the possibility of iron interaction with carbon or oxygen of initial carbonyl ligands.

And the last but not the least, it was preliminary shown that the introduction of extremely active nano-particles open a perspective to simplify and depreciate the curing stage of ceramic polymer precursors.

Table 2. Characteristics of chemical bonds.

BOND NATURE	BOND ENERGIES, kJ/mol	BOND LENGTHS, pm
C - H	414	
C - C	334	154
Si - H	314	
Si - Si	196	234

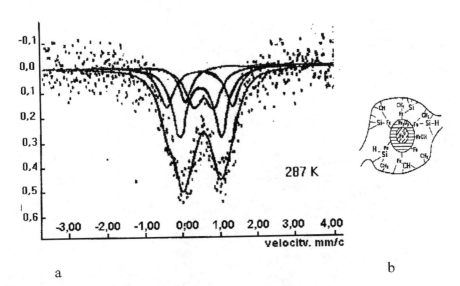

a   b

Figure 6. NGR-spectrum of NPFeCS (a) and scheme of nano-particle status (b).

This study was supported by the International Scientific and Technological Center, Project No. 582.

References

1. C.Y. Ho, S.K. El-Rahaiby, ASSESSMENT OF THE STATUS OF CERAMIC MATRIX COMPOSITES TECHNOLOGY IN THE UNITED STATES AND ABROAD - Ceramic Engineering & Science Proceedings, Vol. 13, 1992, No 7–8, pp. 3–17.
2. E.G. Bulter, NEEDS AND MARKET PROSPECTS FOR CERAMIC FIBRES IN THE AEROENGINE INDUSTRY- ECCM-6, Euro–Japanese Colloquium on Ceramic Fibres, Woodhead Publishing, Ltd., Abington Hall, England, 1993, pp. 81–108.
3. A.M. Tsirlin, TIRANNO AND SILICON NITRIDE FIBERS WITHOUT SUB-STRATE - Soviet Advanced Composite Technology Series, Vol. 1, Ch.5, pp. 455-580, CHAPMAN & HALL, London, 1995.
4. P. Greil ACTIVE-FILLER-CONTROLLED PYROLISIS OF PRECERAMIC POLYMERS- J.Amer.Ceramic Soc. Vol. 78, No. 4, pp. 835-848 (1995).
5. R. Riedel, G. Bill, A. Kienzle, BORON-MODIFIED INORGANIC POLYMERS-PRECURSOR FOR SYNTHESIS OF MULTICOMPONENT CERAMICS - Appl. Organometallic Chem. Vol. 10, pp. 241-256 (1996).
6. S.P. Gubin, I.D. Kosobudskii, METALLICHESKIE KLASTERY V POLIMERNYH MATRITSAH - Zhurnal Uspehi Himii, Vol. LII, Vyp. 8, pp. 1350-1364 (1983).
7. A.V. Kozinkin, O.V. Sever, S.P. Gubin, A.T. Shuvaev, 1. INVESTIGATION ON THE COMPOSITION AND STRUCTURE OF IRON-CONTANING CLUSTERS IN A FLUORINATED POLYMERIC MATRIX – Inorg. Materials. Vol. 30, No. 5, pp. 634-640 (1994).

Effect of thermal cycling and thermal exposure on the mechanical properties of particulate reinforced MMC.

J. D. Lord*, B. Roebuck*, P .Pitcher**

*   Centre for Materials Measurement and Technology, National Physical Laboratory, Teddington, Middlesex, TW11 0LW, UK

**   Structural Materials Centre, DERA, Farnborough, Hampshire, GU14 6TD, UK

Abstract

Some of the key markets for particulate reinforced MMCs are for applications in the aerospace and automotive industries, replacing conventional materials and offering improved performance via weight saving, higher specific strength and stiffness, wear resistance and improved thermal stability. In many cases the components are subjected to thermal exposure or thermal cycling during operation, and such conditions can have a detrimental effect on material performance. In this work, tests have been carried out on a variety of particulate reinforced MMCs and matrix alloys to evaluate the test methodology and materials performance after thermal cycling between 25-200°C and long term exposure for up to 10,000 hours at 100, 150, 200 and 260°C. Room temperature and elevated temperature tests have been used to characterise material performance, and a novel miniaturised test rig has been developed which was able to distinguish clear differences in the dimensional stability, thermal cycling, fatigue and creep performance of the MMCs over a range of temperatures and conditions.

At present, no standards exist specifically for the mechanical testing of particulate MMCs. A draft test procedure has been developed for room temperature tensile tests based on the results and analysis of data from a number of intercomparison exercises, and this has been extended to cover elevated temperature tests. Results are presented for a variety of MMCs, with some comment on the strain measurement and data analysis methods.

Results showed that MMCs with heat treatable 2000-series aluminium matrices offered the greatest stability after long term thermal exposure. Thermal cycling from room temperature to 200°C had little effect on strength values and general performance. The stability is a function of the maximum temperature, exposure time and the number of heating cycles and, for MMCs, is keenly affected by the properties of the reinforcement and matrix and the state of heat treatment. Tests showed that the Young's modulus of MMC reinforced with Al_2O_3 particulate fell significantly at elevated temperatures compared with the SiC reinforced materials, which remained relatively stable. Also, from tests carried out in the miniaturised test rig, the MMCs could be split into two groups, according to whether the materials softened or hardened during thermal cycling. The trends observed with the miniature test rig compared well with those from the conventional tests.

Effect of the Angle Between Fibres and Tensile Axis on Static Properties of Unidirectional Reinforced Titanium MMC

M.P.Thomas and M.R.Winstone

Structural Materials Centre, Defence Evaluation and Research Agency, Farnborough, Hampshire, GU14 0LX, UK

Abstract

Tensile properties of unidirectionally reinforced Ti MMC are measured, at ambient temperature and 600°C, as a function of the angle (θ) between fibres and the applied stress. With the exception of the ambient temperature proportional limit, all properties drop as θ increases from zero, because of the decreasing influence of the fibres. The ambient temperature proportional limit increases with increasing θ up to 10°, because of the effect of residual stresses on the matrix dominated yielding. It then drops as θ increases beyond 10°, because yielding becomes dominated by interfacial failure. The failure strain follows an inverse trend to the ambient temperature proportional limit. No existing model is able to predict the proportional limit at ambient temperature and the failure strains. Empirical models are proposed to predict these properties.

1. Introduction

Where the applied stress is along the fibre axis, Titanium metal matrix composites (Ti MMCs) exhibit improved specific mechanical properties when compared to monolithic Ti alloys. The mechanical properties of Ti MMCs drop rapidly, however, as the angle (θ) between the fibre axis and the applied stress increases. The mechanical properties of Ti MMCs with fibres at 90° to the applied stress are often lower than for monolithic Ti alloys, because of failure at the weak fibre/matrix interfaces. This presents a challenge to the design of Ti MMC engineering components. It is important, therefore, to build up a database of off-axis properties for current Ti MMCs and to model the behaviour.

2. Materials and Procedure

The MMC is a Ti-6Al-4V alloy reinforced with DERA Sigma SM1240, C/TiB$_x$ coated SiC fibres. The MMC was produced via the fibre-foil route at DERA Sigma, as 8 ply panels with a nominal fibre volume fraction of 0.33. Rectangular specimens, 150mm x 20mm, were electro-discharge machined (EDM) from the panels at a variety of angles to the fibre axis. The specimen edges were ground smooth to remove machining damage. Glass fibre reinforced plastic (GFRP) tabs were bonded to specimen ends to protect them when gripped in the testing machine. Tests were conducted in air, at ambient temperature and 600°C, on a servo-hydraulic machine at a strain rate of 5.5 x 10^{-5} s^{-1}.

3. Results and Discussion

At both temperatures, the tensile strength and Young's modulus decrease as θ increases (Figures 1 and 2), as previously observed [1,2]. The fracture surfaces of Figure 3 clearly show that as θ increases up to 45° the area fraction of failed fibres decreases, whilst the interfacial area present on the fracture surface increases. Thus, the decrease in MMC tensile strength and Young's modulus as θ increases up to 45° can be attributed to the decreasing influence of the high strength and stiffness of the fibres. At θ > 45° the fracture surface morphology changes very little which corresponds to the almost constant value of tensile strength and Young's modulus at these values of θ.

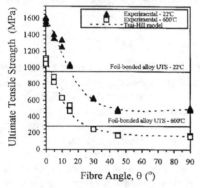 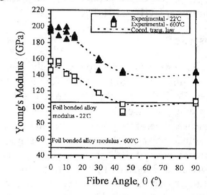

Figure 1: Effect of fibre angle on the ultimate tensile strength

Figure 2: Effect of fibre angle on the Young's modulus

The ambient temperature proportional limit increases as θ increases from 0° to 10°, and thereafter decreases (Figure 5). This is due to two conflicting yielding mechanisms - matrix yield and interfacial failure [1]. At θ>10°, MMC yield is dominated by interfacial failure, and in this regime increasing θ makes interfacial failure easier, resulting in a decreasing proportional limit. At θ <10°, MMC yield is dominated by flow of the matrix, as previously shown in specimens with θ=0° [3-7]. Figure 4 shows the resultant matrix slip bands in a specimen with small θ. Matrix yield at ambient temperature is strongly affected by the residual stresses. No predictions of residual stress in SM1240 reinforced Ti MMCs could be found in the literature. Values of matrix residual stress in Ti-6-4 alloy reinforced with carbon coated fibres [8-11] can be corrected, however, for the change in coating thickness and Young's modulus of the SM1240 fibre [12]. This results in estimates of the matrix residual stresses in the current MMC of 338 MPa in the axial direction (θ = 0°) and -226 MPa in the transverse direction (θ = 90°). The variation of residual stress with θ means that as θ increases, the applied stress required to cause matrix yielding also increases. Thus, at θ <10° increasing θ results in an increasing MMC proportional limit at ambient temperature. At 600°C, however, the matrix residual stress will be negligible and, in any case, the matrix yields almost immediately on load up [3]. Increasing θ is not likely, therefore, to have any effect on matrix yield. At this temperature the proportional limit is affected only by interfacial failure. Thus, it remains static at θ ≤ 10°, where interfacial failure has a negligible effect, and then decreases as θ increases above 10° and interfacial failure becomes important (Figure 5).

At both testing temperatures the failure strain decreases as θ increases from 0° to 15°, increases up to θ = 45° and thereafter remains constant (Figure 6). The point of inflexion seems to coincide with the change in fracture mechanism from fibre dominated to matrix dominated.

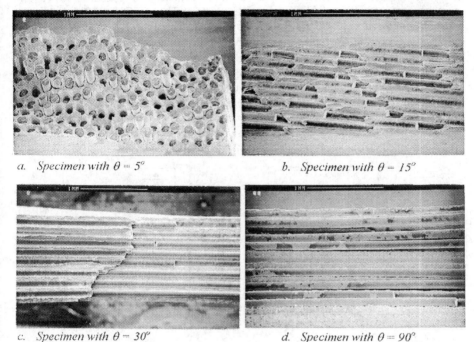

a. Specimen with θ = 5°   b. Specimen with θ = 15°

c. Specimen with θ = 30°   d. Specimen with θ = 90°

Figure 3:   Fracture surfaces of MMC tested at ambient temperature (600°C similar)

For specimens with θ ≥ 45°, tested at 600°C, the failure strains were beyond the limit of the extensometer used (5%) as a result of the onset of creep mechanisms. This problem may be overcome by increasing the strain rate, thus preventing creep.

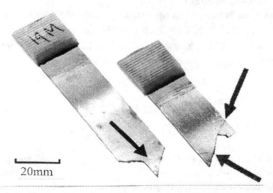

20mm

Figure 4:   Slip bands in the matrix of specimens with θ = 0° (600°C specimen shown)

Recently, measurements have been made of the tensile properties of Ti-6Al-4V produced from alloy foils using the same fabrication parameters as the MMC [3]. These properties are plotted as solid lines in Figures 1,2 and 5. The MMC only offers improved strength compared to foil-bonded alloy at small θ. This is due to the deformation mechanism becoming dominated at large θ by interfacial failure, and the strength of the interface in the MMC [13-15] is significantly lower than that of the foil-bonded alloy [3].

At 600°C, the proportional limit of the MMC is superior to foil-bonded Ti-6Al-4V at all values of θ. Young's modulus of the MMC is consistently superior to that of foil-bonded Ti-6Al-4V due to the influence of the stiff SiC fibres at small values of θ and matrix constraint between fibres at high values of θ. The strain to failure is reduced compared to foil-bonded Ti-6Al-4V, because of the poor failure strain of the SiC fibre and constrained matrix ligaments, which dominate at small and large θ respectively.

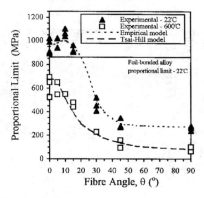

Figure 5: Effect of fibre angle on the proportional limit

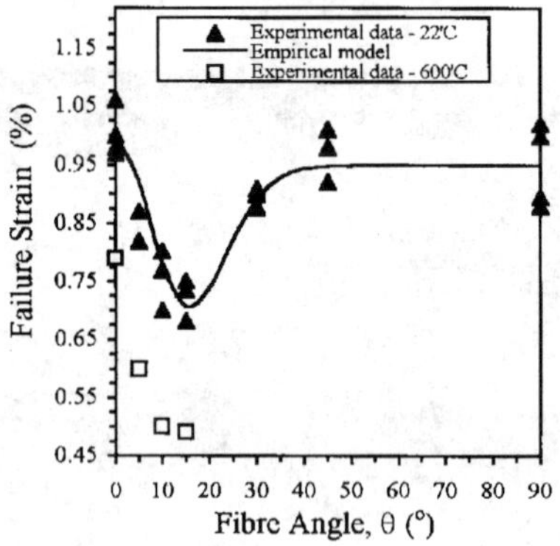

Figure 6: Effect of fibre angle on the failure strain

4. Modelling of Mechanical Properties

4.1 Ultimate Tensile Strength

A number of models have been proposed to predict the tensile strength of MMCs as a function of θ. The models of Cratchley [16,17] and of Sun et al [1,2] both suffer from large inaccuracies at small values of θ. The Tsai-Hill model predicts the strength more accurately over the entire range of θ and is given as [18]:

$$\sigma_\theta = \left[\frac{\cos^2 \theta \left(\cos^2 \theta - \sin^2 \theta \right)}{\sigma_0^2} + \frac{\sin^4 \theta}{\sigma_{90}^2} + \frac{\cos^2 \theta \sin^2 \theta}{\tau_c^2} \right]^{-\frac{1}{2}}$$

Eqn. (1)

where: σ_θ = Tensile strength of MMC at angle θ $\qquad$ σ_0 = MMC tensile strength at θ = 0°

σ_{90} = MMC tensile strength at θ = 90° $\qquad$ τ_o = MMC shear strength

A value for the MMC shear strength can be found from [2]:

$$\tau_c = \frac{\sigma_{90}}{\sqrt{2b_{66}}} \qquad \text{where: } b_{66} = \text{constant} \qquad \text{Eqn. (2)}$$

Although b_{66} is a constant it has been found to vary between Ti MMC systems, and even between batches of nominally the same MMC [1,2]. Using the procedure of Sun et al [1], it was found that $b_{66} = 1.51$ at ambient temperature and $b_{66} = 0.90$ at 600°C for the current MMC. Using these values in equation 2 provides an estimate of the MMC shear strength as 287 MPa at ambient temperature and 123 MPa at 600°C. When these values are substituted into equation (1), the Tsai-Hill model accurately predicts the tensile strengths for the current MMC as shown in Figure 1.

4.2 Young's Modulus
The off-axis Young's modulus can be modelled using the co-ordinate transformation law [1]:

$$E_\theta = \left[\frac{\cos^4 \theta}{E_0} + \left(\frac{1}{G_{12}} - 2\frac{\nu_{12}}{E_0} \right) \sin^2 \theta \cos^2 \theta + \frac{\sin^4 \theta}{E_{90}} \right]^{-1} \qquad \text{Eqn. (3)}$$

where: E_θ =MMC Young's modulus at fibre angle θ E_0 =MMC longitudinal Young's modulus
        E_{90} =MMC transverse Young's modulus G_{12} = MMC shear modulus
        ν_{12} =MMC Longitudinal Poisson's ratio

Mean literature values to be used with this equation are shown in Table 1. Using these data in equation 3 overestimates the Young's modulus at intermediate values of θ. This is due to inaccuracy in the shear modulus value, which has most influence at these values of θ. The value in Table 1 came from data on Ti MMCs reinforced with carbon coated fibres. It is noted that the bond strength in Ti MMCs with TiB_x coated fibres is weaker than in those with carbon coated fibres [13, 14,19-21]. This is likely to result in a lower shear modulus. Equation 3 accurately predicts the Young's modulus if a value of $G_{12} = 54$ GPa is used at ambient temperature and $G_{12} = 39$ GPa at 600°C (Figure 2).

Parameter	Value		Source
E_0 (longitudinal modulus)	Ambient 600°C	197.7 GPa 151.0 GPa	Current testing
E_{90} (transverse modulus)	Ambient 600°C	142.0 GPa 107.0 GPa	Current testing
G_{12} (shear modulus)	Ambient 600°C	59.8 GPa -----	References 1,5,22
ν_{12} ($V_f \approx 0.34\%$)	0.27		References 4,5,19,22

Table 1: Data for use in equation 9 to predict Young's modulus

4.3 Proportional Limit
The unusual shape of the proportional limit curve at ambient temperature (Figure 5) means that none of the current models can be used. Thus, a new model needs to be developed. There are two opposing mechanisms for MMC yield: matrix yield, which dominates at small θ and interfacial failure, which dominates at large θ.

As explained in section 3, matrix yield strength increases with θ due to residual stresses. Any new model needs to account for this change in matrix yield strength. Studies are required, therefore, to examine how the matrix residual stress alters with θ.

It is also noted that the interfacial failure stress will change with θ. At low values of θ interfacial failure occurs in shear, whilst at large values of θ interfacial failure occurs in tension. The shear and tensile strength of interfaces in Ti MMCs are likely to be significantly different [20-21, 23-25]. Whilst the interfacial shear strength in Ti-64 / SM1240 has been measured (with values between 80MPa to 103MPa) [13,14,19], no work has yet quantified the interfacial tensile strength. This area must be investigated and a function derived to account for variation of the interfacial failure strength with θ.

In the absence of detailed data for the above parameters, the experimental data may still be modelled in an empirical form. Using the values of matrix residual stress obtained in section 3, and assuming a linear variation in residual stress with increasing θ, an equation can be fitted to the resulting data set to give the value of residual stress at any angle:

$$\sigma_{\theta ym} = \sigma_{0y} + (6.27 * \theta) \qquad \text{Eqn. (4)}$$

where: $\sigma_{\theta ym}$ = MMC proportional limit at angle θ, where matrix yielding is the only yield mechanism

The contribution that matrix yielding makes to the MMC proportional limit decreases rapidly from 100% at θ = 0° to 0% at θ ≥ 30°. Interfacial failure does not occur during MMC yield at small values of θ, but as θ increases beyond 10° it becomes increasingly important until at θ > 45° interfacial failure is the only yield mechanism. Thus, an equation using trigonometric functions can be derived to model each of these effects:

effect of angle on contribution of contribution of
matrix yield strength matrix yield interfacial failure

$$\sigma_{\theta y} = \left\{ \left[\sigma_{0y} + (6.27 * \theta) \right] \left[\left(-0.5 * \tanh \frac{\theta - 24}{11} \right) + 0.5 \right] \right\} + \left\{ \sigma_{90y} \left[\left(0.5 * \tanh \frac{\theta - 24}{11} \right) + 0.5 \right] \right\} \qquad \text{Eqn.(5)}$$

This equation fits the experimental data to a satisfactory degree (Figure 5).

At 600°C, the residual stresses in the MMC matrix are almost completely relaxed, and matrix yield requires the same applied stress for all values of θ. The variation in MMC proportional limit with θ is due only to the change in the relative amounts of matrix yield and interfacial failure. Thus, at this temperature the Tsai-Hill equation can predict the proportional limit, if the tensile strengths in equation 1 are replaced by the corresponding proportional limits, as shown in Figure 5.

4.4 Failure Strain

A simple macromechanical model has been suggested [1,2] to predict the failure strains, such that:

$$\varepsilon_{\theta} = 100 \left(\frac{\sigma_{\theta y}}{E_{\theta}} \right) + \left[\left[\frac{3}{2} \left(\sin^4 \theta + 2a_{66} \sin^2 \theta \cos^2 \theta \right) \right]^{\frac{1}{2}} \right]^{m+1} \alpha \sigma_{\theta}^{m} \qquad \text{Eqn. (6)}$$

where a_{66}, m and α are constants.

Using the procedure of Sun et al to determine these constants, it is found that a_{66} cannot be assigned a single value for the current MMC, but rather varies with fibre angle. The model is of limited use, therefore, in the current case. Equation 6 also predicts no plastic strain for specimens with $\theta = 0°$, which is not the case in most Ti MMCs [3-7].

A new model needs to be developed to predict the failure strains of MMCs in this testing regime. The failure strain is determined by two processes: fibre failure, which dominates at small θ and matrix ligament failure, which dominates at large θ (Figures 3). It is noted that fibre failure strain will alter with θ. At small values of θ, fibre failure occurs in a tensile mode, whilst the fracture appearance of broken fibres at large values of θ suggests that they fail in shear i.e. no longitudinal splitting of fibres is seen. It is likely that the shear failure strain of SiC fibres will be lower than the tensile failure strain, which would account for the drop in MMC failure strain as θ increases up to 15° (Figure 6). The tensile and shear failure strains of SiC fibres must be determined and a function derived to account for the variation in fibre failure strain with θ.

In the absence of specific data, the current experimental results may be modelled empirically using trigonometric functions to model the relative importance of matrix ligament failure and fibre failure:

contribution of   contribution of matrix
fibre failure   ligament failure

$$\varepsilon_\theta = \varepsilon_0 \left(\cos^{40}\theta\right) + \varepsilon_{90} \left(1 - \cos^{19}\theta\right) \qquad \text{Eqn. (7)}$$

where: ε_θ = Failure strain for fibre angle θ ε_0 = MMC longitudinal failure strain
      ε_{90} = MMC transverse failure strain

This equation fits the experimental data very well (Figure 6).

5. Conclusions

1. The mechanical properties of unidirectional fibre reinforced Ti MMCs are highly dependent on the angle between the fibre axis and tensile axis. The MMC only offers improved proportional limit and tensile strength over foil-bonded, monolithic alloy in the regime $\theta \leq 15°$

2. The ambient temperature proportional limit follows a trend not previously reported, as a result of matrix residual stress.

3. Failure strains decrease as θ increases up to 15° and then increase as θ increases to 45°. The point of inflexion coincides with the change in failure mechanism from fibre dominated to matrix ligament dominated.

4. The ultimate tensile strength and Young's modulus of the MMC at both temperatures and the proportional limit at 600°C can be modelled using previously developed macromechanical models.

5. New models are required to predict the ambient temperature proportional limit and the failure strain at both testing temperatures.

Acknowledgements

British Crown Copyright 1998/DERA. Published with the permission of the Controller of Her Britannic Majesty's Stationary Office.

References

1. C.T.SUN, J.L.CHEN, G.T.SHA and W.E.KOOP, J.Comp.Mat. Vol.24 (1990) pp.1029
2. C.T.SUN, J.L.CHEN, G.T.SHA and W.E.KOOP in "Mechanics of Composites at Elevated and Cryogenic Temperatures", AMD-Vol.118 (ASTM, Philadelphia, 1991) pp.119
3. M.P.THOMAS and M.R.WINSTONE, Scripta Materialia Vol.37 (1997) pp. 1855
4. S.MALL and P.G.ERMER, J.Comp.Mat. Vol.25 (1991) pp.1668
5. S.JANSSON, H.E.DEVE and A.G.EVANS, Met.Trans. Vol.22A (1991) pp.2975
6. J.GAYDA and T.P.GABB, Int.J.Fat. Vol.14 (1992) pp.14
7. P.C.WANG, S.M.JENG, H.P.CHIU and J.M.YANG, J.Mat.Sci. Vol.30 (1995) pp.1818
8. J.B.BRAYSHAW and M.J.PINDERA, J.Eng.Mat.Tech. Vol.116 (1994) pp.505
9. J.F.DURODOLA and C.RUIZ, in "Advanced Composites '93: International Conference on Advanced Composite Materials". Edited by: T.Chandra and A.K.Dhingra (TMS, Pennsylvania, 1993) pp.1133
10. G.F.HARRISON, B.MORGAN, P.H.TRANTER and M.R.WINSTONE, in "Characterisation of Fibre Reinforced Titanium Matrix Composites", AGARD Report No.796 (NATO - AGARD, Neuilly Sur Seine, France, 1994) pp.14.1
11. S.M.THOMIN, P.A.NOEL and D.C.DUNAND, Met.Trans. Vol.26A (1995) pp.883
12. D.UPADHYAYA, D.M.BLACKLETTER and F.H.FROES, in "Titanium '92: Science and Technology", Vol.III. Edited by: F.H.Froes and I.Caplan, (TMS, Pennsylvania, 1993) pp.2537
13. M.C.WATSON and T.W.CLYNE, in "Titanium '92: Science and Technology". Edited by: F.H.Froes and I.Caplan (TMS, Pennsylvania, 1993) pp.2569
14. Z.X.GUO and B.DERBY, in "Titanium '92: Science and Technology". Edited by: F.H.Froes and I.Caplan (TMS, Pennsylvania, 1993) pp.2633
15. A.VASSEL, M.C.MERIENNE, F.PAUTONNIER, L.MOLLIEX and J.P.FAVRE, in "Proceedings of the Sixth World Conference on Titanium". Edited by: P.Lacombe, R.Tricot and G.Beranger (Les Editions de Physique, Les Ulis, France, 1988) Vol.II pp.919
16. D.CRATCHLEY, A.A.BAKER, P.W.JACKSON in "Metal Matrix Composites" ASM STP 438 (ASTM, Philadelphia, 1968) pp.169
17. M.TAYA and R.J.ARSENAULT in "Metal Matrix Composites: Thermo-mechanical Behaviour" (Pergamon Press, Oxford, 1989) pp.74
18. T.W.CLYNE and P.J.WITHERS in " An Introduction to Metal Matrix Composites", (Cambridge University Press, 1993) pp.227
19. M.C.WATSON and T.W.CLYNE Composites, Vol.24 (1993) pp.222
20. Y.LE PETIT-CORPS, T.MACKE, R.PAILLER and J.M.QUENISSET in "Developments in the Science and Engineering of Composite Materials - Proceedings of ECCM-3". Edited by: A.R.Bunsell, P.Lamicq and A.Massiah (Elsevier Applied Science, London, 1989) pp.185
21. L.MOLLIEX, J.P.FAVRE, A.VASSEL and M.RABINOVITCH J.Mat.Sci. Vol.29 (1994) pp.6033
22. G.M.NEWAZ and B.S.MAJUMDAR, Eng.Fract.Mech. Vol.42 (1992) pp.699
23. J.M.YANG, S.M.JENG and C.J.YANG, Mat.Sci.Eng. Vol.A138 (1991) pp.155
24. D.B.GUNDEL, B.S.MAJUMDAR and D.B.MIRACLE, Scripta Materialia Vol.33 (1995) pp.2057
25. S.G.WARRIER, B.S.MAJUMDAR, D.B.GUNDEL and D.B.MIRACLE, Acta Materialia, Vol.45 (1997) pp.3469

High Temperature Ductility of
P/M Aluminium-Lithium Composites

M.J.Tan, X.Zhang
School of Mechanical & Production Engineering
Nanyang Technological University, Singapore 639798.

Abstract

Aluminium-Lithium (Al-Li) alloys along with aluminium-based metal-matrix composites (MMCs) have been the subject of much study due to high specific strength and modulus. However, their inherent brittle nature makes processing of these materials difficult using conventional metal forming techniques thereby limiting their applications. This problem could be overcome by making use of superplastic forming techniques. Here, a study of aluminium-lithium composites using powder metallurgy is presented together with comparisons with the matrix material. The aim of this work is to study the deformation behaviour under isothermal testing conditions to study the potential of superplasticity in this composite system.

1. Introduction

The demand for high specific and improved properties to satisfy modern aerospace material applications has led to the development of light weight Aluminium-Lithium alloys [1,2] and metal-matrix composites [3-5]. The superiority of these new advanced materials in terms of their physical and mechanical properties has enabled the design of more and demanding aerospace vehicles in recent years. However, their development has been hamstrung by the lack of sufficient property data and characterisation, suitable fabrication technology and most importantly, cost. Discontinuously reinforced MMCs based on powder metallurgy (P/M) has the advantage of reduced cost, isotropic properties and improved wear resistance [6-7] as compared to continuously reinforced MMCs.

Superplasticity is used to describe extraordinary high ductilities obtained during tensile deformation of polycrystalline materials. This phenomenon was first observed over 60 years ago[8] and only in the past 40 years or so has considerable effort been devoted to exploit its potential due to the large manufacturing cost and weight savings (through redesign) which can be gained by switching from conventional to superplastic forming. The limited amounts of ductility at room temperature for most MMCs also render superplastic forming an attractive manufacturing option, if their high temperature deformation behaviour is as such.

2. Experimental Details

Rapid solidification Aluminium powder, AA8090 of composition 2.54 weight % Li, 1.49% Cu, 0.91% Mg, 0.13% Zr (size <105μm) and SiC particles (mean size 4μm) were obtained from Sumitomo Light Metals Co., Japan and Lonza Ltd Co., Switzerland respectively. The mean size of the matrix powder is 22µm with a density of 2.539 g/cm³. The SiC particles are α-type with a density of 3.252 g/cm³ and the volume fraction of reinforcement was chosen as 8 volume % (10 weight %).

Both the original matrix powder and SiC particles had been dried first at 120°C in a low vacuum oven for 5 hours before they were dry mixed for another 5 hours. The as-mixed powder were then unidirectionally hot pressed at a temperature of 545°C under a pressure of 130MPa for 20 minutes. Finally, the as-pressed billets were extruded at 375°C using a reduction ratio of 16:1. To avoid fir-tree cracking (speed cracking), a novel extrusion technique, namely, front pad extrusion was developed [9]. Mechanical tests were done using an Instron tensile testing machine within a chamber-type furnace.

3. P/M Aluminium-Lithium

Isothermal tensile tests of the matrix were conducted in a temperature range of 525°C to 555°C at a constant initial strain rate of $2 \times 10^{-4} s^{-1}$ to $1 \times 10^{-1} s^{-1}$. The results of elongation to fracture are given in Figure 1.

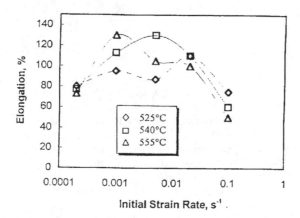

Figure 1. Elongation to fracture of matrix as a function of initial strain rate

Within a wide strain rate range of $2 \times 10^{-4} s^{-1}$ to $1 \times 10^{-1} s^{-1}$ the elongations to failure of the matrix alloy varied between 50% to 130%. At the temperature and strain rate of 540°C and $5 \times 10^{-3} s^{-1}$, or 555°C and $1 \times 10^{-1} s^{-1}$, the matrix alloy showed an optimum elongation in percentage of 130%. This optimum elongation value, and other elongations obtained in the testing range are significantly higher than the ductility value at room temperature, which is 4.3%.

The stress versus strain relationship of the matrix in logarithmic scale is given in Figure 2. For the matrix, the relationship of stress versus strain rate does not seem to have a conventional sigmoidal shape that represents the three regions with distinguishable

NON VALE COME SCONTRINO FISCALE

CARTASI NC VISA

RISTORANTE IL DAVID - RI
P.za Signoria 2/r
S/E-CE 7340508

C4929905382095   0199

OPER 016 08/06/98 17:34/

AUT   080268- BATCH 242

TRANSAZIONE: 042420016

AMOUNT/IMP: 18,480

C/M SIGNATURE-FIRMA

J.R Powell

TRANSAZIONE ESEGUITA

COME SCONTRINO FISCALE

mechanisms. It is likely that the same deformation mechanism dominates within the
temperature and strain rate range.

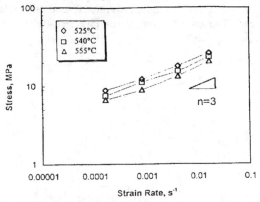

Figure 2. Logarithmic stress versus strain rate relationship of matrix.

It is suggested that for powder metallurgy aluminum alloys and composites the concept of
a threshold stress is commonly used to account for higher apparent stress exponent [10-
12]. The threshold stress values in superplastic materials are usually determined
graphically by performing the same plot using a stress exponent (n) of 2, i.e. strain rate
sensitivity $m=0.5$ ($=1/n$) [10,13]. The stress exponent value which not only gives a best
linear fit but also extrapolates positive threshold stresses is believed to be able to represent
the dominant deformation mechanism. Negative threshold stress does not have any
physical meaning.

A series of plots were made from n=1 to n=5 and the best linear fit occurs at n=3 as given
in Figure 3 with threshold stresses varying in a range of 2 to 5 MPa. Stress exponent of 3
is usually found in those solution alloys with coarse grains, named as Class I [14]. The
climb of dislocations whose slips are dragged by solute atoms is the dominant mechanism
here. For the matrix alloy, within the testing temperature range, almost all of the alloying
atoms of lithium, magnesium and copper became solute atoms, of which the total reached
about 4wt.%. It is likely that these solute atoms play an important role during deformation
at high temperature.

Another possible explanation for a stress exponent of 3 is that continuous recrystallization
may dominate the deformation process. Continuous recrystallization commonly occurs at
the initial stage of superplastic deformation in Zr containing aluminum alloys. Blackwell
and Bates [15] reported that in superplastic deformation of an AA8090 (Al-Li-Cu-Mg-Zr)
alloy the strain rate sensitivity index m increased with strain from 0.3 at a strain of 0.1 to
more than 0.5 at a strain approaching unity at a strain rate of 10^{-4} s^{-1}. This was attributed
to dynamic recrystallization. Almost the same result was reported by Hamilton *et al* [16]
in a Al-Li-Cu-Zr alloy. Therefore, $m = 0.3$, i.e. n = 3, seems to be one of the consequence
of continuous recrystallization. All the micromechanisms suggested for continuous
recrystallization are due to the activities of subgrains (the misorientations of which are low
angle) such as Boundary Sliding (Sub)grain Rotation model, Dislocation Glide (Sub)grain

Rotation model and (Sub)grain Neighbor Switching model etc. [18] such that the strain rate sensitivity index m can not reach 0.5. With increase of the number of high angle boundaries by continuous recrystallization, the conventional grain boundary sliding mechanism becomes dominant. That is why the strain rate sensitivity index m was increased to 0.5 at large strains.

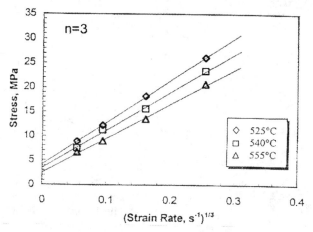

Figure 3. Threshold stress analysis for matrix alloy at stress exponent n=3.

However, if dislocation movements in subgrains control continuous recrystallization, due to the pinning of ultra-fine second phase particles, e.g. Al_3Zr in Zr containing aluminum alloys, continuous recrystallization is actually quite similar to the solute drag mechanism. It implies that continuous recrystallization and solute drag mechanism may affect the process simultaneously.

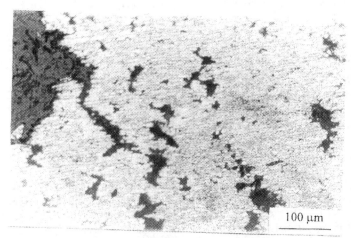

Figures 4. Microstructures of matrix after isothermal tensile test.

The microstructure observation of the matrix alloy after tensile test in Figures 4 clearly shows that continuous recrystallization occurred in the isothermal tensile test. Most of areas are recrystallized. The observations support the view that continuous recrystallization is the main cause of a stress exponent of 3.

4. Isothermal Tensile Testing of MMC

Isothermal tensile tests of the MMC(4μm, 10wt% SiCp) were performed within a temperature range of 525°C to 555°C and at an initial strain rate range of $2x10^{-4}$ s^{-1} to $1x10^{-1}$ s^{-1}. The results of elongation in percentage to fracture are shown in Figure 5. All the elongations to fracture were in the range of 50% to 113%. At the strain rate less than $5x10^{-4}s^{-1}$ faster strain rate can give better elongation value. As strain rate exceeds $1x10^{-3}$ s^{-1}, the elongation values begin to decrease. The optimum condition is at a strain rate of $5x10^{-4}$ s^{-1} and a temperature of 555°C with a maximum elongation to fracture of 113%. These results do not show any sign of high strain rate superplasticity [10]and when strain rate goes to $1x10^{-1}$ s^{-1} the elongation to fracture drops to 50%.

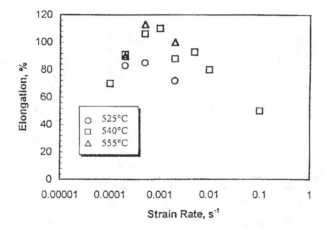

Figure 5. Effect of temperature and strain rate on elongation to fracture of MMC.

Chokshi et al [17] have mentioned that a commercial extruded bar of a 2124 aluminum composites reinforced with 20 vol.% SiC whiskers was not superplastic due to its limited reduction ratio. After further thermo-mechanical processing into thin sheet form (most possible by rolling) the material exhibited superplasticity. So the amount of working in processing plays an important role in refining the matrix grain particle shape and size. When grain size is reduced to much less than 10μm, high strain rate superplasticity and good elongation became feasible in aluminum-based MMCs [10]. A series of reports by Mabuchi et al. [19-21] showed that greater amounts of work (100:1 extrusion compared to 44:1) lead to a faster optimum strain rate, lower stress, and greater elongation in Si_3N_4 whisker composites. The main deformation processing of this research is extrusion with reduction ratio of 16:1 at 375°C. The reduction ratio is insufficient to obtain fine grains to achieve high strain rate superplasticity. Therefore, further thermo-mechanical processing such as extrusion and/or rolling is desirable.

The optimum elongation value of 113% is already much higher than the ductility at room temperature which is only 3.4%, and close to the optimum elongation value of 130% in the matrix alloy (Figure 1). Considering that there are few reports on superplasticity of Al-Li/SiCp composites up till now, to do detailed analysis about the deformation characteristics and microstructural evolution is helpful to find out potential of superplasticity in this material.

The stress versus strain rate relationship of the MMC in logarithmic scale is shown in Figure 6. The relationship does not show a clear "S-shape". The stress exponent n is about 3, which means the strain rate sensitivity m is around 0.3. However, a strain rate sensitivity of 0.3 does not bring about high elongation values. In fact, the elongation in percentage starts to decrease when strain rate goes beyond 5×10^{-4} s^{-1} as shown in Figure 1.

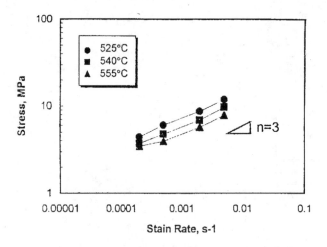

Figure 6. Logarithmic stress versus strain rate curve of MMC.

The regressions of stress versus (strain rate)$^{1/n}$ with different stress exponents are given in Figure 7 and the threshold stress extrapolated and square of efficiencies of correlation are tabulated in Table 1. It is observed that when stress exponent is 2 and 3 the data give good linear fit. Stress exponent of 2 implies the possibility that deformation is controlled by conventional superplastic mechanisms such as grain boundary sliding, climbing of dislocations along grain boundary etc. Good regression result at stress exponent of 3 also imply that continuous recrystallization or/and solute drag mechanism are of importance.

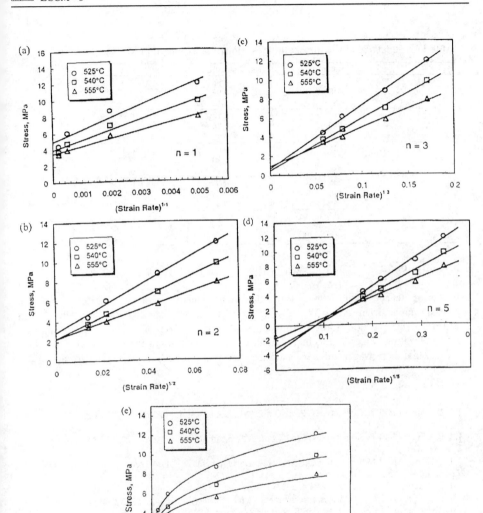

Figure 7. A threshold stress analysis for the MMC at different stress exponents.

Table 1 Dependence of the threshold stress on stress exponent in MMC.

	Temperature, °C	Threshold Stress, MPa	R^2
n=1	525	4.98	0.9478
	540	3.97	0.9715
	555	3.50	0.9803
n=2	525	2.88	0.9941
	540	2.28	0.9988
	555	2.25	0.9984
n=3	525	0.69	0.9974
	540	0.50	0.9941
	555	0.93	0.9891
n=5	525	-3.77	0.9938
	540	-3.11	0.9839
	555	-1.75	0.9747
n=3.51	525	0	0.9969
(power law)	540	0	0.9941
	555	0	0.9786

5. Conclusion

Isothermal tensile testing results show that at a temperature of 525 to 555°C and at a constant initial strain rates of $2x10^{-4}s^{-1}$ to $1x10^{-1}s^{-1}$ the MMCs have elongation of 50% to 113% whereas the elongations of the matrix alloys range from 50% to 130%. Continuous recrystallization was found to be the main micromechanism but incomplete, this is due to the fact that there was insufficient materials deformation in the material prior to testing.

References

[1] E.A.Starke Jr, T.H.Sanders (eds) Proc. Conf. "Aluminium-Lithium Alloys II", TMS-AIME, Warrendale, PA(1984)

[2] E.A.Starke Jr, T.H.Sanders (eds) Proc. Conf. "Aluminium-Lithium Alloys V", MCE Publications, Warley, UK (1989)

[3] B.Terry, "Metal-Matrix Composites - Current Developments & Future Trends in Industril Research & Applications", Elsevier, UK (1990)

[4] S.J.Harris, Mat. Sc. & Tech. Vol 4, 231, (1988)

[5] S.V.Nair, J.K.Tien, R.C.Bates, International Metall. Review Vol 30 No 6, 275 (1985)

[6] P.Martineau, M.Lahaye, R.Pailler, R.Naslain, M.Couzy, F.Cruege, J. Mat. Sci. Vol 19, 2749, (1984)

[7] T.G.Nieh, Metall. Trans A, Vol 15, 139 (1984)

[8] C.E.Pearson, J.Inst. Metals, Vol.54, 111 (1934)

[9] X.Zhang, M.J.Tan, Materials & Manuf. Processes, Vol 11, No 5, 749 (1996)

[10] R.S.Mishra, T.R.Bieler, A.K.Mukherjee, Acta Metall., Vol 43, 877 (1995)

[11] R.S.Mishra, A.K.Mukherjee, Scripta Metall. Vol 25, 271 (1991)

[12] R.S.Mishra, T.R.Bieler, A.K.Mukherjee, Scripta Metall., Vol 26, 1605 (1992)

[13] F.A.Mohamed, J.Materials Sci. Vol 18, 582 (1983)

[14] O.D.Sherby, P.M.Burke, Progress in Materials Science, Vol 13, 325 (1969)

[15] P.L.Blackwell, P.S.Bate, Metall. Trans. Vol 24A, 1085 (1993)

[16] C.H.Hamilton, B.A.Ash, D.Sherwood, H.C.Heikkenen, in "Superplasticity in Aerospace" (eds. H.C.Heikkenen, T.R.McNelley, TMS,Warrendale, PA, 29 (1988)

[17] A.K.Chokshi, T.R.Bieler, T.G.Nieh, J.Wadsworth, A.K.Mukherjee, *ibid*, 229 (1988)

[18] M.T.Lyttle, J.A.Wert, J. Materials Sci., Vol 29, 3342 (1994)

[19] T.Imai, M.Mabuchi, Y.Tozawa, M.Yamada, J.Mater. Sci. Lett., Vol 9, 255 (1990)

[20] M.Mabuchi, T.Imai, J.Mater. Sci. Lett., Vol 9, 761 (1990)

[21] M.Mabuchi, K.Higashi, S.Tanimura, T.Imai, K.Kubo, Scripta Metall. Vol 25, 1675 (1991)

Strength, deformability and failure of swaged metallic composites under quasistatic and high strain rate loading

L. W. Meyer[*], L. Krueger[*], I. Faber[*], P. Woidneck[**], K. Buehler[***]

[*]   TU Chemnitz, Materials and Impact Engineering, 09107 Chemnitz, Germany,
        http://www.wsk.tu-chemnitz.de/lwm/index.html
[**]   WIWEB, Landshuter Straße 70, 85345 Erding, Germany
[***]   IFAM, Lesumer Heerstraße 36, 28717 Bremen, Germany

Abstract

To improve the ballistic performance of kinetic energy penetrators, high strength tungsten-fiber reinforced metal matrix composites were manufactured by melt infiltration. The content of tungsten fibers were varied in the range of 25 % to about 91 % tungsten. Cu-2 %Be - alloy was used as matrix material. The material behavior of the composites were investigated under quasi-static and dynamic tensile and compressive loading under variation of the matrix content considering the anisotropy of the composite material. In order to characterize the failure behavior mechanical tests were performed in fiber direction as well as across the tungsten fibers.

1. Introduction

Some requirements for materials used for kinetic energy penetrators are high yield stress and strength, high density and a certain amount of deformability. Fiber reinforced composites with high density can be proper candidates as kinetic energy penetrators [1]. Refractory metal wires or fibers in general have been reported to exhibit a good balance of high strength and ductility, owing to drawn grain structure and alloying additions [2].

In order to achieve the aim of developing penetrators with remarkably increased ballistic efficiency, both the investigations of strength and failure mechanism at high rates of strain and the optimization of the relation between strength and deformability between fiber and matrix are needed.

Because of the limited information about the material behavior of fiber reinforced materials under high strain rate impact loading, basic investigations are required. Another important goal is to improve finite element calculations of composite materials by introducing the true material behavior. Essential to the development of constitutive equations used in numerical simulations is information on dynamic material properties. This task is further complicated by the directional dependence of properties due to the anisotropy of the material.

The present paper reports results from studies on means to reinforce by tungsten wires. The composites were produced by melt infiltration technique. In a first step to get access to suitable adjustability of strength and failure between matrix and fiber, the amount of fiber was varied. Quasi-static as well as dynamic tests were performed.

2. Materials and Methods

2.1 Processing

Commercially available tungsten wires with 2 wt.-% ThO_2 and 1 mm in diameter were put longitudinal into a crucible. Cu-2Be was melt infiltrated at 1250 °C. This process took 1 hour. After infiltration of matrix material, the composite was confined by a steel tube with an outer diameter of 30 mm. The following swaging process was performed in two steps. At first the material was heated at 300 °C for 30 minutes and swaged to 23 mm in diameter. After preheating at 300 °C the second swaging process was performed to a final diameter of 17 mm. After this process the tungsten wires had a diameter of 0.7 mm.

Three types of composites with different amount of matrix-material were manufactured. For the highest packing density of tungsten wires (90.7 % of the total area) a density of 17 g/cm^3 were achieved, Fig. 1. Additionally, composites with 50 % as well as 75 % amount of matrix were produced.

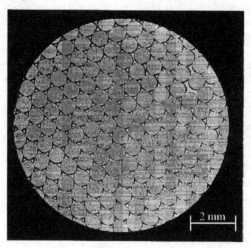

Fig. 1: Transverse section of a swaged compound of tungsten fibers infiltrated with CuBe (W 91-CuBe)

2.2 Experimental

2.2.1 Test set-up for tensile tests

In order to measure the material behavior under a wide range of strain rates, different experimental methodologies were employed. Under quasi-static loading conditions, a servo-hydraulic testing machine was used. Dynamic tension tests were conducted using a rotating wheel test setup. The force was measured directly at the specimen. Additional details of the experimental methodology has been reported elsewhere [3].

In order to measure material properties in transverse direction of the reinforcement, small specimens were manufactured by electrical discharge machining. The test apparatus for 10 mm long specimen and the specimen geometry for the tension experiments are shown in Fig. 2.

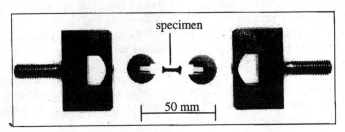

Fig. 2: Tensile loading device for 10 mm long specimens, splitted to pieces with tension specimen

2.2.2 Test set-up for compression tests

The compression specimens have a cubic form with 6 x 6 x 6 mm^3 and 8 x 8 x 8 mm^3 when tested in transverse direction, Fig. 3. In the axial direction disc shaped specimens with heights of 2 - 3 mm and 5 - 10 mm in diameter for quasi-static tests and heights of 1 - 2 mm and 5 - 10 mm in diameter for dynamic tests were used. The quasi-static experiments were performed using a servo-hydraulic testing machine. For dynamic tests at strain rates in the range of 10^2 to 10^3 1/s a drop weight test set-up was used.

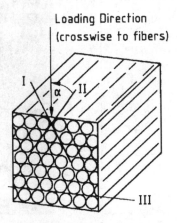

Fig. 3: Cubic form of compression specimen crosswise to fibers with three planes of layers (I, II, III). The smallest angle „α" is used to characterize the position of the loading direction to the orientation of the planes of layers.

3. Results

3.1 Material behavior under tension - W91-CuBe

The tension behavior of W91-CuBe under quasi-static strain rates of $\dot{\varepsilon}=5\text{x}10^{-4}$ 1/s as well as dynamic strain rates up to $\dot{\varepsilon}=10^3$ 1/s is shown in Fig. 4. Under quasi-static loading, a 0.2 %- yield strength of about 1630±63 MPa and an ultimate tensile strength of about 1750±96 MPa was measured. The elongation varied between 0.1 % and more than 2.5 %.

Increasing the strain rate to $\dot{\varepsilon}=10^0$ 1/s, the deformability is reduced and reaches 0.35 % to 0.65 %. The yield stress as well as the ultimate tensile strength are strain rate sensitive, compared with quasi-static loading.

At the higher strain rates of $\dot{\varepsilon}=10^2$ 1/s and 10^3 1/s the composite fails by brittle fracture in the elastic range of the stress-strain curve. At the highest strain rates, ultimate tensile strengths of 1300 to 1430 MPa were measured.

Experiments under quasi-static loading of specimens taken crosswise to the fibers have shown early failure in the elastic part of the stress - strain behavior at about 80 MPa. Ground and polished specimens have shown similar failure behavior to the eroded specimens. The failure was initiated in the W- fiber at the grain boundaries, Fig. 5.

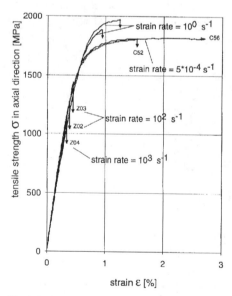

1000 µm

Fig. 4: Stress - strain behavior under tensile loading as a function of strain rate, W91-CuBe, loading direction parallel to the fibers

Fig. 5: Tensile specimen, fracture surface transverse to the fiber direction

3.2 Material behavior under compression - W91-CuBe

The compression behavior of W91-CuBe as a function of strain rate is quite different to the tensile behavior, Fig. 6. The yield stress of the composite is highly strain rate sensitive. For example, at 5 % plastic deformation about 500 MPa increase in flow stress is measured at 10^7 1/s enhanced strain rates. Under dynamic loading of $\dot{\varepsilon}=10^3$ 1/s a compression strength of 2300 MPa is achieved. The material softening with strain leads to a stress plateau of about 2000 MPa at a plastic strain of 25 %.

At higher strains, the flow stress is increasing again. Two explanations can be named. On the one hand, the strain and strain rate hardening dominates again over thermal material softening. On the other hand, the compression tests were conducted using disk shaped specimens with a height to diameter ratio of 0.2 to 0.3. This leads to an increased influence of friction, compared with specimens of height to diameter ratio of 1, respectively.

Therefore, the measured flow stress was corrected assuming that a friction coefficient of about μ=0.06 is possible, when the lubrication is sufficient. As reported in the literature, friction coefficients of μ=0.02 to 0.07 are found [4].

The correction was performed by using equation 1 [5].

$$R_{p,c} = \frac{R_{p,measured}}{1 + \dfrac{D \cdot \mu}{3 \cdot h \cdot \sqrt{3}}} \qquad \text{equation 1}$$

D .... diameter, h ... height

Furthermore, metallographic examination of tested specimens with 1 mm in height have shown no signs of buckling of the W-fibers. Hence, buckling is not supposed to be the reason for the strain hardening behavior during deformation at higher plastic strains.

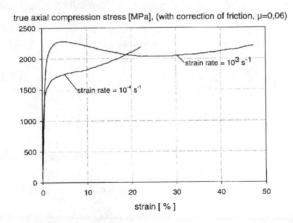

Fig. 6: True axial compression stress versus compression strain at two rates of strain, W91-CuBe

With an increasing angle between the possible easiest glide (I or II) and loading direction, a decrease of deformability and hence of compression strength takes place, Fig. 7 and Fig. 8. The total compression strain reduces from ε_t = 3.5 to 0.5 %. The compression strength is reduced from σ_{max} = 1500 to 900 MPa.

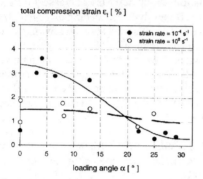

Fig. 7: Total compression strain transverse to fibers under quasistatic and dynamic loading as a function of the loading angle α, W91-CuBe

Fig. 8: Comparison of compression strength transverse to fibers as a function of loading angle α under quasistatic and dynamic loading, W91-CuBe

The failure is occurring by shear processes in the matrix material along its slip planes between the fibers. The reason for the failure at low overall deformability when tested transverse to the fiber direction is the low thickness of about 4 to 8 μm of matrix material between the tungsten fibers. Under dynamic loading transverse to the fibers, the deformation is drastically reduced. By increasing the loading angle α to more than 20 °, the total strain is found to be less under quasi-static loading than that under dynamic strain rates, Fig. 7.

The behavior of compression strength as a function of loading angle and strain rate is given in Fig. 8. As can be seen additionally from Fig. 9, the flow stresses and strengths are reduced, if the loading angle is increased. In all cases fracture occurred, after the maximum in stress was reached.

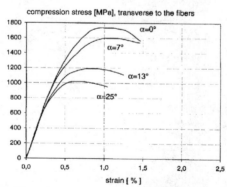

Fig. 9: Typical stress - strain behavior of W91-CuBe under dynamic loading of $\dot{\varepsilon}$ =200 1/s, compression stress transverse to fibers

3.3 Material behavior of W50-CuBe and W25-CuBe

The compression behavior in the direction along the fibers of W50-CuBe is presented in Fig. 10. Under dynamic loading of $\dot{\varepsilon}=10^3$ 1/s, a region of nearly constant stress of about 1650 MPa between 5 % and 30 % plastic deformation is measured. The strain hardening rate starts with the highest value of 350 MPa at the beginning of plastic deformation and decreases with increasing strain because of the lack of strain hardening at high strain rates between 5 % and 30 % strain.

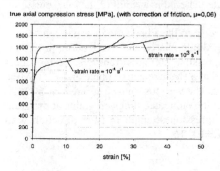

Fig. 10: Comparison of true compression stress - strain behavior at two strain rates, W50-CuBe

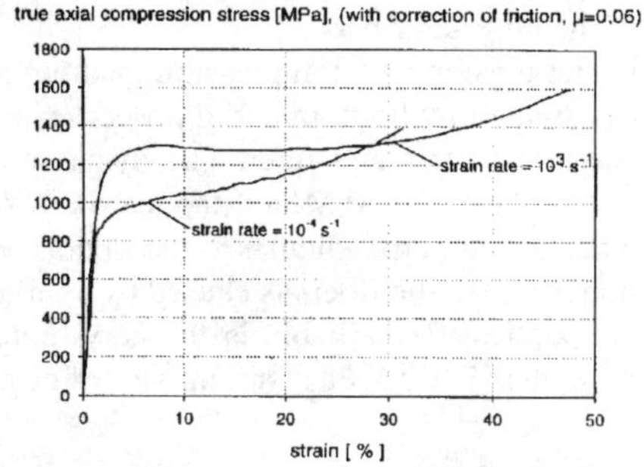

Fig. 11: Comparison of true compression stress-strain behavior at two strain rates, W25-CuBe

With further enhanced matrix content, the flow stress of the W25-CuBe alloy is reduced due to the lower strength of the matrix, Fig. 11. The measured maximum strain hardening rate at 3 % to 5 % plastic deformation is about 250 MPa.

With compression loading in fiber direction, even at high strains beyond 30 % deformation, no delaminations between matrix and fibers are observed. Under transverse loading to the fibers, however, the spreading of the matrix material leads to delaminations, preferred in the 45 ° plane to a shear failure in the matrix, Fig. 12 and Fig. 13. These failures develop before reaching 10 % compression strain, where the dynamic loading has been interrupted. Cracks are observed with a length of up to 3 mm, Fig. 13. Furthermore, numerous delaminations of the fibers and cracks are visible, Fig. 12 and Fig. 13.

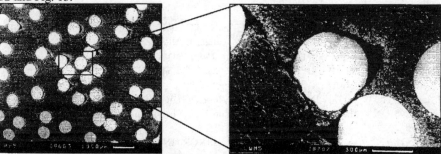

Fig. 12: Transverse compressed W25-CuBe after $\varepsilon = 10$ % deformation with delaminations and shear cracks

Fig. 13: Detail of Fig. 12

4. Discussion

A comparison of the material behavior of composites with different volume fraction of tungsten fibers show differences in flow stresses, strengths and deformabilities. As an example, the yield stress at 2 % plastic deformation along the fibers as a function of tungsten content and strain rate enhances with fiber content and strain rate. The material W91-CuBe has the highest strain rate sensitivity of the flow stress of about 500 MPa, compared with about 350 MPa for W50-CuBe and 200 MPa at W25-CuBe. The results suggest that the rate sensitivity of the composite material is a result of the rate sensitivity of the tungsten fibers.

Under tension and compression loading transverse to the fibers, the material behavior is determined by both, material properties of the single phases and the bonding properties between matrix and fiber. The dynamic compression strengths transverse to the fibers varies between 950 MPa if the matrix is loaded at a shear sensitive plane and 1750 MPa, when tested perpendicular to the fibers ($\alpha=0$ °). The very low tensile strength of 80 MPa transverse to the fibers is caused by a failure of the elongated tungsten grains at the grain boundaries. Remarkable is the very high compression strength in fiber direction with more than 2200 MPa at strain rates of nearly $\dot{\varepsilon}=10^3$ 1/s.

5. Conclusion

The dominant content of closed-packed tungsten fibers leads to excellent axial quasi-static tensile and compressive as well as dynamic compressive strength of the composite W91-CuBe combined with an also excellent formability in the same direction excluding a strain rate embrittlement under tensile loading. The strength behavior under both tensile and compressive loading crosswise to the fibers is determined by the low strength of the matrix, respectively the combination of the matrix and fiber strength and is therefore relatively low. An elevation of the matrix content leads to an improvement of the deformability only at very high contents, but under loss of the good strength behavior. Because of the low influence of the matrix in improvement of the ductility, the matrix content in the composite should decreased in order to use the high strength of the tungsten fibers. Future work should be directed to prevent the early failure between the grain boundaries of the tungsten fibers.

References

[1] Tweed, J. H.; Loosemore, G. R.: FEASIBILITY OF TUNGSTEN-BASED METAL-MATRIX COMPOSITES - Tungsten and Tungsten Alloys - 1992, p. 471-478

[2] Ritzert, F. J., Dreshfield, R. L.: PROGRESS TOWARD A TUNGSTEN ALLOY WIRE/HIGH TEMPERATURE ALLOY COMPOSITE TURBINE BLADE - Tungsten and Tungsten Alloys - 1992, MPIF, p. 455-462

[3] Meyer, L. W.; et. al.: DYNAMIC BEHAVIOR OF HIGH STRENGTH STEELS UNDER TENSION - Shock Waves and High-Strain-Rate Phenomena in Metals, Plenum Press, 1981, p. 51-63

[4] Gorham, D. A.; et. al.: SOURCE IN ERROR IN VERY HIGH STRAIN RATE COMPRESSION TESTS - 3rd Conf. on the Mechanical Properties at High Rates of Strain, Oxford, 1984, p. 151

[5] Avitzur, B.: Israel J. Techn. 2 (3), 1964, p. 295-304

Effects of Damage and Mechanical Behaviour on Aluminium -Matrix Composites

E. Gariboldi, M. Vedani

Dipartimento di Meccanica - Politecnico di Milano
Piazza L. da Vinci 32, I-20133 Milan, Italy

Introduction

During the last decade the mechanical properties of discontinuously reinforced Metal Matrix Composites (MMC's) have been extensively investigated. It was found that they were strongly affected by damage. The present paper is aimed at giving a review of the experimental investigations carried out on damage of aluminium-matrix composites differing for matrix alloy, reinforcement volume fraction and heat treating conditions. The materials were experimentally investigated either in uniaxial room temperature tension and compression, or in fatigue and high temperature tension and creep.

Materials and experimental procedures

The composites investigated were 2014 and 6061 Al alloys reinforced with nominal volume fractions of 10 and 20% Al_2O_3 particles. The materials, supplied as extruded bars, were studied either in T6 or T4 conditions.

Microstructural observations carried out on the composites showed that the reinforced was made up of blocky-shaped Al_2O_3 particles uniformly distributed in the aluminium matrix. The increase in reinforcement volume fraction of the 20% Al_2O_3 composite was achieved by using coarser particles (the particle size was 9,9 μm and 20,6 μm for the 10% and 20% composite, respectively). Correspondingly, the mean free path between particles was calculated to be 35 μm and 47 μm. For both materials the matrix grain size corresponded to grade 7 of the ASTM E112 standard classification.

Damage in composites was measured by monitoring the variation in the elastic modulus and by the acoustic emission technique in tension and compression testing. Details of the experimental procedures where given in previous papers [1-3]. Fatigue and creep properties were also measured and the correlation with the damage behaviour of the materials was discussed.

Results

Figure 1 shows the damage behaviour of a 2014 20% Al_2O_{3p} T6 composite as a function of plastic strain in tension. Damage was evaluated by the parameter $D=1-Eo/E(\varepsilon_p)$. In the formula, Eo is the initial elastic modulus and $E(\varepsilon_p)$ is the elastic modulus measured after a plastic strain ε_p. The graph shows that damage is clearly active even after small plastic strains and that it progresses almost linearly during deformation. It was also found that damage was strongly affected by matrix composition and temper as well as reinforcement shape and volume content. In any material condition, microstructural and fractographic analyses allowed to state that the decrease in load bearing capacity (i.e. damage) was related to reinforcement fracture.

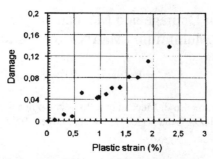

Figure 1. Damage evolution of a 2014 20% Al₂O₃ₚ T6 composite.

Acoustic emission was measured by positioning specific probes on the specimens during testing. The acoustic activity was more intense during the initial stages of plastic straining when coarse particles failed in a brittle manner. Figure 2 compares the acoustic emission energy measured during tension and compression tests on a 6061 20% Al_2O_{3p} T6 composite. It is shown that the damage is more evident and localised in tension and that it occurs mainly before macroscopic yielding. On the contrary, in compression damage occurs over the whole straining range, albeit at significantly lower energy levels.

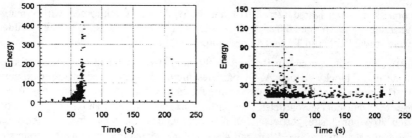

Figure 2. Acoustic emission energy in a 6061 20% Al_2O_{3p} T6 composite loaded in tension (left) and compression (right).

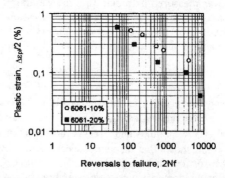

Figure 3. Low cycle fatigue resistance of 6061 matrix 20% and 10% Al_2O_{3p} T6 composites.

The differences in damage behaviour as affected by reinforcement size and volume fraction strongly modified the material behaviour under fatigue loading condition. The diagram given in figure 3 shows the different behaviour of 6061 matrix 20% and 10% Al₂O₃ₚ T6 composites. The 10% Al₂O₃ material featured a smaller size of the reinforcement particles that revealed to be more resistant to failure. This aspect resulted in increased fatigue life either under low or high cycle regimes.

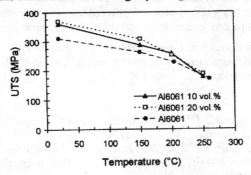

Figure 4. Ultimate tensile strength vs. temperature for the 6061 Al₂O₃ₚ T6 composites.

Tensile tests were carried out on 6061 matrix 20% and 10% Al₂O₃ₚ composites at temperatures up to 250 °C [4]. The corresponding unreinforced alloy was also considered for comparison purposes. The resulting tensile strength is given in Figure 4. The improvement in strength brought about by particle addition at room temperature was gradually lost when temperature reached the highest values.

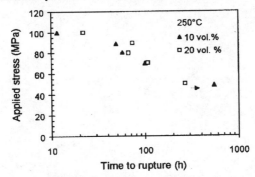

Figure 5. Stress vs. time to rupture for the 6061 Al₂O₃ₚ composites creep tested at 250°C.

A similar behaviour of the two particulate composites was also observed under constant load creep tests at 250°C. Times to rupture vs. initial stress are given in Figure 5. The creep stress exponent in classical Norton's law were similar for the two composites. The only factor affected by volume fraction and particle size appeared to be ductility. SEM observations of crept specimens confirmed the presence of two different microstructural damage mechanisms acting at this temperature: void nucleation occurred mainly by particle-matrix debonding in the 6061-10% Al₂O₃ₚ composite, while particles cracked and acted as the main site for the process of void nucleation and growth in the material with the higher particle volume fraction. Thus, the results suggested that creep at the investigated temperature was mainly controlled by aluminium matrix deformation. The reinforcement assumed a remarkable effect only during the last stage of creep deformation, when the occurrence of microstructural damage led to the final fracture.

Conclusions

The experimental tests carried out on several particulate reinforced composites showed that damage mechanism and its extent was affected by many microstructural factors such as matrix composition and temper, reinforcement volume fraction and size.

The differences in damage rate and behaviour conditioned the material performance under load in simple tension and compression and determined the fatigue life and creep ductility of the composites.

References

1. M. Vedani, E. Gariboldi. Damage and ductility of particulate and short fibre Al-Al$_2$O$_3$ composites. Acta Materialia, 44, 8 (1996) 3077-3088
2. E. Gariboldi, C. Santulli, F. Stivali, M. Vedani. Evaluation of tensile damage in particulate reinforced MMC's by acoustic emission. Scripta metallurgica et Materialia, 35, 2 (1996) 273-277
3. E, Gariboldi, C, Santulli, F. Stivali, M. Vedani. Evaluation of damage in MMC's by acoustic emission technique. Mater. Sci. Forum, vol.217-222 (1996) 1461-1466
4. E. Gariboldi, M. Vedani. Behaviour of aluminium matrix composites in the temperature range 200-250°C. Proc. Int.Conf. "Advencing with Composites '97", Milan, 1997, pp. 29-40

The Microstructure and Mechanical Properties at High Temperatures of Nextel 720 Fibres

F. Deléglise, M.H. Berger, A.R. Bunsell
Centre des Matériaux de l'Ecole des Mines de Paris, BP 87. 91003. Evry cedex,France

Abstract

The thermomechanical properties of Nextel 720 fibres have been studied. Their microstructures were examined in order to establish the micromechanisms which limit these fibres. The two-phases microstructure which was observed increased resistance to creep but subcritical crack growth was found to become dominant at high temperatures.

I. Introduction

The development of ceramic matrix composites capable of operating at high temperatures above 1100°C in air requires fibre reinforcements which are stable under these conditions. Oxide based fibres do offer the possibility of long term use in air above 1400°C. However at temperatures above around 1000°C single phase polycrystalline alumina fibres have been found to suffer from creep and above 1150°C weakening of the grain boundaries results in major strength loss [1]. The Nextel 720 fibre is the latest of a series of fibres produced by 3-M of which have been based on alumina or mullite, and has an original two phase microstructure of alumina and mullite [2].

II.Material

The Nextel 720 fibre is composed of 85% Al_2O_3 and 15% SiO_2. A sol-gel method is used to produce a fully crystallised fibre with an approximate composition of 55% mullite and 45% α-alumina. The most important innovation compared to other alumina fibres is its two-phase microstructure consisting of mosaic and elongated grains.

III. Experimental procedure

The mechanical behaviour of the fibre has been studied at room and high temperatures up to 1200°C. Creep and tensile tests at different strain rates have been carried out on as-received and heat-treated fibres (3 hours at 1200°C in air). Single filaments were tested on a horizontal tensile machine with a maximum displacement speed of 1.5mm/s. [3]. Before each test the diameter of the fibre was measured using a Watson ocular mounted on an optical microscope. A slotted electrical resistance oven allowed the fibres to be tested at high temperatures with extremely rapid heat-up rates. The jaws remained cold and the hottest part of the furnace was 25 mm long.
The fibre fracture surfaces were observed using a LEO DSM 982 Gemini SEM operating at a low voltage (1 to 3 keV). The effects of heat treatment on the microstructure were investigated using a PHILIPS EM 430 TEM-STEM with an acceleration voltage of 300 kV. X-ray diffraction was used to analyse the compositions of the different phases and the influence of heat treatments on the crystal structures.

IV Results
1. Room temperature characterisation
1.1. Microstructure

The external appearance of the fibre is shown in Figure 1. The fibre is circular in cross section. The average diameter is 12.5 μm with a standard deviation = 0.55 μm. The fracture surface shows "grains" of around 0.5 μm the nature of which will be explained by TEM observations.

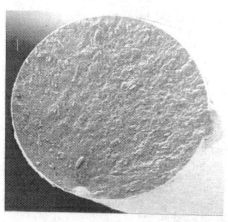

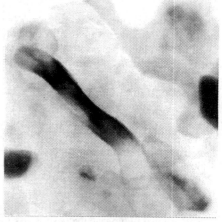

5 μm
Figure1: SEM micrograph of N720 fibre.

100 nm
Figure2: Elongated grain with an aspect ratio of 6.

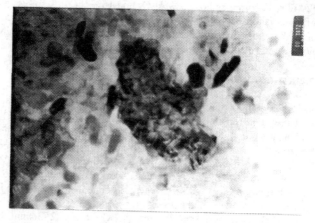

200 nm

Figure 3: Large mosaic grain of about 650 nm composed of several subgrains and surrounded by many grains with aspect ratios of 3 to 4.

The fibre can be seen to be composed of three types of entities. Observation at low magnification showed large diffracting grains of 0.5 μm or more, producing the granular aspect of the fibre surface observed in SEM. More thorough observations indicated that these "grains" were in reality composed of several grains. Alumina and mullite grains of 50 to 100 nm in size with curved interfaces constitute the bulk of the fibre. Many of these grains were grouped by two or three and diffracted together indicating crystallographic orientation relationships. The fibre enclosed elongated grains of 20 to 50 nm in width with a length to width ratio of 3 or 4 in general but grains with aspect ratio of 6 were also found (Figure 2). The Al:Si ratio in these grains was more than 9 in weight which is more alumina rich than the average composition of Nextel 720 (Al:Si=7).

1.2. Mechanical properties

The room temperature strengths were determined at different gauge lengths between 5 and 250 mm. At least 35 tests were performed at each gauge length.

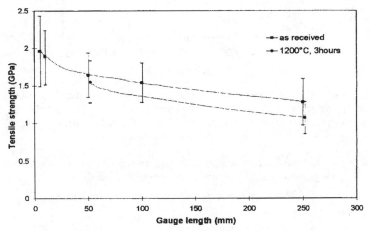

Figure 4: tensile strength as a function of gauge length.

Strains to failure were small and they never exceeded 1%. The Young's modulus was seen to be 251 GPa with a standard deviation of 12 GPa. No influence of strain rate was found on room temperature properties. A Weibull analysis of the results permitted a value of 4.9 for the m modulus to be calculated.

1.3. Fracture analysis at room temperature

The fibres showed linear elastic behaviour with brittle failure. The fracture surfaces of all fibres broken at room temperature had the same appearance and were planar. Crack propagation was intra or intergranular. Figure1 shows the fracture surface of an as-received N720 fibre broken at room temperature with a failure strength of 1.30 GPa.

2. High temperature characterisation
2.1 Tensile tests

Up to 1000°C the fibres showed perfectly linear elastic behaviour. At higher temperatures (1100°C to 1200°C) fibres broke very quickly at loads less than 2g. No plasticity was observed during high temperature tensile testing and the fibres still exhibited brittle failure surfaces.

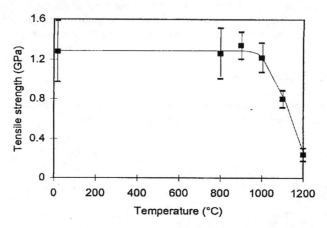

Figure5: Tensile strength at a constant strain rate of 6.10^{-5}/s as a function of temperature. For each temperature at least 35 fibres were tested. Around 1000°C the strength suddenly dropped as shown in figure5. At 1200°C the failure strength was 0.24 GPa which is only 14% of the room temperature failure strength.

Figure 6 shows that the fracture strengths depend on the strain rates which indicate that subcritical crack growth (S.C.G.) occurs in the Nextel 720 fibre.

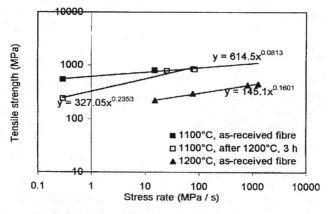

Figure 6: Failure strengths at 1100°C and 1200°C of as-received and heat treated fibres as a function of strain rate.

2.2 Creep tests

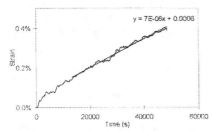

Figure 7: Creep curve at
0.23 GPa and 1100°C.

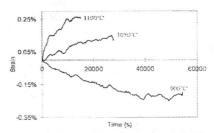

Figure 8: Creep curves at 0.3 GPa
and at several temperatures.

Many creep tests were conducted at different stresses and temperatures. It was very difficult to observe steady state creep as the fibres rapidly broke because of subcritical crack growth. At high temperature or stress (above 0.2 GPa at 1100°C) catastrophic failure during primary creep occurred in less than 3 hours. Only one fibre was found to creep and survive after 40000s. For this test at 1100°C and 0.23 GPa a creep rate of 7.10^{-8} / s was calculated between 15000s and 45000s (Figure 7).
Negative creep was observed from 800°C to 1000°C in Figure 8 resulting from a densification of the fibre at high temperature.

2.3 High temperature microstructure
2.3.1 Microstructural evolution after heat treatment

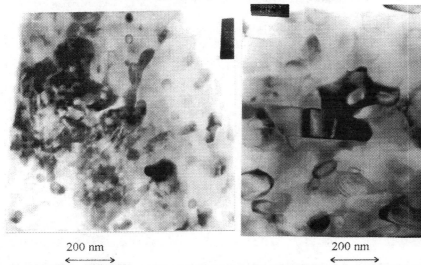

Figure 9: TEM. micrography after
a 5 hours heat treatment at 1200°C.

Figure 10: TEM. micrography after
a 5 hours heat treatment at 1400°C.

After a heat treatment at 1200°C the microstructure seemed to be almost unchanged compared to the as-received fibre. The same phases with the same sizes as before the treatment were observed. After heat treatment at 1400°C the microstructure showed the same entities but their sizes had increased. Grains were bigger (100 to 300 nm) but they had retained their characteristics except for a change from rounded shapes to angular forms.

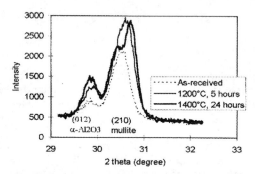

Figure 11:XRD analysis of fibre before and after heat treatments.

After heat-treatment the increase of the α-alumina peak and the doubling of the mullite peak (Figure 11) indicate a change in the mullite structure. At room temperature the alumina-silica ratio in the mullite is close to 2:1 and its structure is tetragonal. After heat treatment the mullite recristallises in the orthorombic system while losing alumina and the ratio is closer to 3:2

2.3.2. Fracture analysis

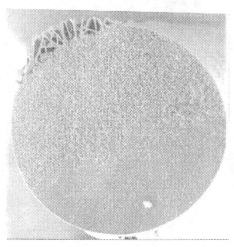

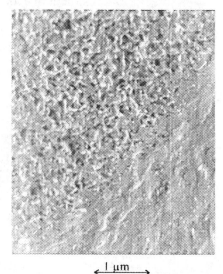

Figure 12: Tensile fracture surface at 1200°C

Figure 13: Detail of the center of a tensile fracture surface at 1200°C

Up to 1000°C, the fibre fracture surfaces have the same appearance as those at room temperature however from 1100°C the fracture surfaces are drastically modified. Failure was seen to be always initiated at a large defect (up to 1.5μm) consisting of alumina-rich needles. As can be seen in Figure 12 the fracture surfaces present an area which fans out symmetrically from these needles and consists of two zones. The fracture surface is at first rough due to intergranular failure and then smother due to intragranular failure (see Figure 13).

V. Discussion
1. Fracture mechanisms

At room temperature and up to 1000°C, the fracture mechanisms do not change. From Figure 1 the failure strength has been used to calculate K_{IC} from the Griffith's relation applied to a semi-elliptical surface crack:

$$K_{IC} = \sigma Y \sqrt{a\pi} \qquad \text{with } Y = 0.67 \ [4] \qquad \longrightarrow \qquad K_{IC} = 1.3 MPa\sqrt{m}$$

From this value the critical defect size can be calculated assuming that the shape is constant. At 250 mm gauge length, 1 GPa $< \sigma_r < 1.6$ GPa so 0.5 μm $<a_c<1.2$ μm.
The granulometry of the fibres is fine and smaller than the critical defect size so that, as observations of the fracture surfaces have shown, failure is always initiated at defects, essentially voids, created during the manufacturing process.

From 1100°C the fracture is provoked by slow crack growth (SCG) initiated at large needle shaped defects created by the combination of high temperature and stress. The fracture surface shows two areas. The first corresponds to an intergranular SCG. When the stress concentration at the tip of the crack is great enough K_I reaches K_{IC} and crack propagation becomes catastrophic. This corresponds to the intragranular area.
This phenomenon occurs even at very high stress rates so SCG can dominate failure even at very fast loading rates. SCG is very detrimental for the high temperature properties of the fibre and its effect is aggravated by a temperature increase. Crack velocity V can be represented by the following Paris expression which relates it to the applied stress intensity factor K_I [5]: $V = V_0 \left(\dfrac{K_I}{K_IC}\right)^N$

were N is the crack growth exponent which indicates susceptibility to SCG.
The N values can be calculated from the slope of the curve represented in Figure 6: At 1100°C N = 10.3 for as received fibres and N = 3.2 for heat-treated fibres. At 1200°C N= 5.2 for as received fibres. A N value of 17 at 1000°C was calculated by Goring and Schneider [6]. The crack growth exponent, N, decreases clearly with increasing temperature which shows that the fibre becomes more and more susceptible to SCG when temperature increases. At 1200°C and in the best conditions (high stressing rates), the fibre retains only 30% of its room temperature strength. These results are lower than those announced by the manufacturer [2] and may be due to differences in experimental techniques.

2. Creep behaviour

The creep resistance of Nextel 720 is better than this of pure alumina fibres. Indeed the creep rate at 1100°C under 0.23GPa is seven times smaller than that of FP fibres under the same conditions (5.10^{-6}/s) [1]. These improved properties are due to the two-phase microstructure. However SCG occurs from 1000°C during creep tests and causes premature failure of the fibre before reaching the secondary creep stage.

3. Influence of a 3hours, 1200°C, heat treatment

After the heat treatment N at 1100°C is even smaller than N at 1200°C of non treated fibres. A 3 hour heat treatment at 1200°C had a detrimental effect on susceptibility of the fibre to SCG and on its residual room temperature properties even though it had no detectable influence on the general microstructure. Such a heat treatment had a localised effect on the microstructure. It can be assumed to have created point defects.

VI. Conclusion

The two phases microstructure of the Nextel 720 fibre has been shown to confer greater creep resistance than that found with other polycrystalline oxide fibre. However this advantage is limited by premature fracture due to subcritical crack growth.

References

[1].V.Lavaste, M.H. Berger and A. R. Bunsell, MICROSTRUCTURE AND MECHANICAL CHARACTERICS OF ALPHA-ALUMINA-BASED FIBRES ,Journal of Materials Science, Vol. 30, 1995, p. 4215-4225.
[2] D.M.Wilson, S.L. Lieder and D.C. Luenburg, MICROSTRUCTURE AND HIGH TEMPERATURE PROPERTIES OF NEXTEL 720 FIBERS, Ceramic engineering and Science Proceedings, Vol. 16, N° 5, 1995, p. 1005-1014.
[3] A.R. Bunsell, J.W.S. Hearle, AN APPARATUS FOR FATIGUE-TESTING OF FIBRES, J. Phys. E Sci. Instrum. 4 (1971) 868
[4] STRESS INTENSITY FACTORS HANDBOOK, 1987, Vol. 2, p 654-655, Ed. Y.Murakami, Pergamon Press
[5] S.M. Wiederhorn, SUBCRITICAL CRACK GROWTH, Concise Encyclopedia of advanced ceramic Materials, 1991, p. 461-466, Ed. R.J. Brook, Pergamon Press.
[6] J. Goring and H. Schneider, CREEP AND SUBCRITICAL CRACK GROWTH OF NEXTEL 720 ALUMINO SILICATE FIBERS AS RECEIVED AND AFTER HETA TREATMENT AT 1300°C, Ceramic engineering and Science Proceedings, Vol. 18, N° 3, 1997, p. 95-104.

Neutron diffraction measurements of matrix stresses in planar random fibre aluminium matrix composites

B. Johannesson[∘], T. Lorentzen and O.B. Pedersen

Materials Research Department, Risø National Laboratory
DK-4000 Roskilde, Denmark
[∘] Engineering Research Institute, University of Iceland,
Hjardarhaga 2-6, IS-107 Reykjavik, Iceland

Abstract

The neutron diffraction technique is used to study matrix strain components during tensile loading of aluminium matrix composites with planar random alumina fibres. Analysis of the mean stress hardening rate shows that measurements based on the permanent softening underestimates the mean stress and the hardening rate.

1. Introduction

Optimization of the thermoplasticity of metal matrix composites requires an understanding of the influence of reinforcement geometry on the residual stresses produced by thermoplastic deformation. An introduction to recent analytical and numerical modelling of residual stresses for different reinforcement geometries can be found in [e.g. 1]. Since Orowan[2] and Wilson's [3] studies of residual stresses the Bauschinger experiment has been used widely to obtain estimates of the mean matrix stress produced with a wide variety of reinforcement geometries [4-6], ranging from equiaxed particles to aligned continuous fibres. Fundamental to the Orowan-Wilson approach is the association of "permanent softening" [2] with the mean matrix stress. Recently, Johannesson and Ogin [7,8] introduced a further modification of the Orowan-Wilson approach, which is applicable to the industrially important squeeze-cast aluminium matrix composites with planar random Saffil fibres. As in the case of earlier modifications of the approach it is essential to calibrate the measurements using direct measurements of residual elastic strains by a diffraction technique. The need for such calibration is quite strongly emphasized by recent studies [9-11] , and the purpose of this communication is to present and discuss the result of such a calibration of Johannesson and Ogin's modification.

The present calibration utilizes *in-situ* neutron diffraction measurements of the mean matrix stress evolving during tensile elongation of the squeeze-cast composites. The advantage of neutron diffraction, rather than X-ray diffraction, lies in the large penetration depth of the neutron beam into aluminium, which allows the mean matrix stress to be sampled over the entire volume of the composite specimens.

2. Experimental details.

The composites were prepared by squeeze infiltration [e.g. 1] of Al_2O_3 (Saffil) fibre preforms [12] with commercial purity (99.98%) aluminium. After infiltration the aluminium solidifies under a pressure of about 32 MPa. Fibre volume fractions were f = 10%, 15% and 20%, the average fibre length is about 500 μm and the average fibre diameter about 3 μm. The fibre orientation distribution is random in the plane of the preform (x_2x_3) and the angles between the fibre axes and the x_1 axis are distributed close to 90° [13]. In order to get a statistically representative measure of internal strains with the neutron diffraction technique, it is important that the number of diffracting grains is as large as possible [14]. Therefore, specimens for the neutron diffraction experiments were cut from the bottom of the composite billets, where the matrix grains are uniformly small (mean grain size about 100-300 μm). A problem with the composites with f = 10% is that large columnar matrix grains (about 0.5-1 mm diameter and several mm long) extend to the region of the billet from which the specimens were cut, so that only about half of each specimen has the small matrix grains.

Neutron diffraction experiments were carried out at the DR-3 reactor at Risø in a loading device which can be mounted on the neutron spectrometer [15]. Aluminium powder was used as a stress free reference. The aluminium {111} reflection was used to monitor the mean elastic lattice strain in the matrix during controlled stepwise loading of test specimens. A reorientation of the loading device by 90° allows for measurements of the transverse lattice strain components. One lattice strain component is measured during each loading experiment and hence different strain components are measured in different specimens. Repeated diffraction measurements in any one specimen showed that the uncertainty in each strain measurement is $\pm$ $0.9 \cdot 10^{-4}$. The specimens had a gauge length of 50 mm and a cross section of 6×6 mm^2. For f = 15% the large columnar grains extend about 1 mm into the specimens. The specimens with f = 20% contain only matrix with small grains.

3. Cyclic Bauschinger experiments

The diffraction experiments are analysed in section 4.2 and compared with results from cyclic Bauschinger experiments [7, 8] on specimens with large columnar matrix grains of diameter 0.5 mm and several millimeters long. The question then arises if the matrix grain size affects the hardening rate in these composites. To answer this question, cyclic Bauschinger experiments were carried out on 6 composite specimens with the small matrix grains. Specimens with fibre volume fractions of 10%, 15% and 20% were cycled to obtain hysteresis loops at plastic strain amplitudes between 0.05% and 0.4%, with loops recorded every 0.05%. The hysteresis loops were then analysed to obtain the permanent softening values $\Delta\sigma^T$ and $\Delta\sigma^C$, which were used to obtain an estimate of the plastic mean stress and the mean stress hardening rate (for a detailed description of how $\Delta\sigma^T$ and $\Delta\sigma^C$ are evaluated and used in the analysis, see refs. [7,8]). The results are presented and discussed in section 5.

4. Results.

Analysis of stress-strain hysteresis loops [7, 8] using Johannesson and Ogin's modification of the Orowan-Wilson approach gives information on both the residual mean stress and the plastic mean stress in the matrix. In this section we present neutron diffraction measurements of the matrix strain in the planar random fibre aluminium matrix composites and compare them with results of analysis of the permanent softening.

4.1. Residual matrix strain.

All three components of the (thermal) residual strain, $<e_3>_M^R$, $<e_2>_M^R$ and $<e_1>_M^R$ were measured in virgin specimens and the results are summarised in fig. 1. For the composites with fibre volume fraction of 10%, only about half of each specimen contains the fine matrix grains necessary for neutron diffraction, the other half consisting of coarse columnar grains. This was clearly reflected in the considerable scatter in the results and these results are, therefore, not included in fig. 1. The residual strain components were measured for five specimens with f = 15% and ten specimens with f = 20%. The scatter in the results is similar to the intrinsic uncertainty of the neutron diffraction method, showing that specimen to specimen variation is minimal. The average value of each strain component is small and no clear distinction can be made between the in-plane components ($<e_3>_M^R$, $<e_2>_M^R$) and the normal component ($<e_1>_M^R$).

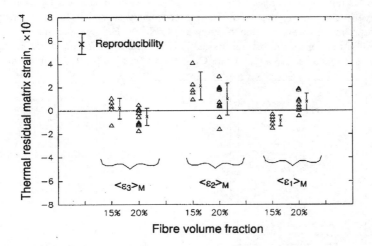

Figure 1. Neutron diffraction measurements of the three thermal residual strain components in virgin specimens with fibre volume fraction 15 and 20%. The x-es with error bars are the mean value and standard deviation for each group of specimens. The reproducibility gives the intrinsic error in each strain measurement, $\pm 0.9 \cdot 10^{-4}$.

4.2. Plastically induced mean stress.

The three components of the elastic matrix strain, $<e_1>_M$, $<e_2>_M$ and $<e_3>_M$, were measured with the neutron diffraction technique during tensile loading of composite test specimens. One component was measured during each loading experiment and simultaneously the composite total strain and load was monitored. Three specimens were tested for each component and the mean value of each of the three measurements for f = 20% is shown in fig. 2. The error bars give an indication of the specimen to specimen variation (one standard deviation). By comparing the scatter in fig. 2 with the intrinsic uncertainty of the neutron diffraction technique ($\pm 0.9 \cdot 10^{-4}$) it is clear that specimen to specimen variation is similar to the uncertainty of the neutron measurements. The neutron diffraction method reveals clearly the differences between longitudinal ($<e_3>_M$) and transverse ($<e_2>_M$ and $<e_1>_M$) strain components during loading of composite test specimens. Elastic matrix strain in the longitudinal direction increases with applied load, as expected, until saturation is reached, because of plastic flow in the matrix. Strain in the transverse directions decreases with applied load because of Poisson contraction. It is noteworthy that, as for the thermal residual strain (fig. 1), no clear distinction can be made in fig. 2 between in-plane ($<e_2>_M$) and normal ($<e_1>_M$) transverse strain components. The measured strain components shown in fig. 2 can be used to obtain the plastic mean stress in the composites. We begin by writing the general equation describing the average matrix stress, σ_M [16, 1]

$$\overline{\sigma}_M = \sigma^A + <\sigma>_M^P + <\sigma>_M^A + <\sigma>_M^R \tag{1}$$

where σ^A is the applied stress, $<\sigma>_M^P$ and $<\sigma>_M^A$ the plastic and elastic mean stresses respectively, and $<\sigma>_M^R$ the residual stress (originally caused by thermal mismatch). The elastic mean stress can be written as $<\sigma>_M^A = B\sigma^A$, where the B matrix is delivered by the variable constraint model [17,18]. Rearranging eq. 1 and using $\sigma_M = C_M<e>_M$ and $<\sigma>_M^R = C_M<e>_M^R$ where C_M is the stiffness matrix for aluminium, $<e>_M$

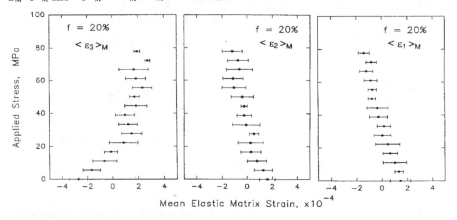

Figure 2. The mean elastic matrix strain measured with the neutron diffraction technique during loading of composite test specimens with f = 20%. For this fibre volume fraction, three specimens were tested for each strain component and the error bars are calculated from the scatter in the results.

the average matrix strain, $<e>_M^R$ the average residual strain and I the identity matrix, the plastic mean stress can be written as

$$<\sigma>_M^P = C_M(<e>_M - <e>_M^R) - (I+B)\sigma^A \qquad (2)$$

The right hand side of eq. 2 can now be evaluated using the experimental results of fig. 2. In order to compare the neutron measurements with analysis of permanent softening [7, 8], we need to know the plastic mean stress of eq. 2 as a function of composite plastic strain, e_{pc}. For that purpose, e_{pc} was estimated from the composite stress-strain curve for each point in fig. 2. The basis for a comparison of the plastic mean stress measured with the neutron diffraction technique on the one hand, and analysis of permanent softening on the other hand, is eq. 13 in ref. [7]. The results of the neutron diffraction experiments are shown as a function of composite plastic strain in fig. 3 (filled circles) for f = 20%. The first measurement is taken at zero applied load, in which case the mean elastic matrix strain, $<e>_M$, is equal to the residual strain, $<e>_M^R$ (caused by thermal mismatch). Hence, eq. 2 shows that the plastic mean stress, $<\sigma>_M^P$, is zero at zero applied load. Fig. 3 shows that the plastic mean stress in the matrix increases linearly with plastic strain in the range tested (the line is a best fit to the data set). Also shown in fig. 3 are measurements of the plastic mean stress made on the same composite system using Johannesson and Ogin's modification of the Orowan-Wilson approach (left hand side of eq. 13 in ref. [7]).

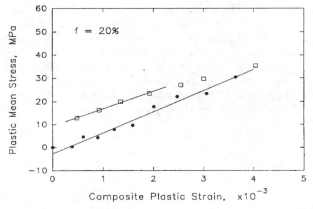

Figure 3. The development of the plastic mean stress in the planar random fibre aluminium matrix composites with fibre volume fraction 20%. The filled circles represent results of the neutron diffraction experiments as analysed by eq. 2, and the squares are results based on Johannesson and Ogin's modification of the Orowan-Wilson approach [7, 8].

5. Discussion

Johannesson and Ogin [7, 8] studied the hardening behaviour in the planar random fibre aluminium matrix composites using their modification of the Orowan-Wilson approach. They fitted a line to the first four points of each plastic mean stress curve (e.g. fig. 3) and took the slope of that line as as measure of the mean stress hardening rate. Extrapolation of the mean stress curves to zero plastic strain results in a non-zero intercept which increases with fibre volume fraction [7]. We find it likely that the non-zero intercept is related to an increased effect of the roundedness of the reverse flow curve, which has been shown to increase with fibre volume fraction [4]. At high fibre volume fractions and small plastic strains, the roundedness of the reverse flow curve also makes the determination of the permanent softening more difficult. It is noteworthy that the problem with the non-zero intercept is avoided when the hardening rate is considered. The neutron diffraction measurements of the mean stress hardening rate are shown in fig. 4 for fibre volume fraction of 15% and 20%. The diffraction measurements with greatest statistical significance were made for f = 20% (more specimens tested and a larger number of grains in the gauge volume). Compared with results of the neutron diffraction experiments, it turns out that Johannesson and Ogin's modification of the Orowan-Wilson approach underestimates the hardening rate by about 25-35%. If we express the mean value of $\Delta\sigma^T$ and $\Delta\sigma^C$ [7] as $<\Delta\sigma>$, the equation relating the permanent softening to the mean matrix stress measured in the neutron diffraction experiment can be written as

$$\frac{1}{2}<\Delta\sigma> = a\left\{\frac{\left|<\sigma_3>_M^P - <\sigma_1>_M^P\right|}{1+B}\right\}_{ND} \tag{3}$$

where the numerical value of the factor a is 0.71 ± 0.04. ND stands for neutron diffraction measurements. The ratio of $<\Delta\sigma>$ to the plastic mean stress is therefore not two, but $2a = 1.42 \pm 0.08$. This is in general agreement with the experimental work and analysis of Wilson and Bate [9] and Prangnell et al. [10] who showed that the permanent softening must be less than twice the mean matrix stress. As described in section 3, the neutron diffraction measurements were carried out on composite specimens with small matrix grains, but the material used in the Bauschinger experiments [7, 8] consisted of large columnar matrix grains. To check if the matrix grain size affects the hardening rate, cyclic Bauschinger experiments were carried out

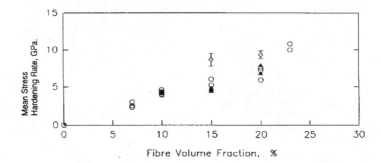

Figure 4. Mean stress hardening rate as a function of fibre volume fraction for the planar random fibre aluminium composites. The circles are based on measurements of permanent softening values for composite specimens with large columnar matrix grains [7, 8]. The filled triangles are obtained by the same method, but on the composites with small matrix grains used in the neutron diffraction experiments. The two squares with error bars represent the hardening rates obtained by analysing the results of the neutron diffraction experiments (slope of the curve in fig. 3).

on fine grained composite specimens designed for the neutron diffraction experiments. The results are given in fig. 4, which shows that for the two matrix grain sizes tested, the mean stress hardening rate is not affected by the matrix grain size. The neutron diffraction experiments, carried out on fine grained material, can therefore safely be compared with measurements made by the permanent softening method on material with large matrix grains.

6. Conclusions

1. Neutron diffraction measurements of thermal residual lattice strains in virgin specimens showed that all strain components were small and close to zero. No significant difference was observed between strain components in the plane of the fibres ($<e_3>_M$ and $<e_2>_M$) and normal to the plane of the fibres ($<e_1>_M$).

2. During tensile loading, the matrix strain in the loading direction increases, but saturates due to plastic flow in the matrix. The two transverse strain components decrease with increasing applied load because of Poisson contraction. No difference was observed in the behaviour of the two transverse strain components.

3. Analysis of cyclic Bauschinger experiments, based on measurements of permanent softening, showed that for the grain sizes tested here, the matrix grain size does not affect the mean stress hardening rate.

4. Analysis of neutron diffraction measurements of the three lattice strain components showed that the plastic mean stress is zero at zero plastic strain and increases linearly with plastic strain in the range tested. The non-zero intercept of the mean stress as analysed with the permanent softening is not a feature of the results of the neutron diffraction measurements.

5. The mean stress hardening rate measured from the results of the neutron diffraction experiments provides a calibration of Johannesson and Ogin's modification of the Orowan-Wilson approach. The modification is found to underestimate the hardening rate by about 25-35%. This means that the ratio of the permanent softening (or $<\Delta\sigma>$) to the mean stress is not two, but $2a = 1.42 \pm 0.08$.

Acknowledgements

This work was carried out within the Engineering Science Centre for Structural Characterization and Modelling of Materials while BJ was a postdoctoral researcher at Risø, supported financially by the Icelandic Council of Science, the Engineering Science Centre and NorFA. During the preparation of this paper, BJ was supported financially by the Icelandic Council of Science.

References

[1]. T.W. Clyne and P.J. Withers: INTRODUCTION TO METAL MATRIX COMPOSITES. Cambridge University Press, 1993.

[2]. E. Orowan: CAUSES AND EFFECTS OF INTERNAL STRESSES. Internal stresses and Fatigue in Metals, Ed. by G.M. Rassweiler and W.L. Grube, Elsevier, London, 59, 1959.

[3]. D.V. Wilson: REVERSIBLE WORK HARDENING IN ALLOYS OF CUBIC METALS Acta Metall. **13**, 807, 1965.

[4]. J.D. Atkinson, L.M. Brown and W.M. Stobbs: THE WORK-HARDENING OF COPPER-SILICA IV. THE BAUSCHINGER EFFECT AND PLASTIC RELAXATION, Phil. Mag., **30**, 1247, 1974.

[5]. H. Lilholt: HARDENING IN TWO-PHASE MATERIALS-I. STRENGTH CONTRIBUTIONS IN FIBRE-REINFORCED COPPER-TUNGSTEN. Acta Metall., **25**, 571, 1977.

[6]. O.B. Pedersen: THERMOELASTICITY AND PLASTICITY OF COMPOSITES-II. A MODEL SYSTEM. Acta Metall., **38**, 1201, 1990.

[7]. B. Johannesson and S.L. Ogin: INTERNAL STRESSES IN PLANAR RANDOM FIBRE ALUMINIUM MATRIX COMPOSITES-I. TENSILE TESTS AND CYCLIC BAUSCHINGER EXPERIMENTS AT ROOM TEMPERATURE AND 77 K. Acta Metall. Mater., **43**, 4337, 1995.

[8]. B. Johannesson and S.L. Ogin: INTERNAL STRESSES IN PLANAR RANDOM FIBRE ALUMINIUM MATRIX COMPOSITES-I. MEAN STRESS HARDENING AND RELAXATION. Acta Metall. Mater., **43**,4349, 1995.

[9]. D.V. Wilson and P.S. Bate: REVERSIBILITY IN THE WORK HARDENING OF SPHEROIDISED STEELS, Acta Metall. **34** (6), 1107, 1986.

[10]. P.B. Prangnell, W.M. Stobbs and P.J. Withers: CONSIDERATIONS IN THE USE OF YIELD ASYMMETRIES FOR THE ANALYSIS OF INTERNAL STRESSES IN METAL MATRIX COMPOSITES, Mat. Sci. Eng. **A159**, 51, 1992.

[11]. P.B. Prangnell, T. Downes, P.J. Withers and T. Lorentzen: AN EXAMINATION OF THE MEAN STRESS CONTRIBUTION TO THE BAUSCHINGER EFFECT BY NEUTRON DIFFRACTION Mat. Sci. Eng. **A197**, 215, 1995.

[12]. T.W. Clyne, M.G. Bader, G.R. Cappleman and P.A. Hubert: THE USE OF δ-ALUMINA FIBRE FOR MMCs. J. Mater. Sci. **20**, 85, 1985.

[13]. D.v. Hille, S. Bengtsson and R. Warren: QUANTITATIVE METALLOGRAPHY IN A SHORT ALUMINA FIBRE REINFORCED ALUMINIUM ALLOY, Comp. Sci. and Techn., **35**, 195, 1989.

[14]. B. Johannesson, T. Lorentzen and O.B. Pedersen: THERMAL RESIDUAL STRESSES IN PLANAR RANDOM FIBRE MMCs in Proc. of ICSMA-10, 21-26 Aug., Sendai, Japan, Oikawa et al. (eds.), p. 377, 1994.

[15]. T. Lorentzen and N.J. Sørensen: A NEW DEVICE FOR IN-SITU LOADING OF SAMPLES DURING NEUTRON DIFFRACTION STRAIN MEASUREMENTS, Proc. 12th Risø International Symposium on Materials Science, Ed. by N. Hansen et al., Risø National Laboratory, Roskilde, Denmark, 489, 1991.

[16]. P.J. Withers, W.M. Stobbs and O.B. Pedersen: THE APPLICATION OF THE ESHELBY METHOD OF INTERNAL STRESS DETERMINATION TO SHORT FIBRE METAL MATRIX COMPOSITES, Acta Metall. **37** (11), 3061, 1989.

[17]. B. Johannesson and O.B. Pedersen: ANALYTICAL DETERMINATION OF THE AVERAGE ESHELBY TENSOR FOR TRANSVERSELY ISOTROPIC FIBRE ORIENTATION DISTRIBUTIONS. Acta Mater. (in press), 1998.

[18]. O.B. Pedersen and B. Johannesson. APPLICATION OF FIBRE ORIENTATION DISTRIBUTIONS IN ANALYTICAL MODELLING OF COMPOSITE MATERIALS. Acta Mater., in preparation.

Mimicking the layered structure of natural shells as a design approach to the fiber-matrix interfaces in CMCs

R. Naslain, R. Pailler, X. Bourrat, F. Heurtevent
Laboratory for Thermostructural Composites
UMR-5801 (CNRS-SEP-UB1), University of Bordeaux, 3 Allée de La Boétie,
33600 Pessac, France

Abstract

Nacre, a tough material present in seashells, is composed of calcium carbonate (aragonite) and organic alternating layers with thicknesses in the range 250-500 nm and 10-50 nm, respectively. Synthetic $(PyC-SiC)_n$ multilayered interphases mimicking the structure of nacre, are designed and produced by pressure pulsed CVD/CVI, from propane and MTS/H_2 precursors, at 900-1000°C and under low pressures, with a computer-controlled apparatus. Layers with thicknesses ranging from a few nm to a few μ are deposited by adjusting the number and duration of reactant pressure pulses. These interphases are used in model Nicalon (or Hi-Nicalon)/SiC microcomposites. The materials are characterized from structural and mechanical standpoints. A comparison is worked out between natural nacre and $(PyC-SiC)_n$ multilayers, the two materials displaying many common features despite the fact that the elementary sequence is different and they are produced according to very different growth mechanisms.

1. Introduction

Seashells (such as abalone, nautilus or pearl oyster) are tough materials although mainly consisting of brittle calcium carbonate, $CaCO_3$. This unique behavior is related to the fact that the inner portion of seashells, the nacre, is actually a multilayered structure in which the brittle layers of calcium carbonate (usually present as aragonite) are separated from one another by organic thin films of proteins and polysaccharides, acting as crack arresters [1-11].

Ceramic matrix composites (CMCs), such as SiC fibers/SiC matrix composites, are inverse composites which is to say that the failure strain of the matrix is much lower than that of the fibers, with the result that the matrix fails first under load. The composites display a tough or brittle behavior depending on whether the fiber/matrix (FM) interfaces have the capability of deflecting or not the matrix microcracks. When the FM is too strong, the microcracks propagate (mode I) through the fibers, the composites being brittle. In contrast, when the FM bonding is weak enough, the FM interfaces act as mechanical fuses, deflecting the microcracks (mode II) and protecting the fibers from an early failure [12]. The control of the FM bonding is achieved through the use of a thin layer of a compliant material with a low shear strength, the interphase, which is usually deposited on the fibers before embedding them in the matrix [13, 14].

The design of the FM interfacial zone in CMCs used at high temperatures is difficult inasmuch as the functions of the interphase are often contradictory. The best mechanical behavior is achieved when there is a good balance between the mechanical fuse function and the load transfer function. Further, the interphase may also act a diffusion barrier and should be compatible with the fiber, the matrix and the atmosphere, at high temperatures. The most commonly used interphase materials in SiC-matrix composites are pyrocarbon and hex-BN. Both have a layered crystal structure and can be cleaved parallel to the basal plane. When deposited with the layers parallel to the fiber surface and strongly bonded to the fibers, they can act intrinsically as mechanical fuses, the matrix microcracks being deflected in a diffuse manner within the interphase itself and not simply at the fiber surface as often observed in composites with a too weak FM bonding [13, 15].

A more subtle and flexible way of interphase design in CMCs might be through mimicking the layered microstructure of seashell nacre, by extending the concept of layered interphases from the atomic scale (as for pyrocarbon or hex-BN) to the nanometer or micrometer scales. The interphase would actually be a sequence of a stiff material X (analogous to calcium carbonate) such as SiC, associated with a compliant material Y (comparable to the protein/polysaccharide film), such as pyrocarbon, hex-BN or layered oxides, the X-Y sequence being repeated n times. $(X-Y)_n$ interphases can be in principle engineered (or tailored), the adjustable parameters being the nature of X and Y, the sequence number n, and the thickness of each sublayer [16, 17]. Finally, the concept can also be further extended to the matrix itself, leading to a new class of tough and oxidation-resistant synthetic materials [18, 19].

The aim of the present contribution is to show how $(PyC-SiC)_n$ interphases displaying some common features with nacre, have been designed for use in SiC (Nicalon or Hi-Nicalon)/SiC composites and produced according to a specific version of the chemical vapor deposition/infiltration process referred to as P-CVD/CVI.

2. Background on structure, properties and formation of nacre

Nacre is a biocomposite with a multilayered architecture comprising layers of brittle aragonite platelets (typically, 5-10 µm in size and 0.25-0.50 µm in thickness) and layers of an organic material (proteins and polysaccharides), arranged in a "brick and mortar" structure (fig. 1) [1, 2]. The structure of nacre displays very specific features which are responsible for its remarkable mechanical properties in terms of toughness and strength. First, the volume fraction of the inorganic phase, i.e. the aragonite platelets, is extremely high (typically, 95-98%). As a matter of fact, the thickness of the organic layers is much lower (typically, 10-50 nm) than that of the calcium carbonate layers (from 250 to 500 nm and even more, depending on the species of the seashells). Second, the aragonite platelets exhibit the same preferred orientation (with their [001] axis perpendicular to the layer) within a given layer and from layer to layer. Third, the organic layers comprising both polysaccharides (such as chitin) and proteins, are highly compliant but strongly bonded to the aragonite layers [1, 2, 6].

From a mechanical standpoint, nacre displays high performances with respect to monolithic aragonite, which is weak and brittle (with a value of K_{IC} as low as 0.25 MPa $m^{1/2}$ [7]). Nacre exhibits a failure strength ranging from 170 to 220 MPa, a K_{IC} of the order of 10 MPa $m^{1/2}$ (20 to 40 times that of monolithic aragonite) and a strain energy release rate as high as 1000 - 1200 J/m^2 (almost 3000 times that of unreinforced aragonite [5]). This increase in toughness has been assigned to various mechanisms including crack blunting or deflection in the organic layers, platelets sliding and pull-out, platelets debonding and crack bridging by ligaments in the organic layers [1, 2, 4]. Further, a self-constraint mechanism, in which the weakest brittle layers are constrained by their stronger neighboring layers (which supposes a strong enough interlayer bonding) has been proposed to explain the increase in failure strength observed when moving from monolithic aragonite to nacre [7].

The formation mechanisms of nacre is still a subject of research and controversy. It has been depicted on the assumption that each aragonite layer is nucleated on the thin interlamellar organic sheet acting as template, or on the proposition that the aragonite crystals continuously grow from one layer to the next, through pores in the interlamellar organic sheets [1, 9].

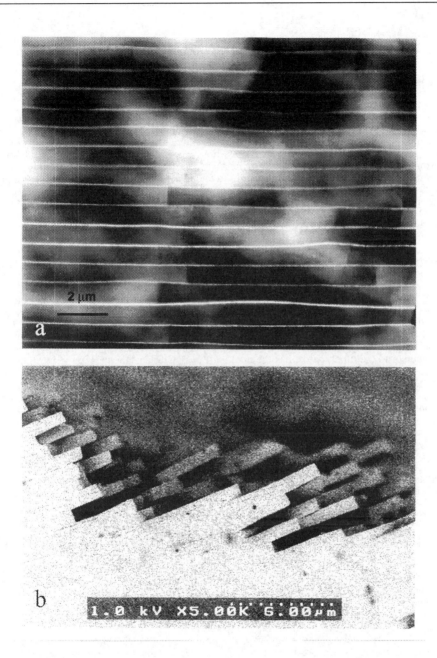

Fig. 1 : (a) Nacre as seen by TEM in contrasted brightfield mode showing occurrence of aragonite platelet layers (dark) and organic matter layers (bright), (b) pull-out of the platelets visible on the fracture surface of dry nacre obtained from a three point bending test (high resolution SEM on polished surface).

3. Bioinspired (PyC-SiC)$_n$ multilayered interphases

3.1. Experimental

(PyC-SiC)$_n$ multilayered interphases that mimic to some extent the microstructure of nacre, with an overall thickness lower than 1 μm, $10 < n < 30$, and elementary layers in the PyC-SiC sequence that could be as thin as a few nm, were deposited by P-CVD/CVI, mainly on straight Nicalon or Hi-Nicalon monofilaments, according to a process which has been depicted elsewhere [17]. Pyrocarbon was deposited from propane, and SiC from a mixture of methyltrichlorosilane (MTS) and hydrogen, at relatively low temperatures (900 - 1000°C) and pressures (a few kPa) in order to achieve low deposition rates (fig. 2). The fibers were mounted on a carbon holder with a high temperature cement and set in the hot zone of the deposition chamber. Each constituent of the PyC-SiC sequence was deposited through the repetition of short pressure pulses of the corresponding gaseous precursor, as shown schematically in fig. 3. Each pulse can be analyzed on the basis of four domains : the injection of the reactants in the previously evacuated deposition chamber (domain I), the deposition period or residence time (domain II), the evacuation of the unreacted precursor and reaction products (domain III) and finally, a period during which the deposition chamber is maintained under vacuum (domain IV). The deposited thickness per pulse being known for pyrocarbon and SiC (through calibration preliminary experiments), i.e. from a few Å to about 10 nm per pulse (depending on the T, P conditions and residence time), the thickness of each constituent in the elementary PyC-SiC sequence is fixed by the numbers of propane and MTS/H$_2$ pulses. A computer was used to control the opening and closing of the pneumatic valves, the lengths of domains II and IV within a pulse, the switching from one gaseous precursor to the other, as well as the number n of PyC-SiC sequences in the interphase. The fibers were used either as-desized or after a pre-treatment in order to change the nature of the surface [17]. The first layer deposited onto the fiber surface was usually pyrocarbon. After deposition of the interphase, a SiC-matrix was deposited from the same MTS/H$_2$ precursor, yielding *microcomposites* with a fiber volume fraction of about 50%.

Longitudinal thin foils prepared according to a technique described elsewhere [20] were observed by transmission electron microscopy (TEM). Mechanical tensile tests were performed at room temperature (gauge length : 10 mm) and fatigue tests (static or dynamic loading) run at 700°C in air, in order to assess lifetime, according to procedures depicted elsewhere [17].

3.2. Results and discussion

A few preliminary experiments were run to ascertain the flexibility of the deposition process and its ability to produce (PyC-SiC)$_n$ multilayers mimicking "thick" layers of nacre. An example of such multilayers deposited on a flat substrate (SiC-infiltrated graphite) and displaying an overall thickness of 17 μm, is shown in fig. 4a. The multilayer is composed of 200 (PyC-SiC) sequences, the PyC layers (5 nm) being as thin as the protein/polysaccharide layers in nacre. Conversely, by changing the deposition parameters, a multilayer of similar thickness (21 μm) but with n = 10 only was also deposited, in which the thickness of the layers were e(SiC) = 800 nm and e(PyC) = 1350 nm. These data clearly show that (PyC-SiC)$_n$ ceramics exhibiting a multilayered structure either at the nm-scale or at the μm-scale (and including thus structures with layer thicknesses similar to those of seashell nacre) can be produced in a very flexible manner by P-CVD/CVI.

(PyC-SiC)$_n$ multilayered interphases were then deposited on SiC-based fibers. An example of such interphases is shown in fig. 5a. It is composed of 10 (PyC-SiC)

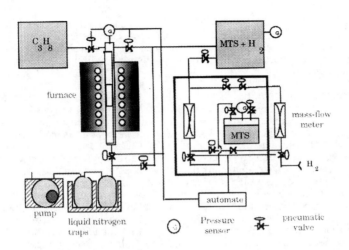

Fig. 2 : Apparatus for the deposition of (PyC-SiC)$_n$ multilayered interphases by P-CVD/CVI, according to ref. [17].

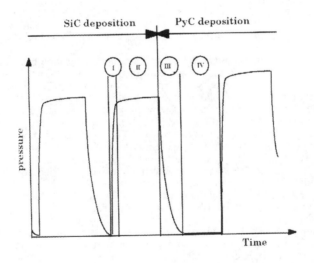

Fig. 3 : Example of pressure pulses (schematic) used in the deposition of (PyC-SiC)$_n$ multilayered interphases, according to ref. [17].

sequences with e(SiC) = 40 nm and e(PyC) = 10 nm, and it displays an overall thickness of 500 nm, a value which is common in Nicalon/SiC composites. It is surrounded with the brittle SiC matrix. Such an overall structure is also observed in seashells where the tough nacre internal portion of the shell is surrounded with brittle calcium carbonate (usually as calcite).

In the multilayers shown in fig. 4a and 5a, the brittle SiC-layer in the elementary PyC-SiC sequence is highly crystalline, the β-SiC crystals (blende-type) growing with their [111] direction perpendicular to the substrate. The layer consists of pure SiC (C/Si at. ≈ 1). It is polycrystalline, with a crystal size which can be as large as the layer thickness, as shown in fig. 4b. Further, the growth of SiC-crystals of large size and displaying a facetted morphology, yields rough interfaces between the pyrocarbon and the SiC layers which may render more difficult crack deflection. It is noteworthy that aragonite crystals in nacre also grow with a preferred orientation, their size being much larger than that of the SiC-crystals in the (PyC-SiC)$_n$ multilayers. However, the crystalline systems and the crystal growth mechanisms are totally different. Finally, the PyC layers when deposited from propane under appropriate conditions, exhibit themselves a strongly anisotropic layered microtexture referred to as rough laminar (RL), the graphene atomic sheets being roughly parallel to the substrate [13-16, 21]. In nacre, the organic protein/polysaccharide layers also display an anisotropic texture.

In a further design step, the highly crystalline pure SiC-layers were replaced by diphasic SiC + C nanocrystalline layers, in order to impede the growth of large size SiC-crystals and hence, to smooth the PyC/SiC interfaces. This can be easily achieved in P-CVD/CVI by changing the chemical composition of the SiC-precursor, i.e. the MTS/H$_2$ ratio [17]. An example of a [PyC-(SiC+C)]$_{10}$ multilayered interphase with smooth internal interfaces, deposited on a Hi-Nicalon fiber, is shown in fig. 5b. Incidentally, replacing stiff SiC (E ≈ 400 GPa) by a SiC+ C mixture, decreases the stiffness of the layer (the Young modulus of the SiC+C Hi-Nicalon fiber, with C/Si at. = 1.39, is only 270 GPa). In terms of stiffness, the SiC+C layers are thus more compliant than their pure SiC counterparts but probably still stiffer than aragonite layers (E = 70 GPa) [5].

A last important parameter in the design of an interphase, is the strength of the interfaces which are present. When an interphase comprises one or several sublayers with a cleavable layered crystal structure (such as pyrocarbon), the interfaces between them (or between the fiber and such a sublayer) should be strong in order : (i) to induce *diffuse* crack deflection between atomic planes (here, the graphene sheets) and (ii) to permit a dense microcracking of the brittle matrix [13-16]. It is noteworthy that the bonding between the aragonite and organic layers in seashell nacre, has also been reported to be strong [1]. The bonding between PyC and SiC layers deposited from the gas phase is believed to be strong. Conversely, that between a SiC-based fiber and the first PyC-layer of the interphase can be very weak when the fiber surface is highly contaminated with oxygen, the silica/RL pyrocarbon interface being extremely weak [13]. Thus, the fibers were usually pretreated before the interphase deposition in order to change their surface composition and to strengthen the fiber-interphase bonding.

From a mechanical standpoint, all the SiC/SiC microcomposites with a multilayered interphase displayed a non-brittle behavior with an extended non-linear stress-strain domain and matrix multiple microcracking, when tensile tested at room temperature. The matrix microcracks were actually deflected in the multilayered interphase acting as a mechanical fuse (fig. 5b), protecting the fiber from an early failure, as nacre in seashells protects the animal against damage occurring in the brittle outer portion of the shell. Moreover, the higher failure stresses were achieved, in microcomposites with interphases exhibiting a given PyC-SiC sequence number (e.g. n = 10), for specific layer thickness values, i.e. e(SiC) ≈ 30 nm and either very thin (3 nm) or relatively thick (20-40 nm)

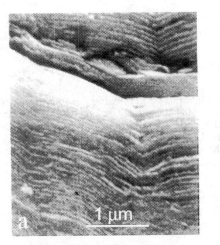
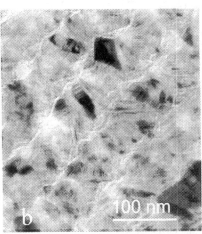

Fig. 4 : (PyC-SiC) multilayers : (a) on flat substrate (n = 200 ; e (SiC) = 80 nm e (PyC) = 5 nm), (b) roughness induced by SiC growth [17].

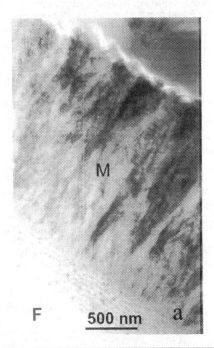
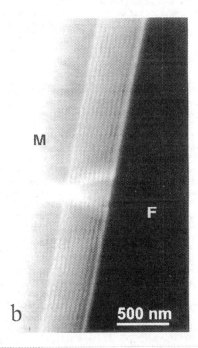

Fig. 5 : (PyC-SiC)$_n$ multilayered interphases : (a) deposited on a Nicalon fiber (n = 10 ; e (SiC) = 40 nm, pure SiC ; e (PyC) = 10 nm), (b) deposited on a Hi-Nicalon fiber (n = 10 ; e (SiC) = 30 nm, nanocrystalline ; e (PyC) = 10 nm) [17].

PyC layers [17]. Finally, the lifetime tests performed in air at 700°C under load showed that the best results were obtained with the multilayers comprising thin PyC layers (3 nm).

4. Conclusion

Model SiC/SiC composites with multilayered interphases mimicking to some extent the layered microstructure of seashell nacre, have been successfully designed and fabricated by P-CVD/CVI. The synthetic and natural layered materials display many common features. They are composed of stiff and compliant alternating layers strongly bonded to one another, the compliant layers being much thinner (in the range of a few nm) than their stiff counterparts. Both materials are tough and deflect cracks in a very tortuous manner within the soft layers. Although the growth mechanisms are different, the crystals in the stiff layers exhibit a preferred orientation and the soft layers are also anisotropic. However, the organic material is nacre is probably much more compliant than the RL pyrocarbon in the synthetic interphases. The concept of layered materials mimicking nacre is being extended to other X-Y sequence, to the matrix and to more complex fiber architecture.

Acknowledgements

This work has been supported by the French Ministry of Education Research and Technology, by SEP division de SNECMA and by Conseil Regional d'Aquitaine. It is part of a long range program of research on interphases in non-oxide composites. The authors are indebted to S. Goujard from SEP for valuable discussion, M. Alrivie for TEM sample preparation, M. Saux and J. Forget for assistance in preparing the manuscript.

References

[1] M. Sarikaya, J. Liu, I.A. Aksay, NACRE : PROPERTIES, CRYSTALLOGRAPHY, MORPHOLOGY, AND FORMATION, Biomimetics : Design and Processing of Materials (M. Sarikaya, I.A. Aksay, eds.), pp. 35-90, AIP Press, American Institute of Physics, Woodbury (NY), USA, 1995.
[2] K.E. Gunnison, M. Sarikaya, J. Liu, I.A. Aksay, STRUCTURE-MECHANICAL PROPERTY RELATIONSHIPS IN A BIOLOGICAL CERAMIC-POLYMER COMPOSITE : NACRE, Mater. Res. Soc. Symp. Proc., Vol. 255, pp. 171-183, Materials Research Society, Pittsburgh, 1992.
[3] J. Liu, M. Sarikaya, I.A. Aksay, A HIERARCHICALLY STRUCTURED MODEL COMPOSITE : A TEM STUDY OF THE HARD TISSUE OF RED ABALONE, Mater. Res. Soc. Symp. Proc., Vol. 255, pp. 9-17, Materials Research Society, Pittsburgh, 1992.
[4] M. Sarikaya, K.E. Gunnison, M. Yasrebi, I.A. Aksay, MECHANICAL PROPERTY - MICROSTRUCTURAL RELATIONSHIPS IN ABALONE SHELL, Mater. Res. Soc. Symp. Proc., Vol. 174, pp. 109-116, Materials Research Society, Pittsburgh, 1990.
[5] A.P. Jackson, J.F.V. Vincent, R.M. Turner, THE MECHANICAL DESIGN OF NACRE, Proc. Roy. Soc. Lond., 1988, B234, pp. 415-440.
[6] S. Weiner, Y. Talmon, W. Traub, ELECTRON DIFFRACTION OF MOLLUSC SHELL ORGANIC MATRICES AND THEIR RELATIONSHIP TO THE MINERAL PHASE, Int. J. Biol. Macromol., Dec. 1983, Vol. 5, pp. 325-328.
[7] X.F. Yang, A SELF-CONSTRAINT STRENGTHENING MECHANISM AND ITS APPLICATION TO SEASHELLS, J. Mater. Res., Jun. 1995, Vol. 10, N° 6, pp. 1485-1490.
[8] N. Koyama, K. Suzuki, K. Date, STRUCTURE AND DEFORMATION OF SHELL, Proc. Annual Meeting JSME/MMD, Iwate Univ. (Morioka), Aug. 23-24, 1995, Vol. A, N° 95-2, pp. 119-120.
[9] T.E. Schäffer, C. Ionescu-Zanetti, R. Proksch, M. Fritz, D.A. Walters, N. Almqvist, C.M. Zaremba, A.M. Belcher, B.L. Smith, G.D. Stucky, D.E. Morse, P.K. Hansma, DOES ABALONE NACRE FORM BY HETEROEPITAXIAL NUCLEATION OR BY GROWTH THROUGH MINERAL BRIDGES?, Chem. Mater., 1997, 9, 1731-1740.
[10] G.A. Ozin, H. Yang, I. Sokolov, N. Coombs, SHELL MIMETICS, Adv. Mater., 1997, Vol. 9, N° 8, pp. 662-667.

[11] *I.A. Aksay, D.M. Dabbs, J.T. Staley, M. Sarikaya*, BIOINSPIRED PROCESSING OF CERAMIC MATRIX COMPOSITES, Proc. 3rd Euro-Ceramics (P. Duran, J.F. Fernandez, eds.), 1993, Vol. 1, pp. 405-418, Faenza Editrice Iberica S.L., Spain.

[12] *J. Aveston, G.A. Cooper, A. Kelly*, A SINGLE AND MULTIPLE FRACTURE, Proc. Conf. Properties of Fibre Composites, NPL, Nov. 4, 1971, pp. 15-26, Guildford, IPC Science and Technology Press, UK, 1971.

[13] *R. Naslain*, FIBER-MATRIX INTERFACES AND INTERPHASES IN CERAMIC-MATRIX COMPOSITES PROCESSED BY CVI, Composites Interfaces, 1993, Vol. 1, N° 3, pp. 253-286.

[14] *R.R. Naslain*, INTERPHASES IN CERAMIC MATRIX COMPOSITES, Ceram. Trans., 1996, Vol. 79, pp. 37-52.

[15] *R. Naslain*, THE CONCEPT OF LAYERED INTERPHASES IN SiC/SiC, Ceram. Trans., 1995, Vol. 58, pp. 23-29.

[16] *C. Droillard*, PREPARATION AND CHARACTERIZATION OF COMPOSITES WITH PyC/SiC MULTILAYER INTERPHASES, PhD Thesis, Univ. Bordeaux-1, June 19, 1993.

[17] *F. Heurtevent*, $(PyC-SiC)_n$ NANOMETER-SCALE MULTILAYERED MATERIALS : APPLICATION AS INTERPHASES IN CERAMIC MATRIX COMPOSITES, PhD Thesis, Univ. Bordeaux-1, March 29, 1996.

[18] *W.J. Lackey, K.L. More*, LAMINATED MATRIX COMPOSITES : A NEW CLASS OF MATERIALS, Ceram. Eng. Sci. Proc., 1996, Vol. 17, N°4, pp. 166-173.

[19] F. Lamouroux, S. Bertrand, R. Pailler, R. Naslain, A MULTILAYER CERAMIC MATRIX FOR OXIDATION RESISTANT CARBON FIBERS-REINFORCED CMCs, Proc. HT-CMC-3, Osaka, Sept. 6-9, 1998 (to be published).

[20] *S.Jacques, A. Guette, F. Langlais, X. Bourrat*, CHARACTERIZATION OF SiC/C(B)/SiC MICROCOMPOSITES BY TRANSMISSION ELECTRON MICROSCOPY, J. Mater. Sci., 1997, Vol. 32, pp. 2969-2975.

[21] *P. Dupel*, PYROCARBON PULSED CVD/CVI : APPLICATION TO THERMOSTRUCTURAL COMPOSITES, PhD Thesis, Univ. Bordeaux-1, May 24, 1993.

Inorganic matrix composite materials: peculiarities, modelling, testing

Yu. Dimitrienko

NPO Mashinostroenia Corporation, Gagarina 33a, Reutov, Moscow region 143952, RUSSIA, Phone: +7(095)528-0450, E-mail: dimitnw.math.msu.su

Abstract

In this work, a mathematical model of inorganic-matrix composites (IMCs) and processes of their manufacturing is developed. New advanced IMCs on ceramic and glass fibres have been synthesizied, which can work for a long time under temperatures 1200 - 1400 °C in an oxidizing environment, at temperatures 1200 - 1400 °C the IMCs have a higher specific strength than C/C and C/C-SiC and are essentially less expensive. The developed model allows us to forecast the peculiarities of IMCs under high temperatures such as an increase of strength at temperatures 400-500 °C, a change of the character of deforming from plastic to embrittling and then to plastic, the appearance of plastic residual deformations, the dependence of strength on deforming rate etc.

1. Introduction

At present composite materials based on inorganic matrices (aluminophosphate, magnesium-phosphate, chromephosphate etc.) are advanced for creation of large-scale thermoloaded structures working under temperatures up to 1200... 1500 °C in oxidizing environment [1, 2].

Deficiencies of the up-to-date high-temperature materials are well known: heat-resistant steels (HRS) have higher specific weight (σ/ρ) (strength/density), carbon-carbon composites (C/C) can not work for a long time in oxidizing environment even with antioxidizing coatings, ceramic composites are well resistant against oxidizing under high temperatures but up to now they do not allow to create large-scale thin-walled shell structures, in addition they are very expensive.

Inorganic matrix composites (IMC) have substantional advantages as compared with the material classes mentioned above: within the temperature interval 1200...1400 °C their specificstrength (σ/ρ) is higher than for carbon-carbon composites (see Figure 1) and heat-resistant steels; they are essentially less expensive than for ceramic composites, and they allow to create large-scale shell structures. Moreover, IMC, unlike carbon-carbon composites, are radiotranslucent, that permits to apply them successfully for structures of antenna fairing under high temperatures.

2. Synthesis

Technology of synthesis of IMC does not differ practically from the manufacturing technology for polymer-matrix composites (PMC) and consists of the following

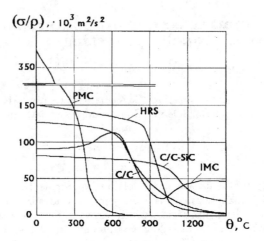

Figure 1: Dependence of specific strength (σ/ρ) on temperature θ of heating in oxidizing environment for different composites

stages: preparation of a liquid binder in the form of the mixture of aluminophosphate binder and fine-dispersed powder of solidifier, impregnation of fibres by the binder, laying them into a mould, and then moulding and heat treatment in a furnace under temperature not more than 350 °C and pressure not more than 2 MPa. The manufacturing technology for large-scale structures of the type of thin-walled and netted shells and strengthened panels made of IMC proves to be less expensive by 3 - 5 times than the one for carbon-carbon composite and ceramic composite structures for which temperature up to 2000 °C and pressure up to 40 MPa and also a substantional number of repeating technological cycles are usually used.

3. Properties

Inorganic matrices under heating up to temperatures 1400 °C are not thermostable: there occur several subsequent solid-phase transitions (practically without the appearance of gas phases) therein, which lead to nonmonotonous changing physico-mechanical properties of the materials (Fig.1).

For example, a sequence of phase transformations in aluminochromephosphate (ACP) matrix can be presented as follows:

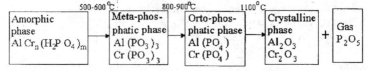

Within the temperature interval 400 - 600 °C one can observe the growth of strength and elastic characteristics of IMC by approximately 30 - 40 %. Falling down the strength within the interval 600 - 900 °C is caused by lowering the strength of reinforcing ceramic, glass or carbon fibres in the oxidizing environment. Inorganic matrix holds its mechanical properties and is just strengthened within the interval 900 - 1300 °C and then is stabilized up to 1500 °C. Thus, within the temperature interval 1200 - 1400 °C the specific strength of IMC proves to be higher than for C/C-SiC composites in air (Fig.1).

Interaction of inorganic matrix with glass and ceramic fibres under normal and elevated temperatures occurs not only in the adhesion way but also due to chemical reactions. As a result, between the matrix-fibre interface appears, which consists of solid dispersed particles almost not connected one with another. This interface sharply diminishes the composite strength under high temperatures. Therefore, one of the main directions of further works is to increase the interface strength. Adhesion with carbon fibres is lower, for its improvement silicon-organic coatings of fibres are applied.

As with PMC, composites IMC are sensitive to chages of technological regimes and parameters: pressure of moulding, temperature and regime of solidification etc., therefore, in manufacturing of complex structures with curvilinear surfaces, ribs, elements of strengthening etc. the properties of IMC can essentially change. To forecast properties of IMC in constructions more accurately, a mathematical model of inorganic-matrix fabric textile composite is developed.

4. Modelling

The model [3 - 5] connects final elastic and strength properties of textile IMC with properties of its fibre and separate phases of its matrix and also with geometrical parameters of a structure of the composite: content of fibres, their thickness, curving, misalignment, breakage etc. In addition, the model describes phase transformations in the matrix and its chemical interaction with fibre under high temperatures.

A composite material is considered to be a multilevel structure in the overwhelming majority of cases, with six structural levels (Figure 2). Each n-th structural level is represented by a collection of repeating structural elements being periodicity cells of the n-th type (PCn) n=1,...6. For a textile composite, its periodicity cell consists of s layers placed in the orthogonal direction to the same Ox_1 direction, where each the layer represented by the element of the second structural level is formed by the system of curved threads of unidirectional material and consists of a collection of PC5. PC5, in its turn, consists of a continuous collection of components characterized by curving angle ϑ_m in the certain plane Ox_1x_3 or Ox_1x_2. Each component from the continuous collection PC5 being the element of the third structural level is represented by unidirectional material. Unidirectional material being, in essence, a thread of separate fibres consists of repiodicity cells PC4. According to the model developed in the paper this cell is formed by two components: microcomposite with a destroyed fibre and microcomposite with intact fibre. Part of the microcomposite with destroyed monofibre increases during loading the composite and breakage of separate monofibres. Microcomposite with a destroyed fibre is a collection of periodicity cells PC3, the component of which consists of a destroyed monofibre, surrounding matrix and the fibre-matrix interface. Microcomposite with an intact fibre consists of PC3, the component of which consists of an intact fibre, matrix

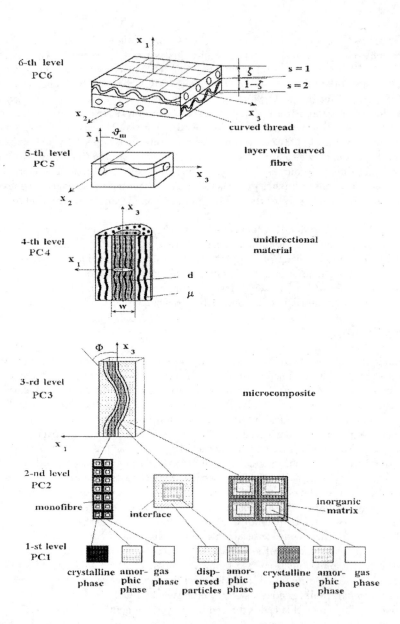

Figure 2: A scheme of six-level structure of IMC

and the fibre-matrix interface, being elements of the 2nd structural level. And finally a matrix undergoing phase transformations during heating consists of two solid phases, amorphic and crystalline phases, and also voids.

Intermediate phases (meta- and orto-phosphate) are considered as sequential stages of one continuous transformation of the matrix from its amorphic phase state to crystalline.

With the help of the described structural scheme, we can determine all the elastic moduli and strengths of composite in different directions, in particular for the strength in tension in the reinforcing direction of textile inorganic matrix composite we have

$$\sigma_{3T} = \zeta \left(\frac{1}{\sigma_{uT}'^2} + \frac{4}{3} \left(\frac{\vartheta_m^2}{\sigma_{mS}^0 \tilde{a}_3} \right)^2 \right)^{-1/2},$$

$$\sigma_{uT}' = \sigma_f^0 a_f(\theta) H_0 \left(\frac{\varphi_f}{1 - \varphi_f} \right)^r \left(\frac{\sigma_{fS}^0 \tilde{a}_m(\theta)}{\sigma_f^0 a_f(\theta)} \right)^\omega,$$

$$\tilde{a}_3 = \tilde{a}_m \left((1 - \delta_f)\delta_f + \frac{1 + \nu_m}{2} \delta_f^2 \frac{E_f^0}{E_m^0} \frac{\tilde{a}_f(\theta)}{\tilde{a}_m(\theta)} \sin^2 2\Phi \right),$$

where σ_f^0 is the strength of monofibres at $\theta_0 = 20°C$; σ_S^0 and σ_{fS}^0 are the strengths of the matrix in shear and adhesional strength at θ_0, respectively; $\tilde{a}_m$ and $\tilde{a}_f$ are the functions of varying strengths of matrix and fibre during heating (they depend on temperature θ and phase content of the matrix and fibre [4]); E_f^0 and E_m^0 are the elastic moduli of fibres and matrix at θ_0; H_0, r and ω are constants depending on the spread in monofibre strengths in the thread [4]; φ_f is the content of fibres in the composite; ζ is the part of fibres in the Ox_3 direction; Φ is the angle of monofibre misalignment in the thread [4]. Figure 3 shows a dependence of strength σ_{3T} on temperature θ for composite with ACP matrix and glass fibres. Comparison of experimental and theoretical data exhibits their good agreement and sufficient accuracy of modelling of the high-temperature behaviour of the composite.

Comparison with experimental data showed that modelling of properties of IMC and technological processes of synthesis is quite adequate.

5. Testing

Phase transformations occurring in inorganic-matrix composites under high temperatures are the causes of the appearance of specific features of the composites, some of which have been indicated above. To investigate the thermomechanical behaviour of glass-fibre composites with APC matrix, physico-mechanical testing was conducted for plane specimens under high temperatures in air. We investigated the following properties: a) peculiarities of the diagram stress (σ_{3T}) – strain (ε_3); b) the effect of deforming rate $\dot{\varepsilon}_3$ on the strength; c) the presence of residual deformations after unloading.

At normal temperatures, IMC have a nonlinear graph of deforming $\sigma_{3T}(\varepsilon_3)$, under off-loading there appears a hysteresis of residual deformations. Under high temperatures up to 1200 °C, IMC are embrittled, however, the diagram of their deforming

also has a nonlinear character, and, finally, under limit temperatures 1500 - 1600 °
IMC become anew plastic, flow under loading and become melted (Figs. 4 and 5).

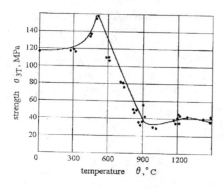

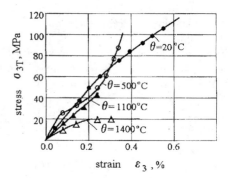

Figure 3. Strength σ_{3T} vs temperature θ for composite with ACP matrix and glass fibres

Figure 4. Stress-strain diagram of IMC at different temperatures

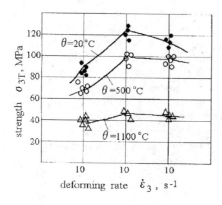

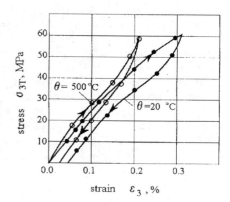

Figure 5. Strength σ_{3T} vs deforming rate $\dot{\varepsilon}_3$

Figure 6. Stress-strain diagram of IMC at different temperatures

At normal temperatures, the deforming diagram $\sigma_{3T}(\varepsilon_3)$ has an up-convexity, that indicates the presence of plastic (residual) deformations. This is also verified by the measurement of deformations after unloading (Figure 4). At temperature 500 °C the diagram $\sigma_{3T}(\varepsilon_3)$ considerably changes: at high stresses σ_{3T} there appears a section of the diagram with down-convexity, that indicates embrittlement of the material. Plastic deformations decrease there (Figure 6).

Strength of IMC σ_{3T} considerably depends on deforming rate $\dot{\varepsilon}_3$ of a specimen (Figure 5), and this dependence is non-linear: while $\dot{\varepsilon}_3$ increases, the strength σ_{3T}

first grows and then decreases. This dependence holds for all the investigated temperatures from 20 to 1100 °C. The strength reaches maximum at $\dot{\varepsilon}_3 = 10^{-4}$ s^{-1}. Maximum difference in values of σ_{3T} for different $\dot{\varepsilon}_3$ is approximately 30 %.

6. Conclusions

- New advanced IMC on glass and ceramic fibres were synthesizied, which can work for a long time under temperatures 1200 - 1400 °C in an oxidizing environment.

- Under temperatures 1200 - 1400 °C IMC have a higher specific strength than C/C and C/C-SiC and are essentially less expensive.

- The mathematical model of IMC has been developed which describes its thermomechanical behaviour under high temperatures.

- Thermomechanical testing was conducted for specimens of IMC based on APC matrix and with glass fibres. The testing showed that IMCs have a number of specific peculiarities: a non-linear diagram of deforming, transition from viscoplastic character of deforming to embrittling and then anew to plastic character with temperature increasing, the presence of residual deformations, unmonotonous dependence of the strength on temperature etc.

- IMCs have a wide application in different fields, such as aerospace industry, civil fire protection industry. However, applications of IMCs in structures requires a thorough studying their thermomechanical behaviour under actual conditions.

References

[1] Dimitrienko Yu.I., Epifanovski I.S., Shells of reinforced ceramic composite materials in airspace structures. *Proceedings of the 18-th International Symposium on Space Technology and Science,* Kagoshima, Japan, 1992, 104.

[2] Dimitrienko Yu.I., Epifanovski I.S., Investigation of high-temperature deformations of composites on inorganic matrices. *Proceedings of Moscow Conference on Composites,* 1991.

[3] Dimitrienko Yu.I., Mathematical modelling of composite materials manufacturing processes. *Proc. of the IMACS Symposium on Mathematical Modelling,* 1994, Austria, 5, 878-881.

[4] Dimitrienko Yu.I., Thermomechanical Behaviour of Composite Materials and Structures under High Temperatures. 1. Materials and 2. Structures. *Composites. Part A: Applied Science and Manufacturing,* 1997, 28A, 453 - 471.

[5] Dimitrienko Yu.I., Modelling of Mechanical Properties of Composite Materials under High Temperatures. Part 1. Matrix and Fibres and Part 2. Properties of Unidirectional Composites. *Applied Composite Materials,* 1997, 4, N 4, 219-261.

Roles of Interfacial Microstructure
on Interfacial Shear Strength of SiC/SiC

T. Hinoki*, W. Zhang*, Y. Katoh*, A. Kohyama* and H. Tsunakawa**

*Institute of Advanced Energy, Kyoto University,
Gokasho, Uji, Kyoto 611-0011, Japan
&
Crest, Japan Science and Technology Corporation,
4-1-8 Honmachi, Kawaguchi, Saitama 332 Japan
**Engineering Research Institute, The University of Tokyo,
Yayoi, Bunkyo-ku, Tokyo 113, Japan

Abstract

In order to improve fracture toughness of SiC/SiC composites, interfaces between fibers and matrices were analyzed. SiC/SiC composites with various carbon coating thickness were fabricated by chemical vapor infiltration. Interfacial shear stress between fiber and matrix was evaluated by fiber push-out test of thin specimens. Interfacial microstructure was analyzed by scanning electron microscopy (SEM) and transmission electron microscopy (TEM). TEM specimens were prepared with a focused ion beam processing device. Microstructures of debonded (after the push-out test) interfaces were examined. High resolution TEM images of interfaces with different carbon coating thickness were compared each other. The effect of interfacial microstructure on interfacial shear stress of SiC/SiC was discussed.

1. Introduction

Ceramic matrix composites (CMCs) are expected to apply for not only airplane and spaceplane but also advanced energy system such as high thermodynamics efficiencies of gas cycles and fusion system because of their high temperature strength and inherently low induced radioactivity [1].

The importance of shear strength of fiber-matrix interfaces on mechanical properties of CMCs has long been emphasized [1, 2]. The authors' group has been also concentrating its efforts in the relationship between mechanical properties and shear stress of fiber-matrix interface [3-5]. Interfacial shear stress of SiC/SiC composites depends on process and thickness of fiber coating, however, the effect of interfacial microstructure has not clearly been understood mainly because of difficulty in preparing thin film specimens for microstructural examination by TEM.

The objective of this work is to evaluate the effect of carbon coating characteristics on interfacial shear stress of SiC/SiC composites and to clarify the relationship between interfacial shear stress and microstructure. In order to prepare thin film specimens for microstructural examination by TEM, new device, focused ion beam (FIB) device was introduced. Site of debonding during push-out tests was identified by means of TEM. High resolution TEM images of interface between fiber and matrix were obtained and roles of interfacial microstructure on interfacial shear stress of SiC/SiC were discussed.

2. Experimental Procedure

2. 1. Materials Used

Materials used in this study were SiC fibers (Nicalon CG and Hi-Nicalon) reinforced SiC matrix composites. SiC/SiC materials were fabricated by chemical vapor infiltration (CVI) method, following fiber coating with carbon of various thicknesses ranging from 0.05 to 3.5 μm. They were used to see the effect of carbon coating on interfacial shear stress and interfacial microstructure.

2. 2. Fiber Push-out Test

To evaluate the fiber-matrix interfacial shear stress, push-out tests of single fibers [4-10] were carried out with constant displacement rates by scanning electron microscope with in-situ micro-indentation test capability (SEMITEC) and an ultra-micro indentation test machine. Schematic illustration of SEMITEC is shown in Fig. 1. The specimens for push-out tests were sliced about 500 μm-thick by automatic dicing saw. Then the specimens were planed under 80 μm-thick by mechanical polishing. Specimens of 40-80 μm-thick were placed on a specimen holder made of tungsten carbide with a groove of 50 μm-width, and fibers above the groove were loaded by triangular pyramidal diamond indenter with the maximum load of 1 N. The semi-apex angle of indenter was 68 degrees. The specimen surface was examined by scanning electron microscopy and surface profilometry prior to and after indentation tests.

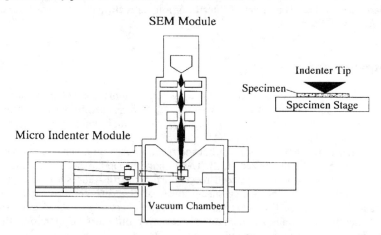

Fig. 1. Schematic illustration of SEMITEC

2. 3. Microstructural Examination by TEM

Specimens for TEM examination were sliced about 250 μm-thick and tip of the specimens were processed about 50 μm-thick by means of dicing saw. Thin films under 0.1 μm-thick for TEM examination were prepared from 50 μm-thick parts with a

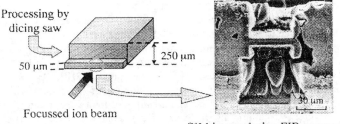

Processing by dicing saw

50 μm

250 μm

Focussed ion beam

SIM image during FIB processing

30 μm

Fig.2. Schematic Flow of TEM examination specimen preparation by FIB device

focused ion beam processing device as shown in Fig. 2. Thin films of debonded interface areas after push-out tests were also selected and prepared by FIB device. Microstructures of both undebonded (prior to the push-out test) and debonded (after the push-out test) interfaces were examined by TEM.

3. Results and Discussions

3. 1. Interfacial Shear Stress

3. 1. 1. Definition of Interfacial Shear Stress

Typical examples of indentation curve for fiber push-out tests are shown in Fig. 3. At first indenter made inroads into fiber during process A. The gradient change labeled 'push-in' in Fig. 3 (a) is considered to indicate the initiation of interfacial debonding. These should be basically the same results with those in refs. 7-9, although the shape of indenter was not flat-bottom but triangular pyramid. Crack between fiber and matrix propagated and fiber was pushing in and sliding during process B. A larger step, following the interfacial debonding, appeared to correspond to the fiber push-out, because fiber push-out was observed whenever this behavior was detected. Indenter

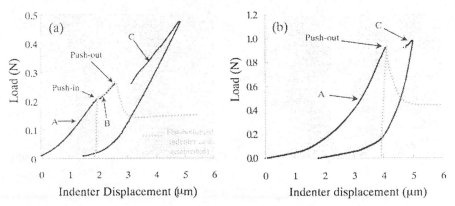

Fig. 3. Indentation curve under fiber push-out test of SiC/SiC specimen

touched matrix and load increased again during process C.

Fig. 3 (b) was indentation curve in the case of relatively thin specimen. Comparing with (a), the gradient change of push-in and process B was not seen in Fig. 3 (b). The reason of the difference is specimens used in this experiment were so thin that push-out load was smaller than push-in load in the case of thick specimen as shown in Fig. 3 (a). This trend is quite similar with the case of C/C composites [10]. As a preliminary interpretation, it is considered that push-out in the case of (b) occurs when pushing stress overcomes interfacial shear stress. The meaning of load at push-out in the case of (b) is quite different from one in the case of (a). Push-out load in (a) is used to propagate crack and for friction by clamping stress and Poisson expansion and to progress Poisson expansion. From this, the load at push-out in the case of (b) was used to determine interfacial shear stress. Interfacial shear stress was defined as:

$$\tau_{is} = P / \pi D t \qquad (1)$$

where τ_{is}; interfacial shear stress, P; load at push-out,
        D; fiber diameter, t; specimen thickness.

3. 1. 2. Effect of Carbon Coating Thickness on Interfacial Shear Stress

In order to study the effect of fiber coating on interfacial shear stress, SiC/SiC composites with various carbon coating thicknesses were fabricated at National Research Institute for Metals (NRIM) under the collaboration with Dr. Noda and Mr. Araki. In these materials, flat woven fabrics of Hi-Nicalon were used and fiber coating of carbon was applied by the CVI method with the variations in coating thickness from 0.2 to 3.5 μm, prior to the CVI processing to infiltrate SiC as the matrix. Fiber push-out tests of these specimens were carried out.

The resultant effect of carbon coating thickness on interfacial shear stress is shown in Fig. 4. Each carbon coating thickness in the specimens was measured from SEM images. Interfacial shear stress drastically decreased with increasing of carbon coating thickness. The relationship between interfacial shear stress and bending strength was discussed in our previous paper [5]. Effect of interfacial microstructure on interfacial shear stress is provided in following section.

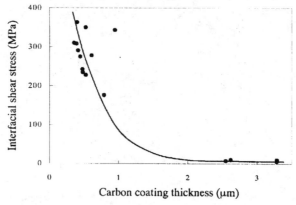

Fig. 4. Relationship between carbon coating thickness and interfacial shear strength

3. 2. Microstructure

3. 2. 1. Interfacial Microstructure

Clear TEM images of interface between fiber and matrix were obtained by means of FIB device. Fig. 5 shows interfaces between fiber and matrix. The specimen of (a) was coated about 0.2 µm-thick carbon on fibers and 0.5 µm-thick carbon was coated in the case of (b). Carbon characteristics of both specimens such as grain size and interface between carbon and matrix were also different because deposition rate of (a) was faster than one of (b). Interfaces between carbon coating and matrix were rougher than another sides. High resolution TEM images between fiber and carbon coating were also obtained. Basically both fiber and carbon coating are amorphous structures although β-SiC grains also exist in fiber. Sublayer characterized by large amount of oxygen as shown in the case of Nicalon fiber [11] was not confirmed because low-oxygen fibers, Hi-Nicalon were used in these examinations. However another layer was confirmed inside of carbon layer adjacent to fiber. In those layers carbon grain direction was parallel to fiber surface and different from the other carbon grain direction. Thickness of those layers was about 10 nm in both cases of (a) and (b).

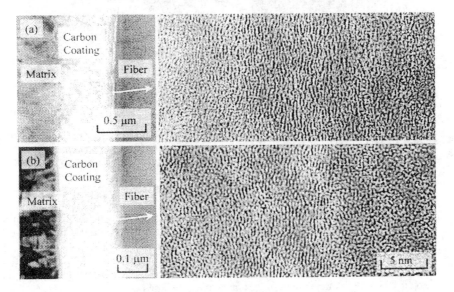

Fig. 5. Interfacial Microstructure of SiC/SiC Composites

3. 2. 2. Debonded Interface

In order to understand the dependence of interfacial shear stress upon carbon coating thickness, microstructure of debonded interface was examined by TEM. Fig. 5 shows the microstructures of the fiber-matrix interface of the specimen with 50 nm-thick fiber coating after a push-out test. Site at which the debonding occurred during the push-out test was identified as the carbon layer adjacent to the fiber. One of the reasons debonding occurs not in interface between carbon coating and matrix but in this site was considered that matrix side was rougher than another side as mentioned in previous section. Another reason was the layer confirmed inside of carbon layer adjacent to fiber. This site was weaker than the other parts because of grain direction. Preliminary investigation suggested that carbon coating thickness effects change of microstructure of this layer. However the results of high resolution TEM examination did not show conclusive difference. From this results it is considered that the other part of carbon coating is issue to decide interfacial shear stress.

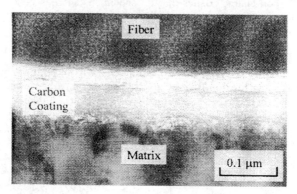

Fig. 5. Debonded interface after push-out test

4. Conclusions

In order to understand roles of interfacial microstructure on interfacial shear stress of SiC/SiC, SiC/SiC materials with various carbon coating thickness were fabricated. Shear stress between fibers and matrices were evaluated and microstructures of SiC/SiC were examined by TEM.
The conclusions are;

(1) A definition of interfacial shear stress as obtained from the push-out test of thin specimen is proposed. From this definition, effect of carbon coating thickness on interfacial shear stress was evaluated.
(2) The layer whose grain direction was parallel to fiber surface and different from the other grain direction was confirmed in carbon coating. Thickness of the layer was about 10 nm and independent of carbon coating thickness.
(3) Site of debonding during push-out tests was identified as carbon coating adjacent to fibers. However interfacial shear stress was independent of microstructure in this part.

Acknowledgement

The authors would like to express their sincere appreciation to Dr. T. Noda and Mr. H. Araki of National Research Institute for Metals for their kind supports to this work. This work was partly supported by Core Research for Evolutional Science and Technology (CREST) program under the title of "R & D of Environment Conscious Multi-Functional Structural Materials for Advanced Energy Systems".

References

[1] *L.L. Snead, R. H. Jones, A. Kohyama and P. Fenici,* **Status of Silicon Carbide Composites for Fusion**, Journal of Nuclear Materials, Vol. 233-237, 1996, pp.26-36.

[2] *D. H. Grande, J. F. Mandell and K. C. C. Hong,* **Fiber-Matrix Bond Strength Studies of Glass, Ceramic, and Metal Matrix Composites**, Journal of Materials Science, Vol. 23, 1988, pp.311-328.

[3] *W. Zhang, T. Hinoki and A. kohyama,* **Crack Initiation and Growth Characteristics in SiC/SiC under Indentation Test**, Proc. ICFRM-8(1997), in print for J. Nucl. Mater.

[4] *T. Hinoki, A. Kohyama, W. Zhang, H. Serizawa and S. Sato,* **Effect of Interfacial Shear Strength on Mechanical Property of SiC/SiC**, SCI. Rep. RITU A45, 1997, pp. 133-136.

[5] *T. Hinoki, A. Kohyama, S. Sato and T. Noda,* **Effect of Fiber Coating on Interfacial Shear Strength of SiC/SiC by Nano-indentation Technique**, Proc. ICFRM-8, 1997, in print for J. Nucl. Mater. .

[6] *D. B. Marshall,* **An Indentation Method for Measuring Matrix-Fiber Frictional Stresses in Ceramic Composites**, Journal of the American Ceramic Society, Vol. 67, 1984, pp.C-259-C-260.

[7] *E. Lara-Curzio, M. K. Ferber and A. R. Lowden,* **The Effect of Fiber Coating Thickness on the Interfacial Properties of a Continuous Fiber Ceramic Matrix Composite**, Ceramic Engineering and Science Proceedings, Vol. 15 [5], 1994, pp. 989-1000.

[8] *E. Lara-Curzio, M. K. Ferber and T. M. Besmann,* **Fiber-Matrix Bond Strength, Fiber Friction Sliding and the Macroscopic Tensile Behavior of a 2D SiC/SiC Composite with Tailored Interfaces**, Ceramic Engineering and Science Proceedings, Vol. 16 [5], 1995, pp. 597-612.

[9] *E. Lara-Curzio, M. K. Ferber and T. M. Besmann,* **Interfacial Characterization of SiC/SiC Composites with Multilayered Interphases**, High-Temperature Ceramic-Matrix Composites I, Vol. 57, 1995, pp. 311-316.

[10] *K. Watanabe, A. Kohyama, S. Sato, H. Serizawa, H. Tsunakawa, K. Hamada and T. Kishi,* **Evaluation of Interfacial Shear Strength of C/C Composites by means of Micro-Indentation Test**, Materials Transactions, JIM, Vol. 37, No. 5, 1996, pp. 1161-1165.

[11] *F. Doreau, J. Vicens and J. L. Chermant,* **TEM and EDX Investigations of Experimental SiCf-YMAS Composites**, High-Temperature Ceramic-Matrix Composites II, Vol. 58, 1995, pp. 355-360.

Determination of the Interlaminar Shear Strength of Ceramic Matrix Composites at High Temperatures

W. Geiwiz, M. Schiebel, D. Madani, F.J. Arendts

Institute of Aircraft Design (IFB), University of Stuttgart
Pfaffenwaldring 31, D-70569 Stuttgart, Germany
Phone: +49-711-658-2408
Fax: +49-711-685-2449
Email: geiwiz@ifb.uni-stuttgart.de

Abstract

The interlaminar shear strength is an important characteristic value for fibre reinforced materials. This paper describes the investigation of this value for a fibre reinforced ceramic material. The behaviour of this material was investigated by temperatures ranging from room temperature up to 1600°C according to the field of application of such materials. After the development of a suitable testing method similar to the short beam bending test extensive test series at different temperatures were performed. The results of these tests show a significant increase of the interlaminar shear strength with increasing temperature.

1 Introduction

Within the scope of the special research program SFB 259 'High Temperature Problems of Reusable Space Transportation Systems' a new quality of high temperature material will be developed for use in hot structures. For these applications it is necessary exactly to know the thermomechanical and thermophysical behaviour of this material. These properties have to be determined for the whole field of applications of this material (from room temperature up to 1600°C in oxidative atmosphere).

This material belongs to the group of the fibre reinforced ceramics. For fibre reinforced materials made of layers with different properties the interlaminar shear strength is a very important characteristic value. This value is a measure for the fibre matrix bonding. It is already known, that the fibre-matrix bonding for this material is relatively high in comparison to the bending strength. Therefore most of the classical test methods for the interlaminare shear strength can't be used and it is necessary to develop a new test method.

2 Experiments

2.1 Test Material

The C/C-SiC material was produced by the DLR Stuttgart[1] in a liquid infiltration process [1]. The C/C-SiC is a ceramic matrix composite (CMC). The reinforcement fibres are 2D-carbon-woven plies, the matrix is SiC. The strong fibre-matrix bonding is typical for this material. This material combines the high temperature and oxidation resistance of the monolithic ceramic with the good mechanical properties of the reinforcement fibres.

2.2 Specimens and testing device

C/C-SiC is very different in its thermomechanical properties from other fibre reinforced materials or monolithic ceramic. The material properties are significantly influenced by the strong fibre-matrix bonding. This means, that the shear strength is relatively high in comparison with the flexural strength. Because of this unfavourable ratio most of the specimens failure in the classical short span bend test by bending and not by shear stresses.

There are different standards to determine the interlaminar shear strength of CMC-materials. A good overview is given by B. Thielicke [2]. The best results at room temperature and also at elevated temperatures were received with the compression shear test (see fig.1.a).

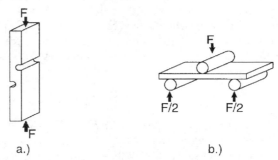

a.)   b.)

Figure 1 : a.) Compression shear test - b.) short span bend test

In our case, we have chosen the very simple short span bend test (see fig.1.b) and have modified the shape of the specimen as shown in fig.2 [4]. Because of the shape of the cross section, we get a strong shear stress increase in the middle of the specimen, which will lead to an interlaminar failure of the specimen.

---

[1] German Aerospace Centre, Institute for Structure and Design, Pfaffenwaldring 38-40, D-70569 Stuttgart, Germany

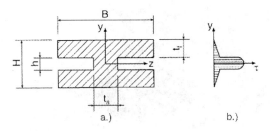

Figure 2: a.) Specimens geometry - b.) Shear stress

Using the ETB leads to the following equation for the ILSF

$$\tau_{ILSF} = \frac{3}{4} \cdot F \cdot \frac{BH^2 - bh^2}{(B-h)(BH^3 - bh^3)}$$ (1)

with τ_{ILSF} :   interlaminar shear strength [MPa]

         B :   specimen width [mm]

         H :   specimen thickness [mm]

         h :   strap height [mm]

         F :   failure load [N].

This equation could be simplified with the following approximation

$$\tau_{ILSF} = \frac{3}{4} \cdot \frac{F}{t_s \cdot H}$$ (2)

with t_S :   strap width [mm].

By variation of the different test parameters - such as thickness of the specimen H, width of the strap t_s or support span l_a - an optimum test geometry could be found, which leads to the desired failure mode at all temperatures.

2.3 Test procedure

For the tests the specimens were put in the normal short span beam test fixture made of steel for room temperature (see fig.3) or SiC-ceramic for high temperature (see fig.4).

Figure 3 : Flexure test fixture for room temperature

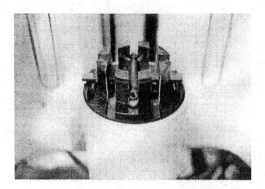

Figure 4 : Ceramic fixture for tests at high temperatures

All tests were carried out on an INSTRON multiaxial testing machine in position control mode with a displacement rate of 0.5 mm/min. The furnace system is a single zone furnace with resistance heating elements and a maximum heating rate of 12°C/min.

As it was expected most of the specimens failed between the two layers in the middle section (see fig.5). When bending failure occurred, it could be easily excluded by controlling the specimens after the test.

Figure 5 : Different specimens with the significant interlaminar failure

3 Results

3.1 Results of the pilot tests

3.1.1 Comparison of the different testing methods

Large series of tests were performed to compare the result of our new modified short beam bending test with the other methods. Some of them didn't show the right failure mode. For the classical short beam bending test all specimens failured because of tension forces on the bottom side. Responsible for this failure is the unfavourable ratio between interlaminar shear strength and the bending strength.

The measured shear strength in the compression shear test is a little bit smaller than in the modified short beam bending test. But this is already known from other fibre reinforced materials.

The pilot tests were done with two different materials. The first series were done with glass fibre reinforced plastic, because an interlaminar failure in this material could be observed very easily. After this first series the test were continued with C/C-SiC. The problem of the C/C-SiC material was the varying material quality from plate to plate. So we have to concentrate our investigations on a few important parameters to make all specimens for one parameter variation out of the same plate.

3.1.2 Parameter investigations

When the examination started, there were no definitions about the test parameter. To explain the influence of the test parameters a lot of test had to be done. The following parameters were investigated:

- Strap with t_S
- Specimen thickness
- Support span
- Stacking sequence

3.1.2.1 Strap width t_S

Critical by the production of the specimens was the determination of the strap width. The problem could be solved with the construction of a special measuring equipment. The strap width directly influences the failure load and was chosen as small as possible. The minimal strap width depends on the width of the carbon fibre roving. In fig. 6 the relation between strap width and shear strength is shown. There is only a small increase with decreasing strap width, but in our case for t_s=1,75mm the ILSS is nearly constant.

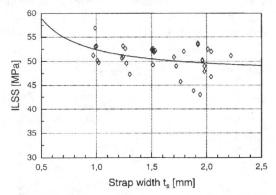

Figure 6 : Influence between strap width and ILSS

3.1.2.2 Specimen thickness H

The specimen thickness is necessary for the calculation of the ILSS and although of the bending strength. An increase of the specimen thickness effects an increase of the failure loads. When all other dimensions are constant the critical bending load increases more than the critical interlaminar load. That means with greater specimens thickness the interlaminar failure will occur clearly before the bending failure.

3.1.2.3 Support span l_a

The support span was varied from 15 to 50 mm (see fig. 7). With increasing support span the value of the shear strength approximates asymptotically to a limit. For our material this limit is reached for l_a=40 mm.

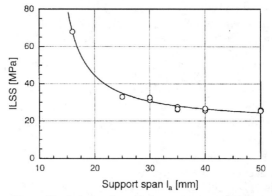

Figure 7 : Relation between support span and ILSS

3.1.2.4 Stacking sequence

The examined composite material is made of orthotropic layers and its mechanical properties are closely related with the stacking sequence. An exception is the ILSS, which is normally independent from the stacking sequence. This could be shown in the tests (see fig. 8).

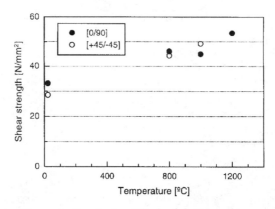

Figure 8 : Results for different stacking sequences

3.2 Results of the temperature tests

The tests were performed at temperatures ranging from room temperature up to 1400°C in inert gas atmosphere. Fig. 9 shows some typical stress deflection curves.

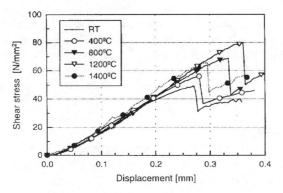

Figure 9 : Typical stress-displacement-curve

The curve shows a significant decrease when the interlaminar failure occurred. Because the specimen is not totally damaged after this descent the curve increases again until flexure failure is attained.

The determination of the failure mode for the tests above 1200°C is a little bit difficult, because the shape of the curve is not clear sometimes. In these cases the examination of the broken specimens is necessary to find out the real failure mode.

Temperature [°C]	Increase of the ILSS [%]
21	0
400	13,6
800	34,1
1000	37,8
1200	60,4
1400	38,8

Table 1 : Relative increase of the interlaminar shear strength for C/C-SiC measured in the modified short span bend test

The measured interlaminar shear strength increases with increasing temperature (see table 1). The maximum shear strength with an increase of more than 60% was measured at 1200°C. This phenomena is already known and could also be observed by other test methods [3].The absolute values are slightly higher then the ILSS measured in the compression shear test. Maybe this depends on the influence of the normal stress in the unsymmetrical compression shear specimens, which will lead to failure shortly before the maximum shear strength is reached. But the largest uncertainty is the tested material. The quality of this material shows a large variation in the interlaminar forces. So the interlaminar shear strength varies from 30 to 50 MPa at room temperature and from 50 to 85 MPa at 1200°C. But the relative increase is always the same.

4 Conclusions

The presented test method shows an easy way to determine the interlaminar shear strength of fibre reinforced ceramic. The preparation of the specimens is very simple and the existing bending device can be used. The method was successfully tested for temperatures up to 1400°C. To compare the results with the other standard methods it is necessary to test enough specimens from the same material quality, in the same environment and the same furnace system.

Acknowledgements

This work was supported by the DFG (German Research Foundation) within the scope of the SFB 259. Also the author would like to thank all supporters and co-workers.

References

[1] R. Kochendörfer; *Liquid Silicon Infiltration - a Fast and Low Cost CMC-Manufacturing Process*, ICCM 8; Honolulu 1991

[2] B. Thielicke, U. Soltesz; *The Interlaminar Shear Strength of Carbon-Fibre reinforced Carbon - Differences between Various Experimental Methods*; Composites testing & Standardisation (ECCM-CTS); EACM; Bordeaux 1992; page 297-296

[3] B. Thielicke, U. Soltesz; *The High Temperature Strength of C/C and C/C-SiC Composites under Shear Loading*, High-Temperature Ceramic-Matrix Composites 1; The American Ceramic Society; Ceramic Transactions, Volume 57; page 401-406

[4] A. Theuer; *Zur Bestimmung der interlaminaren Scherfestigkeit an flüssiginfiltrierter Faserkeramik*, Festigkeitsseminar über keramische Verbundwerkstoffe, September 1993; Fortschrittsberichte der Deutschen Keramischen Gesellschaft; Heft 4, Band 9 (1994); page 134-1411

Composites with structure-unstable bindings: the formation of micro- and mesostructure under different type of loading

S.N.Kulkov

Institute of Strength Physics and Materials Sciences
Russia Academy of Sciences, Siberian Branch
2/1, pr. Academicheskii, Tomsk, 634021, Russia
e-mail: ksn@red.ispms.tomsk.su

Abstract

There carried out a study of a regularities of deformation and fracture of the composite with structure-unstable binder, the peculiarities of inelastic behavior of the matrix having structural phase transformation.
It was shown that there was formed the ultra-fine structure with typical size of crystallites less 10 nm having high plasticity and high capacity to hardening. This structural state of the matrix lead to the effective transmission of external loading to hardener and a dislocation slides even in typical brittle particles, for example, titanium carbide on a microlevel and rotation of the carbide particles and on a mesolevel result in multiple cracking of a plastically deformed particles and, finally, to the high value of the fracture toughness. The material fragmentation on the fracture surface is founded, and the presence of transformation leads to amorphisation of the fracture surface.

1. Introduction

The composites with disperse hard particles being in a relatively soft metal matrix represent a special class of materials - the so called hard alloys, which are widely used in engineering both as structural and cutting tool materials with hard particles increasing strength and hardness and a plastic matrix giving rise to the high toughness and plasticity to the whole material. At low content the particles promote the higher yield stress of plastic deformation of a materials (as a carbides in steels for example). In other case plastic matrix gives some plasticity and toughness to a brittle material (carbide for example).

From all the existing models only some of them [1] pay a sufficient attention to a binding phase. Nevertheless, the problem becomes apparently principal firstly for physics of the deformation process of a such class of materials to be understood for correct modeling of mechanical behavior of such composites, and for increasing of the properties of the materials and a new-generation composites with the highest properties to be worked out. The fact is that non-uniformity state of stress in the disperse-hardened composite stipulates considerable mechanical constraint of deformation playing the leading part in the higher properties of the hard alloys. Taking into account a rather small (less then 1-2 μ) size of the interparticle interlayers and the higher yield stress of a matrix as a result of the lower thickness of its interlayers, it's difficult to expect that dislocation sliding should be effective under these conditions.

So, it is necessary to look for new materials as the binding phases that should provide the effective deformation of the composite under strained conditions and preserve its fracture as a result. In our previous work [2,3] it have been shown, that the usage in a composite as binder of an alloy with structural transformation permits the essentially to increase its mechanical properties. The alloys with thermoelastic martensitic transformation may be taken to materials of such a class owing to their crystal structure instability with respect to shear, for example, NiTi [3].

Our results have been demonstrated the efficiently higher plasticity of an alloy can be achieved at the constant level of strength due to the transformation of the structure-unstable binder, it being not important, in what a way this is achieved - due to changing either of its composition or the deformation temperature. In any case the specific energy of plastic deformation increases approximately by a factor of 3.

The main purpose of the work is to study a structures at a various scale levels, the phase composition, the deformation and fracture of TiC - TiNi composites with a structure-unstable binding phase.

2. The experimental procedure and materials

The TiC-TiNi composite with martensitic transformation in binder was investigated. The material was obtained by powder metallurgy methods and was a cylinder form with the size 10*10 (mm), which were used for different types of loading - for compression tests with quasi-static loading and as a flyer for penetration to the half-infinite targets prepared from aluminium alloy. The speed of flyers was varied from 500 up to 2500 m/sec. After loading analysis of its microstructure by the X-ray and TEM was carried out.

3. Results and discussions

On Fig.1 are shown stress-strain curves for quasi-static loading for stable and unstable states of binders in composites. As one can see, in the second case the plasticity is higher then in stable state of binder.

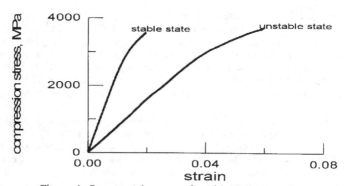

Figure 1. Stress-strain curves for TiC-TiNi composite

On Fig.2 are shown the X-ray patterns, obtained in initial condition and from fracture surface of a flyer. As one can see, after loading to high strains X-ray reflections, belonging to a binding phase are not practically visible, while reflections of carbide has a broadening. Such kind of X-ray lines permits to make the conclusion, that in volume of a binder material the amorphous structure are formed.

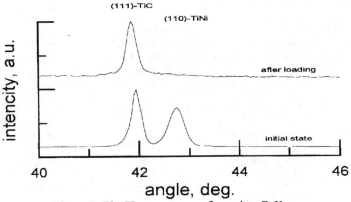

Figure 2. The X-ray patterns of samples, CuKα.

For analysis of an internal structure of such material after loading the researches on thin foils, cutting from fractured samples were carried out the TEM investigations.

The investigations of deformation and fracture processes of TiC-NiTi alloys with the structure transformation showed for NiTi deformation to be accompanied by considerable changes of structure state when being lost shearing stability of its lattice. Already in the non-deformed TiC-NiTi samples the NiTi structure is inhomogeneous to a high degree. One can observe the characteristic rippled contrast at the light-field electron-microscopic patterns what testifies to NiTi pre-transition state. Under loading still in a region of the composite elasticity NiTi the microstructure changed from a disperse domain into a banded contrasting one that is characteristic for the intermediate shear structure. Firstly, one can observe generation of diffusion peaks in the electron diffraction patterns and then extra-reflexes both in commensurable and incommensurable positions with the different parameter of commensurability in the different directions of axes of the reciprocal lattice, what testifies to several variants of the martensite domains each being generated with its own real structure. A such character of NiTi transformation is caused by highly inhomogeneous state arising near the hard particles of the composite under loading. Under the conditions of high stress gradients appearing in a matrix, the directions of atomic displacements in microzones stipulating local losses of B2 structure stability are determined by stressed states arising under loading at a moment in a given microvolume of the binding. These conditions determine the orientation of newly generating martensite domains too. NiTi transformations of such a character results in simultaneous decrease of the peak and the integral intensities of B2-phase lines in X-ray patterns not being accompanied by growing or arising new martensite peaks. One can only observe the diffusive of the most intensive lines of monoclinic NiTi in unidiffusive peak, what is characteristic when being generated a fine-disperse structure.

Under deformation with the higher velocity then yield strength in zones of the binding being under the most stressed conditions, there appears a disperse structure consisting of disoriented fragments of B2 phase and martensite domains. The electron diffraction patterns, taken from these zones have a characteristic ring shape of a different kind, mainly rings of point peaks and separate arcs arranged in one azimuth range of the wide (110) B2 ring against the weak diffuse (110)-B2 background. One can see the wide, highly intensive ring sharply standing out the others and being 0.201-0.240 nm in width corresponding to an interval of the interplanes distances. One can see the most intensive (002), (111), and (020) peaks of a monoclinic phase and (110) peak of B2 cubic structure. Moreover, one frequently meets with electron diffraction patterns being rings of point peaks with reflections of B2 and martensite structures against the background of diffuse (110) ring of B2. Second-order peaks are weekly distinguishable. Under deformation to considerable degrees there are observed in separate zones of the composite binding (preferentially in the vicinity of intercarbide boundaries and with continuous (110) diffuse rings of B2 and rare arcs in the main azimuth directions. Diffraction of a such character corresponds to quasiamorphous state.

The binding phase on a micro-level is in two-phase state - on figures one can see both an austenitic phase, and martensitic one in a form of a plate. As it is possible to see, it has places in volume of a binding phase, and at approach to carbide particles closely adjoin to border of section "binder-carbide", Fig.3. The area of contact "martensitic plate - carbide grain" different for carbides with the different size. On Fig.4 the angles measured on the various images of contact "martensitic plate - carbide grain" are shown. It is visible, that than it is less size of a carbide grain, that it is more angles of contact. It, as appear, testifies that in a loading process there is the rotation of the carbide grains that greater, than it is less their size.

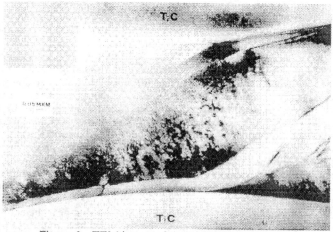

Figure 3. TEM image of the TiC-TiNi composite.

Such rotation of the carbide grains cannot occur without formation of specific internal structure of a binding phase. On Fig.5 data on measurement of a azimuth angles of binder fragments from degree of plastic deformation are shown. It is visible, that the high-grained structure (for TEM) in initial condition is broken with growth of strain into fragments mis-orientation the friend concerning friend, and the angle of a mis-orientation is increased exponential with growth of deformation. It is characteristic, that in carbide grains there is the increase of dislocation density is observed.

Thus in conditions shock loading the binding phase is in condition of all-round compression and shift - in it conditions, close to conditions in the Bridgmen chamber - pressure plus shift are realized.

So, it may be realized the following transformation scheme in the binder phase of the composite under deformation: B2 => B2 + B19' => B2 + "quasiamorphous state", with formation of fine-grain, highly disoriented structure less then 10 nm in grain size, characterizing by high plasticity and strengthening and stipulating external loading effectively transferred onto a strengthener, causing simultaneous dislocation slipping even in typically brittle TiC particles. Subsequent loading increase results in multiple cracking of plastically deformed particles and finally in high value of fracture toughness, continuity of the material being preserved.

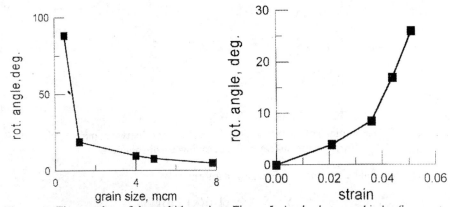

Figure 4. The rotation of the carbide grains vs. its sizes

Figure 5. Angles between binder fragments vs. degree of plastic deformation of composite

On Fig.6 are shown the optic microscopic images of a surface of cross-section of a flyer after its interaction with target. The analysis of microstructure of a samples shows, that in material a many microcracks, concentrating in a form of strip-lines or "tracks", taking place through the whole sample are observed. The fact of occurrence of such "tracks" testifies to special condition of a binder at the moment of impact. Moreover, these "tracks" were formed at initial stage of fracture and only has then taken place a spall. As appear, it is connected to presence in composite of a disperse carbide phase. Really, if to construct a distribution of a carbides on its sizes of particles and distribution of distances between cracks, it is possible to see the certain law.

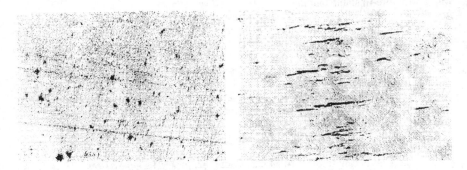

Figure 6. Mesocracks in form of "tracks".

First of all on distribution N(h) is displayed at least three maxima at $h=2$, 4 and 8 μ, Fig.7, while the average size of a carbide grain is equal 1 μ. The similar picture is observed for materials with other size of a carbide grain.

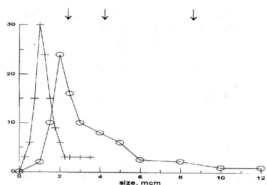

Figure 7. Distributions of carbide graines (+) and distance between mesocracks (o). Arrows are indicated three maxima on distribution.

This law can be presented analytically in a form:

$$Y = \Sigma \, A_i \exp[-\alpha(x + 2^i d)]$$

Where d - average size of a carbide grain, and A_i - function, describing change of intensity of maxima. According to the experimental investigations it can be presented in a form:

$$A_i = I_i \, (x + b_0)^2 .$$

This function depending on b_0 in positive area of argument can be or decreasing, or increasing. In a first case b_0 is very great, in second - it is very small. Experimental data show, that b_0 - is great and, as appear, it is connected to the length of a critical crack.

4. Conclusions

Thus, it is possible to point out a number of the main features of behavior of a composite with a structure-unstable binding.

It was shown that the physical meaning of usage of the structure-unstable bindings in the composites is that to lower a scale of structure levels of plastic deformation and fracture owing to formation of a micro-crystal structure in the binding phase under inhomogeneous loading.

Inhomogeneous deformation of the binding phase able to change a structure under external loading causes its transformation. It should be noted for the latter to occur to the different degrees following the deformation degree, the higher inhomogeneity of stressed state and plasticity of the binding material owing to transformation, the higher dispersion of the binding structure.

References

[1] Almond E.A. Strength of Hard Metals. *Metal Science*, #12, pp.587-592, 1978.
[2] Kulkov S.N., Poletika T.M., Chuhlomin A.Y. and Panin V.E. The Influence of the Phase Content of TiC-TiNi alloys on a Fracture Behavior and Mechanical Properties. *Poroshkovaya Metallurgiya*, v.8 (260), pp.88-92,1984.
[3] Kulkov S.N., Poletika T.M. Heterophase Materials with Structure Instabilities: Structure Levels of Deformation and Fracture. In: *Structure Levels of Deformation and Fracture*, Novosibirsk, Nauka, pp.187-203, 1990.

Prediction of the fatigue behaviour of Metal Matrix Composites using a micro-macro approach

E. Le Pen, D. Baptiste

Laboratoire de Microstructure et Mécanique des Matériaux - CNRS URA 1219
ENSAM - 151, Boulevard de l'Hôpital - 75013 PARIS - France

Abstract

Using very heterogeneous materials in structural parts submitted to cyclic loadings, leads us to present an elastoplastic micromechanical model. After some recalls on the homogenisation principle based on a mean field theory, a non-linear kinematic and isotropic strain hardening is introduced in the matrix. Validation is made on a Al-3.5% Cu / SiC particles and the case of an A356 / Al$_2$O$_3$ fibers is treated as a first application.

Introduction

Introducing reinforcement in structural parts can improve fatigue resistance of aluminium alloys. The shape, percentage and orientation of these inclusions has an influence on the mechanical behaviour of the composite and using these kind of materials in automotive or aerospace industry require prediction tools during low cycle fatigue in respect with the microstructure. Several analytical methods have been used to obtain an "homogenised" behaviour based on the microscopic description of the constituents. First of all, here is presented the micromechanical model, taking into account hardening in the matrix through an incremental scheme. It involves morphological properties of the reinforcement and provides mean stress and strain fields in the constituents. Then, an application to a SiC particles reinforced Al-3,5%Cu shows the accuracy of the prediction in full-reverse cycling, compared with FEM calculations. Finally, a A356 / Al$_2$O$_3$ 3D-distributed fiber composite is studied. The microstructure is analysed and allows elastoplastic predictions.

The elastoplastic model

The composite behaviour follows a classical approach in the elastic range, based on the works of T. Mori & K. Tanaka[1]. In order to change from the macroscopic scale to the microscopic scale in each constituent, one needs to describe the material on the whole : every heterogeneity is integrated in a Representative Elementary Volume, which undergo a uniform stress field. The local stress and strain fields in the matrix are averaged and the effect of the heterogeneity is calculated following Eshelby's equivalent inclusion principle[2]. In our case the comparison material is made of pure matrix.

The stiffness tensor of the composite is obtained after homogenisation of the stress and strain tensors on the REV :

$$L = L_0 \left[I + \sum_1^N f_r Q_r \left(I + \sum_1^N f_r (S_r - I) Q_r \right)^{-1} \right]^{-1}$$
(1)

where L_0 is the matrix stiffness tensor, I the identity tensor and S_r the Eshelby tensor of the r- inclusion family which have the same stiffness tensor L_r and volume fraction f_r.
Note that

$$Q_r = \left[(L_0 - L_r) S_r - L_0 \right]^{-1} (L_r - L_0)$$
(2)

and L integrate the stiffness and geometrical (through S_r) properties of the constituents.

The microscopic scale is determined with help of localisation tensors A_0 and B_0 in the matrix and A_r and B_r in the r-inclusion, such that average strain and stress fields in the matrix are

$$\varepsilon_0 = A_0 E, \quad \sigma_0 = B_0 \Sigma$$
(3)

$$\text{with } A_0 = \left[I + \sum_1^N f_r S_r Q_r \right]^{-1} \quad \text{et} \quad B_0 = \left[I + \sum_1^N f_r Q_r (S_r - I) \right]^{-1}$$
(4)

and average strain and stress fields in the r-inclusion are

$$\varepsilon_r = A_r E, \quad \sigma_r = B_r \Sigma$$
(5)

$$\text{with } A_r = (I + S_r Q_r) A_0 \quad \text{et} \quad B_r = L_r A_0 L^{-1}$$
(6)

This model has been chosen because of the materials of interest, composite materials with volume fractions from 0 to 20% of discontinuous reinforcements. Requirements in implementation aptitude and calculation velocity make us balance for this facing the widely used self-consistent scheme.

Reinforcements remain elastic and plasticity is considered within the matrix. Instead of a secant moduli method (Tandon & Weng[3] , N. Bourgeois[4]) the extension of the elastic equations above is achieved through the tangent moduli (G.Hu[5]). The behaviour is described with an incremental method, based on the differentiated equations.

$$\dot{\sigma}_r = L_r \dot{\varepsilon}_r, \quad \dot{\sigma}_0 = L_0' \dot{\varepsilon}_0 \quad \dot{\Sigma} = L' \dot{E}$$
(7)

In the Eshelby's principle, inclusion stresses perturbation is now $\sigma_r^{pi} = L_0' (S_r' - I) \varepsilon *$ with S_r' the Eshelby tensor deduced from the tangent properties of the matrix.

The unknown tangent stiffness tensor of the matrix has to be determined, considering plastic flow properties.

$$L_0' = L_0 - \frac{(L_0 \, n) \otimes (n \, L_0)}{\dfrac{h}{\sigma_{eq}} + n \, L_0 \, n}$$
(8)

with n normal to the yield surface, h plastic modulus and $\sigma_{eq} = \sqrt{\frac{3}{2}(s_0 - x_0'){:}(s_0 - x_0')}$ the equivalent stress. s_0 and x'_0 are deviatoric stress and kinematic hardening tensors.

An isotropic non-linear kinematic law as described by Chaboche[6] is chosen, to traduce hardening or softening phenomena and reach stabilised cycles in fatigue (compared with a power law). The yield surface is written

$$f(\sigma_0, R, x_0) = \sigma_{eq} - R(p) \qquad (9)$$

The hardening isotropic component evolves with the cumulative plastic strain p as $R(p) = R_0 + Q(1 - e^{-bp})$, while kinematic hardening has an instantaneous definition $\dot{x}_0' = \frac{2}{3}C\dot{\varepsilon}_0^p - Dx_0\,\dot{p}$, where $\dot{\varepsilon}_0^p$ is the plastic strain rate of the matrix.

Finally, when plastic flow occurs, solving the system $\left(f = 0, \quad \dot{f} = 0 \quad \text{et} \quad \dot{\varepsilon}_0^p = n\dot{p}\right)$ in the matrix gives the tangent stiffness tensor L'_0 (see 8). This tensor is not isotropic and the homogenisation strictly needs to be made with an non-isotropic calculation of the Eshelby tensor to obtain the overall stiffness tensor of the composite. However, as a first approximation it will be assumed to be isotropic.

Computational topics

The full reverse cycling fatigue loading is set by a strain increment on the REV. The first estimate of the local stress state in the matrix is find by an elastic localisation.

$$\begin{cases} \sigma_0^{est} = \sigma_0(t) + L_0\,\dot{\varepsilon}_0 \\[2mm] n^{est} = \frac{3}{2}\dfrac{s_0^{est} - x_0'(t)}{\sigma_{eq}(t)} \end{cases} \qquad (10)$$

The plastic differentiated equations are solved by iterations to determine $\dot{p}$ and the plastic strain at t+dt. It is achieved following a local integration with a radial return method (Burlet & Cailletaud[7]). Indeed, from the elastic estimate, tensors at the instant t+dt are calculated along the perpendicular direction to the estimated yield surface :

$$\begin{cases} \dot{x}_0' = \frac{2}{3}C\dot{\varepsilon}_0^p - Dx_0\,\dot{p} \\[2mm] R(p) = R_0 + Q(1 - e^{-h\cdot p}) \\[2mm] \dot{p} = \frac{3}{2.h}(s_0 - x_0')\dot{\sigma}_0 \\[2mm] h = C - \frac{3}{2}Dx_0(s_0 - x_0') + b(Q - R(p)) \end{cases} \qquad (11.\text{a-d})$$

The tangent stiffness tensor of the REV is deduced with (1), but the macroscopic strain direction has to be balanced by iterations to satisfied a unidirectional stress state.

Evaluation on a Al-3,5%Cu SiC$_p$-reinforced composite

A first example is presented in this section, based on results of FEM calculations carried out by Llorca, Suresh & Needleman[8]. The material is made of an Al-3,5%Cu matrix with 6, 13 and 20 per cent (vol.) SiC particles. The matrix is assumed to be homogeneous isotropic with a Young modulus of E_0=72 GPa and Poisson's ratio of ν_0=0,33. Particles are considered as spheroids of diameter 3,5 μm and elastic properties E_1=450 GPa et ν_1=0,17.

The strain-hardening parameters of the matrix are identified on the simulated hysteresis loops :

	R0	b	Q	C	D
Al-3,5%Cu matrix :	165	89.5	140	20000	2300

The behaviour is well described until stabilisation (5th cycle) as shown on fig.1. Good convergence is obtained for strain increments up to 5.10^{-2} % (for example 20 points per cycle if $\Delta E/2 = 0,5\%$), but when tangent modulus tends to zero (for higher strains and no strain-hardening) the isotropic approximation leads to instabilities. Calculations could be improved by a anisotropic form of L'_0 combined with the Eshelby tensor for spheroidal inclusions in an anisotropic media (Mura[9]).

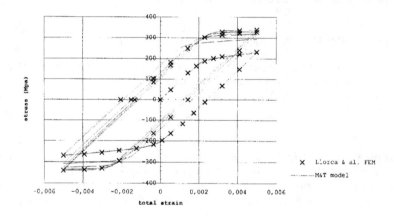

Figure 1 : hysteresis loops for 13% SiCp with ΔE/2=0,005

Tensile tests simulations (fig.2a) reveal the influence of volume fraction on the monotonic behaviour of the composite. Stress-strain curves are in good agreement but for reinforcement rates close to 20%, the Mori & Tanaka tangent moduli approach under-estimate material's strength. This may be due to an overall plastic strain in the matrix, while local plastic strain remains localized at the top of the particles in FEM calculations.

2a)   2b)

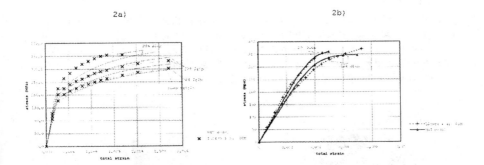

Figure 2 : monotonic (2a) and cyclic (2b) strain hardening for different SiCp volume fractions.

However, cyclic stress - strain curves for both 13% and 20% composites (fig.2b) show that this gap vanish under stabilised cycles.

Predictions for a A356 reinforced with Al2O3 fibers

Since the micromechanical model takes into account several reinforcement families, it is of interest to predict the behaviour of a more complex composite. As part of automotive applications, we study a short fibre reinforced metal matrix composite manufactured by indirect squeeze-casting.

Processing and microstructure

Alumina fiber preforms (ICI-Saffil RF grade) are impregnated by an A356 alloy under speed and pressure control. After a T6 heat treatment, the microstructure is very heterogeneous and made of α-Al dendrites, Al-Si eutectic phase including 1μm diameter Si precipitates (gray) and small $MgAl_2O_4$ spinels concentrated at the fibre-matrix interface (fig.3). A few porosity have been seen on clusters of fibres, resulting from bad impregnation.

The passage real microstructure - VER is accomplished, considering only two constituents : a homogeneous isotropic matrix , and Al_2O_3 fibers. Indeed, matrix mechanical

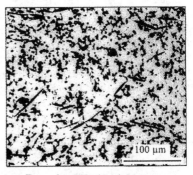

Figure 3: composite microstructure

properties are deduced from an A356-T6 alloy which almost contains the same heterogeneity. These results must be handled with care because of a different mould geometry for the pure alloy specimen in which heterogeneity distribution is different.

	E_{young} (GPa)	v	R0	Q	b	C	D
A356-T6	72	0,33	200	30	10	58000	680
Al_2O_3f	300	0,25	-	-	-	-	-

Fiber orientation distribution is measured by quantitative image analysis. Even if preform processing (Fukunaga[10]) should lead to 2D randomly oriented phase, experimental measurements are not so clear (fig.4a). If θ is the angle to loading direction (3) and ϕ to the observation plane (2,3), orientations can be divided in families characterised by (θ_i, ϕ_i) and volume fraction f_i (fig.4b).

4a)   4b)

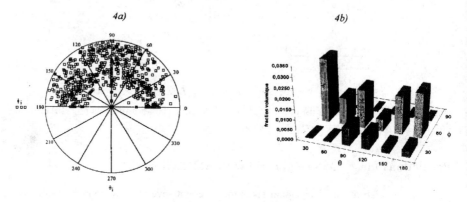

Figure 4: fiber orientation distributions as measured (4a) and rearranged for the model (4b)

Numerical results

Low cycle fatigue calculations are carried out on a composite REV involving 13 fiber families with total volume fraction 13.34%. Stress-strain curves are similar to those obtained for a 2D randomly oriented system ($\phi =0°$), and results are presented for this simple REV (fig.5). Higher stiffness and yield strength for the composite compared to aluminium alloy leads to a better general strength, but total plastic strain amplitude is increased with fiber volume fraction. Stabilised hysteresis loops in the matrix for different composites, as shown on fig.6, gives the confirmation of extra cumulated plastic strain for 13.34% and 20% composites.

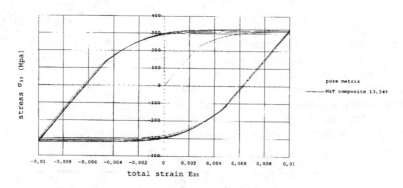

Figure 5: hysteresis predicted loops for the composite and for the aluminium alloy (pure matrix)

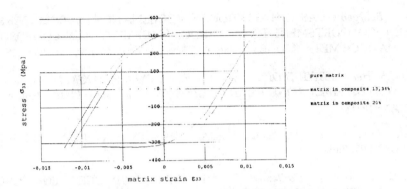

Figure 6: influence of fiber volume fraction on the matrix loading within the composite

Experimental tests are to be conducted to confirm the general tendencies, but for higher cycle fatigue predictions, damage mechanisms will be included to maintain a correspondence between microstructure and mechanical properties.

Conclusion

A micromechanical model has been proposed to predict the behaviour of composites under cyclic loadings. The main difficulty is to combine numerical solving of plasticity in the matrix with a micro-macro transition. However hysteresis loops are in good agreement with FEM calculations for a particle reinforced aluminium composite. It can be used for multiple Al_2O_3 fiber orientations with a high interaction between microstructure and macroscopic response. Accuracy needs to be completed by different strain amplitude experiments, which will permit integration of the matrix cracking damage observed in this former material.

References :

1. *T. Mori, K. Tanaka*, AVERAGE STRESS IN MATRIX AND AVERAGE ELASTIC ENERGY OF MATERIALS WITH MISFITTING INCLUSIONS - *Acta Metall.*, 1973, vol. 21, pp. 571-574

2. *J.D. Eshelby*, THE DETERMINATION OF THE ELASTIC FIELD OF AN ELLIPSOIDAL INCLUSION, AND RELATED PROBLEMS - *Proc. R. Soc. London*, 1957, vol. A241, 376-396 (1957)

3. *G.P. Tandon, G.J. Weng, A THEORY OF PARTICLE-REINFORCED PLASTICITY - Trans. ASME, 1988, vol. 55, pp. 126-135*

4. *N. Bourgeois*, CARACTÉRISATION ET MODÉLISATION MICROMÉCANIQUE DU COMPORTEMENT ET DE L'ENDOMMAGEMENT D'UN COMPOSITE À MATRICE MÉTALLIQUE : AL/SICP - *Thèse de l'Ecole Centrale de Paris*, 1992

5. *G.K. Hu, A METHOD OF PLASTICITY FOR GENERAL ALIGNED SPHEROIDAL VOID OR FIBER REINFORCED COMPOSITES* - *Int. Journ. Plast.*, 1996, vol. 12, pp. 439-449

6. *J. Lemaître, J.L. Chaboche*, MÉCANIQUE DES MATÉRIAUX SOLIDES - DUNOD, Paris, 1985

7. *H. Burlet, G. Cailletaud*, NUMERICAL TECHNIQUES FOR CYCLIC PLASTICITY AT VARIABLE TEMPERATURE - *Eng. Comput.*, 1986, vol. 3, pp. 143-153

8. *J. Llorca, S. Suresh, A. Needleman*, AN EXPERIMENTAL AND NUMERICAL STUDY OF CYCLIC DEFORMATION IN METAL-MATRIX COMPOSITES - *Met. Trans.*, 1992, vol. 23A, pp. 919-934

9. *T. Mura, MICROMECHANICS OF DEFECTS IN SOLIDS*, 2nd edn. *Martinus Nijhoff, Dordrecht*

10. *H. Fukunaga, SQUEEZE-CASTING PROCESSES FOR FIBER REINFORCED METALS AND THEIR MECHANICAL PROPERTIES*, Cast reinforced metal composites, Proceedings of the international symposium on advances in cast reinforced metal composites, Ed. Fishman & Dhingra, ASM publication, 1988, pp. 101-107

Creep Rupture Time Prediction of   Al/Al$_2$O$_3$/C Hybrid Metal Matrix Composites

H.W.Nam*, K.S.Han*

*Department of Mechanical Engineering - Pohang University of Science & Technology - San 31 Hyojadong, NamKu, Pohang, Kyungbuk, Republic of Korea

Abstract

The creep characteristics of a Al/Al$_2$O$_3$/C composites were studied. The mechanical properties and creep rupture time of Al/Al$_2$O$_3$/C composites were compared with matrix alloy and Al/Al$_2$O$_3$ composites. The tensile strength of Al/Al$_2$O$_3$/C composites was higher than that of Al/Al$_2$O$_3$ composites, but the creep rupture time of Al/Al$_2$O$_3$/C composites was shorter than that of Al/Al$_2$O$_3$ composites. This is due to the formation of brittle aluminum carbide(Al$_4$C$_3$). The stress exponent and activation energy of Al/Al$_2$O$_3$/C composites were found to be 7.1 and 167.5kJ/mol.
The new method was proposed for the assessment of creep rupture time. This method was based on the conservation of the creep strain energy. The theoretical predictions were compared with those of the experiment results and a good agreement was obtained. It was found that the creep life is inversely proportional to the (n+1)th power of the applied stress.

1. Introduction

Discontinuously reinforced metal matrix composites(MMCs) are attractive for many structural applications because materials exhibit unusual combinations of mechanical physical and thermal properties. These properties include high modulus and strength, good wear resistance, good heat resistance and low thermal expansion.
In these composites, carbon fiber reinforced aluminum alloys are expected as aerospace and automobile materials, such as cylinder liner of internal combustion engine, because of light specific weight, high wear resistance. Particularly, carbon fibers serve as solid lubrication in order to achieve good wear resistance of high temperature.
However, carbon fibers are difficult to wet by molten aluminum alloys and react with an alloy to form brittle aluminum carbide(Al$_4$C$_3$) at high temperature[1]. The Al$_4$C$_3$ make an effects on mechanical properties of Al/Al$_2$O$_3$/C composites[2]. Generally, Al/Al$_2$O$_3$/C composites were used in high temperature condition. In these condition, various properties of Al/Al$_2$O$_3$/C composites are changed. So, for the engineering application, it is important to understanding high temperature exposed mechanical properties of Al/Al$_2$O$_3$/C composites.
In this research, high temperature tensile and creep properties of Al/Al$_2$O$_3$/C composites were compared with those of Al/Al$_2$O$_3$ composites. Also, the creep rupture time of Al/Al$_2$O$_3$/C composites are studied using Norton's equation and the failure strain energy

equation. The experimental results and theoretical predications obtained are compared with modified Hoff's law.

2. Prediction of creep rupture time

Many researchers have been studied the creep rupture life of commonly used metal and alloy. Lason and Miller[3] first introduced the concept of a time-temperature grouping in the form of T(K + log t), which was based on the earlier Hollomon-Jaffee expression for tempering of steel.

Another approach to predict the creep rupture life is to use Hoff's law[4], and in that Norton's law is used as a constitutive equation

$$\dot{\varepsilon}_s = \dot{\varepsilon}_{sc} \left(\frac{\sigma}{\sigma_c} \right)^n \tag{1}$$

Where $\dot{\varepsilon}_s$ is the velocity of secondary creep, and $\dot{\varepsilon}_{sc}$ is strain rate due to the characteristic stress σ_c. Hoff assumed that the volume of the specimen is constant regardless of the creep state and the relationship between applied stress and rupture time was derived.

$$t_R = \frac{1}{n \dot{\varepsilon}_{sc} \left(\dfrac{\sigma_o}{\sigma_c} \right)^n} \tag{2}$$

The proposed method assume that the failure strain energy is conserved for all the creep test conditions. And for the simplicity, it is assumed that the secondary creep is dominant in all creep stage then the creep failure strain energy can be simply expressed by using Norton's equation. The creep rupture life prediction concept is shown in Fig.1

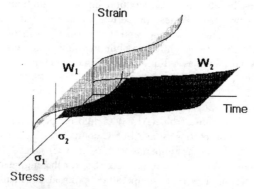

Fig.1 Concept of creep life prediction

Assuming that a failure strain energy equilibrium condition is achieved the following equation.

$$W_1 = W_2 \tag{3}$$

where W_1, W_2 is the strain energy when the applied stress is σ_1, σ_2, respectively. The strain energy can be calculated by integrating the applied stress multiply the infinitesimal strain.

$$W = \int_0^{\varepsilon_u} \sigma d\varepsilon \tag{4}$$

Equation (4) is a function of stress and strain. To represent the strain energy equation as a function of stress and temperature, Equation (5) was used.

$$\frac{d\varepsilon}{dt} = A\sigma^n \exp\left(-\frac{Q}{RT}\right) \tag{5}$$

Using Eq(4), Eq.(5) the following expression is obtained.

$$W = \int_0^t \left[A\sigma^{n+1} \exp\left(-\frac{Q}{RT}\right)\right] dt = A\sigma^{n+1} \exp\left(-\frac{Q}{RT}\right) t \tag{6}$$

Using the failure strain energy equilibrium condition, Eq.(6) can be expressed as:

$$A\sigma_1^{n+1} \exp\left(-\frac{Q}{RT_1}\right) t_1 = A\sigma_2^{n+1} \exp\left(-\frac{Q}{RT_2}\right) t_2 \tag{7}$$

Rearranging Eq.(7), a simple creep rupture life prediction equation is obtained which can be used to predict the creep rupture life by only knowing the stress exponent 'n' and activation energy 'Q'.

$$\frac{t_1}{t_2} = \left(\frac{\sigma_2}{\sigma_1}\right)^{n+1} \exp\left(-\frac{Q}{R}\left(\frac{1}{T_2} - \frac{1}{T_1}\right)\right) \tag{8}$$

3. Experimental procedure

3.1 Materials
For the fabrication of the composite material, a wrought alloy of AC2B Al alloy was used. The alumina fibers, SAFFIL RF Grade, from ICI Co. and carbon fibers, Kreca, from Kureha Chemical Industry Co. were used. The composite materials were fabricated by a squeeze casting method, and for the composite materials a 15%(vol.) self-made preform was used.

3.2 High temperature tensile test and Creep test
Test specimens were machined from cast ingots parallel to the applied pressure direction. In high-temperature tensile test, 4 specimens were used for composite materials. Creep testing was conducted in air using a SATEC lever arm creep machine with horizontal self-alignment capabilities. Strain was monitored using a linear variable differential transformers(LVDT) with the output registered by an electric pen strip chart recorder. The testing temperature was monitored using a thermocouple placed at the center position of the specimen gage length. Temperature fluctuations were found to be less than 2°C of the designated value. Creep test specimens with a 6.25mm diameter and a 25 mm gage length were machined perpendicular to the direction of the applied pressure. The load applied was equal to 40~70% of the high temperature(250°C) tensile strength of the material and the testing temperature was in the range of 230~280°C under a

constant load condition.

4.Results and Discussion

4.1 High temperature tensile strength

Fig. 2 shows yield and tensile strength of matrix alloy and composites at 250°C. The yield strength of Al/Al$_2$O$_3$/C composites is not much increased compared with the matrix alloy. But the tensile strength of Al/Al$_2$O$_3$/C composites is increased about 17% compared with the matrix alloy. Tensile strength of Al/Al$_2$O$_3$/C composites is much increase than yield strength because, in case of composite, yield strength much depend on matrix alloy but tensile strength depend on both matrix alloy and fiber.

The tensile strength of Al/Al$_2$O$_3$/C composites is higher than that of Al/Al$_2$O$_3$ composites. During the fabrication of Al/Al$_2$O$_3$/C composites, matrix alloy react with a carbon fiber to form brittle aluminum carbide(Al$_4$C$_3$). The Al$_4$C$_3$ considered to be behavior like fine precipitates and make a precipitation hardening effect.

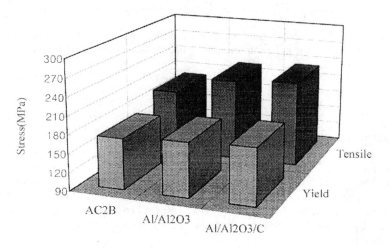

Fig. 2 Yield and Tensile stress of various materials at 250°C

4.1 Creep behavior and life prediction

Most of the creep test were conducted until failure occurred. The constant temperature creep experiments were conducted to determine the stress exponent. Figs.3 illustrate the relationships between the steady state creep strain rate and the applied stress of the Al/Al$_2$O$_3$/C composites

The stress exponents of Al/Al$_2$O$_3$ and Al/Al$_2$O$_3$/C composite are found to be 12.3 and 7.1, respectively. Also, constant load(126Mpa) creep experiments were conducted to found the activation energy. The activation energy of Al/Al$_2$O$_3$ and the Al/Al$_2$O$_3$/C composite is found to be 550.5kJ/mol and 167.5kJ/mol respectively. In Figs. 4, the relationships between the steady state creep strain rate and the test temperatures are illustrated..

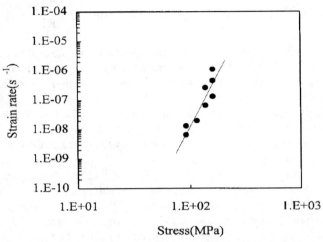

Fig. 3 Creep strain vs. Applied stress for Al/Al$_2$O$_3$/C composite at 250°C

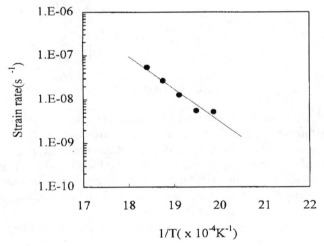

Fig. 4 Creep strain rate vs. temperature for Al/Al$_2$O$_3$/C composite at applied stress 126MPa

Norton's equation is expressed as follow

$$\text{Al/Al}_2\text{O}_3 \text{ composites} : \quad \varepsilon_s = 11.3 \times 10^{-62} \exp\left(-\frac{550.5\text{kJ/mol}}{RT}\right)\sigma^{12.3}$$

$$\text{Al/Al}_2\text{O}_3\text{/C composites:} \quad \dot{\varepsilon}_s = 21.6\text{x}10^{-52} \exp\left(-\frac{167.5\text{kJ/mol}}{RT}\right)\sigma^{7.1}$$

Dragone[5] conducted steady state creep experiments on 26 vol.% alumina fiber reinforced Al-5% Mg alloy MMC. The measured stress exponent obtain by Dragone range from 12.2 at 200°C to 15.5 at 400°C. In this study a stress exponent of 12.3 was obtained for the Al_2O_3 reinforced MMC.

The result of this study is similar to that was obtained by Dragone, despite the fact that the matrix material used was different. This suggests that the high stress exponent and the activation energy are not so much dominated by the nature of the matrix, but rather very much governed the reinforcement phase.

The activation energy and stress exponent of $\text{Al/Al}_2\text{O}_3\text{/C}$ composites are much lower than those of $\text{Al/Al}_2\text{O}_3$ composites. As previously mentioned, the matrix alloy react with the carbon fiber to form Al_4C_3 particles. When a $\text{Al/Al}_2\text{O}_3\text{/C}$ composites is exposed at high temperature, the particles coalesce into larger ones and some of the benefit is lost, the particles being bad effects on the property of $\text{Al/Al}_2\text{O}_3\text{/C}$ composites. This is illustrated in Fig. 5

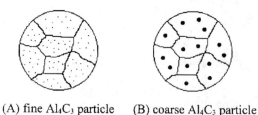

(A) fine Al_4C_3 particle   (B) coarse Al_4C_3 particle

Fig. 5 Precipitation hardening effect of Al_4C_3 particle

Fig. 6 shows creep rupture time of various materials. Is spite of high tensile strength , creep rupture time of $\text{Al/Al}_2\text{O}_3\text{/C}$ composites is shorter than that of $\text{Al/Al}_2\text{O}_3$ composites. This results also support that the Al_4C_3 made a bad effects on the creep properties.

Creep rupture life prediction experiment were conducted under two conditions. The first condition is a fixed temperature condition, and the second is a fixed stress condition. The experiment results were compared with those obtained by using the proposed creep life prediction equation and modified Hoff's equation.

In the case, where the test temperature is constant , Eq.(8) can be simplified as follows

$$\frac{t_1}{t_2} = \left(\frac{\sigma_2}{\sigma_1}\right)^{n+1} \tag{9}$$

In using Eq.(9) it is possible to predict the creep rupture life only by knowing the stress exponent n. This equation shows that the creep life is inversely proportional to (n+1)th power of the applied stress. The proposed equation shows that if materials have a high stress exponent then the creep rupture life will be long and the opposite will be found true if the stress exponent is low.

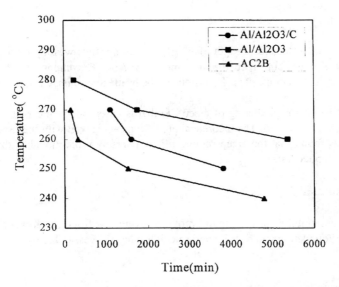

Fig. 6 Creep rupture time of various materias at126MPa

Figs.7 shows the prediction of the creep rupture life in a constant temperature condition. It was found that predictions obtained from using Eq.(9) agree well with the experimental results.

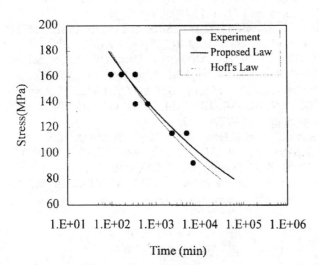

Fig. 7 Creep rupture time prediction of Al/Al$_2$O$_3$/C composite at 250°C

5. Conclusions

1) High temperature tensile strength of $Al/Al_2O_3/C$ composites was higher than that of Al/Al_2O_3 composites, but creep rupture time of $Al/Al_2O_3/C$ composites were shorter than that of Al/Al_2O_3 composites. This is due to the brittle aluminum carbide(Al_4C_3).

2) A new equation for predicting of creep life that is based on the creep strain energy conservation was proposed. The theoretical predictions agree well with the experimental results. It was found that the creep rupture life is inversely proportional to the (n+1)th power of the applied stress.

Acknowledgment

The authors are grateful for the support provided by a grant from The Korea Science & Engineering Foundation and Safety and Structural Integrity Research Center.

References

[1] *M.F. Amateau*, PROGRESS IN THE DEVELOPMENT OF GRAPHITE ALUNIMUM COMPOSITES USING LIQUID INFILTRATION TECHNOLOGY J. Comp. Mat, pp. 279-296, 1976,   Vol. 10
[2] *J.I.Song, H.D.Bong and K.S.Han*, CHARACTERIZATION OF MECHANICAL AND WEAR PROPERTIES OF $Al/Al_2O_3/C$ HYBRID METAL MATRIX COMPOSITES - Scripta. pp.1307-1313,1995. Vol.33, No. 11
[3] *R. Viswanathan*, DAMAGE MECHANISMS AND LIFE ASSESSMENT OF HIGH TEMERAUTRE COMPONENTS, ASM INTERNATIONAL, 1989.
[4] *N. J. Hoff* THE NECKING AND RUPTURE OF RODS SUBJECTED TO CONSTANT TENSION LOADS J. Appl. Mech. pp.105~108, 1953, Vol. 20
[5] *T. L. Dragone and W. D. Nix* STEADY STATE TRANSIENT CREEP PROPERTIES OF AN ALUMINUM ALLOY REINFORCED WITH ALUMINA FIBERS Acta metall., pp.2781~2791, 1992, Vol.40.
[6] *J.I.Song, K.H.Jung, H.U.Nam, K.S. Han*, EVALUATION OF DYNAMIC FRACTURE TOUGHNESS OF $Al/Al_2O_3/C$ COMPOSITES BY INSTRUMENTED IMPACT TESTS Proc. KSCM Conf., pp.204~209, 1995.
[7] *S. Suresh, A. Mortensen, A. Needleman*, FUNDAMENTAL OF METAL MATRIX COMPOSITES Butterworth-Heinemann, 1993

Creep properties of Ni-NiO in-situ composite

H. Takagi*

* Department of Ecosystem Engineering - The University of Tokushima -
2-1 Minamijosanjima-cho, Tokushima 770-8506 Japan

Abstract

Ni-NiO in-situ composites with NiO fiber spacing of 2-6 μm have been prepared by a unidirectional solidification process. The relationship between solidification rate and microstructure of the composite has been investigated with SEM. Creep tests were conducted at intermediate temperatures. Within the range of testing temperature steady state creep was observed. The activation energy of 300 kJ/mol and stress exponent of 5 obtained for steady state creep suggest that power-law creep was dominate. Grain boundaries, which were originally parallel to stress axis, migrated during creep. This grain boundary migration leads to fracture initiation sites.

1. Introduction

In-situ composites, in which fibers or lamellae are aligned in metal matrix by directional solidification of alloys of near eutectic composition, have significant mechanical properties in terms of room temperature and high temperature strength [1-3]. Thus the in-situ composites have been recognized as advanced metal matrix composites for next generation.

In this paper, the creep behavior of Ni-NiO in-situ composites grown at several different rates was studied. The purpose of this paper is to examine the dominant creep deformation mechanism at intermediate temperatures and the creep rupture characteristics of Ni-NiO in-situ composites.

2. Experimental details

Powders of Ni (99.6% pure) and NiO (99.9% pure) were mixed to obtain a predetermined composition. This mixture was then compacted, sintered and induction melted in an argon atmosphere. The resultant ingots were machined to 11.5 mm diameter by 170 mm long.

Directionally solidifying was conducted in an argon atmosphere by means of the apparatus shown schematically in Fig. 1. The molten sample, held in an alumina crucible, was withdrawn at a controlled velocity through a graphite susceptor heated by a five-turn induction coil attached to a 10 kW induction heater operating at 50kHz. Using the withdrawal rod, the sample was lowered into a water-cooled copper chill pipe. The temperature of the graphite susceptor during directional solidification was held within 2 K by a programmable PID controller. The maximum melting temperatures were about 100 K above the eutectic liquidus temperature.

The temperature gradient, determined with a Pt-6Rh/Pt-30Rh thermocouple, was approximately 100 K/cm. This gradient was the slope of the cooling curve measured on the eutectic liquidus temperature. Because of the difficulty of measuring the actual growth rate of the specimen, it was assumed that that rate was equal to the lowering rate of the crucible, i.e. the solidification rate.

The specimens were tested in air at constant load by using a lever-type testing machine. The stress range covered was from 28 to 73MPa; the temperature range was from 973 to 1123K. Temperature was held within ± 2K of the testing temperature. The elongation of the specimen was measured by using the differential-transformer transducer(accuracy of 5μm) which was subjected to connect to extension rods fixed to the grips of the testing machine.

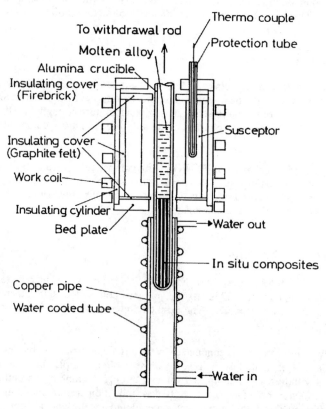

Figure. 1. Schematic diagram of unidirectional solidification rig.

3. Results and discussions

3-1 Creep curves
Figure 2 shows creep curves for the Ni-NiO in-situ composite solidified at 5 and 60 cm/h obtained at 56 MPa and 1073 K. Creep curve for the dispersion-strengthened Ni-NiO composite obtained at the same condition is also shown in this figure for reference. The creep curve consists of three parts that is primary, secondary and tertiary creep region. The in-situ composites have considerably large fracture strain, in addition the percentage of primary creep region is comparatively large. It is obvious that the creep life for the in-situ composite increases with solidification rate.

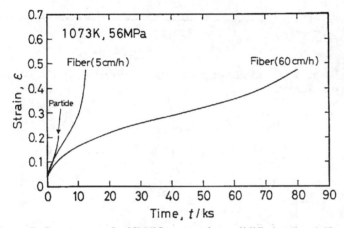

Figure. 2. Creep curves for Ni-NiO composites solidified at 5 and 60 cm/h.

3-2 Stress dependence of steady-state creep rate
Steady-state creep rates of Ni-NiO in-situ composite solidified at 5 cm/h are plotted on a log-log scale as a function of the modulus-compensated stress as shown in Fig. 3. Shear modulus G can be evaluated from Young's modulus of Ni[4], E by $G=E/2(1+v)$, where v is Poisson's ratio of 0.312[5]. A few data with symbol * showing extremely lower steady-state creep rates than expected were obtained. Metallographic observations revealed that such abnormal specimen have many grain boundaries parallel to growth direction. With a few exception, steady-state creep rates present a linear relationship with the stress exponent of 9.5. For the Ni-NiO composite solidified at 60 cm/h shows similar stress dependence with the stress exponent of 10.5.

3-3 Temperature dependence of steady-state creep rate
Plots of the steady-state creep rate at constant modulus-compensated stress versus reciprocal absolute temperature are shown in Fig. 4. In the case of 5 cm/h, the calculated values of activation energies are given in Table 1. These activation energies are higher than that for lattice self diffusion in Ni (284kJ/mol[6]). It is noteworthy that this in-situ composite shows high activation energies for creep.

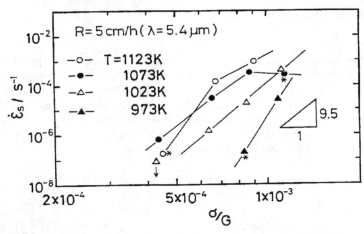

Figure. 3. Stress dependence of steady-state creep rates for Ni-NiO composites solidified at 5 cm/h.

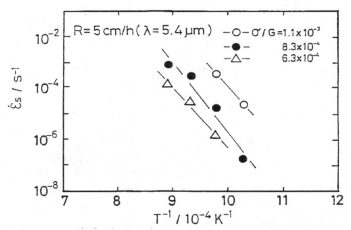

Figure. 4. Temperature dependence of steady-state creep rates for Ni-NiO composites solidified at 5 cm/h.

Table. 1. Activation energies for Ni-NiO composites solidified at 5 cm/h.

Modulus–compensated Stress, σ/G	Activation Energy, Q_c (kJ/mol)
6.3×10^{-4}	360
8.3×10^{-4}	390
1.1×10^{-3}	370

3-4 Dominate creep mechanism

It is well established that particle-strengthened alloys have both high stress exponent, n and high activation energy Q derived from interaction between dislocation and particle[7]. With consideration of existence of internal stress, the steady-state creep rates were related by

$$\dot{\varepsilon}_s = A \left(\frac{\sigma - \sigma_{th}}{G}\right)^{n'} \exp\left(-\frac{Q'}{RT}\right) \tag{1}$$

where A is a constant, σ_{th} is internal stress, G is a shear modulus, n' is a stress exponent, Q' is activation energy, R is gas constant and T is absolute temperature. Using an internal stress proposed by Ansell and Lenel[8], n' and Q' in equation (1) were evaluated. Figures 5 and 6 show stress and temperature dependence, respectively. From equation (1) we have n' = 5 and Q' = 300 kJ/mol; thus, creep behavior for this composite is controlled by a power-law creep.

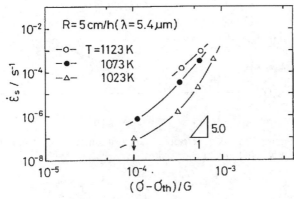

Figure. 5. Stress dependence of steady-state creep rates for Ni-NiO composites solidified at 5 cm/h.

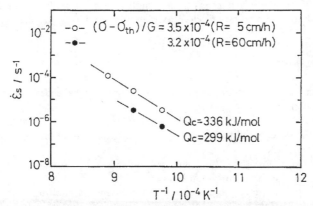

Figure. 6. Temperature dependence of steady-state creep rates for Ni-NiO Composites solidified at 5 and 60 cm/h.

3-5 Fiber spacing dependence of steady-state creep rate
Figure 7 shows the relationship between steady-state creep rate and fiber spacing. The
steady-state creep rate increases with increasing fiber spacing. There exists the linear
relationship between the logarithmic form of steady-state creep rate, ε_s and fiber spacing,
λ, that is,

$$\dot{\varepsilon}_s \propto \lambda^{4.5}. \tag{2}$$

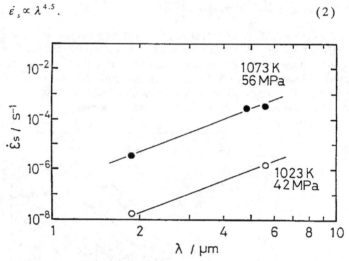

Figure. 7. Steady-state creep rate as a function of the fiber spacing
for Ni-NiO in-situ composites.

3-6 Fracture behavior
Optical micrograph of the longitudinal section near the fracture surface of the specimen
tested at 56MPa and 1073K is shown in Fig. 8. As is obvious from this figure, the
grain boundary morphology changes from straight to zigzag. Many round-type cracks
are observed at the grain boundaries approximately normal to the tensile axis.

4. Conclusions

Steady-state creep behavior for Ni-NiO in-situ composites have been investigated with
the following conclusions:
(1) Steady-state creep rate decreases with decreasing the fiber spacing. Structural
    control by unidirectional solidification has significant effect on creep properties for
    in-situ composites.
(2) Creep behavior for this composite is controlled by a power-law creep with
    consideration of existence of internal stress.
(3) Grain boundaries, which were originally straight and parallel to stress axis, migrate
    during creep. Cavities are preferentially formed at the grain boundaries. This
    cavities at grain boundary leads to fracture initiation sites.

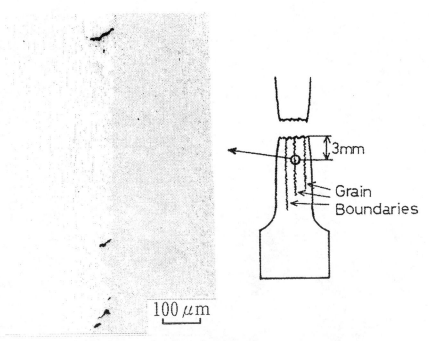

Figure. 8. Optical micrograph of the longitudinal section of the specimen crept at 56 MPa and 1073 K.

References

[1] F. D. Lemkey, E. R. Thompson, J. C. Schuster, H. Nowotny, THE QUATERNARY SYSTEM FE-CR-MN-C AND ALIGNED FERROUS SUPERALLOYS, In Situ Composites IV, 31-50, 1982

[2] T. Ishii, D. J. Duquette, N. S. Storoff, THE LOW CYCLE FATIGUE BEHAVIOR OF THREE ADVANCED NICKEL-BASE EUTECTIC COMPOSITES, In Situ Composites IV, 59-67, 1982

[3] M. McLean, DIRECTIONALLY SOLIDIFIED MATERIALS FOR HIGH TEMPERATURE SERVICE, Metal Society, 1983

[4] P.E.Armstrong,H.L.Brown, DYMANIC YOUNG'S MODULUS MEASUREMENTS ABOVE 1000C ON SOME PURE POLYCRYSTALLINE METALS AND COMMERCIAL GRAPHITES, Trans. Met. Soc. AIME, 230, 962-966, 1964

[5] C.J.Smithells, SMITHELLS METALS REFERENCE BOOK, 6th ed., Butterworths, London, 15-3, 1983

[6] K. Monma, H. Suto, H. Oikawa, J. Japan Inst. Metals, 28, 188, 1964

[7] K. Kucharova, A. Orlova, H. Oikawa, J. Cadek, CREEP IN AN ALUMINIUM ALLOY STRENGTHENED BY AL4C3 PARTICLES, Mat. Sci. and Eng. A, 102, 201-209, 1988

[8] G. S. Ansell, F. V. Lenel, Acta Met., 8, 612, 1960

Investigation of the Corrosion Behaviour of Aluminium Reinforced with Coated Carbon Fibres

B. Wielage, A. Dorner

Institute for Composite Materials and Surface Technology,
Technical University of Chemnitz, Germany

Abstract

Coatings on carbon fibres influence the electrochemical corrosion behaviour of MMCs significantly. The electrochemical corrosion is known to be a very complex system property. The following contribution describes results from corrosion tests on MMCs containing uncoated and coated carbon fibres in 3,5 wt% NaCl or 3,5 wt% Na_2SO_4 solution. Metallic fibre coatings were completely dissolved due to the fabrication process of the composites. Although the presence of copper and nickel causes a shift of the pitting potential E_{PITT} in the anodic direction and copper leads to a much lower i_{CORR} in both electrolytes, minor anodic polarization leads to a remarkable increase of i_{CORR}. The ceramic SiC-coating on the fibres lowers i_{CORR} in both solutions over the whole range of the polarization curve in comparison with MMCs containing uncoated carbon fibres. This suggests that the applied SiC-coating could be exploited as a corrosion protection. The generated pyC-coating is found to decrease the corrosion resistance remarkably. A probable reason for that may be an enhanced surface activity of the pyrolytical carbon-coating (pyC-coating), which could be reduced by changing the synthesis parameters of the pyC-coating for example. Finally, after exposure in both electrolytes microcrevices around the fibres are detected on MMCs with local chemical interactions only. In that case the hydrolytic Al_4C_3-needles obtain a length of maximum 0,5 μm and grow on some points from the fibre surface into the aluminium. If there are a thick Al_4C_3 layer around the fibres or an interface without visible Al_4C_3-needles, no microcrevices could be observed after exposure in both electrolytes.

1 Introduction

Metal matrix composites possess interesting qualities like considerably improved mechanical properties connected with low weight, high creep or wear resistance. However, the whole entity of the materials behaviour has to be taken into consideration. For the practical application of MMCs the corrosion behaviour should be known. Until now little attention was paid on the corrosion behaviour of MMCs. Although there are some important scientific investigations of the corrosion sensibility of carbon fibre reinforced aluminium [1-3], there are numerous open questions. The reason for that is the extreme sensibility of the corrosion behaviour. There are a lot of overlapping effects influencing the electrochemical processes. So finally, the corrosion behaviour of a definite MMC material is a very complex property determined by a wide range of factors.

In most cases investigations indicate that the corrosion resistance of reinforced metal suffers from essential degradation [4]. Corrosion of MMCs can occur preferable with anodic dissolution of the metallic matrix, the reinforcement or both, the matrix and the reinforcement. If the metal and the reinforcement build a galvanic couple with a remarkable potential difference a heavy accelerate electrochemical reaction can take place in presence of an electrolyte. Besides, the interface between the components plays an important role. Reaction products could cause a further potential difference for instance.

Last but not least, the fibre coating can determine the electrochemical behaviour of the MMC in different ways. The following paper presents results from electrochemical corrosion tests done on unidirectionally reinforced aluminium, which contains coated carbon fibres.

2 Experimental Methods

The selected materials are pure aluminium for the metal matrix and carbon fibres. Unidirectional and endless reinforced aluminium has to be fabricated by using uncoated and coated fibres. The chosen fibre coatings are a SiC-, a pyC-, a copper- and a nickel-coating. The ceramic SiC-coating and the pyC-coating are deposited on endless carbon fibres by means of thermally-activated chemical vapour deposition. The metallic coatings result from chemical or electrolytic depositions. Subsequently, preforms with unidirectional orientation of the fibres are made. Finally, the MMCs selected for this study are prepared by conventional Squeeze Casting using constant and optimized process parameters for each infiltration. By means of the described processing method the fibre content of the MMCs amounts to 70 vol%. An expectation is made by the MMCs containing fibres with metallic coatings, which obtain a fibre content of 40 vol% only. Because of the extraordinary importance of the interfacial circumference, composites samples are investigated in the Transmission Electron Microscope (TEM) mainly to prove the wetting, the build up of reaction products and the condition of the fibre coatings. For the examination in the TEM, (2 x 3) mm discs are prepared with perpendicular fibre orientation in the plane of the piece. This specimens are prepared to a thickness of approximately 100 to 120 μm and mechanically thinned to around 10 μm. Subsequently, specimens are thinned to perforation in an ion beam milling apparatus. Thinned composites are investigated in the 200 kV-TEM HITACHI H8100.

The specimens for corrosion tests are polished to an 1 μm finish. For the potentiodynamic tests samples are embedded in an epoxy resin before polishing and contacted to ensure current flow. All corrosion tests described in this paper are carried out in neutral 3,5 wt% NaCl solution and in neutral 3,5 wt% Na_2SO_4 solution. The electrolytes are not moved or aerated. During the potentiodynamic tests the MMCs are allowed to stabilize at their corrosion potential for at least 1 hour. After this time the polarization starts from the corrosion potential in the cathodic direction with 0,1 mVs^{-1}. Finally, after reaching -1500 mV the polarization is turned in order to record the anodic branch until - 500 mV are reached. The discussed measurement results represent an average of at least three tests. Further information about the electrochemical corrosion behaviour are given by simple immersion tests and the assessment of the corroded MMCs in the Scanning Electron Microscope (SEM).

3 Results and Discussion

TEM Studies

By means of TEM excellent wetting of uncoated and coated carbon fibres is evident. The liquid aluminium was able to penetrate into tight spacing between the fibres during the infiltration process. Furthermore, it is evident from the TEM studies that the interfacial interactions are heavy determined by the applied fibre coating. In general, three kinds of interfaces are developed during the infiltration process.

First, only local Al_4C_3-needles on the interface are formed. Such local chemical interactions are noted on the interface between uncoated carbon fibres and aluminium (Fig. 1) as well as between pyC-coated carbon fibres and aluminium (Fig. 2). The slim carbide needles obtain a maximum length of about 0,5 µm and exhibit around the same quantity in both MMCs. Results from the TEM reveal that the pyC-coating on the fibres wasn't harmed during the infiltration process. So the pyC-coating serves as an effective fibre protection and chemical interactions take place between the pyC-coating and the aluminium only. From all that observations excellent load transfer into uncoated and pyC-coated carbon fibres can be expected. However, the presence of the pyC-coating influences the micro- and macromechanical behaviour of the composites, which is pointed out on another place [5].

Second, if the carbon fibres obtain a Cu- or Ni-coating before the infiltration with liquid aluminium, thick layers of Al_4C_3-needles are formed during the Squeeze Casting process. The metallic fibre coatings are resolved completely in the liquid aluminium and intermetallics are formed. In addition, the carbon fibres are heavy destroyed by the extensive build up of Al_4C_3. There are large notches visible on the fibre surface (Fig. 3), which documents a strongly decreased fibre strength. Furthermore, carbide needles are detected some micrometers away from the fibres completely surrounded by aluminium. During the infiltration process it seems this needles are tired off the interface and at the same time new needles are formed. Although the same Squeeze Casting parameters are chosen for all MMCs, heavy build up of Al_4C_3 on the interface only takes place between metallic coated fibres and aluminium. The presence of the Cu- and Ni-coating seems to promote the formation of Al_4C_3.

Third, there are no chemical reaction products formed at the interface. This is the case if the fibres are surrounded by a SiC-coating. The SiC-coating acts in analogy to the pyC-coating as an excellent fibre protection by preventing chemical interaction between the carbon fibre and the aluminium. The SiC-aluminium-interface shows no sign of chemical interactions (Fig. 4).

$$3 \; SiC + 4Al \rightarrow Al_4C_3 + 3 \; Si \tag{1}$$

The presence of a Si-O-layer could be detected between SiC-fibre coating and the aluminium matrix. Probably, this oxide layer serves as a barrier and prevents the thermodynamically preferred build up of Al_4C_3 and elementary Si according equitation 1.

Electrochemical Corrosion Tests

Tab. 1 and 2 reveal results from the potentiodynamic corrosion test in Na_2SO_4-solution and in the pitting corrosion generating NaCl electrolyte. No significant shift of the corrosion potential E_{CORR} of the MMCs is remarkable in comparison with unreinforced aluminium in NaCl. However, the corrosion current density i_{CORR} is affected. In general, the reinforcement with carbon fibres leads to a significant increase of i_{CORR}. Moreover, the corrosion current density exhibits different changes in dependence on the applied fibre coating. As visible, the largest enhancement of i_{CORR} is found on MMCs containing pyC-coated fibres. A minor increased i_{CORR} reveals the MMCs with formerly Ni-coated fibres. MMCs with SiC- and formerly Cu-coated fibres show reduced i_{CORR} in comparison with composites containing uncoated carbon fibres. Finally, the pitting potential E_{PITT} is moved due to the presence of copper in the MMC in anodic direction, whereas on the other MMC is no significant shift of E_{PITT} recognizable.

Tab.1: Results from the potentiodynamic polarization tests in 3,5 wt% NaCl

MATERIAL	E_{CORR} [mV]	i_{CORR} [μA cm^{-2}]	E_{PITT} [mV]
unreinforced aluminium	-752 ±6	1,0 ±0,3	-739 ±10
C/Al-composite	-792 ±39	20,3 ±9,6	-768 ±11
C(pyC)/Al-composite	-754 ±3	100,4 ±1,1	-759 ±7
C(SiC)/Al-composite	-765 ±7	7,1 ±3,9	-772 ±8
C(Cu)/Al-composite	-708 ±7	12,3 ±7,5	-679 ±2
C(Ni)/Al-composite	-793 ±4	25,2 ±1,8	-748 ±8

All corrosion potentials are detected to be more positive in Na_2SO_4 solution than in NaCl. In addition, a clearly shift of the corrosion potentials in the anodic direction due to the integration of uncoated and coated carbon fibres into aluminium is found in the Na_2SO_4 electrolyte. Once more, the presence of the pyC-coating on the fibre-matrix interface has caused the largest increase of i_{CORR}, followed by the nickel containing MMCs. The MMCs containing SiC-coated fibres reveal a lower i_{CORR} than composites reinforced with uncoated fibres again. However, it is realized that the current density in the presence of copper or nickel is significant increased at more positive potentials. It is clear from that observation, that a heavy accelerate anodic solution of the metallic matrix occurs under anodic polarization.

Tab. 2: Results from the potentiodynamic polarization tests in 3,5 wt% Na$_2$SO$_4$

MATERIAL	E_{CORR} [mV]	i_{CORR} [μA cm^{-2}]
unreinforced aluminium	-528 ±18	2,1 ±0,7
C/Al-composite	-352 ±45	4,9 ±4,8
C(pyC)/Al-composite	- 418 ±32	7,4 ±0,8
C(SiC)/Al-composite	-310 ±21	3,4 ±0,2
C(Cu)/Al-composite	-224 ±26	3,1 ±0,5
C(Ni)/Al-composite	-462 ±32	6,9 ±3,4

SEM studies of the corroded MMCs exhibit typical features of pitting corrosion in NaCl solution. In addition, composites with local formation of Al_4C_3 show microcrevices at the interface, which documents enhanced anodic dissolution around the fibres (Fig. 5). This is observed in NaCl and Na_2SO_4 solution. The strongest matrix dissolution reveals the MMCs with pyC-coated fibres. The anodic dissolution extends to the whole matrix and isn't limited around the fibres especially. Only massive corrosion products and fibres maintain on the composites surface after 30 days immersion in NaCl (Fig. 6). Whereas without an external polarization the MMCs containing copper or nickel show the lowest corrosion sensibility (Fig. 7). In addition, on copper and nickel containing MMCs are found elementary segregations of this metals on the surface of the MMCs after the corrosion tests.

4 Conclusions

(1) As reported by other researchers [1-3] the corrosion resistance of aluminium is decreased significantly due to the reinforcement with carbon fibres. This is probably due to the fact, that the carbon fibres act as effective microcathodes.

(2) Fibre coatings influence the corrosion sensibility remarkably.

(3) The presence of the applied pyC-coating decreases the electrochemical corrosion resistance. A reason for that may be the surface activity of different kinds of pyrolytical carbon, which for example could be changed by careful choice of the carbon precursor or processing parameters of the pyC-coatings [6].

(4) The applied SiC-coating improves the electrochemical corrosions behaviour. Probably, this ceramic coating acts like a kind of isolator between the excellent conductible carbon fibres and the aluminium, which obtains an outstanding electrical conductivity as well.

(5) The chosen metallic coatings enhance the current density under anodic polarization of the MMCs. Due to that observation, a use of copper or nickel can't be recommended. Probably, galvanic coupling between the intermetallics formed during the Squeeze Casting and other components causes accelerate anodic dissolution. Furthermore, the observed segregations of elementary nickel or copper can lead the heavy galvanic corrosion.

(6) The circumference of the fibre-matrix-interface seems to be an important aspect for the corrosion sensibility and the dominant corrosion mechanism. The amount and geometry of the reaction products is decisive for the existence of crevices around the fibres, which are caused by preferable anodic matrix dissolution at the interface. It seems, crevice corrosion takes place only, when minor chemical interaction has led to the build up of small spacing on the surface of the carbon fibres. Apparently, thick carbide layers or an interface without the hygroscopic Al_4C_3 do not create appropriate conditions for crevice corrosion.

References

1. S. L. Coleman, V. D. Scott, B. McEnaney, Corrosion behaviour of aluminium-based metal matrix composites, J. Mater. Sci., 29 (1994) 11 2826-2834

2. L. H. Hihara, R. M. Latanison, Galvanic Corrosion of Aluminum-Matrix Composites, Corrosion 48 (1992) 546-552

3. I. Dutta, L. R. Elkin, J. D. King, Corrosion Behavior of a P130x Graphite Fiber Reinforced 6063 Aluminum Composite Laminate in Aqueous Environments, Electrochem. Soc. 138 (1991) 3199-3209

4. A. Turnbull, Review of corrosion studies on aluminium metal matrix composites, British Corrosion Journal 27 (1992) 27-35

5. B. Wielage, A. Dorner, Korrosionsuntersuchungen an faserverstärktem Aluminium, in: K. Friedrich (ed.), Verbundwerkstoffe und Werkstoffverbunde, DGM Informationsgesellschaft Verlag, Oberusel, 1997, 545-550

6. B. Wielage, A. Dorner, Einfluß einer Faserbeschichtung aus pyro-Carbon auf das elektrochemische Korrosionsverhalten von Kohlenstoffaser/Aluminium-Verbunden, accepted to be published in: Werkstoffe und Korrosion, contribution 3240 (97-11-23)

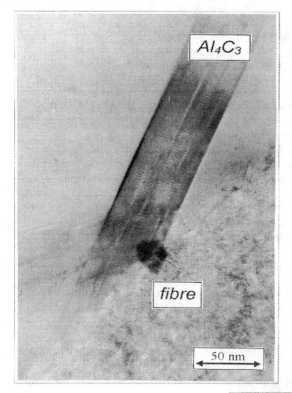

Fig. 1:

Uncoated carbon fibre in aluminium showing excellent wetting and local build up of the reaction product Al_4C_3, TEM

Fig. 2:

PyC-coated carbon fibre in aluminium showing excellent wetting and local build up of the reaction product Al_4C_3, TEM

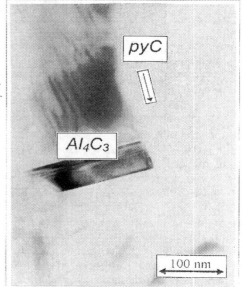

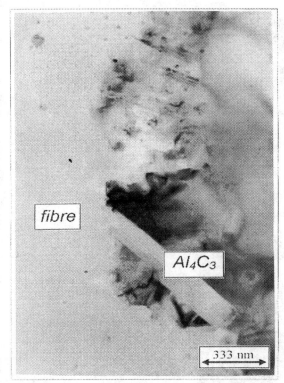

Fig. 3:

Formerly Ni-coated carbon fibre in aluminium showing heavy fibre destruction due to intensive chemical reaction on the interface (Al_4C_3-needles), TEM

fibre

Al_4C_3

333 nm

Fig. 4:

SiC-coated carbon fibre in aluminium showing excellent wetting and no build up of reaction products, TEM

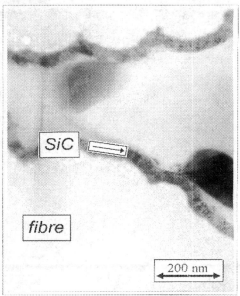

SiC

fibre

200 nm

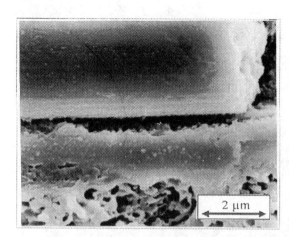

Fig. 5:

MMC (uncoated carbon fibres in aluminium) after exposure for 30 days in 3,5 wt% NaCl:

1) Microcrevice on the fibre-matrix-interface

2) Pitting corrosion of the aluminium matrix, SEM

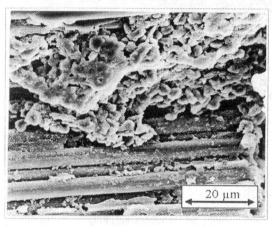

Fig. 6:

MMC (pyC-coated carbon fibres in aluminium) after exposure for 30 days in 3,5 wt% NaCl showing heavy electrochemical corrosion, SEM

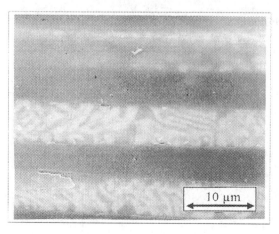

Fig. 7:

MMC (Ni-coated carbon fibres) after exposure for 30 days in 3,5 wt% NaCl showing minor anodic dissolution of the matrix, SEM

Processing and interfacial reaction of Nb/MoSi2 laminate composites and their effects on impact properties

S. P. Lee*, W. J. Park**, M. Yoshida*, G. Sasaki* and H. Fukunaga*

* Dept. of Mechanical Engineering, Hiroshima University, Kagamiyama 1-4-1,
  Higashi-Hiroshima, 739-8527, Japan
**Dept. of Mechanical & Ship Engineering, Gyeongsang National University, 445,
  In-Pyeng Dong, Tongyeong, Gyeongnam, Korea

Abstract

Nb/MoSi2 laminate composites have been successfully fabricated by hot pressing in a graphite mould. Lamination of Nb foil and MoSi2 layer showed a sufficient improvement in the absorbed impact energy comparing to that of monolithic MoSi2 material. The impact value of Nb/MoSi2 laminate composites obviously reduced when sintered at temperatures higher than 1523K, even if the composite density contributing to impact load increased along with fabricating temperatures. Impact value of laminate composites was also drastically decreased with the growth of reaction layer after the heat treatment. However, it was effective in the improvement of the impact value to increase the pressure at the same sintering temperature.

1. Introduction

Molybdenum disilicide($MoSi_2$) is considered to be an attractive candidate for future gas turbine and high performance engines in aerospace vehicles as well as various industrial applications. $MoSi_2$ has an excellent oxidation resistance comparing to most of other intermetallic compounds at the elevated temperature above 1273K and its density(6.3 g/cm³) is lower than that of nickel based superalloy. Furthermore, $MoSi_2$ has considerable potentials for the improvement of mechanical properties due to its excellent chemical stability with many kinds of ceramic reinforcements[1]. However, practical applications of $MoSi_2$ have still been restricted by its pest behavior, the insufficient fracture toughness at the room temperature and the reduced strength at higher temperature than 1473K. Several attempts have been focused on composite process in order to improve the critical damage tolerance of $MoSi_2$. Recent works on the addition of SiC whisker, TiB_2 particle and TiC particle to $MoSi_2$ have shown small improvements in the fracture toughness at room temperature[2, 3, 4]. It has been also found that $MoSi_2$ based composites contained Nb short fiber has shown a sufficient improvement in static fracture energy comparing to that of monolithic $MoSi_2$[5]. However, such a microstructural variation of $MoSi_2$ material has only shown the limited improvement effect in the fracture toughness and the fracture energy. Therefore, it is necessary to improve the damage tolerance of $MoSi_2$ through structural configurations. Lamination strategy is considered as another way to improve the fracture energy at room temperature, because it can delay the propagation crack through plastic deformation of component material and interfacial delamination. In order to apply $MoSi_2$ as high temperature structures, it is required to estimate impact properties under dynamic load as well as fracture toughness at room temperature. Unfortunately, there have been few studies to investigate impact properties of $MoSi_2$ based composites.

The primary purpose of the present work is to investigate the effect of fabricating condition on impact properties and interfacial reaction layer of Nb/MoSi₂ laminate composites. The secondary goal is to estimate the influence of interfacial reaction layer between MoSi₂ and Nb on impact properties of laminate composites after heat treated at different temperatures. In addition, interfacial reaction product between MoSi₂ and Nb is analyzed by EPMA, and then the fracture mechanism depending on the growth of interfacial reaction layer is discussed.

2. Experimental details

2.1 Fabrication of Nb/MoSi₂ laminate composites

By alternating MoSi₂ powder layer with four layers of Nb foil, and then hot pressing in a graphite mould, Nb/MoSi₂ laminate composites were fabricated. The matrix material in this experiment was a commercial MoSi₂ powder supplied by Japan New Metal Corporation with an average particle size of 2.8μ m. The thickness of 99.99% Nb foil in this system was 0.2mm. Table 1 shows fabricating conditions for Nb/MoSi₂ laminate composites. The dimension of as-pressed laminate composites was $8\times20\times80$mm³.

Table 1 Fabricating conditions for Nb/MoSi₂ laminate composites.

Volume fraction of Nb foil	(%)	10
Consolidation temperature	(K)	1473, 1523, 1573, 1623, 1773
Consolidation pressure	(MPa)	20, 30, 40
Consolidation time	(ks)	0.9, 1.8, 3.6
Vacuum pressure	(Pa)	1.33×10^{-2}

2.2 Instrumented Charpy impact test

Impact properties for monolithic MoSi₂ material and Nb/MoSi₂ laminate composites were evaluated at the room temperature by an instrumented Charpy impact test machine. The test velocity and the span length of the specimen were 3.3 m/sec and 40mm, respectively. Figure 1 shows the geometry and the dimension for impact specimen. The U shaped notch was introduced with EDM. The impact test was carried out on the flat wise specimen, and each load-displacement curve was directly monitored by the oscilloscope. Charpy impact value(E_c) was determined from the absorbed impact energy calculated with the area under load-displacement curve, divided by the fracture area of the notch of specimen (4.0×6.8 mm².).

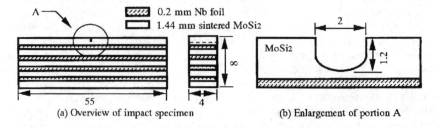

(a) Overview of impact specimen   (b) Enlargement of portion A

Figure 1. Geometry and dimension of the impact specimen

2.3 Interfacial reaction zone and fracture mechanism analyses

The microstructure constituent of the interfacial reaction zone between Nb foil and MoSi2 was analyzed with the JEOL JXA-8900RL WD/ED Combined Microanalyzer. The thickness and the composition of the reaction region were estimated by WDS(Wave dispersive spectrometer) line analysis and semi-quantitative analysis processes. Moreover, the thickness of reaction layer produced by the heat treatment was measured from WDS line analysis profile and then the effect of its thickness on impact value of Nb/MoSi2 laminate composites fabricated at 1773K was investigated. The heat treatment was conducted at 1873K and 1973K for 18ks. In addition, the plastic deformation of Nb foil and the interfacial delamination were macroscopically observed to explain the variation of absorbed impact energy.

3. Results and discussion

3.1 Density depending on process conditions

Sintered densities of Nb/MoSi2 laminate composites fabricated with a variety of process conditions are shown in Table 2. Theoretical density($6.53Mg/m^3$) of laminate composites was calculated with the rule of mixture. The density of laminate composites increased along with consolidation temperature, consolidation pressure and consolidation time because of the high densitification of MoSi2 powder. It can be seen that the densest laminate composite is one fabricated at 1773K. Therefore, it is regarded that the optimum fabrication condition for Nb/MoSi2 laminate composites may be 1773K, 30MPa and 3.6ks, considering sintered density and matrix strengthening.

Table 2 Sintered density and impact results of Nb/MoSi2 laminate composite fabricated with various conditions

Consolidation temperaature (K)	Consolidation time (ks)	Consolidation pressure (MPa)	Average density (Mg/m^3)	Relative density (%)	Impact load (N)	Displacement (mm)
1423			5.18	79	405.5	2.31
1523			5.65	86	469.9	2.75
1573	3.6	30	5.91	90	471.0	2.02
1623			5.97	91	474.5	1.87
1773			6.17	94	324.0	1.43
	0.9		5.61	86	445.2	1.81
1623	1.8	30	5.95	91	466.2	1.85
	3.6		5.96	91	474.5	1.87
		20	5.85	89	423.9	1.92
1623	3.6	30	5.95	91	474.5	1.87
		40	6.07	93	481.0	1.87

3.2 Interfacial reaction zone

The interfacial microstructure of Nb/MoSi2 laminate composites fabricated at 1773K and 30MPa for 3.6ks is shown in Figure 2. It was found that two obvious phases differing from composition proportion of Mo, Si and Nb formed at the interfacial region. The line analysis profile also shows that Si diffused far deeper into Nb region than Mo. This is due to the high diffusion rate of Si relative to Mo at 1623K. The diffusion coefficient of Si and Mo in Nb was 6.16×10^{-13} and 8.37×10^{-18} m²/s, respectively, which was calculated from the available literature data[6]. Interfacial reaction products for Nb/MoSi₂ laminate composites fabricated at this temperature were dominated by the diffusion of Si into Nb region, which resulted in forming other intermetallic compounds adjacent to Nb, such as (Nb, Mo)Si₂ and Nb₅Si₃.

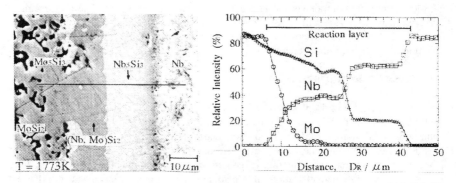

Figure 2. SEM observation and WDS analysis for interfacial reaction layer of Nb/MoSi₂ laminate composites fabricated at 1773K

Effect of consolidation temperature and heat treatment condition on the thickness of the reaction layer between Nb and MoSi₂ in Nb/MoSi₂ laminate composites is shown in Figure 3. The thickness of reaction layer is defined as the region from beginning of Nb element to Nb amount being zero. This layer increased along with the consolidation temperature. The reaction layer created at 1773K was about $35\,\mu$m in thickness, being about four times compared to that of 1473K. In addition, its thickness increased to about $60\,\mu$m after the composite was heated at 1973K for 18ks. This is believed due to the high diffusion rate of Si at a higher fabricating temperature.

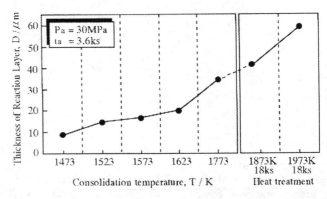

Figure 3. Effect of consolidation temperature and heat treatment condition on the growth of the reaction layer for Nb/MoSi₂ laminate composites

3.3 Impact behavior and absorbed impact energy

Figure 4 shows the effect of consolidation temperature on impact behavior for Nb/MoSi$_2$ laminate composites. As a comparison, an impact behavior of the monolithic MoSi$_2$ material sintered at 1623K is shown in this figure. The monolithic MoSi$_2$ material displays a typical brittle behavior, that is, impact load catastrophically drops down at the maximum load. By contrast, the fracture behavior of laminate composites exhibits a stable crack propagation stage beyond the maximum load. Especially, this ductile behavior of laminate composites was obviously revealed in the case of lower consolidation temperature, together with the reaction layer reduction. The maximum impact load, fracture displacement and sintered density for Nb/MoSi$_2$ laminate composites fabricated with different conditions are summarized in Table 2. The lamination of Nb foil with MoSi$_2$ powder shows an obvious improvement in the maximum load and the fracture displacement, comparing to those of the monolithic MoSi$_2$ fabricated at 1623K. The maximum impact load increased to a peak value at 1623K and went dramatically down at 1773K. The fracture displacement of laminate composites has a decrease tendency at a temperature higher than 1523K. Therefore, it can be found from Figure 3 and Table 2 that the maximum load of Nb/MoSi$_2$ laminate composites depends on the sintered density and the thickness of the interfacial reaction layer, whereas the fracture displacement is mainly dominated by the reaction layer constraining the plastic deformation of Nb foil.

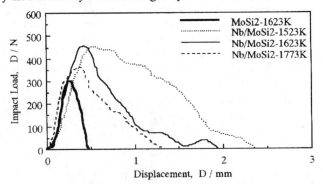

Figure 4. Impact behavior for monolithic MoSi$_2$ material and Nb/MoSi$_2$ laminate composites depending on consolidation temperatures

Figure 5 shows the absorbed impact energy for Nb/MoSi$_2$ laminate composites fabricated at different temperatures. The absorbed impact energy is divided into the crack initiation energy and the crack propagation energy. The former corresponds to the area under load-displacement curve till maximum load and the latter is one behind the maximum load. By laminating with Nb foil, the crack initiation energy and the crack propagation energy increased more than three times and seven times, respectively, comparing to those of monolithic MoSi$_2$ sintered at the same temperature(1623K). In laminate composites, the crack initiation energy has an analogous level with the increase of consolidation temperature, but the crack propagation energy remarkably reduces when fabricated at a process temperature higher than 1523K. This is because laminate composites fabricated at 1523K has a larger impact load and fracture displacement compared to those of laminate composites at 1773K in the crack propagation behavior as shown in Figure 3, even if the displacement corresponding to the maximum load has the same level. It can be considered from this figure that the deformation behavior of Nb foil and the interfacial delamination mainly contributes to the crack propagation energy of Nb/MoSi₂ laminate composites.

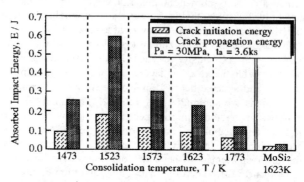

Figure 5. Absorbed impact energy for monolithic MoSi₂ material and Nb/MoSi₂ laminate composites depending on consolidation temperatures

3.4 Charpy impact value of Nb/MoSi₂ laminate composites

Figure 6 represents effects of consolidation temperature and heat treatment condition on the Charpy impact value of Nb/MoSi₂ laminate composites. The impact value increased more than five times by laminating Nb foil, comparing to that of monolithic MoSi₂ material sintered at 1623K. However, impact values of Nb/MoSi₂ laminate composites rapidly reduced at a process temperature higher than 1523K.

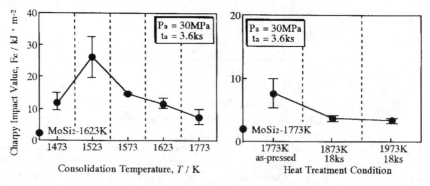

Figure 6. Effect of consolidation temperature and heat treatment condition on the Charpy impact value of Nb/MoSi₂ laminate composites

It could be seen that impact values of laminate composites were independent on composite density shown in Table 2, even if the density increased along with consolidation temperatures. In detail illustrations, the impact value of the laminate composites fabricated at 1773K was 7.7kJ · m⁻², decreasing to one third compared to 26.2kJ · m⁻² for the composite at 1523K. It has been found that the variation of the impact value was influenced by the growth of the interfacial reaction layer related to the fabricating temperature, since this layer constrains the deformation behavior of Nb foil and the interfacial delamination and results in the reduction of the fracture displacement for laminate composites(see Table 2). The effect of reaction layer thickness on the impact value obviously reveals in laminates composites heated at high temperatures. The impact value of the laminate composites fabricated at 1773K significantly decreased after the heat treatment and displayed the same level as that of monolithic matrix material. Therefore, it is essential to suppress the growth of the reaction layer between MoSi₂ and Nb foil.

Effect of consolidation pressure and consolidation time on the Charpy impact value of Nb/MoSi₂ laminate composites is shown in Figure 7. Impact values of laminate composites have an increase tendency along with consolidation pressure, but its impact value accorrding to consolidation time is nearly constant as about $12kJ \cdot m^{-2}$. This is because the sintered density of the laminate composites contributing to the maximum load increases, as shown in Table 2.

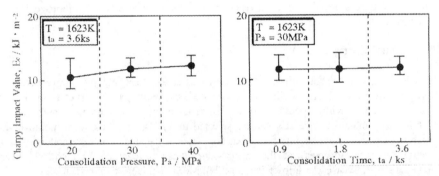

Figure 7. Effect of consolidation pressure and consolidation time on the Charpy impact value of Nb/MoSi₂ laminate composites.(Consolidation temperature:1623K)

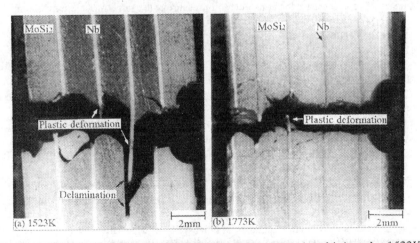

Figure 8. Typical fractographs for Nb/MoSi₂ laminate composites fabricated at 1523K and 1773K.

3.5 Fracture mechanism

From macroscopic observation of $Nb/MoSi_2$ laminate composites fabricated at 1523K and 1773K as displayed in Figure 8, it can be seen that both brittle fracture of $MoSi_2$ layer and ductile fracture of Nb foil coexist in fracture profiles of laminate composites. The laminate composites fabricated at 1523K dominantly exhibits the plastic deformation of Nb foil and the interfacial delamination, whereas the laminate composites at 1773K displays the straight crack propagation in front of notch and have a smaller deformation of Nb foil compared to that at 1523K. Such a different fracture mode is resulted from the reaction layer growth according to process temperatures, since the interfacial constraint reduces the plastic deformation of Nb foil, which leads to the interfacial delamination. In addition, it can be illustrated from the view point of fracture profiles that the difference of absorbed impact energy is related to the extent of Nb deformation and interfacial delamination depending on consolidation temperatures. Therefore, these results suggest that the suppression of interfacial reaction layer is very effective to improve impact properties .

4. Summary

The lamination of Nb foil with $MoSi_2$ powder was an excellent strategy to improve impact properties for monolithic $MoSi_2$, but represented an anisotropic properties in the impact value associated with process temperatures. The optimum processing temperature of $Nb/MoSi_2$ laminate composites can be selected to 1773K from the view point of the sintered density, but the interfacial reaction layer created at this temperature was a detrimental factor to deteriorate the impact value. It can be regarded that a critical process condition to increase the impact value of $Nb/MoSi_2$ laminate composites was 1523K, 30MPa and 3.6ks in this study, considering the ability of the plastic deformation of Nb foil and the interfacial delamination. In addition, it is effective in the improvement of the impact value to increase the consolidation pressure at the same process temperature.

Reference

[1] P. J. Meschter and D. S. Schwartz, SILICIDE-MATRIX MATERIALS FOR HIGH-TEMPERATURE APPLICATION - JOM, 1989, Vol. 42, No. 11, pp.52-55

[2] D.H.Carter, J.J.Petrovic, R.E. Honnell and W.S. Gibbs, SiC-MoSi2 COMPOSITES - Ceram. Eng. Sci. Proc., 1989, Vol. 10, pp.1121-1129

[3] R. Tiwari, H. Herman and S. Sampath, VACUUM PLASMA SPLAYING OF MoSi2 AND ITS COMPOSITES - Mater. Sci. Eng., 1992, A155, pp.95-100

[4] J. M. Yang, W. Wai and S. M. Jeng, DEVELOPMENT OF TiC PARTICLE-REINFORCED MoSi2 COMPOSITES - Scripta Metall., 1989, Vol. 23, pp.1953-1958.

[5] D. E. Alman and N. S. Stoloff, The EFFECT OF NIOBIUM MORPHOLOGY ON THE FRACTURE BEHAVIOR OF MoSi2/NB COMPOSITES- Metall. Trans. A., 1995, Vol. 26, pp.289-303

[6] L.Xiao and R. Abbaschian, INTERFACIAL MODIFICATION IN NB/MoSi2 COMPOSITES AND ITS EFFECTS ON FRACTURE TOUGHNESS - Mater.Sci.Eng., 1992, A155, pp.259-145

New metall matrix composites for superior wear resistance at elevated temperatures

H. Berns*, S. Koch*

*Institute for materials, chair of materials technology, Ruhr-University Bochum,
IA2/152, D-44780 Bochum, Germany

Abstract

A metal matrix composite (MMC) with NiCr20Al4Si3 as metal matrix (MM) and
30 vol% of hard WC/W_2C particles (HP) was tested in sliding abrasion up to 900°C.
The abrasive particles (AP) flint and corundum lead to a high wear resistance of the
MMC under argon atmosphere caused by the hardness ratio $H_{HP}/H_{AP} > 1.2$ and a
strengthened wear surface by embedded AP. Silicon carbide result in $H_{HP}/H_{AP} < 1$ and
the mechanism microindentation changes to microcutting. Examining MMC under an
oxidizing atmosphere above 600°C the wear resistance decreases drastically caused by
primarily oxidizing the HP. A better wear behaviour in an oxidizing atmosphere is
expected of HP of the type Cr_3C_2.

1. Introduction

At a given content of hard particles (HP) in a metal matrix (MM) of particle reinforced
metal matrix composites (MMC) a dispersion of the HP is beneficial in respect to wear
resistance and toughness. A net-like arrangement of small HP around large MM grains
is detrimental to both properties. The powder metallurgical (PM) manufacturing of
MMC requires a certain range of size ratio d_{HP}/d_{MM} and volume ratio f_{HP}/f_{MM} [1].
The HP themselfes are mostly effective in reducing wear, if they are larger than the
groove width, harder than the abrasive particles (AP) and of higher fracture toughness
than the AP [2].
Previous results showed that the resistance to sliding abrasion increases with
temperature due a self-protecting layer of AP up to a critical temperature which is a
nearly equal to the temperature of recrystallization in the MM (fig. 1a) [2]. A dispersion
of eutectic tungsten carbide improves the wear resistance (fig.1b, c). In conclusion of
this investigation the most heat resistant alloy was selected as MM material for the
present study.
The aim is to investigate the influence of different AP thus variing H_{HP}/H_{AP} and of an
inert versus an oxidizing atmosphere up 900°C.

2. Testing procedure

The MMC were hot isostatically pressed (HIP) for 3 hours under a pressure of 180MPa
at 1070°C. They were tested in a three-body sliding abrasion test up to 900°C (fig. 2).
Body (disc) and counter body (ring) are of the same material. Flint, corundum and
silicon carbide were used as interfacial media and were from 63 to 100µm in size. The

AP enter the interface through radial slots in the ring from a reservoir in a hollow ring holder. Argon or dry air atmospheres were used as enviroment in the tribological system. During the experiments the velocity and the surface pressure (p) were held constant. After a testing distance of 50m the mass loss was determined and the wear resistance was calculated.

Additionally oxidation tests were carried out in a horizontal tube furnace. The increase of mass (Δm) in a slowly flowing dry air current is recorded online continuously using cubic specimen (40x20x5mm) of the MMC.

Vickers-hardness (H) of MM and HP was investigated by microindentation. Specimen and indenter were heated in a vacuum chamber up to 900°C. A computer-controlled normal load was applied by a piezotranslator. Further, Palmqvist cracks initiated by indents in HP were used to calculate their fracture toughness (K_{Ic}) at room temperature.

3. Results and discussion

Fig. 3 shows a microstructure of dispersed WC/W_2C in an NiCr20Al4Si3 MM. A good bonding of HP to the MM is provided by a diffusion rim consisting of M_6C and $M_{12}C$.

This MMC is worn by different AP in sliding abrasion under argon atmosphere (fig. 4). The wear resistance decreases with an increasing hardness of the AP. Using flint and corundum a hardness ratio $H_{HP}/H_{AP} > 1.2$ and a good wear resistance is achieved. Microindentation instead of microcutting takes place and WC/W_2C protrudes above the wear surface (fig. 5a). Silicon carbide leads to a hardness ratio < 1 and allows microcutting (fig. 5b), which removes more material from the surface and leads to lower wear resistance (fig. 4).

During sliding abrasion the largest AP are ground and small particles are embedded in the wear surface forming laminates and dispersions. A process comparable to mechanical alloying creates a dispersoid strengthened MM layer. Wear resistance increases up to the recrystallization temperature. Above this the softening MM cannot support the protecting layer any longer and the wear resistance decreases (fig. 4). At 900°C the slope of the corundum curve is steeper than the one of flint, depending on the hardness ratio H_{MM}/H_{AP}. The hardness ratio of flint reaches values > 1, while corundum achieves values < 1.

The MMC behaviour under dry air atmosphere is shown in fig. 6 and 7. Fig. 6 depicts the increase in mass of the MMC specimen with temperature. The curves for different time periods during oxidation tests demonstrate that WC/W_2C is primarily oxidized. In the first minutes the oxidation rate increases with temperature corresponding to the thermal activation. In the second period there is a maximum of increase at 900°C. Above 900°C the HP at the surface are almost fully oxidized and the MMC builds up a protective oxide layer. This statement is supported by the oxidation rate of the MM. At 1100°C the mass increase of the MM after 3 h is smaller than of the MMC in the first 7.5 minutes. The photograph in fig. 6 shows a section through an oxidized HP. The reaction ($WC + 2 O_2 \rightarrow CO + WO_3$) produces a porous blooming oxide.

Consequently sliding abrasion reveals a drastically decreasing wear resistance of the MMC in dry air atmosphere at elevated temperatures (fig. 7). A photograph of the wear surface shows a oxidized and fragmented HP after sliding abrasion at 700°C. The oxidation of the surface reduces the implantation of AP in the wear surface so that the wear resistance is lowered by a gradually decreasing hardness of the MM up to 700°C.

Above 700°C more stable oxides of the alloy like Al_2O_3 and $NiCr_2O_4$ may be formed at the surface and dispersed in the near surface zone besides the AP, thus enhancing the wear resistance.

4. Conclusions and future prospects

Up to 800°C under argon atmosphere the MMC NiCr20Al4Si3 + 30 vol% WC/W_2C has a superior resistance to sliding abrasion against corundum and similar AP. But under oxidizing atmospheres temperature above 600 °C lead to a reduced wear resistance based on primary oxidation of WC/W_2C. In this case a new HP of equivalent mechanical properties and a much better resistance for high-temperature corrosion is required.

Earlier investigations revealed that CrB_2 tends to microcracking and that NbC oxidizes readily [3]. Cr_3C_2 seems to offer a compromise between chemical and mechanical properties. Chromium builds a dense Cr_2O_3 oxide layer so that the HP is protected from further oxidation. On the mechanical side the HP is superior to corundum as far as toughness and hardness are concerned (fig. 8). Finally the bonding of HP to MM has to be taken into consideration. It can be expected that Cr_3C_2 generates a diffusion zone of M_7C_3 in NiCr20Al4Si3. An examination of the new MMC will reveal whether it is suitable in sliding abrasion at elevated temperature in air.

Applications are seen in crushing of hot sinter or in compacting of hot sinter iron.

References

[1] *H. Berns, C. Nguyen*, A NEW MICROSTRUCTURE FOR PM TOOLING MATERIAL, Met. Phys. Adv. Tech. (1996) 6, pp. 61-71

[2] *H. Berns, S. Franco*, EFFECT OF COARSE HARD PARTICLES ON HIGH-TEMPERATURE SLIDING ABRASION OF NEW METAL MATRIX COMPOSITES, wear 203-204 (1997), pp. 608-614

[3] *H. Berns (ed.)*, HARTLEGIERUNGEN UND HARTVERBUNDWERKSTOFFE, Springer-Verlag Berlin/Heidelberg, 1998

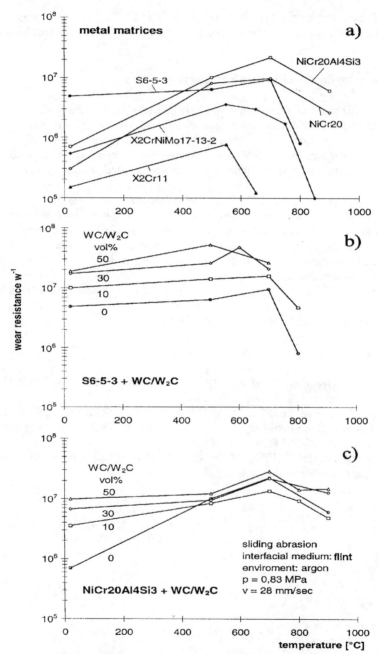

Figure 1: Effect of testing temperature on the wear resistance of MMC in three-body sliding abrasion (see fig. 2)

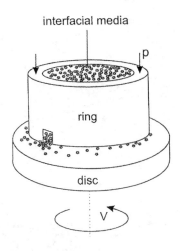

interfacial media

wear resistance

$$w^{-1} = L \cdot \varrho \cdot (\Delta m_s/A_s + \Delta m_r/A_r)^{-1}$$

L = distance
ϱ = density
Δm = mass loss (r = ring, d = disc)
A = contact area (r = ring, d = disc)

abrasives: flint, corundum, silicon carbide
(size 63 - 100µm)

environment: argon, dry air

parameter: v = 28 mm/s, p = 0.83 MPa

Figure 2:   Schematic representation of sliding abrasion test (ring on disc)

Figure 3:   Microstructure of the MMC NiCr20Al4Si3 + 30 vol% WC/W$_2$C

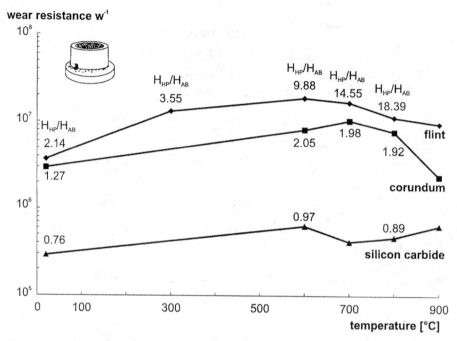

Figure 4: Wear resistance of NiCr20Al4Si3 + 30 vol% WC/W₂C against temperature in sliding abrasion test (s.fig. 2) with different abrasives under argon atmosphere

a)

b)

Figure 5: Wear surface of NiCr20Al4Si3 + 30 vol% WC/W₂C after sliding abrasion (s. fig. 2) under argon atmosphere
a) protruding WC/W₂C after abrasion with flint at 800°C,
b) worn WC/W₂C after abrasion with silicon carbide at 600°C

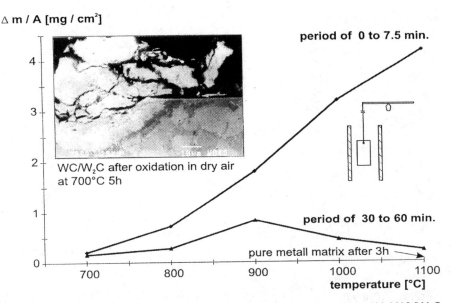

Figure 6:   Increasing mass of the MMC NiCr20Al4Si3 + 30 vol% WC/W$_2$C
versus temperature during oxidation in different periods of the test

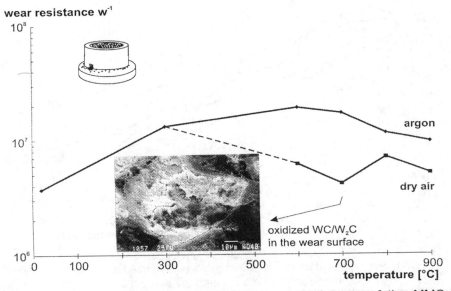

Figure 7:   Influence of enviroment on the wear resistance of the MMC
NiCr20Al4Si3 + 30 vol% WC/W$_2$C in sliding abrasion test with flint

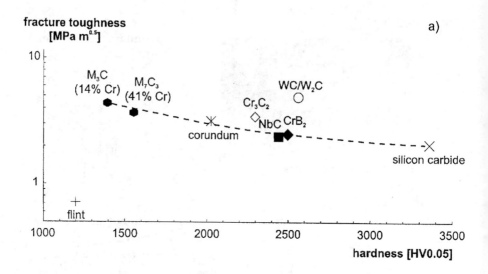

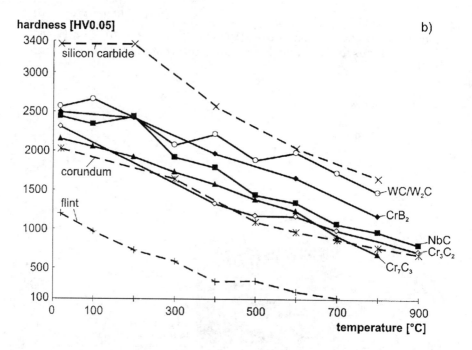

Figure 8: Properties of abrasives and hardphases derived by microindentation:
a) fracture toughness versus hardness at room temperature
b) hardness versus testing temperature

Gradient ceramic composite coatings

A.I. Mamayev, P.I. Butyagin and S.Yu. Tarassov

DTI RITC SB RAS, ISPMS SB RAS, Tomsk, Russia

Introduction

Various kinds of ceramics are greatly resistant nonorganic protective coatings at high temperatures and in aggressive media. However, the technique of obtaining such coatings is complicated and labor - intensive while adhesion with substrates is often unsatisfactory.

We have developed a technique based on microplasma processes which allows to obtain laminar gradient ceramic coatings of different compositions on aluminum, titanium, zirconium, niobium, tantalum and their alloys. Coatings formed in the above way are close to ordinary ceramics in their phase and chemical compositions as well as their mechanical properties.

Model of formation of laminar gradient ceramic coating

Ceramic coatings are produced from water solution of salts in the microplasma regime. A treated sample (anode) and a metal counter electrode (cathode) are immersed into a bath with an alkali solution to which an electric current is applied. The barrier film (5 μk thick) is formed on the surface of the anode. This film is a dielectric. Due to the fact that the anode surface is insulated with the barrier film, the decrease in the current and the increase in the voltage occurs. When the voltage achieves the value ≈ 300 V, there occurs a breakdown in the barrier film and a microarc regime of treatment of the anode surface begins.

The effect of the microplasma process on water is that water vaporizes into a gaseous state. In case the process time is a small enough value, the volume within which the water is dissociated is taken as a constant and the microplasma discharge results in a gas-filled cavity having the density of water. The gas-filled cavity is composed of oxygen and hydrogen atoms and if the electrode under investigation is of the positive polarity (anode), then the formation of oxygen atoms occurs at the interface x=0 and they diffuse into the base metal.

The partial pressure of the oxygen will be equal to 688 atm. The oxygen concentration under such conditions is 0.88 gr/cm3. The boundary problem then takes the form

$$\frac{dC^{A}(x,t)}{dt} = D_{A} \frac{d^2 C^{A}(x,t)}{dx^2} \qquad (1)$$

$$C^{A}(x,o) = 0 \qquad (2)$$

$$C^{A}(0,t) = C^{A} \qquad (3)$$

$$C^{A}(,t) = 0 \qquad (4)$$

where C(x,t), C0- are, correspondingly, the distribution of the oxygen atom concentrations in the metal and the oxygen concentration at the metal-solution interface. x is the distance from the metal-solution interface; F is the Faraday number; T is the temperature; n is the number of electrons involved in the electrochemical reaction; D is the coefficient of diffusion; R is the universal gas constant.

The solution of the boundary problem in case of the oxygen concentration distribution in the layer of aluminum is as follows

$$C_{(t,x)} = C_0^A \exp\left\{\frac{nF}{RT}\eta\right\} erfc\frac{x}{2\sqrt{D_A t}} \tag{5}$$

The oxygen ion distribution at the metal-solution interface and under the boundary conditions of (1), (2), (3), (4) DA=1×10-5=const, if T=298 K; x=0,000001-0,05 cm; t=0,002-9 s is presented in Fig.1. From Fig.1 one can see that the oxygen atom concentration increases with the time of treatment, i.e. the oxygen atoms penetrate into the metal. The oxygen atom concentration decreases with the distance from the surface to deeper layers.

When the temperature rises, the growth in the oxygen concentration and deeper penetration of the oxygen atoms into base metal (Fig.2) are observed due to the temperature dependence of the diffusion coefficient. The calculations were made under the following conditions: x=0,000001-0,05 cm; t= 9 s.; T=293-4000 K.

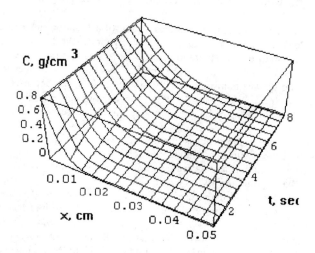

Fig. 1. Variation of oxygen atoms concentration (C, g/cm3) in a   gradient layer vs. the depth of the gradient layer (x, cm) and the treatment time (t, sec).

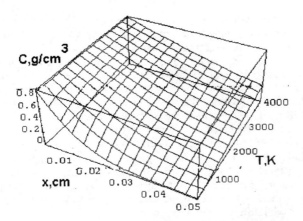

Fig. 2.Variation of oxygen atoms concentration (C, g/cm3) in a gradient layer vs. the depth of the gradient layer (x, cm) and the temperature in microplasma discharge(T, K).

The atoms of oxygen enter into a reaction with the metal to form oxides

$$2Al + 3O = Al_2O_3$$

which is accompanied by variations in composition, density and hardness of the metal (hardness of the oxides is higher than that of the base metal). Since aluminum oxide concentration is a function of the amount of the oxygen atoms, it is evident that both the composition and hardness will vary with the distance away from the surface of the metal.

Results and discussion

As a rule, the material of the ceramic coating is an oxide or a mixture of oxides. X-ray structure analysis showed the obtained coatings consist of metal oxides. Many of the oxides are injected in electrolytes as dispersive powders ($Al2O3$, $ZrO2$, $SiO2$, $Cr2O3$ and others), to form a colloid, i.e. particles of the powders acquire positive or negotive charge, the choice is determined by the charge of ions which are absorbed on the particles' surface. In our case, a source of these ions is the alkaline silicate. Variations in the amount of oxides in the coating with concentration of alkaline silicate in electrolyte are presented in the table 1.

Table 1

	Concentration of Na_2SiO_3 in electrolyte, ml/l	
	50	100
Amount of Al_2O_3 in coating , %	13,42	17,19

The same has been found when the metal-oxide interface particles of the powders are built into coating without changing their structure. It is explained by a short time of influence of the microarc discharges (100-300 μsec). This time is not enough for γ-Al$_2$O$_3$ $\rightarrow$ α-Al$_2$O$_3$ transformation, for example, in spite of the high temperature in the vicinity of the microarc discharge. This phenomenon can be used to design the coatings with different properties. In the picture, shown are the surfaces of the coatings formed from electrolytes contained the different dispersion powders. Injecting powders with high surface area ($\approx$ 150 m^2/g, picture 1a) in the electrolyte allows to increase the total area of the ceramic coatings. These coatings can be used as a substrate for a catalyst. Further analysis of the compositions on those ceramic coatings showed that they can be formed by aluminium oxide (α, γ), zirconium oxide (16 %), chromium oxide (6-14 %), manganese oxide (25 %), cobalt oxide (5 %), copper oxide (4 %). Hence the microplasma treatment in solutions allows to form the coatings which can possess the catalytic activity. In the Fig.3 showed is the variation of microhardness of gradient layers of the coatings with different compositions.

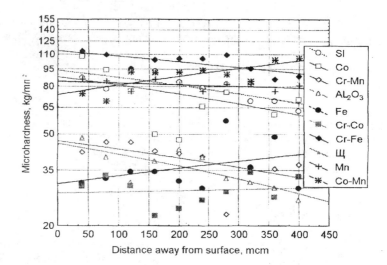

Samples made from the aluminum alloy D16 (Al-Cu-Mg) were treated in the microplasma regime : Ua= 550 V, with the current amplitude varying from 30 A down to 10 A during the whole time of process, δ=200 μsec, the treated area of the sample S=0.08 dm^2 and the time of treatment was equal to 720 sec.
The obtained coatings were 100 - 300 μm thick. Test of the coatings resistant to thermo-shock showed, that they can sustain 40 cycles at temperature of 6000 C without destruction. Microhardness is 600- 2500 kg·mm-2.
Tribotechnical characteristics and wear resistance of the material have been evaluated with the use of pin-on-disk testing procedure on 2168 UMT-1 friction machine. Specimens were 10 mm dia. and 20 mm in length aluminum alloy D16 pins loaded by pneumatic device at 350 N that corresponded to the normal pressure of 8 MPa. Composite coatings were placed on end faces of pins. The tests were carried out in boundary lubri-

cation friction. Prior to the tests specimens were held in kerosene what caused the increase in their weight due to filling of pores with the liquid (see Table 2). Coating porosity was 15 %.

Table 2

Coating component	Dry weight, g	Weight after immersion in kerosene, g	Weight after friction, g	Wear, g	Friction coefficient
Si	15.425.00	15.45800	15.44260	0.01540	0.088
Cr	14.71615	14.73140	14.72085	0.01055	0.08
Cr-Co	14.75015	14.77840	14755.70	0.02270	0.076-0.15
Co	14.66645	14.67860	14.67178	0.00682	0.11-0.2
Co-Mn	14.69440	14.72670	14.71105	0.01565	0.08-0.25
Mn	14.71515	14.74275	14.72060	0.02215	0.05
Fe	14.65050	14.66250	14.65400	0.0085	0.077
Al	13.17180	13.17885	13.17080	0.00805	0.076

The lubricant liquid was also kerosene. Velocity of slide friction was 2 m/s. Duration of each test corresponded to the length of friction path L=7500 m . Friction torque and temperature of the specimen were measured automatically. Wear was measured by loss of weight.

Conclusions

1. It is theoretically substantiated as well as experimentally verified that the microplasma processes may lead to the formation of a gradient layer with gradually varying hardness.
2. The gradient layer thickness is a function of the deposited coating composition and may be as great as 250 μk.
3. The technique, which allows to produce ceramic coatings having catalytic properties was developed.
4. Investigation of the tribotechnical characteristics of composite gradient ceramic coatings showed that the amount of hard phase Al_2O_3 found by X-ray analysis in all coatings has a great effect on friction and wear. Furthermore, porous ceramic coatings being filled with kerosene contribute to the stability of friction maintaining the lubricate ability of the coating.

Tribological behaviour of Al-20%Si/Al₂O₃ high reinforcement content composites produced by a displacement reaction

L. Ceschini°, M.C. Breslin*, G.S. Dahen**, G.L. Garagnani°, G. Poli°

° Inst. of Metallurgy, University of Bologna, V.le Risorgimento 4, Bologna, Italy
* BFD, Inc. 1275 Kinnear Rd, Columbus, OH (USA)
** Dept. of Mat. Sci. & Eng., The Ohio State University, Columbus, OH (USA)

Abstract

The tribological behaviour under dry sliding conditions against a steel countermaterial of a Co-Continuous Ceramic Composite (C^4) with interpenetrating phases (Al-20%Si/Al₂O₃) was studied. This Si-rich C^4 may be a suitable material for use in high-wear environments where structural integrity is desired.

The friction and wear test were carried out under different applied loads and sliding speeds. The wear damage of the composite was investigated by means of microstructural and chemical analyses carried out on both worn surfaces and wear debris. In all the testing conditions the Al-20%Si/Al₂O₃ composite displayed relatively high coefficients of friction (ranging from 0.60 to 0.85) and good wear resistance. The tribological behaviour of the material was explained on the basis of the "third body" composition and its ability to produce partial or full coverage of the wear scars by a compact and mechanically stable iron-oxide transfer layer.

1. Introduction

A new class of composite materials has been recently developed through an *in-situ* displacement reaction, referred to as a *Reactive Infiltration of Shaped Precursors* or RISP reaction which allows near-net shape manufacturing via a very cost-effective process[1,2]. The materials produced by this method, referred to as *Co-Continuous Ceramic Composite* (or C^4) materials, are members of a new class of composite materials known as interpenetrating phase composites (IPCs)[3]. Ideally IPC materials, wherein the constituent phases exist as indipendent, continuous, interpenetrating networks, may feature unusual combinations of properties typically not found in monolithic or traditional composite materials.

The C^4 material used in this study consists of approximately 70 vol% α-alumina, which provides wear resistance and high stiffness, the remaining 30 vol% is an Al-20%Si alloy which acts as a toughening phase and improves thermal conductivity. High levels of Si in Al alloys are known to provide improved wear resistance[4]. In the case of monolithic Al alloys, exceptionally high Si levels produce a structurally weak material. However, in the case of the C^4 material, the presence of the Al₂O₃ structure allows high levels of Si without sacrificing the structural integrity of the composite. Thus Si-rich C^4 may be a suitable material for use in high-wear environments where structural integrity is desired, including piston bore (cylinder) liners, extrusion barrels for the plastic injecton molding industry and as a part of brake rotor assemblies.

In this study, therefore, the friction and wear behaviour of an Al-20%Si C^4 material were investigated under dry sliding conditions against a surface hardened AISI 1040 steel under different applied loads and sliding speeds. The wear damage of the composite was investigated by means of microstructural and compositional analyses carried out on both worn surfaces and wear debris.

2. Experimental

2.1 Material

Co-Continuous Ceramic Composite (or C^4) materials belong to a class of materials known as interpenetrating phase composites (IPCs)[3]. These structures can best be described as multiple constituent composite materials wherein two of the constituent phases exist as independent, continuous, interpenetrating networks. This structure can be contrasted to traditional composite materials (i.e. CMCs and MMCs) where both discrete matrix and reinforcement phases exist.

C^4 materials are generally produced through an *in-situ* displacement reaction between a solid oxide body and a high oxygen affinity molten metal or alloy. Referred to as a *R*eactive *I*nfiltration of *S*haped *P*recursors or RISP reaction, the equation:

$$A_xO_y (s) + M(l) \rightarrow M_iO_j (s) + \underline{A}(l) \qquad (1)$$

will proceed in a forward direction provided that (1) the thermodynamics are favorable (i.e. $\Delta G < 0$), (2) the system exhibits acceptable wetting, and (3) there is a suitable reduction in volume in going from the reactant oxide to the product oxide[1,5]. It is this third condition which is largely responsible for the formation of the co-continuous structure. Since the process is essentially net-shape, the change in oxide volume dictates the amount of metal phase which will be present in the final microstructure.

Of the many potential RISP-formed C^4 materials, Al/Al$_2$O$_3$ C^4 appears to be one of the most promising. In practice, C^4 structures of this type are produced through the preparation of a silica precursor via conventional ceramic processing techniques (e.g. pressing, slip casting, injection molding, etc.). The precursor is formed to the desired shape and dimensions, and subsequently introduced to a molten Al bath and allowed to react according to:

$$SiO_2(s) + Al(l) \rightarrow Al_2O_3(s) + Si_{[Al]} \qquad (2).$$

In this example the Al reacts with the SiO$_2$ precursor, essentially displacing the Si in the ceramic structure. The product Si is left to dissolve in the Al bath (as denoted by the [Al]), thus a surplus of Al is required.

By controlling the ratio of product Si to reactant Al, the final alloy within the C^4 body can be controlled. The resulting composite is comprised of interpenetrating networks of Al$_2$O$_3$ and an Al-Si alloy. During the reaction, the product composite structure maintains the geometry of the precursor with 1% isotropic shrinkage[2]. By carefully controlling the initial Si level in the reaction bath and/or the ratio of product Si vs. reactant Al (i.e. according to Equation 2), the final Si composition of the Al alloy in the C^4 structure can be predicted through simple mass balance calculations.

Thus, it is possible to tailor the properties of the alloy and subsequently the composite structure. Figure 1 is a typical microstructure of the Al-20%Si C^4 material used in this study, consisting of about 70vol.%Al$_2$O$_3$ the remainder being an Al-20%Si alloy. Properties of the Al-20%Si C^4 material are listed in Table I.

Density	Young's modulus	Hardness
g·cm^{-3}	GPa	HRA
3.06	174	62

Table I - Physical and mechanical properties of the Al-20%Si C^4 material.

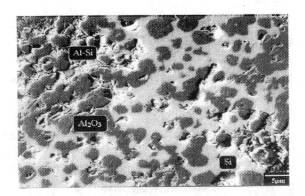

Fig. 1 - SEM micrograph of the Al-20%Si C^4 material produced by RISP processing.

2.2 Test equipment and conditions

Dry sliding tests were carried out using a slider-on-cylinder tribometer[6], where sliders (5x5x70 mm) of the composite under study are pressed against an AISI 1040 carbon steel cylinder (ϕ=40 mm), surface hardened to 62 HRC and finished up to a surface roughness Ra=0.19 μm. The sliders, made of the Al-20%Si C^4 material (hardness 62 HRA), were prepared by metallographic polishing up to a surface roughness Ra=0.2 μm.

The tests were carried out at applied loads of 5, 15 and 30 N (corresponding to maximum hertzian contact pressures about 30, 50 and 70 MPa, respectively) and at sliding speeds of 0.3, 1.2 and 1.8 ms^{-1}, while the total sliding distance was kept constant at 10 km. The tests were carried out at room temperature, in a relative humidity of 40-50%, with at least two repetitions of each test.

During testing both the friction force and total linear amount of wear (i.e. the sum of the amount of wear of the sliders and the cylinder) were continuously measured using a loading cell and a linear variable displacement transducer (LVDT), respectively. After the tests, profiles and maximum depth of the wear tracks, on both slider and cylinder, were determined with a stylus profilometer (pick-up radius of 5 μm).

The worn surfaces were examined by optical and scanning electron microscopy (SEM). Elemental analyses were carried out using Electron Dispersive Spectroscopy (EDS). Wear debris were examined by SEM and by X-ray diffraction.

3. RESULTS AND DISCUSSION

Figure 2 shows the variation of friction coefficient, μ, with sliding distance for the Al-20%Si C^4 material dry sliding against the AISI 1040 surface hardened steel, under applied loads of 5, 15 and 30 N and sliding speeds of 0.3, 1.2 and 1.8 ms^{-1}. The average and standard deviation of the steady state friction coefficients for data collected between 2 and 10 km of sliding distance are reported in Figure 3.

The values of μ are substantially high (never lower than about 0.60) and show little change with the different testing conditions, ranging from a minimum steady state value of about 0.60, at 5 N applied load and 0.3 ms^{-1} sliding speed, up to a maximum steady state value of about 0.85, at 15 N and 1.2 ms^{-1}. For all the applied loads, the coefficient of friction shows a tendency to increase when the speed increases from 0.3 to 1.2 ms^{-1} and then to decrease with a further increase of sliding speed to 1.8 ms^{-1}.

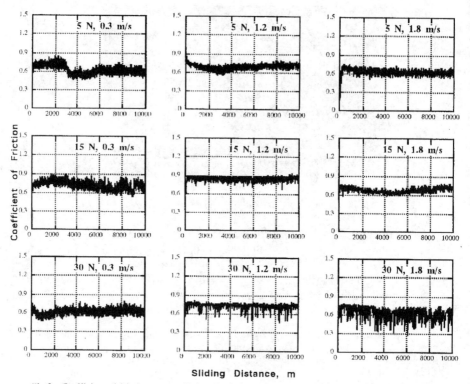

Fig.2 - Coefficient of friction versus sliding distance for the Al-20%Si C[4] dry sliding against a surface hardened AISI 1040 steel.

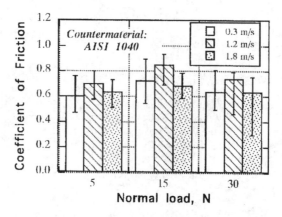

Fig.3 - Average and standard deviations of the steady state coefficient of friction measured at different testing conditions (data collected between 2 and 10 km sliding distance).

The tribological behaviour of the Al-20%Si/Al$_2$O$_3$ C^4 material, dry sliding against a surface hardened AISI 1040 steel, can be mainly explained on the basis of the composition of the "third-body"[7] and of its ability to produce partial or full coverage of the wear scars by a compact and mechanically stable oxide layer[8]. Under all the investigated testing conditions iron was found to be the main component of both wear surfaces and debris (Fig.4).

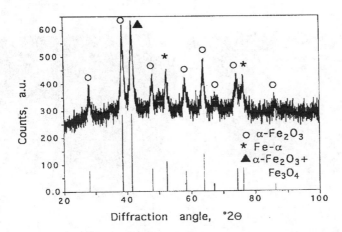

Fig.4 - XRD spectrum of the wear debris collected after dry sliding of the Al-20%Si C^4 material against steel at 30 N and 1.8 m/s.

The source of iron in the "third-body" is obviously the steel cylinder against the Al-20%Si C^4 material slid during the wear tests. In fact, the hard ceramic phase of the composite ploughs into the steel cylinder: this abrasive action, as well as adhesion, results in the formation of debris. During sliding the debris undergo oxidation, as a consequence of the frictional heating generated by the high flash temperature at the asperity contacts and are continuously transferred and compacted onto the C^4 slider surfaces, giving rise to the iron-rich transfer. This layer produces partial or total coverage of the wear tracks, depending on the testing conditions.

On the basis of the "third body" model, therefore, the increase in the friction resistance when sliding speed was increased from 0.3 to 1.2 ms^{-1} can be mainly ascribed to an increase in both the adhesive and abrasive components of the friction coefficient. A further increase in the sliding speed up to 1.8 ms^{-1}, instead, because of a significant increase in oxidation favoured by the frictional heating at the surfaces in contact, gives rise to full coverage of the wear tracks by a thick, well-adherent iron oxide layer (as shown in the SEM micrograph and corresponding Fe-Kα X-ray map in Fig.5). Essentially the same behaviour was also observed by the authors in a similar C^4 material consisting of 70vol.%Al$_2$O$_3$/30vol.%Al[9].

The full coverage of the wear track by the iron-oxide layer, observed at the highest sliding speed (1.8 ms^{-1}), therefore produces a decrease in the coefficient of friction and also limits the wear damage on both slider and cylinder. In fact, as it can be seen from the histograms in Figure 6, showing the values of the maximum depth of the wear scars, measured at the end of each test (i.e. after 10 km of sliding distance) by stylus profilometry on both C^4 sliders (Fig.6-a) and counterfacing steel (Fig.6-b), an increase in sliding speed from 1.2 to 1.8 ms[-1] leads to a decrease in the wear damage on both counterfacing surfaces. Also important is to note that under all the experimental conditions adopted in this study, Al-20%Si C^4 displays good wear resistance with maximum wear scar depths always lower than about 50 μm even after dry sliding at the more severe testing conditions (30 N and 1.8 ms[-1]) for 10 km of sliding distance.

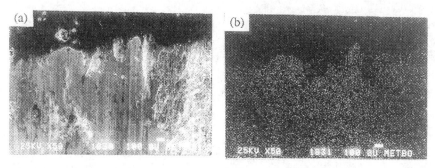

Fig.5 - SEM micrograph (a) and corresponding Fe-Kα X-ray map (b) of the wear scar on the Al-20%Si C^4 material after dry sliding against steel at 1.8 ms[-1] and 30 N.

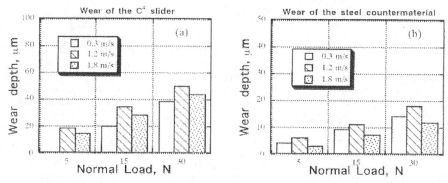

Fig.6 - Maximum wear scar depth measured at the end of the tests (after 10 km of sliding distance) with a stylus profilometer on the C^4 slider (a) and on the surface hardened AISI 1040 countermaterial (b).

Further validation of the previously described theory about the friction and wear behaviour of the tribological couple studied can be derived by an analysis of the shape of the profilometry traces across the wear scars. In fact from Figure 7, showing profiles of representative worn surfaces on both slider (Fig.7-a) and countermaterial (Fig.7-b), it can be seen that an increase in the sliding speed from 0.3 to 1.2 ms[-1] leads to rough surfaces on both slider and countermaterial (as a result of an increase of both the adhesive and abrasive components of the coefficient of friction), while smooth surfaces, typical of iron-oxide covered wear scars, were produced at 1.8 ms[-1].

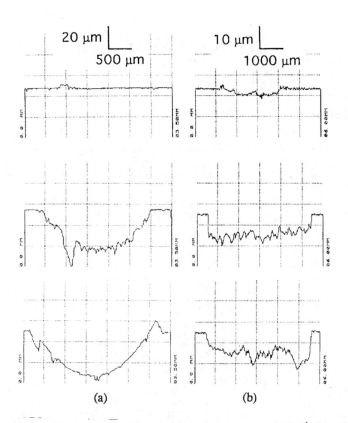

Fig.7 - Profilometer traces of the wear scars measured at the end of the tests on both C^4 sliders (a) and steel countermaterial (b), after dry sliding under different testing conditions.

Conclusions

(1)   A Co-Continuous Ceramic Composite (C^4) with interpenetrating phases was produced through an *in-situ* displacement reaction (referred to as a *R*eactive *I*nfiltration of *S*haped *P*recursors or RISP) between a SiO_2 precursor and an Al bath. The resulting composite is comprised of interpenetrating networks of Al_2O_3 and an Al-Si alloy (Al-20%Si C^4). This Si-rich C^4 may be a suitable material for use in high-wear environments where structural integrity is desired.

(2)   The tribological behaviour of the Al-20%Si C^4 material was studied under dry sliding conditions against a surface hardened AISI 1040 steel at different applied loads and sliding speeds. In all the testing conditions the composite displayed relatively high coefficients of friction (ranging from 0.60 to 0.85) and good wear resistance.

(3)   The tribological behaviour of the material was explained on the basis of the composition of the "third body" and its ability to produce partial or full coverage of the wear scars by a compact and mechanically stable iron-oxide transfer layer.

Acknowledgements

The authors are indebted to M.E. Fuller and A.C. Strange of BFD Inc. for their invaluable contribution in preparing the samples and to N. Mingazzini for his help in carrying out the experimental tests.

References

[1]   M.C. Breslin, J. Ringnalda, J. Seeger, A.L. Marasco, G.S. Daehn and H.L. Fraser, "Alumina/Aluminum Co-Continuous Ceramic Composite (C^4) Materials Produced by Solid / Liquid Displacement Reactions: Processing Kinetics and Microstructures", *Ceramic Engineering & Science Proceedings*, Vol. 15, No. 4, 1994.

[2]   M.C. Breslin, J. Ringnalda, J. Seeger, G.S. Daehn and H.L. Fraser, "Net Shape Processing of Co-Continuous Alumina-Aluminum Composites by the Reaction of Silica and Liquid Aluminum", *Fourth EuroCeramics*, Vol.1, (1995), 413.

[3]   D.R. Clarke, "Interpenetrating Phase Composites," *J. Amer. Ceram. Soc.*, 75, [4], (1992), 739.

[4]   T. Lyman, Ed. "Metals Handbook, 8th Ed. Vol.1: Properties and Selection of Metals", American Society for Metals, Metals Park, Ohio, 1978.

[5]   W. Liu and U. Köster, "Criteria for Formation of Interpenetrating Oxide/Metal Composites by Immersing Sacrificial Oxide Preforms in Molten Metals", *Scripta Materialia*, Vol. 35, No.1, 1996, 35.

[6]   L. Ceschini, G.S. Dahen, G.L. Garagnani, C. Martini, "Microstructure and tribological properties of a Co-Continuous alumina/aluminum composite", Proceed. of "EUROMAT 97", Netherlands Society for Materials Sci., Zvijndrecht, NL, 3, (1997), 319.

[7]   M. Godet, "Third-body in Tribology",*Wear* 136 (1990) 29

[8]   J. Glascott, F.H. Stott, G.C. Wood, "The effectiveness of oxides in reducing sliding wear of alloys", *Oxid. of Met.*, 24 (3-4) (1985), 99.

[9]   L. Ceschini, G.S. Dahen, G.L. Garagnani, C. Martini, "Friction and wear behaviour of C^4 Al_2O_3/Al composites under dry sliding conditions" to be published in *Wear*.

Friction and Wear Characteristics of C(Cu)/C Composite

Yan Zhaowang   Xie Guohong   Fan Zhaohui
(Iron & Steel Research Inst., Bao Steel, Shanghai, 201900, China)

Wu Yuying   Yu Mindong   Zhang Guoding
(Shanghai Jiao Tong University, Shanghai, 200030, China)

Abstract

The pantograph slider is the part used on electric locomotive to get electric current from power line. It is desired that the pantograph slider should not only have good conductivity, mechanical properties and wear-resistance, but also, and may be the most important, have as little wear on power line as possible. In order to meet the demand for high-performance sliders, short carbon fiber reinforced copper-coated-carbon powder composite was developed by cold- and/or hot-press processes. Its friction and wear performance with copper couple was studied. The comparison was also made with some other currently used slider materials. The results shown that the C(Cu)/C composite has so good friction and wear performance that the abrasion of copper couple with the composite is much less than that with other slider materials. The superior friction and wear performance of the composite is benefited by both the carbon fiber reinforcement and the carbon matrix.

1. Introduction

Pantograph sliders is a group of stripe-like parts contacting with the power line, through which electric locomotive gets electric current.

The phenomena of friction and wear between slider and power line arises when locomotive runs. Factors affecting the friction and wear are complicated, including not only power line and locomotive status, materials and properties of the slider, but also some uncertain factors such as the weather. However, dealing with a special railway, we can claim that the key factor is the pantograph slider material exclusively.

Higher performances of slider are required in order to meet the demand of the faster speed of locomotive, and to ensure it to run safely and steadily. In detail, the required performances include good conductivity and mechanical properties, low density, long service life, and most important of all, little wear on the power line [1].

The most commonly used sliders in China are made of carbon or P/M materials. In early 1990s, metalized carbon (carbon dipped with copper alloy) was also tested [2]. It was reported that lots of application researches have been made in Europe and Japan.

All of these mentioned materials have their advantages and disadvantages. The P/M slider has high strength and long service life, but has very serious wear on power line made of copper. The carbon slider has good friction and wear properties, but has poor mechanical properties and relatively shorter service life. The metalized carbon slider has good electrical, mechanical properties, but its properties are unsteady (when unthoroughly dipped with metal), its cost is relatively higher, and its wear on copper line is also serious.

It is a good choice to develop composite slider using carbon as matrix. Composite slider, made up of carbon matrix reinforced with metallic powder, fiber or fibrous net, was developed in Japan [3]. This material has higher mechanical and electric properties than carbon material.

Another composite system made up of copper-coated-carbon powder and chopped carbon fiber was developed in our laboratory. Our research works showed that this material has good comprehensive properties, and can be used to make high quality carbon system slider.

In this paper, the friction and wear between the composite and copper couple was investigated, and comparison with other slider materials was made. This paper aims at representing the good wear and friction characteristic of this composite.

2. Materials and Experimental Procedures

The C(Cu)/C composite was made up of a certain volume fraction of chopped short carbon fiber, carbon powder coated with copper, and a little amount of binder, and it was consolidated by the process of mixing, cold press, hot press and sintering.

Two composite samples(referred to as A, B) were examined. The fiber content of A (2.5%) is slightly different to B (5%). In the consolidation process, mixing is performed in dry condition for sample A, but in wet condition for sample B. Cold press is performed at 270MPa for 1 minute, and hot press at 245MPa, >160°C for 5 or 10 minutes.

In addition, three currently used slider materials, pure carbon (referred to as C), metalized carbon (referred to as D) and P/M material (referred to as E), were tested for comparison.

The wear tests were carried out on MM-200 type wear tester. The following condition was applied to each specimen, load 98N, rotation speed of copper ring 400rpm, dry slide wear, the specimen size 7×7×16mm, copper ring size φ60(outer) × φ40(inner) × 10(width)mm, wear test duration 120 minutes. The contact mode of specimen with copper ring is illustrated as figure 1.

The weight loss of carbon materials after wear test is very small and the wear status is very complicated, so the weight loss of the specimen cannot be used to represent the intensity of abrasion. The wear groove width was used instead (as fig 2 shows).

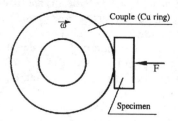

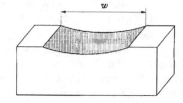

Fig 1. Contact mode of specimen with   Fig 2. Intensity of abrasion is calculated by
        copper ring   measuring wear groove width

material is very steady. It should be noted that the friction coefficient of composite A
and B progressively become smaller. It means that their anti-wear properties increase
with the examining time going on.

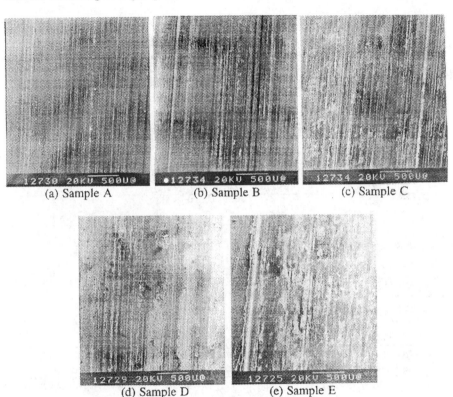

(a) Sample A   (b) Sample B   (c) Sample C

(d) Sample D   (e) Sample E

Fig 3. Surface images of copper rings after wear test

3.3 Analysis of friction and wear properties of composites

The P/M material slider is made by sintering copper or iron based powder.
Unreasonably it will abrase the copper surface very bitterly.

The producing process of metalized carbon material determines that the structure of this
material is not uniform. Metal usually distributes within carbon matrix as lumps or strip,
determined by the original structure of carbon preform. This is the main reason that this
material abrase the copper surface bitterly.

In C(Cu)/C composite, copper-coated-carbon powder works as matrix. A thin layer of
copper coating envelopes carbon particles, after cold or hot press, the copper coating
forms a kind of interconnected network. This copper network adds to increase
conductivity and mechanical properties of composite. Due to its finely and uniformly
distribution, copper has little negative influence on composite anti-wear property. As
the reinforcement, chopped shorted carbon fiber is added mainly to improve

The surface of copper ring after wear test was analyzed on HITACHI S-500 SEM.

3. Results and Discussion

3.1 Materials mechanical and physical properties

The materials mechanical and physical properties of sample A and B are listed in table I. We can see that the density of these two materials is higher than pure carbon slide ($\sim 1.7 \text{g/cm}^3$) and metalized carbon ($\sim 2.4 \text{g/cm}^3$). The conductivity of the composites is superior to carbon($20\text{-}40\mu\Omega\cdot\text{m}$), equivalent to metalized carbon ($\sim 8\mu\Omega\cdot\text{m}$). The bending strength is superior to carbon ($\sim 30\text{MPa}$), inferior to metalized carbon ($\sim 100\text{MPa}$). The compact strength is superior to both carbon($<0.1\text{J/cm}^2$) and metalized carbon($\sim 0.25 \text{ J/cm}^2$).

Table I   The mechanical and physical properties of composite sample A and B

Sample	Density (g/cm^3)	Electric Resistance ($\mu\Omega\cdot$m)	Bending Strength (MPa)	Compact Strength (J/cm^2)
A	2.75	5.45	63.44	0.47
B	2.81	7.67	63.29	0.50

3.2 The results of wear test

The results of wear test are listed in table II. It is shown that the friction coefficient of composite specimens is close to that of currently used slider materials, but their wear loss is less. Considering both friction coefficient and wear loss, the C(Cu)/C composites have superior anti-wear properties.

Table II   The results of wear test

Sample	Friction Coefficient	Wear Groove Width (mm)	Damage of Copper Ring's Surface	Change of Friction Coefficient
A	0.27	2.43	small	steady, progressively become smaller
B	0.26	2.54	small	steady, progressively become smaller
C	0.25	2.68	slightly serious	steady
D	0.24	3.04	serious	unsteady
E	0.29	6.47	very serious	unsteady, progressively become bigger

SEM images of copper rings' surfaces after wear test is shown in figure 3. The wear on copper of each material is much different. P/M material has the most serious damage on copper, metalized carbon the second, pure carbon the third, and the composite samples A and B have very small damage on copper.

Change of friction coefficient during test also varies from one to another. The friction coefficient of P/M material is very unsteady, and become bigger as test proceeds. In other words, P/M material abrases the copper surface more bitterly as test lasts. That is unwanted in electric locomotive running. However, the friction coefficient of carbon

mechanical properties of composite, and is uniformly dispersed in matrix (Figure 4). Carbon fiber also has positive influence on composite's anti-wear property. So the good anti-wear property of C(Cu)/C composite is benefited from both carbon matrix and fiber reinforcement.

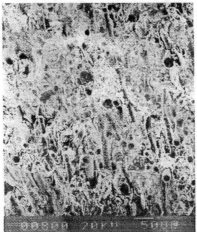

Fig 4.   Uniform distribution of fiber in C(Cu)/C composite

Carbon can form a protective film on the surface of copper during the test. This film can reduce abrasion of copper, and decrease the friction coefficient. From the test results, we know that the C(Cu)/C composite and pure carbon materials can form protective film. But P/M and metalized carbon materials cannot, they roughen the surface of copper, causing the friction coefficient become bigger as test last.

4. Conclusion

1) C(Cu)/C composite has better friction and anti-wear properties than P/M, metalized carbon and pure carbon slider materials.
2) The good friction and anti-wear property of C(Cu)/C composite is benefited from both carbon matrix and carbon fiber reinforcement.
3) Better friction and anti-wear properties, together with its better conductivity and mechanical properties make C(Cu)/C composite as a excellent pantograph slider material.

References

[1] *Jean-Jacques MAILLARD*. Pantographe-catenaire:les materiaux du contact. REVUE GENERALE DES CHEMINS DEFER. No. 718, 1991,17-18
[2] *Liang Roqing*. Proceeding of pantograph slider for electric locomotive in China. Dian Tan (Electro-Carbon). No.1, 1991, 3-8 (in Chinese)
[3] Japanese Patent 1-157464 (1989)

SELF-HEALING GLASSY COATINGS FOR SiCf/SiC COMPOSITES

M. Ferraris, M. Salvo , C. Isola, M. Appendino Montorsi
Politecnico di Torino-Dipartimento di Scienza dei Materiali e Ingegneria
Chimica
corso Duca degli Abruzzi 24 - I10129 Torino

Abstract

This work reports on the synthesis of new self-healing, oxidation resistant glass and glass-ceramic coatings for SiCf/SiC (SiCFil®, FN-Enea, Italy). Glasses and glass-ceramics were applied on the surface of the composite and heated to obtain a coated SiCf/SiC. Some *composite coatings* were obtained on SiCf/SiC by applying a slurry made of SiC particles and glass powder. In some cases, a double coating made of two different coating materials (a glass and a glass-ceramic) was necessary to obtain the self-healing behaviour in the range 700-900 °C. The self-healing property of the some of the coatings was investigated by inducing cracks on them and subsequently by observing the repair after heating them at their working temperature.

1.Introduction

Silicon carbide long fibres reinforced silicon carbide matrix composites (SiCf/SiC) are among the most promising materials for structural high temperature applications (aerospace, automotive, high temperature, high performance engines, turbines,..). They are mechanically reliable at temperatures not obtainable by conventional materials, e.g. super-alloys, and they offer an enormous advantage in terms of specific mechanical properties, compared with metal alloys. Nevertheless, some problems are still to be solved before using SiCf/SiC for all the foreseen applications: they are mainly porous materials and they are subjected to cracks in the silicon carbide matrix under operative conditions.

These features can lead to the oxidation of the fibre/matrix interface (mainly made of carbon) and consequently to the catastrophic failure of the entire material. Moreover, oxygen or other aggressive gases can react with the composite at high temperature and severely damage its structure.

For these reasons, self-healing, oxidation resistant and possibly low cost coating materials, could be useful for a wider employ of SiCf/SiC.

2. Experimental

In some cases (see Table I), the as received SiCf/SiC composite surface was modified by heating it at 1100 °C in Ar flow or by inducing the formation of metal carbides, in order to improve its wettability toward the glasses: slurry of aluminium, magnesium or silicon plus carbon were put on the composite and heated in Ar flow at suitable temperature, in order to obtain the formation of the respective carbides.

Glass (mol.%)	B_2O_3	SiO_2	CaO	Al_2O_3	BaO	Na_2O	MgO	Tg (°C)	Tsoft (°C)	CTE** (10^{-6} °C^{-1})	Surface modifica-tion* for SiCFil®
BMS	62.5	9.6	-	19.3	-	-	8.6	637	670	8.8 (220-420°C)	Mg, Al
BAB	30	-	-	50	20	-	-		680		-
SABB	16.7	77.6	-	1.4	4.3	-	-		960	≅2.5	-
CA	-	-	64.3	35.7	-	-	-	850	1380	9.42 (RT-350°C)	Al, Si+C
SAM	-	57.6	7.4	25	-	2.0	8.0	645	806	4.0 (RT-400°C)	Heating at 1100 °C, Si+C, Al
SNB#	-	60	-	-	20	20	-	465		17.0 (RT-450°C)	

Table I. Composition and thermal characteristics of the coatings

*, the as received composite surface was modified by inducing the formation of metal carbides, in order to improve its wettability toward glasses: slurry of aluminium, magnesium, or silicon plus carbon were put on the composite and heated in Ar flow. A heating in Ar flow at 1100 °C was effective in some cases to obtain the wettability of some glasses on SiC$_f$/SiC.

** CTE SiCFil® = $4 \cdot 10^{-6}$ °C^{-1}
# self-healing layer to be put on SAM

Most of the glass compositions reported in this paper were "designed" especially to fulfil the characteristics of this particular SiC$_f$/SiC (SiCFil®, produced by FN-Enea, Italy, referred such as SiC$_f$/SiC in the paper): its coefficient of thermal expansion (CTE) is about 4 10-6 °C −1 and the XRD on its surface showed the presence of amorphous SiC, together with some α- and β-SiC crystalline phases.

Glasses were prepared by melting of starting products; their thermal properties were measured by Differential Thermal Analysis (DTA, Netzch 4045), Differential Scanning Calorimetry (DSC, Perkin Elmer 7), heating microscopy (Leitz, model II A) and dilatometry (Netzch 4045) (see Table I).

The wettability and the reactivity between SiC$_f$/SiC and glasses were investigated by heating microscopy. Glasses were powdered, sieved and mixed with ethanol to obtain a slurry, then applied on the surface of SiC$_f$/SiC (as received or modified) and heated above the glass melting temperature to obtain a coated SiC$_f$/SiC.

In one case (with SABB glass) some *composite coatings* were obtained on SiC$_f$/SiC by applying a slurry made of SiC particles, SABB powder and ethanol (18 and 45 wt% SiC particles).

In some cases, a double coating made of two different coating materials (a glass and a glass-ceramic) was necessary to obtain the self-healing behaviour in the range 700-900 °C (SAM glass-ceramic/SNB glass).

The coated structures were characterised by scanning electron microscopy (SEM) and compositional analysis (EDS). The self-healing property of the some of the coatings was investigated by inducing cracks on them and subsequently by observing the repair after heating at their working temperature (800 °C, 30 minutes; 700 °C, 60 minutes). Some coated samples were heated 40 hours at 800 °C in a vertical position to investigate the stability of the coating at the working temperature.

3.Results and Discussion

Glasses are known to react with silicon carbide to form gaseous species [1, 2]: nevertheless, the glass composition can be modified to avoid these reactions in the temperature range of interest for the specific application.[2]

Another important factor to be taken into account is the wettability of glass on SiC$_f$/SiC: the contact angles between the molten glass and the composite should be the lowest possible [3-6].

Several glasses and glass-ceramics (borate glasses, silica- and not silica-based) were prepared and their reactivity towards SiC$_f$/SiC will be discussed in this part.

The glasses indicated as BMS, BAB and SABB (upper part of Table I) are borate glasses: they have been effective in giving homogeneous coatings on SiC$_f$/SiC. One of them, BMS, did not wet the as received composite surface: a surface modification by Mg or Al slurries was necessary to obtain the coating (figure 1): this behaviour is still under investigation. The other two glasses, BAB and SABB, showed near to zero contact angles on the as received composite surface (figures 1 and 3).

The SABB glass gave homogeneous, crack free coatings and a polished cross section of a completely coated composite is shown in figure 3.

These three glasses are suitable for applications requiring a self-healing behaviour in the range of 700-800 °C (BAB), 700-900 °C (BMS) and 1000-1100 °C (SABB) . The presence of boron in each composition do not allow their use in fusion reactors, because of the transformation of boron in lithium and helium [7]

The SABB glass was also used to prepare *composite coatings* (18 or 45 wt % SiC particles and SABB matrix): figure 4 (a, b) show the cross sections of the composite coatings on SiC$_f$/SiC. The composite coatings were prepared with the aim of enhancing the temperature at which the SABB coating could be used as self-healing protection for SiC$_f$/SiC. The self-healing property is guaranteed by the softening of the SABB matrix. The encouraging results in terms of wettability and the continuous interface between SiC$_f$/SiC and the composite coating suggested the preparation of *composite joinings* [8].

The glasses and glass-ceramics labelled as CA, SAM and SNB (Table I), were especially prepared to be used in a fusion reactor environment: they do not contain boron oxide and only have low activation materials [9, 10].
Figure 5 shows the polished cross section of a CA-coated SiC$_f$/SiC: the surface of the composite was modified by the formation of aluminium carbide to obtain the CA coating. The surface modification by silicon carbide was less effective mainly for two reasons: first, the temperature and the time used to modify the surface with the formation of silicon carbide could be too severe for the thermo-mechanical integrity of the composite. Aluminium and magnesium slurries were preferred, when possible, as surface modifiers for CA coatings, because their processing temperature could be much lower (about 1100 °C). Secondly, some silicon was found in the CA coating by EDS, probably coming from the silicon carbide modified surface. The presence of silicon (likely as silica in the glass-ceramic) partially vanishes the effectiveness of CA as protection toward lithium silicate ceramic breeder, as discussed below.
The CA glass was prepared with the particular scope of obtaining a protective coating towards the lithium silicate ceramic breeder probably used in the future thermonuclear fusion reactor, which is known to react with the silica normally present on the surface of SiC$_f$/SiC [11]. The CA composition was chosen without silica, boron oxide and lithium oxide, for this fusion application purposes. Preliminary tests of compatibility between a CA and Li4SiO4 ceramic breeder (pellets) was carried out at 800°C for 140 hours under a helium-hydrogen (0.1% vol. of H2) flow: the CA did not show any reaction toward the pellets and protected the composite surface from degradation [9].
The CA coating was not effective as self-healing coating up to 1200 °C: a double coating structure should be developed in this case, as discussed below.

The SAM glass was developed to be used in couple with SNB and to act as self-healing double coating for nuclear fusion applications in the temperature range of 700-900 °C [10].
A glass without boron oxide, having low characteristic temperatures to be self-healing between 700 and 900 C, but with a CTE suitable to coat SiC$_f$/SiC without cracks was very difficult to find. A double coating solution was then chosen: the coating to be in contact with SiC$_f$/SiC (SAM) has a suitable CTE, but can not self-heal at the requested temperatures. A second coating (SNB) has the requested self-healing property and can fill the cracks produced in the composite during thermo-mechanical stresses due to the operative conditions (supposed to be between 700 and 900 C).
Figure 6 shows the polished cross-section of the double coated composite: the upper layer is the SNB glass, containing barium oxide ("bright" in the back scattered electrons image) and acting as self-healing layer in the range of 700-900 °C: the SAM glass-ceramic shows a continuous interface with SNB and the composite. Self-healing properties for this double coated structure was tested at 700 and 800 °C (60 and 30 minutes, respectively) as described in the

experimental: figure 7 (a, b) shows the double coated composite before and after the self-sealing test. The crack made on the coating was completely healed after the test.

The double coated composite was also tested at the same temperatures in a vertical position for 40 hours: the coatings did not flow from the composite and the two coating materials did not mix themselves after this time, as revealed by the EDS analysis.

4.Conclusions

Several glass and glass-ceramic coatings (single coating, double coating, composite coating) have been developed for a specific SiC$_f$/SiC (SiCFil®, FN-Enea, Italy): they have been positively tested in terms of self-healing property and stability at different temperatures.

Each coating described in this paper, could be proposed for several different fields involving SiC$_f$/SiC: i.e. fusion reactors, medium-high temperature applications, etc.

The coating materials and the coating technique are simple and low cost.

Acknowledgements

Many thanks are due to FN-Enea for providing the composites, to Fiat Research Centre for SEM-EDS and to ASP (Associazione per lo Sviluppo scientifico e tecnologico del Piemonte) for supporting this research.

References

[1]. P. Lemoine, M. Salvo, M. Ferraris, M. Montorsi, REACTIONS OF SICF/SIC COMPOSITES WITH A ZINC-BORATE GLASS, Journal of the American Ceramic Society, Vol. 78-6 P.1691 (1995)

[2]. P.Lemoine, M.Ferraris, M.Salvo, M. Appendino Montorsi, VITREOUS JOINING PROCESS OF SIC/SIC COMPOSITES, Journal of the European Ceramic Society, vol.16,n.11(1996), p.1231-1236.

[3]. D.N. Coon, VITREOUS JOINING OF SIC FIBER REINFORCED SIC COMPOSITES, Report n. NTIS DE90-01295. Composite Materials Research Group, Department of Mechanical Engineering, University of Wyoming. 1989.

[4]. C. Isola, M. Salvo, M. Ferraris e M. Montorsi, JOINING OF SURFACE MODIFIED CARBON/CARBON COMPOSITES USING A BARIUM-ALUMINUM-BORO-SILICATE GLASS, in stampa su Journal of European Ceramic Society.

[5]. M.Ferraris, M.Salvo, C.Isola, M.Montorsi, JOINING AND COATING OF SIC/SIC WITH GLASS AND GLASS-CERAMICS, In the Proceedings of "IEA-International Workshop on SiC/SiC Ceramic Composites for Fusion Structural Applications, Ispra (VA),october 28-29 (1996), p. 113.

[6]. I.W. Donald, PREPARATION, PROPERTIES AND CHEMISTRY OF GLASS- AND GLASS-CERAMIC-TO METAL SEALS AND COATING, Journal of Materials Science, 28, (1993)2841.

[7]. E. Lell, N. J. Kreidl, and J. R. Hensler, RADIATION EFFECTS IN QUARTZ, SILICA AND GLASSES, in Progress in Ceramic Science, vol. 4, J. E. Burke Editor, Pergamon Press Ltd., England, 1966, pag.65.

[8]. M. Ferraris, M. Salvo and M. Montorsi, BRUSH JOINTS FOR LONG FIBER REINFORCED COMPOSITES, abstract for the International Conference of Joining of Advanced Materials, Materials Week '98, Rosemont, IL, October 12-15, 1998.

[9]. M. Ferraris, M. Salvo, C. Isola, M. Appendino Montorsi and A. Kohyama, GLASS-CERAMIC JOINING AND COATING OF SIC/SIC FOR FUSION APPLICATIONS, in press on the Journal of Nuclear Materials.

[10]. M. Ferraris, M. Appendino Montorsi, M. Salvo, C. Isola and A. Kohyama, SELF-SEALING MULTILAYER COATING FOR SiCf/SiC COMPOSITES, "IEA-International Workshop on SiC/SiC Ceramic Composites for Fusion Structural Applications, Sendai , Japan, October 1997, in press on the Proceedings.

[11]. T. Sample, P. Fenici, H. Kolbe and L. Orecchia, Journal of Nuclear Materials vol. 212-215 (1994) 1529-1533.

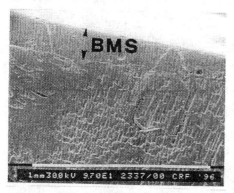

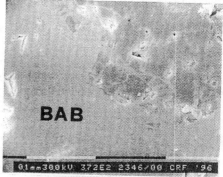

Figure 1. SEM of the polished cross section of a BMS coated SiCf/SiC (composite surface modified by Al slurry).

Figure 2. SEM of the polished cross section of a BAB coated SiCf/SiC: see the infiltration of the glass (bright grey) through the SiC (dark grey).

Figure 3. SEM of the polished cross section of a SABB coated SiC$_f$/SiC.

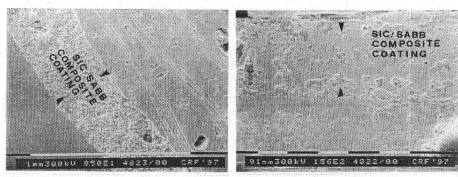

Figure 4. SEM of the polished cross section of two *composite coated* SiC$_f$/SiC: (a) 45 %wt SiC particles in a SABB matrix (b) 18 %wt SiC particles in a SABB matrix.

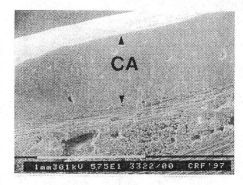

Figure 5. SEM of the polished cross section of a CA coated SiC$_f$/SiC(composite surface modified by Al slurry)

Figure 6. SEM (back scattered electrons) of the polished cross section of a SAM/SNB double coated SiC$_f$/SiC.

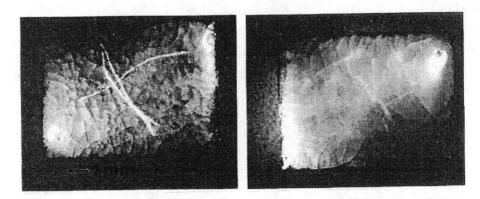

Figure 7. SAM/SNB double coated SiC/SiC before (a) and after (b) self-healing test at 700 C, 60 minutes, in a vertical position.

Friction and wear of composite materials with multilevel shock absorbing structure

A. Kolubaev*, V. Fadin* and V. Panin*

*Institute of Strength Physics & Materials Sciences Pr.Academichesky, 2/1
634021, Tomsk, RUSSIA

Abstract

Results of studies a tribotechnical characteristics of micro- and macroheterogeneous TiC-based composites are present in this article. It is shown that the creation of multilevel structure allows to adjust a process of wear, because can be created the wanted structure of material for given conditions of friction.

1. Introduction

The anti-friction materials, among which there are great number of composite materials based on metals, polymers, and ceramics, are used for manufacturing the components of friction units working under conditions of limited lubrication, high temperature, and high loading. Composite materials have strength properties comparable to pour materials, best running - in, lower friction coefficient, and greater wear resistance. The requirement of raising a power of machines leads to need for use of the composite materials on the metallic base with hard filler (matrix-filled composite), among which there are WC-Co hard alloys and pseudoalloys.

2. Results and discussion

The macroheterogeneous pseudoalloys have been of our main interest, because the loading on their sliding surface is distributed more effective between hard particles and the matrix. In these materials the plastic (protective) thin films forming on the surface of hard particles depend on conditions of external friction. Formation of films are caused by difference between thermal expansion coefficients of the plastic material and the solid particle or are due to mechanical smearing by antifriction component. Load on the surface of a macroheterogeneous material, which are shown in Fig.1, is distributed so that greater contour pressure is acting on the hard area, while the pressure between the antifriction matrix and the mating surface small. This reduces probability of

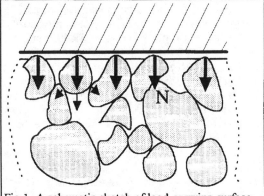

Fig.1. A schematic sketch of load-carrying surface.

fracture of composite material. In our opinion, the size of the hard granules of composite antifriction material with heterogeneous structure should have a comparable size with contour area of the contact, that is more than 300 micrometer.

A required contact of mating surfaces can result from combinations of components of composite material. However there are no substantiated methods to choose the structure of similar composite. Analysis of tribotechnical and strength properties of composite materials shows that the critical pressure can be one of the criterion in choosing the structure and composition of materials, which defines the beginning of plastic deformation, frail fracture or adhesion. Under these conditions the carrying capacity is limited by the elastic limit or strength limit. When increasing the concentration of hard particles, the total area of surface contacts is increasing similarly. This leads to reduction of the contour pressure and thus to an increase of the carrying capacity of sliding surfaces. The best characteristics in this case will have the material which consists of the hard particles.

However, the present consideration ignores character of deformation, changing of the temperature in the friction area, and the adhesive interaction of contacting materials. In their turn these parameters can have a pronounced effect on the correlation of volumes of the solid phase and the anti-friction phase. Thus the question about the optimal size of the hard phase inclusions remains unsolved. It should be solved in any particular case from the requirements of obtaining the best strength and tribotechnical properties.

The results of studies of some physicomechanical and tribotechnical properties of micro- and macroheterogeneous composite materials (CM) are given in this report. These materials belong to the class of matrix-filled composites. For manufacturing the CM is used the multi-level principle of energy damping, which is based on the mesomechanics deformation and fracture of materials. Deciding role in the processes of destruction plays the movement of the volumetric structural elements (meso-volumes). In sliding systems, meso-level of deformation is the main, because it is determined by the interaction between surface irregularities. This interaction is responsible for creating of the local high stress in the short time interval. Relaxation of the local stress in plastic materials may result in rotary modes of deformation which cause the formation of a fragmented microstructure with a submicron size of fragments. Local stresses in the brittle materials are responsible for formation the surface cracks.

The formation of heterogeneous structure must exclude the rotary channel of plastic deformation and brittle fracture of friction surfaces, because the friction energy will be effectively dissipated. It will raise wear resistance. The investigations of the surface structure of microheterogeneous composite materials TiC-CuFe, TiC-NiCr and high-tensile austenitic steel, which were carried out earlier [1,2], confirm really the analysis of process of destroying the surface structure in friction. If heterogeneous structure is saved, the wear, controlled by microplastic deformation, is negligible.

Increase of pressure leads to the intensive heat generation in the friction area of composites TiC-Iå, which causes a dissolution of carbides in the matrix and change a nature of deformation. You can see in Fig.2 that in this case a surface layer forms free from carbides. For austenitic aged steel, however, the formation of the meso-scale structure is not determined by temperature. The hard phase dissolves in the matrix by action of plastic deformation of surface layers. The main types of surface structures are the strong deformed these layers and formation of a fragmented microstructure with a submicron size of fragments. The fragmented microstructure and its microdiffraction pattern are shown in Fig.3. In both cases the deformation on sliding surface takes place at the meso-scale level. The wear becomes catastrophic both for the very hard (TiC-Me) and for comparatively plastic (austenitic steel) material; in these cases the wear particles are comparable to thickness of degraded layer.

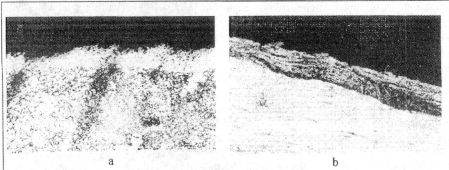

a   b

Fig.2. Microstructures of surface layers under severe friction conditions:
a - TiC - based composite; b - austenitic steel.

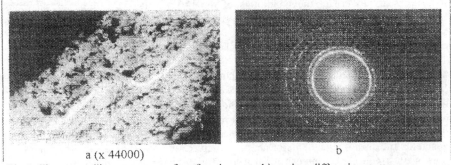

a (x 44000)   b

Fig.3. Fine crystalline structure of surface layer and its microdiffraction pattern.

To raise the wear resistance of heterogeneous materials, it is necessary to increase the heat resistance of the composite components. For this purpose the composites have been synthesized, in which the CoCr alloy was the connecting phase (metal matrix). Indeed, wear resistance and carrying capacity of the TiC-CoCr composites is vastly increased. However, at high (more than 60 MPa) pressure, some samples obtain the cracks at the sliding surface with forming a quasiperiodic structure (Fig.4a), while in the others, which contain the copper in the matrix, the cracks are formed by the action of plastic deformation of surface (Fig. 4b). The cracks were increased and coalesced with time. Finally, this process is characterized by disintegration of the contact area. The disruption of the surface structure can't be explained by compression or extension stresses, because our tribotechnical investigations were carried out at the friction machine with endface sliding bearing. Tangential stresses must be equal in this case and therefore there are no the reasons to generate the periodic cracks.

It is possible that the formation of the cracks in the first case is a result of the generation the elastic waves. It has been shown previously in our article [3], that the amplitude of these waves raise as an exponential function under certain conditions with time. Therefore the explosive stress in the surface area is formed, which creates surface cracking. In the second case the strengthening of surface layer changes a process of deformation and fracture, which becomes dependent from the correlation of strength properties of surface and sub-surface layers. Investigations, which were carried out by V.Panin, have shown that the relaxation of the stress at the boundary between hard surface and sub-surface layers causes the quasiperiodic micro-cracks formation at the surface. In this case the bands of localized deformation are formed in the internal volumes, which accelerate the fracture process of the friction surface.

It is necessary to create the conditions of the energy dissipation to avoid the stress concentrations, which are accompanied by the localization of deformation in the surface layer. This is possible in the case that non-homogeneous structure will be produced. Moreover, the characteristic size of non-homogeneities must be comparable with the scale of plastic deformation in friction. As noted above, this deformation is realized at meso-scale levels. It is necessary to note that the co-ordinated motion of the meso-elements in the system is possible. This motion generates the meso-level of deformation structure of the other order. It is therefore possible to reduce the intensity of wear, if the hierarchy of dissipative structures (multilevel system) will be created. These structures will adequately dissipate friction energy, both under the stationary mode of friction, when the microplastic deformation takes place, and under the catastrophic wear, when the conglomerates of grains are involved in the plastic flow.

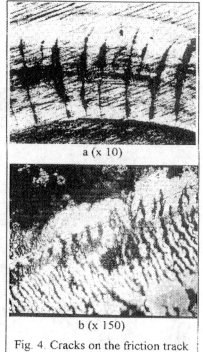

a (x 10)

b (x 150)

Fig. 4. Cracks on the friction track of TiC-CoCr (a) and TiC-Cu-CoCr (b) composites.

The heterogeneous CM with multi-level dissipative structure have been made from the TiC-CuFe and TiC-Cu(100Mn13) granules by the method of impregnation with bronze. Beforehand, the TiC-Me powder was compacted. An additional point to emphasize is that TiC-Me granules have been produced by the self-propagating high-temperature synthesis (SHS) with the following grinding in the ball mill. The resulting hard granules had the micro-heterogeneous structure. The composites contain 40 volume percent of the hard powder (TiC-Me granules) that ensures the elastic contact at the friction surface. Fig. 5 shows the microstructure of these materials, which have the multi-level dissipative structure. It has been found that the composites withstand high compression stresses without noticeable plastic deformation. These stresses are much greater then the friction stresses.

The hard particles of this heterogeneous composite have the elastic contact with mating surface during friction and therefore its wear is similar to the wear of the micro-heterogeneous materials considered in our articles earlier. But some comparative tribotechnical properties of the micro-heterogeneous CM of various TiC content are presented in this report. In the following we use these CM granules as the hard filler.

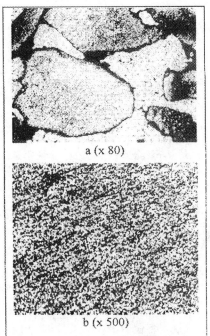

a (x 80)

b (x 500)

Fig. 5. Structure of macroheterogeneous material (a) and microstructure of hard particles TiC-CuFe (b) which contain in this material.

The tests have been carried out with friction sliding of shoe on roller. The specimen was dipped into the oil. In tests, the sliding velocity was 0.5 m/sec, the pressure was varied in steps up to 100 MPa, and the friction way was 700 m. The results are presented in Table 1. It can be seen from Table 1 that the width of the friction track decreases with reduction in the content of copper both for composites with iron and for composites with Gadfild's steel. The width of the friction track decreases with reduction in the content of the connective metal too.

This is because the hardness is increased and therefore the wear is increased. However the wear resistance of more soft composites which contain the iron in the matrix is significantly less then the wear of composites which contain the Gadfild's steel. An investigation of microstructure of the friction surface (Fig.6) has shown that the wear of TiC-Cu(100Mn13) composites is connected with adhesive wear.

Table 1. The main tribotechnical characteristics of micro-heterogeneous CM

№ п/п	Composition, vol.%	Friction coefficient	Width of the friction track, mm	Temperature of oil, K	Hardness, MPa
1	TiC-50%CuFe	0,13	0,910	425	6200
2	TiC-40%CuFe	0,13	0,844	451	8800
3	TiC-50%Cu +(100Mn13)	0,11	0,938	429	7850
4	TiC-40%Cu +(100Mn13)	0,12	0,869	443	10800
5	TiC-35%Cu +(100Mn13)	0,12	0,858	436	12100

Probably, this is connected with low value of shear-stability of Gadfild's steel, which causes the large plastic micro-deformation. Then the deformation goes from micro-level to meso-level, thus increasing the wear. The rough structure of friction surface testifies that the meso-volumes are involved in the wear process. However, test have

confirmed our earlier results about high level of
the tribotechnical characteristics of composites
containing titanium carbide.

An investigation of the relationship between
tribotechnical characteristics and the size of the
TiC-Me granules was carried out to determine the
optimum structure of macro-heterogeneous
composites. Three parties of samples with hard
particles of various sizes - small in size, large in
size, and with all sizes in the interval 0.2 - 1.0 mm,
were prepared. In the latter case, the distribution
of particle's sizes corresponded to a Gauss
distribution. The results of the tribotechnical
investigations of macro-heterogeneous materials
are presented in Table 2. From investigations it
follows that the average temperatures of oil
surrounding friction contact and friction
coefficients are almost the same for all materials.
In addition, these characteristics agree with
characteristics of stellite containing material very
closely. Some differences are observed in the
width of friction tracks. The material with Gauss
distribution of TiC-Me granules, had a smaller
width of friction tracks than materials with
particles of other sizes.

It is probable that a greater wear of the composite
with large particles is determined by the nature of
deformation of plastic component. This
deformation takes place at the high meso-level
due to great sizes of hard particles. The scale
level of plastic deformation decreases with the
decrease of the particle sizes. In this case the

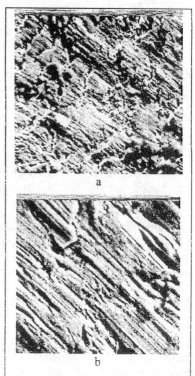

Fig. 6. Friction surfaces of TiC-
Cu(100Mn13) microheterogeneous
material (a) and TiC-
Cu(100Mn13) hard particles (b)
which obtained from this material.
(x 340)

disastrous deformation takes place at high pressures. Besides, the size of hard phase
particles influences the distribution of temperature in surface layer. It exerts some
influence on the strength characteristics, deformation, and wear of material. Decrease of

Table 2. Influence of the granule size to friction parameters of macroheterogeneous
composite

№ п/п	Granule size of TiC-Cu(100Mn13), mm	Friction coefficient	Width of the friction track, mm	Temperature of oil, K
1	0 - 0,4	0,11	0,894 ± 0,023	433
2	0,7 - 1,0	0,11	0,981 ± 0,027	428
3	0,2 - 1,0 (Gauss distribution, without quenching)	0,10	0,613 ± 0,017	426
4	0,2 - 1,0 (Gauss distribution, quenching)	0,11	0,849 ± 0,033	431
5	Composite for drilling bits of usual type	0,10	0,858 ± 0,020	422

size of particles tend to increase an area of the interphase boundaries, resulting in an improvement of heat exchange in macro-heterogeneous system.

It should be added to the aforesaid that the δ - phase, which is contained in the bronze, increases the wear resistance of the macro-heterogeneous material, because the supplementary structural level of deformation is "switch on". It dissipates energy which is supplied to the surface during the friction. Moreover, the tests of macro- and micro-heterogeneous composites showed that the anti-friction properties of composites containing the bronze are the best because the bronze is the solid lubricant.

Notice that the adhesive wear is observed at the TiC-Cu(100Mn13) granules surface (Fig.6b) despite the fact that tribotechnical properties of this macro-heterogeneous materials are very good. It testifies that the wear of the composites with TiC-Cu(100Mn13) particles by the high pressure has the extreme nature. Heavy wear is determined by low shear stability of Gadfild's steel, slow heat removal, and changes of the structure levels of deformation. When the TiC-Cu(Fe) composite are used as a filler the adhesive wear was much less than in the above case (Fig. 7). In this case the wear proceeds on the micro-scale level as judged by the relief of friction surface. Hence the wear tests have shown that the macro-heterogeneous composites containing TiC-Me filler are very promising for

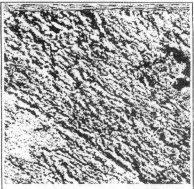

Fig.7. Friction surface of TiC-CuFe macroheterogeneous composite. (x 340)

heavy-duty friction units. Macro-heterogeneous materials of both composites have been processed according to the technology commonly used for the production of plain bearings for the drilling bits. Industrial tests have been made at the plant laboratory and at the oil fields of Tartaria and Western Siberia.

The laboratory stand provided the pressure more then 120 MPa, the sliding velocity of the plain-bearing unit was 0.57 m/sec, the test time was 55 hour. A peculiarity of this tests is that the pressure in the friction zone was variable. It was caused by geometry of bit's surface. A wear resistance of composites was determined by the radial wear of the sliding bearings (Table 3).

Table 3. Wear of drilling bit bearings when testing on stand

Number of the drilling bit section	Wear of the bearing elements			Commentary
	Slide bearing		Journal	
	mm	μm/km	mm	
1	0,12	$1,05 \cdot 10^{-2}$	0,27	There are no any
2	0,02	$1,76 \cdot 10^{-3}$	0,06	volumetric fracture
3	0,02	$1,76 \cdot 10^{-3}$	0,06	

It should be noted that the capacity for work of the experimental drilling bit was preserved after 55 hour whereas drilling bits of usual type were disrupted.

Tests on the oil-fields were carried out under axial loads on drilling bit 140 - 190 kN, that corresponds to a pressure at the interface 70 - 100 MPa, the sliding velocity of bearing was 0.7 - 1.0 m/sec. The conditions of the bit exploitation did not allow to trace a wear of bearings. Stability of drilling bits was therefore defined by the working time of an instrument before wedging one of the bit sections. In Table 4 results of tests of experimental drilling bits in Western Siberia are present. These results show that the average stability and drilling footage of the experimental drilling bits is comparable to the ones of usual type. However the latter contain deficit materials: tungsten, cobalt, and nickel. Smaller stability of the drilling bits containing TiC-Cu(100Mn13) may be stipulated by reasons, about which we spoke above when discussing a wear of composites with different fillers. A similar tests was made in Tartaria.

Table 4. Stability of drilling bits when testing in Western Siberia

N	Composition of CM filler of experimental slide drilling	Average drilling footage, m	Stability of drilling, h
1	TiC-CuFe	241	42
2	TiC-Cu(100Mn13)	101	16
3	Stellite (drilling bits of usual type)	272	39
		222	39
		150	24

Therefore on the grounds of the experimental-industrial tests in different geological conditions, it is possible to conclude that the macro-heterogeneous materials with TiC-CuFe hard granules possess high wear resistance and can be an alternative to the existing materials which contain the deficit components - tungsten, cobalt and nickel.

The using of the composite materials with antifriction matrix and large hard particles having the heterogeneous structure makes possible the redistributing of loads in the material, so that the soft component does not feel the external influences, but gives only lubricating of friction zone. Besides, internal microstructure of hard particles raises a limit of the microplastic deformation and does not allow it to go over to the meso-scale level characterized by the disastrous wear.

3. Conclusions

As is evident from the foregoing, the macroheterogeneous composites can find application in heavy-duty friction units working at pressure up to 100 MPa and in lubricant-deficient conditions. For these materials, the efficient heat-removal is accomplished by the thermal conductivities of both metal matrix and hard granules. Method of self-propagating high-temperature synthesis makes it possible to obtain TiC-Me composites with different phase constitution of metal matrix. Thus, we can obtain the materials with high thermal conductivity and high strength.

References

[1] A.V. Kolubaev, V.V. Fadin, V.E. Panin. AN INVESTIGATION OF THE WEAR RESISTANCE OF COMPOSITES CONTAINING TITANIUM CARBIDE.-Izvestia Vuz. Fizika, 1992, N.12, P. 64-68.

[2] S.Yu. Tarasov, A.V. Kolubaev. STRUCTURE IN SURFACE FRICTIONAL LAYERS ON 36NKhTYu ALLOY. Izvestia Vuz. Fizika, 1991, N.8, P. 9-12.

[3] V.L. Popov, A.V. Kolubaev. THE GENERATION OF SURFACE WAVES IN EXTERNAL FRICTION OF ELASTIC SOLIDS. Pisma v GTF 21, 1995, N.19, P. 91-94.

CREEP OF PARTICLE REINFORCED METAL MATRIX COMPOSITES

R. Pandorf, C. Broeckmann

Institute for Materials, Ruhr-University Bochum, 44780 Bochum, Germany

Abstract

The creep behaviour of the particle reinforced aluminium alloy 6061 is investigated using experimental and numerical techniques. In order to describe the influence of ceramic particles on the overall creep curve of the composite an idealised unit cell model representing a periodic arrangement of particles is generated. The particles are assumed to react linear elastically. Particle cleavage is taking into account. For the matrix an additive viscosity hardening law coupled with an isotropic damage variable is used. The results show that the fracture time is strongly influenced by the presence of particles. Qualitative good agreement with experimental data is found.

1 Introduction

Particle reinforced metal matrix composites (PMMC's) combine advantageous properties of metals like high ductility and toughness with properties known from ceramics like high strength and stiffness. Composites based on aluminium alloys are very attractive in the field of light weight constructions as well as in automotive industries. Some applications are brake disks, drive shafts and pistons. Another benefit of MMC's is their potential for high temperature applications [1,2]. In this context the creep behaviour of PMMC's is of particular interest. This paper deals with the aluminium alloy AA6061, reinforced with Al_2O_3-particles.

2 Experimental Investigations

The MMC under consideration was produced by infiltrating the aluminium melt with Al_2O_3-particles [3]. The ingots were extruded to circular bars of dimension 17x17 mm^2. All specimens were tested in the T6 condition. The chemical composition of AA6061 matrix material and the microstructure of the composite are given in fig.1. The particle's volume fraction is f_P=22 %. The average size of particles was determined to d=14 µm. Uniaxial creep tests under constant stress were carried out at a temperature T=300 °C. The creep curves of composite and pure matrix are characterised by a very short primary creep stage and a large tertiary creep regime. Extensive stationary creep was not observed. Fig. 2 shows the stress rupture curve of the composite and the pure matrix material. The time to fracture of the MMC is up to a magnitude of order shorter as compared to the unreinforced AA6061. In order to find the reason for this effect longitudinal sections of crept specimens were prepared after interrupting the test in the tertiary creep stage. Two different damage mechanisms were found in the reinforced material: damage due to particle cleavage (fig. 3a) and matrix damage by void initiation and growth (fig. 3b).

In the next paragraph a numerical model is presented which is used to simulate the influence of those two types of damage on the overall creep behaviour.

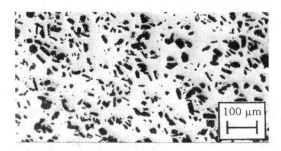

Chemical composition of AA6061 [wt.%]	
Mg	0.93
Si	0.69
Cu	0.28
Cr	0.1
Fe	0.2
Zn	0.02
Mn	0.01
Al	Balance

Fig 1: Microstructure and chemical composition of Al_2O_3-particle reinforced AA 6061

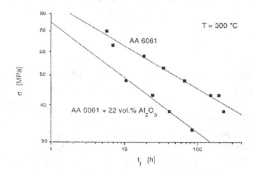

Fig 2: Stress dependence of time to fracture for the reinforced composite and the pure matrix material

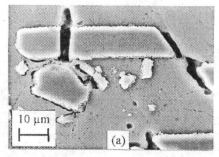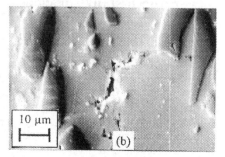

Fig 3: Micrographs showing damage in the particle reinforced AA 6061: (a) broken Al_2O_3-particles, (b) cavities and voids in the metal matrix

3 Numerical Model

The creep behaviour is numerically simulated on a microscopic scale using the finite element program CRACKAN described in detail in [4]. A multiphase model is used for the simulation of the microstructure. The average size of the hard phases in the material under consideration is 14 μm. At this particle size a continuum mechanical description is assumed to be valid [5]. The particles are modelled linear elastically. In order to simulate particle cleavage, a critical continuum mechanical value has to be defined. Most of the Al_2O_3 particles cleave perpendicular to the loading direction along a crystallographic plane (see fig. 3a). Cleavage occurs due to a critical local maximum principle stress.

The theoretical foundations of the viscoplastic material law used for the simulation of the matrix are briefly presented below. The total strain rate is subdivided into an elastic and a plastic part:

$$\dot{\varepsilon} = \dot{\varepsilon}_e + \dot{\varepsilon}_p \tag{1}$$

With the elasticity matrix $D(E,v)$ the relation between the rate of stress and the rate of elastic strain can be described:

$$\dot{\sigma} = D\dot{\varepsilon}_e \tag{2}$$

For the viscoplastic term in equation (1) an additive viscosity hardening law (see [6]) coupled with damage is used:

$$\dot{\varepsilon}_p = \frac{3}{2} \left(\frac{\sigma_{eq} - R(\varepsilon_p) - k}{K_A (1 - \omega(\sigma_{eq}))} \right)^{N_a} \frac{s}{\sigma_{eq}} \tag{3}$$

where k, N_a and K_A are material parameters, σ_{eq} is the effective von Mises stress and s the deviatoric part of the stress tensor. The function $R(\varepsilon_p)$ represents strain hardening and can be understood as a resistance of the material against deformation. The damage parameter ω represents the damage state of the material. In the uniaxial case, the expression $A_0/(1-\omega)$ can be interpreted as the current cross-sectional area A. It is assumed that $\omega = 0$ when the material is in its undamaged state and $\omega = 1$ at rupture. The evolution equation for the hardening term is given by:

$$R(\varepsilon_p) = Q_1 \varepsilon_{eq,p} + Q_2 (1 - \exp(-b\varepsilon_{eq,p})) \tag{4}$$

where Q_1, Q_2 and b are material parameters to be experimentally determined. The equivalent plastic strain is given by:

$$\varepsilon_{eq,p} = \sqrt{\frac{2}{3} \varepsilon_{p,ij} \varepsilon_{p,ij}} \tag{5}$$

In this study it was considered that damage can be neglected during primary creep and strain-hardening is saturated during tertiary creep. Therefore the determination of the material parameters for hardening and damage can be separated.

The parameters for the viscosity hardening law (k, N_a, K_A, b, Q_1 and Q_2) were determined by hardening and relaxation tests under the assumption $\omega = 0$ [6]. An evolution equation for the damage parameter ω was developed in 1958 by Kachanov (see also Ref. [7]):

$$\dot{\omega} = \left(\frac{\sigma_{eq}}{B_0 (1 - \omega)} \right) \tag{6}$$

where B_0 and r are two material parameters. Integration leads to the following equation:

$$-\frac{1}{r+1}(1-\omega)^{r+1} = \left(\frac{\sigma_{eq}}{B_0} \right) t + c \tag{7}$$

where c is the constant of integration. With the boundary conditions ($\omega(t=0)=0$ and $\omega(t=t_f)=1$) it is possible to determine the time to fracture t_f:

$$t_f = \frac{1}{r+1} \left(\frac{\sigma_{eq}}{B_0} \right)^{-r} \tag{8}$$

Inserting of equation (8) in equation (7) leads to the evolution equation for the damage state:

$$\omega = 1 - \left(1 - \frac{t}{t_f(\sigma_{eq})} \right)^{\frac{1}{r+1}} \tag{9}$$

The parameters for B_0 and r were determined by fitting equation (8) against the experimental data given in fig. 2.

4 Numerical Results

In order to investigate the influence of the reinforcement particles on the damage behaviour of the matrix, a periodic array of hard phases was modelled in a uniaxial macroscopic stress field σ_0 (see fig. 5). The hard phases are assumed to react linear elastically and the matrix viscoplastically with isotropic strain hardening coupled with damage. All material parameters used for calculation are given in table 1. All calculations in this paper were considered as plane strain situations. Fig. 6 shows the calculated creep curves for the pure matrix material (AA 6061) in comparison to the composite (AA 6061 + 20% Al$_2$O$_3$). In the primary creep stage the strain rate of the pure matrix material is higher than of the composite. Whereas the pure matrix material has a distinctive secondary stage, the composite shows a very early transition to the tertiary creep region. Time to fracture of the pure matrix is remarkable larger compared to the composite. This agrees well with the experimental observation (see fig. 2).

Constants for elastic behaviour				Constants for damage	
E_{matrix} [MPa]	ν_{matrix}	$E_{hard\ phase}$ [MPa]	$\nu_{hard\ phase}$	B_0 [MPa]	r
51300	0.42	400000	0.22	1026	4.7
Constants for viscoplastic behaviour					
N_a	k [MPa]	K_a [MPa]	Q_1 [MPa]	Q_2 [MPa]	b
4.55	3.0	1712	657.0	12.21	614.1

Tab. 1: Parameters used for numerical calculations

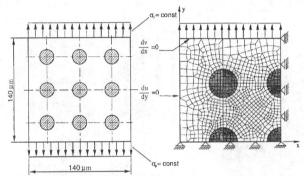

Fig 5: Unit cell with nine hard phases and reduced mesh for numerical calculations

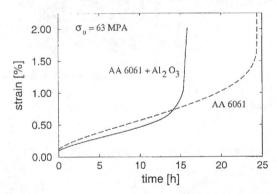

Fig 6: Calculated creep curve for the pure matrix material (AA 6061) in comparison to the composite (AA 6061 + 20% Al₂O₃)

In a second study the influence of particle cleavage on the creep and damage behaviour was investigated. A microstructure containing 19 vol.% hard phases was modelled (fig. 7). The higher the aspect ratio of hard phases is the more they tend to fail due to cleavage. The reason is the stress concentration in the centre part of the particle (fibre effect, see [8]). An aspect ratio of 10 was chosen for the numerical calculations in this study. The stress concentration in the hard phases is due to the higher Young's modulus of the Al_2O_3 and the very local stress fields at dislocation pileups at the matrix/particle interface. This high local stress in the particle is generally underestimated by continuum mechanical methods [9]. Therefore, a relative low critical normal stress for particle cleavage was chosen ($\sigma_{cl,c}$=250 MPa). The evolution of particle failure is monitored in fig. 8. During the creep test the loading of the particles increases until the first hard phase reaches the critical value $\sigma_{cl,c}$. Particle cleavage leads to redistribution of stresses and an increasing loading of the other particles. The effect of the resulting particle cleavage on the stress distribution in the metal matrix was investigated in earlier studies [10].

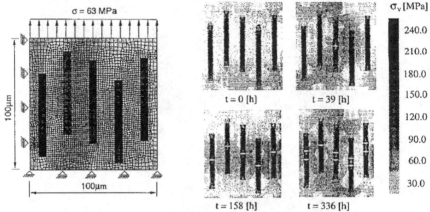

Fig 7: FE-model for a two-phase material with elongated hard phases oriented in loading direction

Fig 8: Distribution of von Mises stress and evolution of particle failure after different time steps

The influence of particle cleavage on the overall creep curve is shown in fig. 9. The comparison of the two curves shows, that particle cleavage leads to an immediate increase of the macroscopic strain. Afterwards the overall strain of the composite is controlled by the viscoplastic behaviour of the matrix.

In fig. 10 the distribution of matrix damage for a calculation without consideration of particle cleavage (fig. 10a) and with particle cleavage is presented (fig. 10b). Damage starts at the end of particles where the stresses are relatively high. In case of particle cleavage regions with high damage are developed between cracked particles and at the ends of adjacent particles.

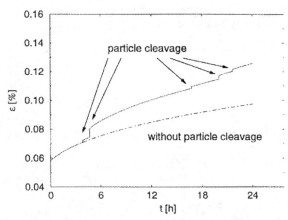

Fig 9: Comparison of calculated creep curves

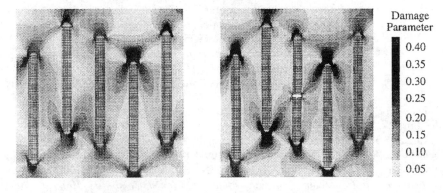

Fig 10: Distribution of local damage

5 Conclusions

- The creep behaviour of particle reinforced aluminium alloy 6061 was modelled. The particles were assumed to react linear elastically, while the matrix was described by a viscoplastic material law taking into account isotropic hardening and damage. Particle cleavage was explicitly simulated.
- A unit cell model was developed, representing a periodic arrangement of globular particles.
- The time to fracture is strongly influenced by the particles. The experimental observation that the composite fails much earlier than the pure matrix is supported by the numerical study.
- A second numerical study shows, that particle cleavage leads to an immediate increase of the macroscopic strain.

- Matrix damage is concentrated at the ends of elongated particles and near microcracks caused by broken particles.

References

[1] Whitehouse, A.F., Clyne, T.W. *Cavity formation during tensile straining of particulate and short fibre metal matrix composites.* Acta Metallurgica, No 6, **41**, (1993) 1701-1711

[2] Corbin, S.F. and Wilkinson, D.S. *Microgeometrical effects on the elastoplastic behaviour of particle reinforced MMCs.* Proc. 9th Int. Conf. on Comp. Mat., Madrid, (1993) 295-302

[3] Degischer, H.P. *Schmelzmetallurgische Herstellung von Metallmatrix-Verbundwerkstoffen,* Erweiterte Vortragstexte eines Fortbildungsseminars der DGM, Institut für Werkstoffkunde und Werkstofftechnik der TU-Clausthal, 1994

[4] Broeckmann, C. *Bruch karbidreicher Stähle - Experiment und FEM-Simulation unter Berücksichtigung des Gefüges.* Dissertation, Ruhr-Universität Bochum, Fortschr.-Ber. VDI, Reihe 18, Nummer 169, Düsseldorf, 1995.

[5] Fischmeister, H.F., Karlsson, B. *Plastizitätseigenschaften grob-zweiphasiger Werkstoffe.* Zeitschrift für Metallkunde, No 5, **68**, (1977) 311-327

[6] Lemaitre, J., Chaboche J.-L *Mechanics of solid materials.* Cambridge University Press, 1994

[7] Leckie, F.A., Hayhurst, D.R. *Constitutive equations for creep rupture.* Acta Metallurgica, **25**, (1977) 1059-1070

[8] Brechet, Y., Embury, J. D., Tao, S., Luo, L. *Damage Initiation in Metal Matrix Composites,* Acta Metall. Mater., No 8, **39**, (1991) 1781-1786

[9] Cleveringa, H.H.M., van der Giessen, E., Needleman, A. *Comparison of Discrete Dislocation and Continuum Plasticity Predictions for a Composite Material,* Acta mater. No 8, **45**, (1997) 3163-3179

[10] Broeckmann, C., Pandorf, R. *Influence of Particle Cleavage on the Creep Behaviour of Metal Matrix Composites* Computational Materials Science, **9**, (1997) 48-55

Application of Mathematical model for Stress Distribution Calculation for Prediction of Thermal Shock Behaviour of Silicon-Based Specimen

T. Volkov-Husović*, R. Jančić*, M. Cvetković*, D. Mitraković*, Z. Popović*

*Faculty of Technology and Metallurgy, Karnegijeva 4, 11000 Belgrade, Yugoslavia

Abstract

Mathematical model for stress distribution calculation in specimen during thermal stability testing was presented in this paper. Mathematical model was applied to silicon-based ceramics: SiC, Si_3N_4 i Si_3N_4-ZT. Fracture resistance parameters, R and R' were also calculated. The obtained results have shown that application of stress distribution calculation combined with fracture resistance parameter R analysis allowed good prediction of the thermal shock behavior of the ceramic specimen.

1. Introduction

Ceramic refractories in furnaces and ovens are exposed to severe thermal conditions and are frequently subjected to rapid changes in temperature that cause high thermal stress. As those materials are brittle, the thermal stress can cause cracking, which degrades the strength of the material, and also leads to increased wear by erosion. Under such temperature conditions it is unreasonable to expect to be able to eliminate cracking altogether, and approach is to attempt to limit the degree of cracking, so that degradation in strength is minimized.

The problem of thermal shock behavior prediction of the material can be solved on two basis : heat transfer conditions and/or fracture mechanic concept. Hasselman [1,2] proposed thermal resistance parameters, based on fracture-mechanic concepts, which could be classified into two types : fracture resistance parameters (R and R') and damage resistance parameters (R''', R'''' and R_{st}). The difference between thermal shock and other loading is the "contactless" generation of transient thermal stress fields inducted by transited temperature fields-hence, non isothermal conditions prevail. Crack propagation occurs when the transient thermal stress field is high enough that the critical stress intensity factor at the tip of at least one of the preexisting crack defects is exceeded. This is the fracture mechanics understanding of thermal shock. Because of the transient and nonisothemal conditions it is not necessarily certain that is measurement of common fracture mechanical properties of ceramics such as fracture toughness, strength and Weibull modulus under pure thermal shock loadings yields the same result as pure mechanic, isothermal measurements.

2. Mathematical model for stress distribution calculation

In order to determine temperature distribution, we start from the heat diffusion equation. Assuming that heat transfer is one-dimensional and that the coefficient of the diffusion is constant, heat diffusion equation can be written in the form [3]:

$$\frac{\partial \theta}{\partial t} = \kappa \frac{\partial^2 \theta}{\partial x^2}, \tag{1}$$

where t is the time, x is the distance, and $\kappa = k/c\rho$ is the thermal diffusivity where k is the thermal conductivity, ρ is the density and c is the thermal capacity. θ denotes relative temperature given by

$$\theta(x,t) = \frac{T(x,t) - T_f}{T_o - T_f}, \tag{2}$$

where $T(x,t)$ is the temperature, T_o is the initial temperature of the heated specimen, while T_f is the temperature of the medium. Boundary conditions of the equation (1) are given by

$$T(x,t) = \begin{cases} T = T_o, & t \le 0, -L \le x \le L, \\ T(-L,t) = T(L,t) = T_f, & t > 0. \end{cases} \tag{3}$$

In our analysis, we have used standard solutions for the temperature distribution in the plain specimen obtained using the Fourier series given by [3]:

$$T(x,t) = T_f + \frac{4}{\pi}(T_o - T_f)\sum_{j=0}^{+\infty} \frac{\sin\left[\frac{(2j+1)\pi}{2}\left(1 + \frac{x}{L}\right)\right]\exp\left\{-\left[\frac{(2j+1)\pi}{2L}\right]^2 \kappa t\right\}}{2j+1}, \tag{4}$$

By knowing the temperature distribution, one can determine the stress distribution using the relation

$$\sigma(x,t) = \frac{\alpha E}{1-\upsilon}\left(-T(x,t) + \frac{1}{2L}\int_{-L}^{+L} T(x,t)\,dx\right), \tag{5}$$

where α is thermal expansion coefficient, υ is Poisson ratio and E is Young modulus. Combining equations (4) and (5), following expressions for the stress distribution can be obtained [4]:

$$\sigma(x,t) = \frac{\alpha E}{1-\upsilon}\left\{-T(x,t) + T_f + \frac{8}{\pi^2}(T_o - T_f)\sum_{j=0}^{+\infty}\frac{1}{(2j+1)^2}\exp\left[-\left(\frac{(2j+1)\pi}{2L}\right)^2\kappa t\right]\right\}. \tag{6}$$

Equations (4) and (6) fully describe the temperature distribution and stress distribution of the specimen, respectively.

3. Materials

Silicon-based ceramics: SiC, Si_3N_4 i Si_3N_4-ZT were used in this paper. Relevant thermal and mechanic properties are listed in Table I.

Table I. Thermal and mechanic properties of materials [5]

Property	A(SiC)	B(Si_3N_4)	C(Si_3N_4-ZT)
Density ρ (g/cm^3)	3.10	3.14	2.45
Young modulus E (GPa)	410	280	165
Strength σ (MPa)	460	500	200
Thermal expansion coefficient α ($^{\circ}C^{-1}$)	4.02×10^{-6}	3×10^{-6}	3×10^{-6}
Heat capacity c (J/kgK)	670	1000	1000
Thermal conductivity k (W/mK)	125	15	1.5
Poisson ratio υ (-)	0.24*	0.21**	0.27*
Medium temperature T_f ($^{\circ}C$)	20	20	20
Specimen temperature T_o ($^{\circ}C$)	1000	1200	1200

* Ref. No. [5], ** Ref. No. [6]

High-frequency thermal fatigue was prodused using a plasma jet [5]. Thermal shock behavior of specimen, measured by number of cycles is presented in Table II.

Table II. Thermal shock behavior of materials [5]

	A (SiC)	B (Si_3N_4)	C (Si_3N_4 -ZT)
Number of cycles	1000	1000	3

As it is presented in Table II. sample A and B, which have a higher thermal conductivity, have good resistance to thermal shock. The specimen of low density, low conductivity silicon nitride (sample C-Si_3N_4-ZT) failed on testing after only 3 cycles [5].

4. Results and Discussion

Presented mathematical model was applied to temperature and stress distribution calculation in silicon based ceramics. Results of the temperature distribution are presented in Fig. 1-3 while the results of stress distribution are presented in Fig 4-6. Location of maximum tensile stress depends on the size of the hottest part. The maximum tensile stress initially locates in the midplane is then located in the bottom part as temperature increase in the hottest part during subsequent cycles. On cooling, the presence of the tensile stress in the impinged zone as a result of restrained contraction of the hottest part, as observed on quench test. Conversely, compressive stresses are produced in the neighboring parts.

Temperature distribution (Fig. 1-3) shows that hotest parts are in the midplane of the samples. Region of the lower temperatures is not similar for the samples. For the sample C "hot" region is much wider than for the samples A and B. Cooling of sample C is the slowest.

Stress analysis shows that stress field includes tensile and shear stress components. The highest values for tensile and shear stress are observed with the A (SiC) specimen. Figures 4 and 5 for the specimen A and B are similar, as their thermal shock behavior.

Results of the stress analysis partially explain the thermal shock behavior of the specimens. Failure of the specimen C is attributed to the presence of the high tensile stresses, whereas survival of the specimen B results from the presence of low tensile stresses when compared with the reference flexural strengths. Location of cracks of the in C specimens is attributed to the presence of maximum tensile stress initially on the inside and then at the surface of the specimens.

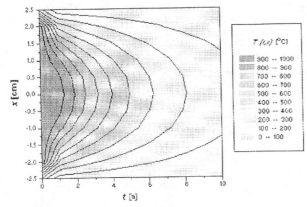

Figure 1

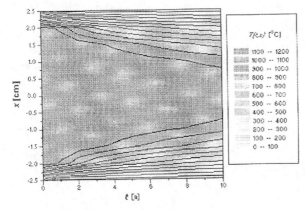

Figure 2

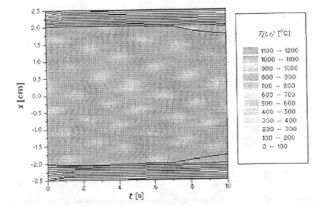

Figure 3

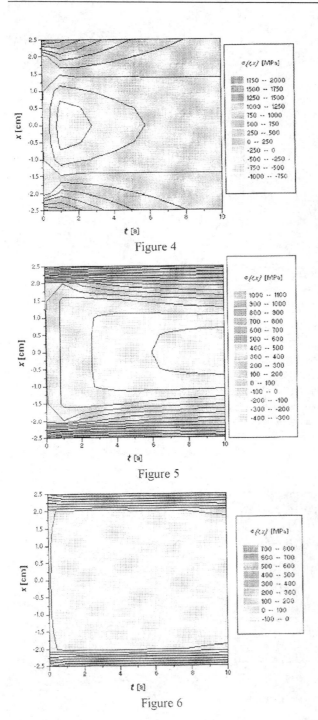

Figure 4

Figure 5

Figure 6

5. Fracture Resistance Parameters

Many theoretical treatments have been presented to describe the thermal shock resistance of brittle ceramic materials in terms of material properties [1,2,7,8]. Resistance parameters can be classified into two types: the thermal shock damage resistance parameter and the thermal shock fracture parameter. In this paper fracture resistance parameters will be used. The thermal shock fracture resistance parameter is based on those material properties which affect the fracture nucleation, and it is represented using equations listed in Table III.

Table III. Fracture resistance parameters

Fracture resistance parameter	Unit	A (SiC)	B (Si_3N_4)	C (Si_3N_4 -ZT)
$R = \sigma(1- \nu)/\alpha E$	°C	212.04	473.76	298.0
$R' = k R$	W/m	26505	7106	477.85

Comparing to the number of cycles (Table II.), results from Table III shows that fracture resistance parameter R' describes shock behaviour more adequate. The lowest value of R' parameter is shown for sample C which had failed after 3 cycles. For the samples A and B fracture resistance parameter have much higher values. From the experimental results [5] specimen A exhibit no significant degradation, and have better thermal shock resistance, then sample B. Fracture resistance parameter for sample A had the highest value, and it is in good correlation with experimental results.

6. Conclusion

A simple mathematical model for stress distribution in ceramic specimen under thermal shock was presented in this paper. Results obtained from the presented model could be useful for :
- location and values of tensile stress,
- ratio of tensile and compressive stress and comparison to flexural strength of material.
Results of the stress analysis partially explain the thermal shock behavior of the specimens. Fracture resistance parameters for chosen materials were also calculated. Analysis of the stress distribution in specimen combined with the calculated fracture

resistance parameter R' can be used for prediction of thermal shock behavior of the ceramic material.

References

[1] *D.P.H. Hasselman*, UNIFIED THEORY OF THHERMAL SHOCK FRACTURE INITATION AND CRACK PROPAGATION IN BRITTLE CERAMICS, *J.Am.Ceram.Soc.* ,Vol. 52 , No.1, 600 (1969)

[2] *D.P.H. Hasselman*, THERMAL STRESS RESISTANCE PAREAMETERS FOR BRITTLE REFRACTORY CERAMICS, A COMPENDIUM, *Am.Ceram.Soc.Bull.* Vol 49, No 12, 1033 (1970)

[3] *R.S. Brodkey, H.C.Hershey*, TRANSPORT PHENOMENA, McGraw Hill, 1988

[4] *T.D.Volkov-Husović, M.Cvetković, D.Mitraković, K.Raić, Z.Popović*, MATHEMATICAL MODELING OF STRESS DISTRIBUTION CALCULATION IN CERAMIC SPECIMEN DURING THERMAL STABILITY TESTING, XLI Yugoslav conference of ETRAN, Zlatibor, 3-6 june 1997., Vol. 4, p. 442-444

[5] *J. Lamon, D.Pherson*, THERMAL STRESS FAILURE OF CERAMICS UNDER REPEATED RAPID HEATINGS, *J.Am.Ceram.Soc.* Vol 74, No 6, 1188 (1991)

[6] *R.W. Davidge*, 1979 Mechanical Behavior of Ceramics, Cambridge University Press, Cambridge, UK, Chap 8, pp 118-31

[7] *T.D.Volkov-Husović, R.M.Jančić, Z.V.Popović*, THERMAL SHOCK STABILITY OF CERAMIC MATERIAL: COMPARISON OF FRACTURE RESISTANCE PARAMETER WITH THE CRITICAL T VALUES $\Delta TC = \Delta TC(BI)$, Meeting of the European Society of Ceramics, Versailles, 22-26 june 1997, Euro Ceramics V, Part 1, Sessions 1A, 1C, 1D, 3, Trans. Tech.Publications, p. 603-607

[8] *T.D.Volkov-Husović, R.M.Jančić, Z.V.Popović*, THERMAL SHOCK STABILITY OF CERAMIC MATERIAL: DAMAGE RESISTANCE PARAMETERS AND CRITICAL FLAW SIZE, Meeting of the European Society of Ceramics, Versailles, 22-26 june 1997, Euro Ceramics V, Part 3, Sessions 6 -12, Trans. Tech.Publications, p. 1778-82.

Fabrication and microstructural evaluation of ZrB₂ ceramics by liquid infiltration

Sang-Kuk Woo, In-Sub Han, Byung-Koog Jang,

Doo-Won Seo, Ki-Suk Hong, Kang Bae, Jun-Hwan Yang

Energy Materials Research Team, Korea Institute of Energy Research,

71-2 Jang-dong, Yusong-Gu, Taejon 305-343, Korea

Abstract

In order to obtain high dense ZrB_2 ceramics, specimens were made by reaction melt infiltration. Reaction behavior, microstructure and mechanical property between infiltrated metals and reactants were investigated. Dense ZrB_2 composites could be fabricated by reaction sintering of molten zirconium with ZrB_2 preform. It has been found that the newly formed ZrB_2 is epitaxially bonded on the original zirconium diboride and ZrB_2 is present as rectangular shapes. When the B was added as reactaent, the average grain size of ZrB_2 was propertionally incresed with the additional content of B. The fracture strength of ZrB_2 composites is approximately 550MPa ~ 400MPa and the fracture toughness is $11.5MPa \cdot m^{1/2} \sim 8MPa \cdot m^{1/2}$ which is more excellent than the value of conventional sintered ceramic materials.

I. Introduction

Zirconium diboride ceramics have a high melting point, high strength, high corrosion resistance, high hardness, high electric conductivity, and excellent needed to attain these favorable properties as the actual component. It is known that their strength and corrosion resistance are very adversely affected by excessive porosity in sintered bodies. But it is diffcult to obtain fully dense body, although ZrB_2 powder is routinely hot-pressed at 2000°C or even lower temperature, with densification aids.[2] In order to overcome these problems, the reaction based processes has been studied. Reaction based processes as an alternative to conventional processing has been of general interest for a long time. Processes based on reactions between a porous solid and an infiltrating liquid phase share the near-net shape and near-net dimension capabilities of gas-phase reaction bonding, as well as the reduced processing temperature relative to solid-state sintering.[3,4] Also, they may have significant advantages in the processing rates and material densities that are achievable.

The prototype of liquid-phase reaction bonded materials is reaction-bonded silicon carbide(RBSC) which is typically fabricated by infiltrating molten silicon into a compacted body of the α-SiC filler and graphite.[5] Recently, the formation of zirconium diboride platelet reinforced ziconium carbide matrix materials by the directed reaction of molten zirconium was reported by Johnson et al.[6]

The purpose of this paper was microstructural evaluation of ZrB_2 composites prepared by reaction sintering. In particular, a shape, distribution and content relationships of the constituent phases will be described. In addition, the mechanical properties of composite specimens were investigated. Both bending and fracture toughness testing were used to characterize the mechanical properties.

II. Experimental Procedure

The raw materials were ZrB_2, B, B_4C(Hermann C. Starck, Berlin,

German) and Zr(Armco Products, Inc.,Ossing, New York, U.S.A). The ZrB$_2$ powders together with reactants and polyethyleneglycol(PEG,#4000) were mixed in a polypropylene bottle for 1hr in an acetone and dried at room temperature. The mixed powders were granulated by using a 60 mesh sieve and die-pressed to form pellets(1" inner diameter). The binder in the preform was burn-out in a electric furnace at 600°C, which removed greater than 99% of binder. The specimens were heated in graphite crucibles in a induction furnace to the 1900°C reaction temperature in vacuum.

The densities of the resulting specimens were measured by the Archimedes principle using water. The phases of specimens was identified using X-ray Diffraction of polished cross sections perpendicular to the growth direction using a Rigaku RTP 300RC. All specimens were examined by both secondary electron and backscattered electron imaging in a Philips XL-30. The flexural strength was measured under 4 point bending(12mm outer span, 6mm inner span) with a #1127 Instron.

III.   Results and Discussion

The optical microscopy images of a etched surfaces of ZrB$_2$ composite fabricated by reaction process were shown in Fig. 1. The images showed that the structure was dense and that there were few or no macro-pores. All materials appeared uniform and isotropic throughout cross sections observed in perpendicular to the growth direction. Figure 2 shows the BEI(back scattered electron image) of the ZrB$_2$-15vol% B and ZrB$_2$-15vol%B$_4$C   preform infiltrated by zirconium.   The white matrix is free zirconium, the gray parts show newly formed ZrC grains which are embedded in free zirconium, and the dark parts show original and newly formed ZrB_2. These final phases are similar to the observed result in the directed reaction of molten zirconium with boron carbide and boron. Also, the BEI micrographs have shown that ZrB_2 is present as rectangulars of 10-20μm in diameter (Fig.2(A,B)) These rectangular shape was different from the platelet shape reported by Johnson et al.[6]

X-ray diffractometry studies were conducted on flat polished sections, and shown in Fig.3. The X-ray analysis showed the presence of three phases, such as ZrB_2, Zr, and ZrC phase, in all the specimens investigated. No significant line broadening was observed in any of the X-ray diffraction peak. That is, the reaction mechanism of this systems seemed to be different from that of the directed reaction of molten zirconium with boron carbide, but seemed to be the same as that of the reaction bonded silicon carbide.[5] So, the reaction mechanism was deduced to precipitation of β-SiC from a solution of graphite in molten silicon. Such as, the reaction begins as soon as the molten zirconium is sufficiently fluid to allow it to rise through the compact, its viscosity being quite high initially but falling with increasing temperature. There is an exothermic reaction of molten zirconium with B and B_4C to form both ZrB_2 and ZrC, as previously reported.[6] This exothermic causes a local temperature rise at the dissolution sites and induces the activity gradient between B, C and Zr. The boron and carbon diffuses to locally cooler sites (original ZrB_2) where it becomes supersaturated in the zirconium. ZrB_2 and ZrC precipitate out to form the products, and these grains grow and deposit epitaxially on the original ZrB_2 by Ostwald ripening during processing.[7]

Table 1 shows the bulk densities and the porosities of ZrB_2 ceramics as a function of reactant contents in the preform. The bulk densities of specimens was $6.18 \sim 6.20 g/cm^3$, and giving a calculated porosity ratio of 0.5% to 0.25%. It suggested that there were almost no pores in the specimens. The average grain sizes of ZrB_2 and ZrC obtained by reaction sintering for 10 min with various contents of B and B_4C are shown in Table 2. As shown in Table 2, the average grain size of ZrB_2 slightly increased with additional content of B. This may be due to the heat of reaction increase upon the reaction Zr and B that might result in the increase of grain grow rate. On the other hand, the average grain size of ZrB_2 slowly decreased with addition of B_4C. It was thought that newly formed ZrC inhibited grain growth of ZrB_2 after all. But in the case of ZrC, the average grain size increased with further increase of B_4C contents.

The flexural strength and fracture toughness of the composites are summarized in Table 3. The strength was increased slightly to addition of 15vol.% reactants and then became dropped gradually. On the other hand, the fracture toughness of ZrB_2 composites slowly decreased with addition of B and B_4C. The flexural strength of ZrB_2 composites is approximately $550MPa \sim 400MPa$ and the fracture toughness is $11.5MPa \cdot m^{1/2} \sim 8MPa \cdot m^{1/2}$ which is more excellent than the value of conventional sintered ceramic materials.

IV. Conclusion

Dense ZrB$_2$ composites could be fabricated by the reaction sintering of molten zirconium with ZrB$_2$-B and ZrB$_2$-B$_4$C preform. Infiltration and the reaction are essentially completed in a few minutes. It was thought that the reaction mechanism was the infiltration followed by the chemical reaction. It was found that the ZrB$_2$ was present as rectangular shapes. It was also observed that three phases of ZrB$_2$, ZrC and Zr coexist when the reactive infiltration of fused Zr was taken place after adding B and B$_4$C to ZrB$_2$. The average grain size of ZrB$_2$ in the reaction-sintered composites increased slightly with an increase in the volume fraction of the B. But in the case of addided B$_4$C, average grain size of ZrB$_2$ is slowly decreased with addition of B$_4$C. The flexural strength and fracture toughness of the composites were showed upto 11.5 MPa $\cdot$ m$^{1/2}$ and upto 550 MPa.

References

1. T.Lundtrom, "Transition Metal Borides" pp351-376, in Boron and refractory borides, ed. Y.L.Matkovich, Springer-Verlag, NY, 1977

2. M.KInoshita, S.Kose and Y.Hamano, "Hot-Pressing of Zirconium Diboride with Binder Metals", J. Ceram. Assoc., 75 [3] (1967)

3. J.S.Haggerty and Y.-M.CHiang, "Reaction-Based Processing Methods for Ceramics and Composites", Ceram. Eng. Sci. Proc., 11[7-8] 757-781(1990)

4. Y.-M.CHiang, J.S.Haggerty, R.P.Messner, and C.Denetry, "Reaction-Based Processing Methods for Ceramic-Matrix", Am. Ceram. Sco. Bull., 68[2], 420-428(1989)

5. P. Popper, "The Preparation of Dense Self-Boned Silicon," Special Ceramics, pp209-19, Heywood, London, 1960

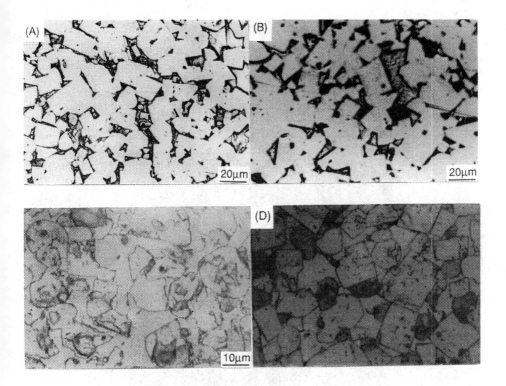

Fig. 1. Optical micrographs of ZrB$_2$ ceramics fabricated by the reaction of ZrB$_2$ preform with Zr melt for 10min. at 1900℃. The preform of specimen contained (A)10vol% boron, (B)20vol% boron, (C)10vol% B$_4$C and (D)20vol% B$_4$C

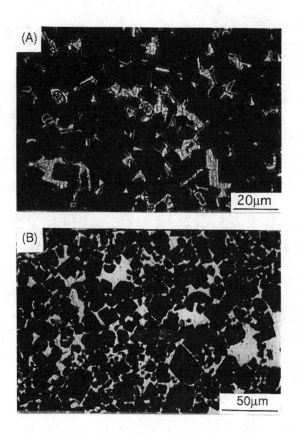

Fig. 2. Backscattered electron images of the ZrB₂ ceramics fabricated by the reaction of ZrB₂ preform with Zr melt. The preform of specimen contained (A)15vol% boron, (B)15vol% B₄C. The dark phase is ZrB₂, the gray phase is ZrC, and the white phase is Zr metal.

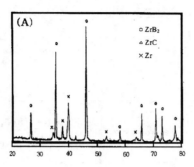

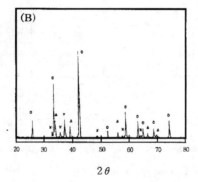

2θ

Fig. 3. XRD patterns of composite specimens fabricated by the reaction of ZrB$_2$ preform with Zr melt for 10min. at 1900℃. The preform of specimen contained (A)15vol% boron, (B)15vol% B$_4$C.

Table 1. Bulk Density and Porosities of ZrB$_2$ Composite as a Function of B and B$_4$C Content in the Preform.

B Content (vol.%)	Bulk Density (g/cm^3)	Porositiy(%)	B$_4$C Content (vol.%)	Bulk Density (g/cm^3)	Porositiy(%)
0	6.24	0.47	0	6.24	0.47
5	6.23	0.35	5	6.21	0.29
10	6.19	0.32	10	6.19	0.29
15	6.18	0.33	15	6.18	0.33
30	6.16	0.33	30	6.15	0.33

Table 2. Average Grain Size of ZrB$_2$ Composite as a Function of B and B$_4$C Content in the Preform.

B Content (vol.%)	Average Grain Size (μm)	B$_4$C Content (vol.%)	Average Grain Size (μm)
0	11.21	0	11.21
5	12.51	5	10.23
10	13.63	10	9.46
15	14.37	15	9.27
30	16.24	30	9.23

Table 3. Flexural Strength and Fracture Toughness of ZrB₂ Composite as
a Function of B and B₄C Content in the Preform.

B Content (vol.%)	Flexural Strength (MPa)	Fracture Toughness (MPa · m$^{1/2}$)	B₄C Content (vol.%)	Flexural Strength (MPa)	Fracture Toughness (MPa · m$^{1/2}$)
0	436	11.5	0	436	11.5
5	477	9.12	5	516	10.0
10	551	8.83	10	561	9.9
15	542	8.48	15	573	8.7
30	430	7.52	30	437	9.5

Dry Sliding Wear Resistance of Toughened ZrO_2-Y_2O_3 and ZrO_2-Y_2O_3-Al_2O_3

N. Savchenko*, S.Tarassov *, A.Melnikov*, S.Kulkov*

*Institute of Strength Physics and Materials Science, Tomsk, 634055, Russia

Abstract

Non-lubricated sliding wear tests were conducted on yttria-stabilized zirconia (Y-TZP) and Y-TZP+Al_2O_3 (3 mol.% Y_2O_3 and 20 or 80 wt%Al_2O_3) ceramic pins rubbed against the steel disc. Wear characteristics of Y-TZP and Y-TZP+Al_2O_3 are represented by three-dimensional wear maps. Different wear regions were identified as a function of applied stress and sliding speed. Microstructural examination of ceramics pins has shown that the wear process is very complex, encompassing many mechanisms which are described.

I. Introduction

Yttria partially stabilized zirconia (Y-TZP) with a potential to martensite transformation toughening is a prominent example of the structural ceramics [1]. Nevertheless, the data on its friction and wear is, to a great extent lacking, in particular this is the case for non-lubricated friction of a metal/ceramics pair. Tribotechnical characteristics of Y-TZP in ceramic/steel pair were related in our previous work [2]. In this work different wear regions were identified as a function of the applied stress (1-10 MPa) and sliding speed (0.2-9.4 m/s). Sudden increases in wear, identified as wear transitions, were found at certain stress and speed values. Below the range of wear mechanism transitions, the wear was mild and the wear mechanism was predominantly plastic deformation and microfracture. Above the wear transition, the wear was severely dominated by brittle fracture. Under high speed, the friction metal transfer film on ceramic pins provides a protection against the wear.

The aim of this paper was to study wear characteristics of Y-TZP+20 wt%Al_2O_3 and Y-TZP+80 wt%Al_2O_3 samples sliding against the steel disk without lubrication within the same range of sliding speed and normal load as reported in [2]. The basic interest was to compare wear structures and wear mechanisms of Y-TZP materials with respect to Al_2O_3 contents.

II. Experimental procedure

ZrO_2 - 3 mol. % Y_2O_3 and Y-TZP+Al_2O_3 (3 mol.% Y_2O_3 and 20 and 80 wt%Al_2O_3) powders have been used to prepare ceramic samples by sintering in vacuum at a temperature of 1750° C. More thoroughly the preparation procedure is described in [3].
. The result of sintering procedure was that zirconia was dominantly presented by its tetragonal modification in all ceramics prepared. Properties of obtained so ceramics are given below in Table I.

Table I. Mechanical and Physical Properties of Y-TZP and Y-TZP+Al$_2$O$_3$ Ceramics

Samples	Density (% theoretical)	Grain size (μm)	Hardness (GPa)	Fracture Toughness (MPa×m$^{1/2}$)	Flexure Strength (MPa)
Y-TZP	98	3.5	11	13	750
Y-TZP +20wt%Al$_2$O$_3$	95	2.5	11.4	7	500
Y-TZP +80 wt%Al$_2$O$_3$	89	4	11.3	4	210

The friction tests have been carried out using the pin-on-disk testing procedure. The normal load has been step-by-step increased during the tests under conditions of non-lubricated friction. Counterbody was a vertically mounted disk made from cast high-speed steel. The samples for tests had a rectangular-shaped cross-section of total area of 40 mm^2. The contact pressure applied to the samples was 1, 2, 5 and 10 MPa, the sliding speed was 0.2, 0.9, 4.7 and 9.4 m/s, respectively. The sliding distance for each of wear tests was 1000 m. Wear measurements were conducted with a micrometer at every combination of load and velocity values. Microstructure of sliding surfaces was studied with the use scanning electron microscopy.

III. Results and Discussion

The three-dimension wear map of the Y-TZP material is shown in Fig. 1(A). Two zones, each having a sufficiently different value of wear rate can be seen there. The first one covers the loading range from 0 to 5 MPa and its marked distinction is a slight load and sliding speed dependence of the wear rate. This zone may be denominated as a stable friction zone. From 5 MPa and on, we have observed not only the increase in wear rate with the contact pressure but the wear rate began to be more sensitive to the variation in the sliding speed. The maximum in wear rate corresponded to the sliding speed value between 0.9-4.7 m/s. Unlike the basic Y-TZP material, Y-TZP+20 wt% Al$_2$O$_3$ and Y-TZP+80 wt% Al$_2$O$_3$ ceramics did not demonstrate a strong sensitivity to the sliding speed, Fig. 1(B), (C). Moreover, the total wear rate of ceramics with alumina is much lower as compared to basic Y-TZP system.

Microstructure and wear mechanisms of Y-TZP material

Within the range of low sliding speeds (0.2-0.9 m/s) and low loads (<5 MPa) a regular grooved wear surface topology can be observed on the samples of Y-TZP being the result of the microcutting occurring on the sliding surface, Fig 2 (A) Adhesion wear controlled by a removal of the large particles occurs at the surface of the ceramics, which have been tested within the range of the medium sliding speed values (4.7 m/s), Fig.2

(B). On testing the samples at a maximal sliding speed (9.4 m/s), the wear surface topology is rather smooth in spite of the clearly seen transversal and longitudinal cracks in
the form of a net, Fig.2 (C). In contrast to the ceramic surfaces tested at 4.7 m/s, there
are practically no spalling and detachment traces on the surfaces of the samples tested at
9.4 m/s.

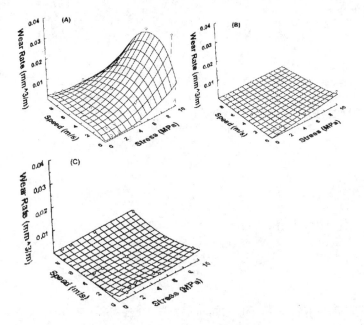

Figure. 1. The three-dimension wear maps for Y-TZP (A), Y-TZP+20 wt%Al$_2$O$_3$ (B)
and Y-TZP+80 wt%Al$_2$O$_3$ (C)

Microstructure, phase contents of sliding surfaces and wear products of Y-TZP were
studied in [2] with the use of X-ray diffraction analysis and scanning electron microscopy attached with X-ray spectral microanalysis. The following conclusions were
drawn: under mild friction (0.2-0.9 m/s and 5 MPa), the wear rate is defined by the action of the transformation toughening and transformation induced plasticity. With the
increase in the sliding speed up to 0.9-4.7 m/s, the transformation toughening does not
occur due to the high temperatures what results in a sharp growth in the contact stresses
with the ensuing thermal brittle fracture in the surface layers and catastrophic wear.
With the increase in the sliding speed up to 9.4 m/s there occurs a fall in the wear rate of
the materials due to the formation of a continuous transfer layer at the surface of the ceramics which closes up the previously formed cracks. The wear rate for a given range of
the studied speed values and loads correlates with the amount of the counterbody
/sample material transferred, Fig.3. The highest wear rate corresponds to the reduced
content of iron in the transfer layer and vice versa.

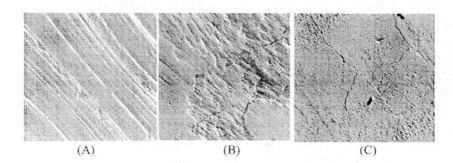

(A)   (B)   (C)

Figure. 2. Sliding wear surfaces on Y-TZP pins after wear tests carried out at 0.9 m/s
and 1 MPa (A), at 4.7 m/s and 10 MPa (B) and at 9.4 m/s and 10 MPa (C). The direction of sliding for (A) and (B) is along with a diagonal of the frame upward left and for
(C) - upward right. The size of all images - 135×135 μm.

Figure.3. X-Ray microanalysis data from the sliding wear surfaces of Y-TZP (From [2])

Microstructure and wear mechanisms of Y-TZP+Al₂O₃ materials

Studying the microstructures of Y-TZP+Al$_2$O$_{3,}$, the emphasis was on tribotechnical tests
with different speed values at a constant contact pressure of 10 MPa. In Fig.4 and Fig.5
shown are the micrographs of worn surfaces. It can be seen that Y-TZP+20wt% Al$_2$O$_3$
demonstrates regular grooved wear surface topology (very much alike that for Y-TZP -
Fig.2 (A)). With the increase in sliding speed, the latter is transformed into a network of
cracks, Fig.4 (B), (C).The difference in surface structure between samples tested at 4.7
m/s and at 9.4 m/s is that the surface of sample tested at 9.4 m/s is fully coated by the
transfer layer metal, whereas there is a discrete type of the transfer metal distribution at
the surface of sample tested at 4.7 m/s. In this case, stripes of the transfer metal alternate
with clean zones of observable initial structure (dark zones - Al$_2$O$_3$; light ones - Y-
TZP). So the system suffers a change in wear surface topology as a function of sliding
speed and very much alike it was the case for pure Y-TZP system.
Approximately the same could be said about the surface wear structures for Y-
TZP+80wt% Al$_2$O$_3$ ceramics. The distinction was that the layer-by-layer fracture had
been observed instead of regular topology for samples tested at low speed,. Fig.5 (A)

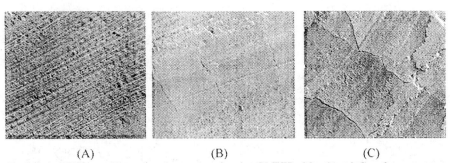

(A)   (B)   (C)

Figure. 4. Sliding wear surfaces on ceramic pins Y-TZP+20wt% Al$_2$O$_3$ after wear carried out at stress 10 MPa and speeds - 0.9 m/s (A), 4.7 m/s (B), 9.4 m/s (C).The direction of sliding - upward right . The size of the image: (A)-400×400 µm (B) and (C) - 135×135 µm

Such a behavior could be likely related to low fracture toughness characteristics of Y-TZP+80wt% Al$_2$O$_3$, which is due to its high porosity. Table 1. It is necessary also to note that during tests, Y-TZP+80wt% Al$_2$O$_3$ ceramics is susceptible to macroscale structural damage even when the overall wear detected by the pin size measurements was low.

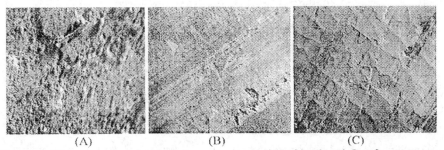

(A)   (B)   (C)

Figure. 5. Sliding wear surfaces on ceramic pins Y-TZP+80wt% Al$_2$O$_3$ after wear carried out at stress 10 MPa and speeds - 0.9 m/s (A), 4.7 m/s (B), 9.4 m/s (C).The direction of sliding -upward right . The size of all images - 400×400 µm

Taking into account the similarity of the wear surface microstructures formed in Y-TZP and Y-TZP+ Al$_2$O$_3$ ceramics, the absence of the catastrophic wear in latter ones at the medium speed range as compared to Y-TZP (Fig.1) is reasonable to connect with their slighter dependence of mechanical characteristics from the martensitic transformation. It is known [1] that Y-TZP+ Al$_2$O$_3$ materials better (being compared to pure Y-TZP) preserve their properties at high temperatures when zirconia's tetragonal phase becomes stable and the transformation toughening is no longer effective. Thus Y-TZP+ Al$_2$O$_3$ materials will provide more resistance to high temperature wear. Another reason behind the absence of the catastrophic wear in latter ones at the medium speed is the following. One cannot but take into account a considerable difference in the heat conductivity values for Y-TZP and Y-TZP+ Al$_2$O$_3$. Thus, as follows from [4], the heat conductivity for ZrO$_2$ and Al$_2$O$_3$ both having zero porosity and at 100°C is 1.95 and 30 Wt/(m×grad), respectively. This distinction can manifest itself in friction in the following manner. Wear of Y-TZP ceramics at 4.7 m/s is accompanied by high temperatures developing at the surface of the samples and being visible as a bright pink light emanating from the friction contact zone. By our estimation, the temperature there may achieve 1000° C. With the increase in the sliding speed up to 9.4 m/s and judging by the emanating light intensity, the temperature achieves 1300-1500° C. Under the same conditions, the intensity of light in the contact zone for Y-TZP+20wt% Al_2O_3/steel is notably lower as compared to Y-TZP, and in case of Y-TZP+80wt% Al_2O_3/steel friction it is not visually observed at all. In connection with this, and due to the poor heat conductivity of the Y-TZP ceramics it is unavoidable the formation of the intense thermal gradients, (much intensive ones as compared to those occurring in Y-TZP+Al_2O_3 materials) which result (in the absence of transformation toughening) in the brittle thermal fracture at the sliding surface. These two reasons , i.e better thermal stability of mechanical properties and higher heat conductivity of Y-TZP+Al_2O_3 (as compared to Y-TZP), define the absence of the catastrophic wear in medium speed range when continuos transfer metal layer is not formed to provide an efficient protection for subsurface ceramic layers.

In our opinion, the network structures arising at the sliding surface after high speed and heavy load friction (Fig.2 (C), 4(C), 5(C)) are worth more thorough look. Similar structures were described in a row of works on wear and friction of ceramics and hard alloys [5,6]. However, nobody has yet made an attempt to find a correlation between those structures' scale and microstructure of materials. We attempted to determine the average size of the fracture block both along and across the sliding direction from the micrographs. Furthermore, the polished sections have been prepared to estimate the depth to which those cracks propagated. The quantitative characteristics of the crack network are given in Table II for the samples tested at 9.4 m/s and 10 MPa

Table II. The quantitative characteristics of the crack network are given in Table II for Y-TZP and Y-TZP+Al_2O_3 ceramic samples tested at 9.4 m/s and 10 MPa

Samples	Average fracture block size along the sliding direction (μm)	Average fracture block size across the sliding direction (μm)	Crack propagation depth (μm)
Y-TZP	65 ±13	105 ±24	15 ±3
Y-TZP +20wt%Al_2O_3	68 ±18	120 ±23	10 ±3
Y-TZP +80 wt%Al_2O_3	49 ±14	75 ±20	15 ±5

Attention is drawn to the high density of cracks in Y-TZP+80wt% Al_2O_3 , which in its turn has the lowest value of the fracture toughness (see Table I). It is reasonable to expect that the total length of cracks formed in subsurface layer will correlate with the fracture toughness under given conditions of friction. From this standpoint it becomes clear why the values of the crack density for Y-TZP and Y-TZP+20wt% Al_2O_3 are comparable, that is, the high temperature fracture toughness values in the friction contact may be in the same order.

The whole situation with cracking may, however, not be so trivial. Many other external and internal reasons should be also considered. It is well-known that severe friction results in formation of defect subsurface layer. Characteristics of friction modified subsurface layer are greatly different from those of initial structure and there may not be any correlation between initial fracture toughness and surface layer cracking. Now we can see two sufficiently different scales in surface fracturing. First one (macroscale) characteristic size is in the order of fracture block) may be defined by distribution of quasistatic component of loading and temperature stresses. The Y-TZP+ Al_2O_3 system suffers a change in wear surface topology as a function of sliding speed very much alike it was the case for pure Y-TZP system. This allows to suggest that formation of macroscale topology is defined by the external loading and microstructure of the samples has practically no effect on the process.

On the other hand, fracture at microscale level (formation of fine wear particles) is defined by local properties of subsurface layer and its dynamic characteristics during friction.

Our further efforts will be aimed at more thorough investigations of energy dissipation in subsurface layers of ceramics.

IV. Conclusion

Unlike the Y-TZP material, Y-TZP+20 wt% Al_2O_3 and Y-TZP+80 wt% Al_2O_3 ceramics did not demonstrate a strong sensitivity to the sliding speed. Moreover, the total wear rate of ceramics with alumina is much lower as compared to basic Y-TZP system.

In particular, wear in studied Y-TZP+Al_2O_3 ceramics is by microcutting and abrasion at 0.9 m/s and 10 MPa. The crack network is formed at the surface of samples with the increase in sliding speed (4.7 m/s). Further increase in sliding speed up to 9.4 m/s results in formation of a continuous transfer layer at the surface of the ceramics which shields the previously formed cracks.

Better thermal stability of mechanical properties and higher heat conductivity of Y-TZP+Al_2O_3 (as compared to Y-TZP), define the absence of the catastrophic wear at medium speed range when continuos transfer metal layer does not provide an efficient protection for subsurface ceramic layers.

In spite of great differences between their wear maps, both Y-TZP+ Al_2O_3 and Y-TZP systems demonstrated similar character of surface topology response to the sliding speed increase. It is plausible that formation of macroscale topology is defined by the external loading and microstructure of the samples has practically no effect on the process.

References
[1] *Nettleship, R. Stevens*, TETRAGONAL ZIRCONIA POLYCRYSTAL (TZP) - A REVIEW - Int. J. High Technology Ceramics. 1987. No.3
[2] *S. N. Kulkov, N. L. Savchenko, S. Yu. Tarassov , A. G. Melnikov*, TRIBOTECHNICAL PROPERTIES OF Y-TZP IN CERAMIC/STEEL COUPLE- Friction and Wear. 1997. Vol. 18, No. 6

[3] *N. L. Savchenko, T. Yu. Sablina, T. M. Poletika, A. S. Artish, S. N. Kulkov*, HIGH-TEMPERATURE SINTERING IN VACUUM OF PLASMA-CHEMICAL ZIRCONIA-BASED POWDERS - Powder Metall. 1994. No.1-2

[4] *W. D. Kingery, J. Francl*, THERMAL CONDUCTIVITY: X, DATA FOR SEVERAL PURE OXIDE MATERIALS CORRECTED TO ZERO POROSITY - J. Am. Ceram. Soc. 1954. Vol. 37, No.2

[5] *A. Libsch, P. C. Becker, S. K. Rhee*, DRY FRICTION AND WEAR OF TOUGHENING ZIRCONIAS AND TOUGHENING ALUMINAS AGAINST STEEL - Wear. 1986. Vol. 110, No. 10.

[6] *S. F. Gnyusov, S. Yu. Tarassov*, FRICTION AND THE DEVELOPMENT OF HARD ALLOY SURFACE MICROSTRUCTURES DURING WEAR - JMEPEG. 1997. Vol.6, No.6

Microstructural Evolution of SiC/SiC Composite under Irradiation

Y. Katoh[1,2], A. Kohyama[1,2] and T. Hinoki[1,2]

[1]Institute of Advanced Energy, Kyoto University
Gokasho, Uji, Kyoto 611, Japan
[2]CREST-ACE, Japan Science and Technology Corporation
Kawaguchi, Saitama 332, Japan

Abstract

Silicon carbide fiber reinforced silicon carbide matrix composite (SiC/SiC composite) is a potential material for use in very severe environment in advanced energy systems. In this work, neutron radiation tolerance of advanced SiC fiber reinforced SiC/SiC composite fabricated through chemical vapor infiltration method was studied by means of dual-beam ion irradiation experiment. The irradiated material was subjected to microstructural examination by transmission electron microscopy following the thin film processing with a focus ion beam device. The result showed the superior radiation resistance of the advanced low-oxygen near-stoichiometric SiC fiber but suggested the need for further study on the synergistic effect of atomic displacement and transmutant helium production.

1. Introduction

Silicon carbide fiber reinforced silicon carbide matrix composites (SiC/SiC composites) are very attractive candidate to materials for advanced energy systems such as future fusion reactors and advanced fission reactors. Beside its potential superior mechanical performance and chemical stability at elevated temperatures, the low induced radioactivity of silicon carbide is especially attractive for heavily neutron loaded applications represented by fusion blanket / first wall structures and high temperature gas reactor core components [e.g. 1-3].

One of the serious concerns of SiC/SiC composite for such nuclear applications is the effect of radiation, especially a synergistic effect of atomic displacement and insoluble gas (i.e. helium) production through nuclear transmutation, on microstructure, physical properties and mechanical properties. Examples of significant radiation effects include the irradiation-induced swelling of crystalline SiC[4, 5], shrinkage of oxygen-containing SiC fibers due to irradiation-assisted oxidation[6], irradiation-induced recrystalization of micro-crystalline fibers[7], the fiber-matrix debonding caused by radiation-induced deformation[8] and irradiation creep of fibers. To avoid or at least minimize these radiation effects, recent trend of SiC fibers is oriented toward lower oxygen content, reduced free carbon and gotten enhanced crystalline structures.

The objective of current work is to determine the synergistic effect of atomic displacement and helium production on microstructure of SiC/SiC composite using the latest SiC fiber which is presently available. Studying the radiation effects in SiC/SiC composites to develop improved severe environment-tolerant materials is among the major purposes of the CREST-ACE program [1].

2. Experimental procedure

For this irradiation study, a SiC/SiC composite material was fabricated using an advanced SiC fiber, namely Hi-Nicalon® Type-S produced by Nippon Carbon Corporation. The Type-S fiber contains reduced amount of oxygen and has near-stoichiometric chemical composition. The representing properties and chemical composition of the Type-S fiber are compared with those of the classic ceramic grade Nicalon (Nicalon-CG) and the more recent Hi-Nicalon fibers in Table 1 [9]. A uni-directionally reinforced composite was made through the matrix forming by chemical vapor infiltration (CVI) method following carbon coating on the fibers also by the CVI processing. The CVI processing was performed at the National Research Institute of Metals (NRIM). The nominal amount of carbon coating was chosen to be 200nm thick.

For the dual-beam ion irradiation experiment, the composite material was square-cut into a $15.0 \times 2.8 \times 2.0$mm bar so that the 2.0mm-long direction be parallel to the fiber direction. The irradiation surface was chosen to be normal to the fiber direction, to enable examination of the depth-dependent irradiated microstructures of the fiber, the matrix and the interphase. The chosen surface to be irradiated was carefully polished and diamond powder-finished. The whole procedure is illustrated in Figure 1.

The ion irradiation was carried out at the dual-beam irradiation station of the High-fluence Irradiation Facility, University of Tokyo (HIT Facility) [10]. 4 MeV Ni^{3+} ions accelerated by a 1MV Tandetron were used to produce atomic displacement damage. Simultaneous implantation of He^+ ions was made, where Van de Graaff accelerator was operated at 1MV and the 1MeV He+ ions were then slowed down using a nickel foil energy degrader. The calculated depth profile of displacement damage, stopped nickel and deposited helium are presented in Figure 2. In this case, average displacement threshold energy of SiC was assumed to be 35eV [11], stoichiometric chemical

	Nicalon-CG	Hi-Nicalon	Hi-Nicalon Type-S
Chemical composition	$SiC_{1.34}O_{0.36}$	$SiC_{1.39}O_{0.01}$	$SiC_{1.05}$
Tensile strength	3.0 GPa	2.8 GPa	2.6 GPa
Elastic modulus	220 GPa	270 GPa	420 GPa
Elongation	1.4%	1.0%	0.6%
Mass density	2.55g/cm³	2.74 g/cm³	3.10 g/cm³
Diameter	14μm	14 μm	12 μm
Specific resistivity	10^3-10^4 Ω·cm	1.4 Ω·cm	0.1 Ω·cm
Max. temperature	1573K	1773K	1873K

Table 1 Comparison of representing properties of Nicalon-CG, Hi-Nicalon and Hi-Nicalon Type-S [9].

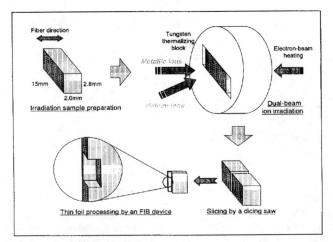

Figure 1   Schematic illustration of specimen preparation procedure for transmission electron microscopy.

composition and 3.0g/cm^3 of mass density were assumed and the calculation was made by TRIM-92 code[12]. Irradiation temperature, displacement damage rate and total dose were 873K, 1×10^{-3} dpa/s and 10 dpa-SiC, respectively.

Following the ion irradiation, the specimen was sliced into 0.2mm-thick thin plate and the irradiated edge region was further thinned to 0.05mm using a dicing saw. The thinned tip was then processed in a focused ion beam (FIB) device to make a thin film, which is parallel to both the fiber and ion bombardment direction, for examination by transmission electron microscopy (TEM). The FIB processing was performed so that a single thin film contains the fiber, matrix and the interphase and each phase contains single-beam irradiated (i.e., no helium deposition), dual-beam irradiated and unirradiated region. Microstructural investigation was performed with JEOL JEM-2010 TEM operating at 200kV.

3. Results and discussion

The CVI-fabricated composite was fairly dense, as usual in uni-directionally reinforced materials, in which the total porosity was approximately 10 percent. The surface of Hi-Nicalon Type-S fibers are thoroughly coated by amorphous carbon and the other surface of the carbon layer was facing the beta-SiC matrix, as observed by secondary ion imaging scheme in the FIB device. The thickness of deposited carbon layer varied in a range of 50 to 300nm while the objected thickness had been 200nm. All the observed pores are typical matrix pores, which were surrounded by the last-deposited matrix SiC.

A low magnification transmission electron micrograph of the thin foil examined in the current study is presented in Figure 3. The mostly bright and black-dotted image on the right side corresponds to the Type-S fiber, that originally consists of variedly oriented beta-SiC micro-crystals. The left hand image represents the matrix where the

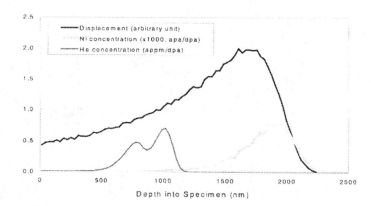

Figure 2   The calculated depth profile of displacement damage, stopped
nickel and deposited helium in SiC. 35eV of displacement threshold energy,
3.0g/cm³ of mass density and stoichiometric chemical composition were
assumed in TRIM-92 calculation.

record of radial growth of beta-SiC grains toward the left side in the micrograph is
observable. The perpendicular thin bright stripe between the fiber and the matrix is a
tyical appearance of carbon interphase. The thin foil's edge in the top of micrograph
corresponds to the surface irradiated by the dual-beam ions. The very edge is covered
with physically deposited tungsten to protect the edge region of the thin foil.

Three different areas of interest were selected from the fiber portion of the thin foil
described above. The first one, 'unirradiated' hereafter, is located about 5 μ m from the
irradiated surface, where is far behind the maximum range of any of the implanted ions.
The second area, or 'single-beam', was selected to be approximately 500nm from the
surface, in which the total deposited concentration of helium should be negligible and
only atomic displacement due to nickel ion irradiation is expected to contribute the
microstructural modification. In the last one, or 'dual-beam' irradiated area,
approximately 1 μ m from the surface, both the atomic displacement and the helium
deposition rate are significant. The average helium deposition rate there was 60 appm

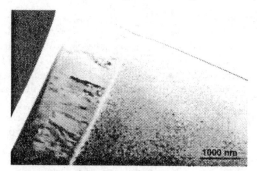

Figure 3   A low magnification transmission electron micrograph of the thin
foil examined (see text). The channel on the left had intentianally been made
to free the internal stress.

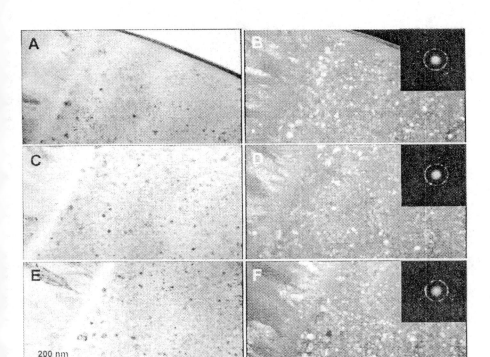

Figure 4   Higher magnification TEM micrographs and the corresponding selected area electron diffraction patterns for single-beam irradiated area (A, D), dual-beam irradiated area (B, E) and unirradiated area (C, F).

(atomic parts per million) / dpa nominal, which corresponds to the typical anticipated helium generation rate in SiC applied to fusion blanket structures.

Higher magnification TEM micrographs and the corresponding selected area electron diffraction patterns for each area are provided in Figure 4. The weak beam dark field images were taken by picking up a part of the <111> diffraction ring of beta-SiC in each condition so that only grains in limited orientation are brightly imaged. The diameter of selected area aperture used for the diffraction imaging was approximately 175 nm. As shown in Figures 4-E and F, the unirradiated Type-S fibers consisted of beta SiC grains of 10 to 50nm in diameter, and such structure did not significantly change after single- and dual-beam irradiation. However, the number of relatively small grains distinctive in the dark field micrographs apparently decreased after single-beam irradiation and this tendency appeared enhanced in dual-beam irradiated area. In the electron diffraction patterns, single-beam irradiation caused the decrease in number of strong reflection spots and instead the increased ring-like feature of the gathered weaker spots, and again this trend were enhanced by the simultaneous helium deposition. In addition, in irradiated fibers and especially in dual-beam irradiated one, number of diffraction spots appeared between <111> diffraction ring and <002> ring. These observations indicate irradiation-induced refining of the micro-crystalline structure and its probable promotion by helium production. The formation of amorphous phase was not confirmed, however. High-resolution transmission electron microscopy will follow this work in near future.

4. Conclusions

A SiC/SiC composite using advanced SiC fiber was fabricated and subjected to dual-beam ion irradiation up to 10 dpa at 873K. Following TEM examination revealed the irradiation-induced refining of micro-crystalline structures in the fiber and its enhancement by simultaneous implantation of helium, nevertheless, the observed microstructural modification due to irradiation was rather insignificant for this level of displacement damage.

Acknowledgement

This work is performed as a part of 'R&D of Composite Materials for Advanced Energy Systems' research project, Core Research for Evolutional Science and Technology (CREST-ACE).The Hi-Nicalon® Type-S fiber used in this study was supplied through the courtesy of Nippon Carbon Corporation. The CVI processing was performed by Drs. H. Araki and T. Noda at the National Research Institute of Metals. The dual-beam ion irradiation experiment was done with the assistance by Drs. H. Shibata and T. Iwai and Messrs. M. Narui and T. Omata at the Research Center for Nuclear Science and Technology, the University of Tokyo. The authors are grateful to these people for their valuable help.

References

[1] A. Kohyama, Y. Katoh, T. Hinoki and W. Zhang, 'Progress in the Development of SiC/SiC Composites for Advanced Energy Systems: CREST-ACE Program', to be presented at this conference.
[2] L. L. Snead, R. H. Jones, A. Kohyama and P. Fenici, J. Nucl. Mater. 233-237 (1996) 26.
[3] G. R. Hopkins, Proc. IAEA Symp. Plasma Phys. and Controlled Necl. Fusion Res., IAEA-CN-33 /s3-3 (1974)
[4] R. J. Price, J. Nucl. Mater. 33 (1969) 17.
[5] R. J. Price, Nucl. Tech. 35 (1979) 320.
[6] L. L. Snead, M. C. Osborne and K. L. More, J. Mater. Res. 10(3) (1995) 736.
[7] A. Hasegawa, G. E. Youngblood and R. H.Jones, J. Nucl. Mater. 231 (1996) 245.
[8] L. L. Snead, S. J. Zinkle and D. Steiner, J. Necl. Mater. 191-194 (1992) 560.
[9] M. Takeda, J. Sakamoto, A. Saeki and H. Ichikawa, Ceramic Engineering & Science Proceedings, 17 (1996) 35.
[10] Y. Kohno, K. Asano, A. Kohyama, K. Hasegawa and N. Igata, J. Nucl. Mater., 141-143 (1986) 794.
[11] H. L. Heinisch, 'Analysis and Recommendations on DPA Calculations in SiC', JUPITER/IEA Joint International Symposium on SiC/SiC Ceramic Composites

for Fusion Structural Applications, Oct. 23-25, 1997, Sendai.

[12] J.P.Biersack and L.G.Haggmark, Nucl. Inst. Meth. 174 (1980) 257.

Effect of Fiber Coatings on Crack Initiation and Growth Characteristics in SiC/SiC under Indentation Test

W. Zhang [1, 2], T. Hiniki[1, 2], Y. Katoh[1, 2], A. Kohyama[1, 2], T. Noda[2,3]

1: Institute of Advanced Energy, Kyoto University, Gokasho, Uji, Kyoto 611, Japan
2: CREST, Japan Science and Technology Corporation, 4-1-8 Honmachi, Kawaguchi, Saitama, 332, Japan
3: National Research Institute for Metals, 1-2-1, Sengen, Tsukuba, Ibaraki 305, Japan

Abstract

SiC/SiC composites are attractive materials for nuclear fusion application and other applications such as heat exchanger and turbines. The objective of this work is to try to find a way to improve the toughness of the materials by investigating the interface microstructure under indentation test. SiC/SiC composites were made by CVI method. Carbon coatings and multiple coatings were applied to be interphase between fibers and matrix for improving the toughness. The effects of the coatings were studied by analyzing crack behavior and fiber debonding behavior after push-in test by use of SEM and TEM. It was found that the fiber coatings had a very important role for improving toughness of SiC/SiC composites by enhancing fibers debond. This work offered some base data for finding an optimum coating thickness on fiber.

1. Introduction

There is a strong demand to develop high performance ceramic matrix composites (CMC) for advanced energy systems, such as nuclear fusion reactors. SiC/SiC composites are considered to be potential candidates for fusion application because of its low induced radioactivity, improved toughness, high temperature strength, a non-catastrophic failure mode, high plant heat efficiency and good irradiation resistance. In addition, SiC/SiC composites are also used for heat exchanger, turbines and so on for their good resistance to corrosion [1]. Same as other composites, fiber-matrix interfacial properties influence mechanical properties of SiC/SiC composites strongly [2,3]. To improve toughness of CMC, like SiC/SiC, a non-bonded or a weakly bonded fiber-matrix interface is known to be desirable as a specific case [4-7] .One approach to modify fiber-matrix interface in SiC/SiC composites is to apply carbon coating on SiC fibers. In this work, Hi-Nicalon fibers coated with varieties of carbon coatings and multiple coatings were used. The objective of the study is to clarify effects of carbon layer thickness and multiple coating thickness on crack initiation and growth behavior in the SiC/SiC composite materials. For the research on interfacial property, it is required to improve performance of SiC/SiC composites to meet each application dependent requirements. The push-in tests were carried out by micro-indentation machine. The microstructures before and after indentation test were inspected by use of a Scanning Electron Microscope (SEM) and a Transmission Electron Microscope (TEM). The foils for TEM observation were prepared by Focused Ion Beam device.

2. Experimental

1) Materials Fabrication

Materials used were 2D woven fabrics of SiC (Hi-Nicalon) reinforced CVI (Chemical Vapor Infiltration) SiC composites. Seven layers of 2D woven fabrics were piled and fixed in 1.8-mm thickness. Carbon coatings on fiber were applied prior to CVI processing to make SiC/SiC composites with variations of thickness. In addition, SiC/SiC composites with multiple layer coatings were prepared, where the first layer was carbon. After fiber coatings, SiC matrix was infiltrated by CVI process from both sides. Each side took about 20 hours for CVI processing.

2) Specimens Preparation

The as infiltrated materials were cut into thin-plate specimens with sizes of $4^{(l)} \times 2^{(w)} \times 0.6^{(t)}$ mm using a low speed diamond saw and a Dicing Saw for micro-indentation

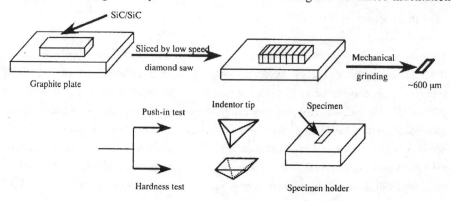

Fig. 1 The Process of Specimens Preparation

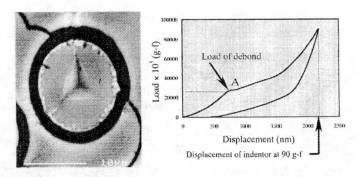

Fig. 2 The Fiber's Image and the Load-displacement Curve after Push-in Test

experiments, of which the process is shown in Fig. 1. Usually a specimen was cut into the thickness of 600 µm for push-in test. Two kinds of diamond paste were used for specimen polishing. TEM specimens were prepared for microstructure observation before and after push-in test. Where two observation directions were selected to be perpendicular and parallel to fiber axis. The thickness for TEM observation was less than 100 nm and the FIB device utilized made it successful to obtain wide and uniform area for observation..

3) Testing

A micro-Vickers hardness-testing machine was used to introduce cracks in the composites and a micro-indentation machine installed in SEM was applied to do push-in test. Specimens were put on a plane copper holder. The fiber diameter selected for the test was about 14 µm. A typical displacement-load curve is in Fig. 2. The load at point A was defined as "load of debond" where interfacial crack was initiated to start debonding. After push-in test, microstructures of the composites were inspected by SEM and TEM.

3. Results and Discussion

1) Hardness Test in Matrix

Cracks initiated in the matrix always stayed in the matrix as far as the indentation load was not too big. And the crack length grew monotonically with increment of the load. Representing images of cracks initiation and growth are shown in Fig.3, where cracks were initiated at the rims of the indentor and propagated, for the most cases, along the direction of rim. Crack could grow across interphase reaching to fibers when the load was big enough which was supposed to be the initiation stage of debonding. The propagating cracks usually deflected at the interface between fiber and coating. Therefore, for thin coating fibers, they were easy to debond. Such crack deflection

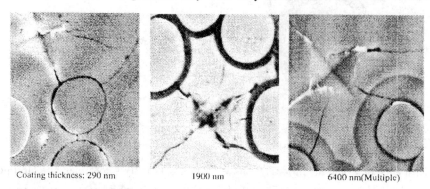

Coating thickness: 290 nm   1900 nm   6400 nm(Multiple)

Fig. 3 Crack Behavior of Different Materials in Hardness Test

behavior is generally thought to be the reason for improving SiC composites toughness. Figure 3 shows crack growth and fiber debonding behaviors for the cases of three different coatings. They were carbon coatings with 290 nm, 1900 nm and multiple coating (carbon, SiC and carbon) with 6400 nm. For the composite with 290 nm coating, the crack initiated by indentation test and debonded multi fibers until the crack was arrested. In the case of fiber with carbon coating thickness of 1900 nm, crack propagated to interface and stopped there without fiber debonding. Similar situation was observed on specimen with multiple layer fibers.

Generally the strength of matrix is lower than that of fibers. The primary function of the matrix is to transfer the load to the fibers while the debonding layer prevents catastrophic failure and provides pseudo-plasticity through fibers debonding and pull out. In the hardness test, thinner coating appeared to produce easy fiber debonding. For improving the toughness of SiC/SiC composite, in the future, composites with coating thickness below 1000 nm will be suggested to be produced for finding the optimum of coating thickness.

2) The fiber debond behavior under push-in test

For single fiber push-in tests on specimens of ~ 600 μm thickness, various debond behaviors were observed. There are two kinds of fibers with a single C-layer and multiple coatings respectively. The indentor load-displacement curve of the single fiber push-in test showed that every fiber had a unique debonding load and the indentor displacement at the maximum load 90 g-f. Considering the debonding load and maximum displacement mentioned the above with the fiber coating thickness together, four figures were drawn out. Fig. 4(a) shows the relation of C-layer thickness with displacement of indentor at 90 g-f under push-in test. With increment of the C-layer thickness the displacement of indentor at 90 g-f increased gradually and then maintained a constant after thickness more than 1000 nm. From the push-in test on multiple coatings (C, SiC, C) specimens the same tendency was seen (Fig. 4(b)). The difference is the different critical thickness from which the displacement didn't change. In addition, the load at which fibers began to debond decreased with the increasing of the multiple coatings thickness (Fig. 4(d)). The result from single coating layer fibers is a little different (Fig. 4(c)). The main tendency of debond load changing fall down. But around the coating thickness of 1000 nm, there is a peak at which the debonding load is the highest.

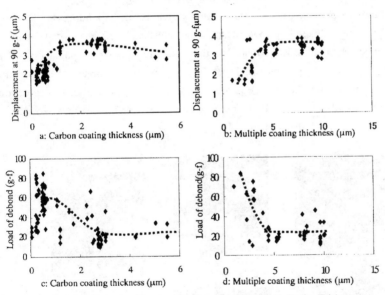

Fig. 4 The Relations of Debond Load and Displacement at 90 g-f with Coating Thickness

From Fig. 4, it appears that thick coatings allowed easier debonding than thin coatings. But the debond load decreased rapidly with increasing coating thickness. If the debond load is too low, the strength of the composite will decrease. Thinner coating has a higher interface shear strength (ISS) and frictional sliding strength (IFS). When ISS is high, the matrix can transfer load to the high strength fibers. However, if the ISS is very high, or IFS is very high, the fibers will not debond which causes two things to happen: 1) Cracks are not tie-up at the interface and the composite fails at low load. And 2)

Unless fiber pull-out and fiber "bridging" occurs the strength at large numbers of fibers are not used. Comparing Fig. 4(b) with Fig. 4(d), it is found that around coating thickness of 3μm the debond load is not so low (~70 g-f) while maximum displacement is not so deep (~2μm). From Fig. 4(c), it looks like hopeful to find a suitable coating thickness which can make fiber debond and has high debond load relatively. Therefore, a suitable coating thickness was suggested to be found in the future work.

3) Observation of crack initiation and propagation under TEM

a) The observation of cracks in the interfaces.

Three kinds of specimens with carbon coating thickness of 1900 nm, multiple coating thickness of 720 nm and 1200 nm were inspected by TEM. The first two kinds were observed from the direction which is perpendicular to fiber axis and the last one from the parallel direction to the fiber axis (Fig. 5(a, b)). In the Fig.5(a), one part of the fibers debonded and debond length is about 3 μm which is corresponding to the load-displacement curve. Debonding took place in the interface of fiber and matrix. In the second fiber, same situation was found. But the crack initiated and propagated between first carbon layer and the second SiC layer. Fig. 5(b) shows the image of a fiber after push-in test from the direction which is parallel to the fiber axis. The fiber also debond partially between fiber and the first layer, and the crack deflected from interface to multiple layers and was arrested there.

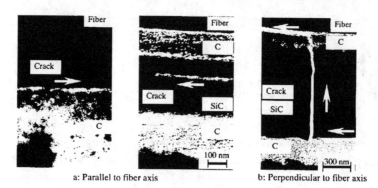

a: Parallel to fiber axis   b: Perpendicular to fiber axis

Fig. 5 Crack Image under TEM

From the three specimens, it is found that the fibers debonded only partially at the load less than 100 g-f. One reason is the load was too low which can not make fiber debonded fully. Other reason is suggested to the deflection of the crack because of some reasons. For the thick coating (single and multiple), typically the crack initiated between fiber and the carbon layer. The crack is possible to deflect to the interface between the second layer and the third layer. It provides more chances to transfer the force from matrix to fiber and at same time the toughness was improved because of the crack propagated at the interface. In the case of thinner multiple coating, crack initiated at the interface between the carbon layer and the SiC layer. It seemed that the carbon coating joined with fiber stronger than that of between carbon coating and SiC layer. Therefore it is possible to obtain a suitable interface shear strength or frictional strength by adjusting the thickness of the multiple layer respectively.

b) Crack initiation and propagation in fiber

In the specimens with multiple coating thickness of 720 nm, crack occurred at the tip of the indentor mark and propagated into the fiber. Fig. 6 shows some TEM foil preparation processes by FIB system. The fiber which has been done push-in test was shown in Fig. 6(a). Indentor mark is seen clearly on the fiber. After several steps by FIB process, thinner layer which the indentor mark still remained is observed(Fig. 6(b)). In the end, on the TEM foil observation area, it is found that the crack initiated at the tip of the indentor mark and propagated into the fiber. The mechanism of the crack propagation will be studied in the future.

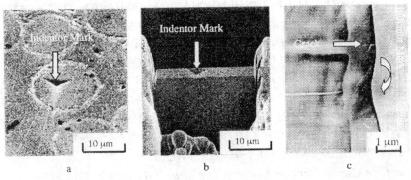

Fig. 6 TEM Foil Preparation Process

4. Summary

1) By hardness test crack behavior and the role of fiber coatings were investigated. In general, cracks initiated in matrix propagated through coating and deflected along fiber surface. Also, in fiber crack deflected at fiber-coating interface. This is the important mechanism for improving toughness of the composite. Within the experimental condition studied, the thinner the coating the toughness was the higher.

2) In push-in test, it was found that the debond load of the fibers decreased with the

increment of the coating thickness. A suitable coating thickness, which will lead to preferable fibers debonding behavior with appropriate interfacial shear strength will be defined together with the microstructural optimization, in the future work.

3) After push-in test the precise TEM observation was carried out to see the relation between crack path and microstructure. Crack deflection in fiber might be related with dispersed crystalline particles in fiber, mostly amorphous in Hi-Nicalon. At interface, crack propagated parallel to graphite basal plane, which was formed along fiber surface epitaxially. In the case of multiple coating Trans-layers crack and crack along interfaces were observed. The analysis of the microstructure is underway and will be provided in the near future.

REFERENCES

[1]*Akira Kohyama*, R & D ACTIVITIES ON ADVANCED COMPOSITE MATERIALS FOR FUSION IN JAPAN, *Materials for advanced energy systems & fission and fusion engineering* '94 pp.34-41.

[2] *A.G.Evans and D.B. Marshall* "THEMECHANICAL BEHAVIOR OF CERAMIC MATRIX COMPOSITES", Acta Metall., 37[10] 2567-83 (1989)

[3] *R.J.Kerans, R. S. Hay, J. Pagano, and T.A. Parthasarathy*, "THE ROLE OF THE FIBER MATRIX INTERFACE IN CERAMIC MATRIX COMPOSITES", Ceram. Bull., 68, [2] (1988) 429-442

[4] *Edgar Lara-Curzio et al*, FIBER-MATRIX BOND STRENGTH, FIBER FRICTIONAL SLIDING AND THE MACROSCOPIC TENSILE BEHAVIOR OF A 2D SiC/SiC COMPOSITE WITH TAILORED INTERFACES. *Ceramics Engineering and Science Proceedings*, Vol.16[5], 1995, pp.597-612.

[5]*Edgar Lara-Curzio, Mattison K. Ferber and Richard A. Lowden*, THE EFFECT OF FIBER THICKNESS ON THE INTERFACIAL PROPERTIES OF A CONTINUOUS FIBER CERAMIC MATRIX COMPOSITE. *Ceramics Engineering and Science Proceedings*, Vol.15[5], 1994, pp.989-1000

[6]*R.H.Jones, L.L.Snead, A.Kohyama, P.Fenci*, RECENT ADVANCES IN THE DEVELOPMENT OF SiC/SiC AS A FUSION STRUCTURAL MATERIAL. ISFNT-4.(Submitted)

[7]*Kenichi Hamada, Shinji Sato, Hideo Tsunakawa and Akira Kohyama, A.*,"INTERFACIAL MICROSTRUCTURE AND MECHANICAL PROPERTIES OF C/C COMPOSITES", *Proceedings of ICCM-10*, Whistler, B.C., Canada, August 1995, Volume VI: Microstructure , Degradation, and Design. pp.423-430

Microstructure and properties of Al_2O_3-$Al_5Y_3O_{12}$ eutectic fibre produced via internal crystallization route

S.T.Mileiko, V.M.Kiiko, N.S.Sarkissyan, M.Yu.Starostin, S.I.Gvozdeva, A.A.Kolchin, and G.K.Strukova
Solid State Physics Institute of the Russian Academy of Sciences, Chernogolovka Moscow district, 142432 RUSSIA

Eutectic oxide fibres look as an attractive reinforcement for a variety of matrices suitable for high temperature service. However, to stimulate a usage of oxide-fibre composites, it is necessary to overcome the problem of a high cost of the fibres. The internal crystallization method (ICM), that is fabricating composites by crystallizing the fibres in channels pre-made in the matrix[1] is now expanding to developing a way to produce fibres by using an auxiliary matrix to be removed after the fibres have been crystallized.

In the present paper, the microstructure and strength of the fibres both extracted from the auxiliary matrix and nested in it are described. Two oxide compositions were used in the crystallization process, the first one, that is 66% Al_2O_3-34% Y_2O_3, corresponds to the eutectic points in the Al_2O_3-$Al_5Y_3O_{12}$ mixture cited in Ref.[2], and the second one, that is 60% Al_2O_3-40% Y_2O_3. To remove a fibre bundle from the auxiliary matrix, a chemical dissolution of molybdenum was used.

Phase composition of the fibres corresponds to observations reported in Ref.[2] The microstructure of the fibres of the first composition is a rather homogeneous one characterized by colonies composed of bent lamellae of $Al_5Y_3O_{12}$ in the Al_2O_3 matrix On the other hand, for a majority of the fibres of the second composition, dendrites are a typical feature.

A highly non-homogeneous microstructure of the ICM-fibres creates a variety of defect sites in the fibre. This yields a rather high strength scatter of the fibres as has been already shown by testing Al_2O_3-$Al_5Y_3O_{12}$-fibre/Mo-matrix composites.[1] In the present work, fibres extracted from the auxiliary matrix were tested by bending them over cylinders of a diameter large enough to neglect end effects. Counting a number of the accumulated breaks and corresponding bending stress yields strength/length dependencies and the strength scatter that allows to calculate the Weibull parameters for bending and tension. The Weibull exponent is between 4 and 5, the scale parameter, that is the fibre tensile strength on a length of 5 mm, changes from 500 MPa for the second composition to about 1000 MPa for the first one. High temperature strength characteristics of the fibres are determined mainly by testing composites with molybdenum matrix.

The work was performed under financial support of International Science and Technology Center, Project # 507-97.

SIMPOSIUM 7

SPATIALLY REINFORCED COMPOSITES

A. KELLY[+] and J.G. PARKHOUSE[*]

[+]Department of Materials Science and Metallurgy, University of Cambridge, Pembroke Street, Cambridge CB2 3QZ, United Kingdom

[*]School of Mechanical and Materials Science, University of Surrey, Guildford, Surrey, GU2 5XH, United Kingdom

Abstract

The possible tightest packing of bars (fibre bundles) of various cross sections has been investigated in detail. The fibre bundles are taken to lie along well defined directions e.g. cube edges, body and face diagonals. The results are used to explore the elastic properties of 3-D composites. In this paper, attention is directed to the possibility of designing a fully isotropic fibrous composite.

1. Introduction

The primary stressed state of a composite is determined by the average stress across the material as a whole. Since stress has six independent components, any stress can be resisted by six sets of rods, each set resisting a portion of the load uniaxially, provided that no more than three of the reinforcement directions are coplanar. Then the primary loading can be resisted solely by the reinforcement. 6-d composites are therefore of particular importance since they are the simplest that do not only rely on their matrices for some primary load transfer. Their matrices are only needed as a local redistributor of load, providing the shear strength necessary to keep the bars bonded together. When the number of reinforcement directions is less than six, even if the reinforcement is rigid, the composite is flexible in all but a few directions, with a flexibility that is determined by the properties of the matrix.

2. Calculation of the Effective Modulus

Fibrous material has a Young modulus of E and fills a proportion of V_f of a volume whose dimensions are large compared to the scale of the pattern. This pattern consists of n sets of rods, all of the same cross-section. Within each set all the rods are straight and parallel, orientated in the direction described by the direction cosines c_i, where $i = 1,2,3$, to the three axes of a Cartesian system.

The proportion of rods in each set are defined by α; the n values of α add up to unity.

Suppose the fibrous pattern is in tension throughout and remains in tension when the volume's boundary moves compatibility with a uniform strain inside the boundary of $\in_{ij}$. This is a strain tensor with nine components. Symmetry requires that there are only six independent components and these will be described by $\in$, a "column vector" where

$$\in^T = \{ \in_{11}, \in_{22}, \in_{33}, 2\in_{32}, 2\in_{31}, 2\in_{12} \} \qquad (1)$$

The components of $\in$ are the three direct strains followed by three engineering shear strains.

The additional strain in any set of rods is a scalar, $\in$, where

$$\in = c_k c_l \in_{kl}$$

This set contributes a scalar stress, σ in its own direction, of

$$\sigma = \alpha E V_f \in$$

Its tensor components with respect to the Cartesian axis system are

$$\sigma_{ij} = c_i c_j \sigma$$

Hence

$$\sigma_{ij} = \alpha E V_f c_i c_j c_k c_l \in_{kl} \qquad (2)$$

Again, symmetry requires that there are only six independent components of σ, which will be described by σ, where

$$\sigma^T = \{\sigma_{11}, \sigma_{22}, \sigma_{33}, \sigma_{23}, \sigma_{31}, \sigma_{12}\} \qquad (3)$$

Equation (2) may now be expressed in matrix from as

$$\sigma = {}_r C \in$$

the prefix indicating that the contribution is from the rth set of rods.

An expansion of ${}_r C$ using equations (1), (2) and (3) gives:

$$
{}_r C = \alpha E V_f
\begin{bmatrix}
c_1^4 & c_1^2 c_2^2 & c_1^2 c_3^2 & c_1^2 c_2 c_3 & c_1^3 c_3 & c_1^3 c_2 \\
 & c_2^4 & c_2^2 c_3^2 & c_2^3 c_3 & c_1 c_2^2 c_3 & c_1 c_2^3 \\
 & & c_3^4 & c_2 c_3^3 & c_1 c_3^3 & c_1 c_2 c_3^2 \\
 & & & c_2^2 c_3^2 & c_1 c_2 c_3^2 & c_1 c_2^2 c_3. \\
\text{symmetric} & & & & c_1^2 c_3^2 & c_1^2 c_2 c_3 \\
 & & & & & c_1^2 c_2^2
\end{bmatrix}
\qquad (4)
$$

These contributions to the total stiffness matrix C are then summed to give

$$C = \sum_{r=1}^{n} {}_r C$$

which may be inverted to give the flexibility matrix S, the preferred form for revealing material properties like the effective Young moduli, $1/S_{11}$ etc., and the Poisson ratios, $-S_{12}, S_{11}$ etc.

3. Results

For instance if a volume fraction V_e of rods of modulus E are distributed equally in the three directions parallel to the edges of a cube then $\alpha = \frac{1}{3}$, and the value of c_i, is either 0 or 1 The array of stiffness constants is of the form

$$\sigma^T = \frac{V_e E}{3} \begin{bmatrix} 1 & 0 & 0 & 0 & 0 & 0 \\ 0 & 1 & 0 & 0 & 0 & 0 \\ 0 & 0 & 1 & 0 & 0 & 0 \\ 0 & 0 & 0 & 0 & 0 & 0 \\ 0 & 0 & 0 & 0 & 0 & 0 \\ 0 & 0 & 0 & 0 & 0 & 0 \end{bmatrix} \in' \tag{5}$$

Similarly if a volume fraction V_d of rods of material of the same cross section are distributed equally parallel to the four directions defined by the body diagonals of a cube then one finds,

$$\sigma^T = \frac{V_d E}{9} \begin{bmatrix} 1 & 1 & 1 & 0 & 0 & 0 \\ 1 & 1 & 1 & 0 & 0 & 0 \\ 1 & 1 & 1 & 0 & 0 & 0 \\ 0 & 0 & 0 & 1 & 0 & 0 \\ 0 & 0 & 0 & 0 & 1 & 0 \\ 0 & 0 & 0 & 0 & 0 & 1 \end{bmatrix} \in' \tag{6}$$

So if we add the arrays together we will have

$$\sigma^T = \frac{E}{9} \begin{bmatrix} Vd+3Ve & Vd & Vd & 0 & 0 & 0 \\ Vd & Vd+3Ve & Vd & 0 & 0 & 0 \\ Vd & Vd & Vd+3Ve & 0 & 0 & 0 \\ 0 & 0 & 0 & Vd & 0 & 0 \\ 0 & 0 & 0 & 0 & Vd & 0 \\ 0 & 0 & 0 & 0 & 0 & Vd \end{bmatrix} \in^T \qquad (7)$$

Isotropy is assured if

$$\frac{1}{2}(C_{11} - C_{12}) = C_{44} \qquad (8)$$

which is satisfied in this case when

$$\frac{3}{2}V_e = V_d \qquad (9)$$

i.e. the proportion of rods lying parallel to cube edges is 2/3 times the proportion lying parallel to cube diagonals. If the total volume fraction of the rods is V_f, then we must set

$$V_e = \frac{2V_f}{5} \quad \text{and} \quad V_d = \frac{3V_f}{5} \qquad (10)$$

Substituting these values in Equation (7) we obtain

$$\sigma^T = \frac{V_f E}{6} \begin{bmatrix} 1.2 & 0.4 & 0.4 & 0 & 0 & 0 \\ 0.4 & 1.2 & 0.4 & 0 & 0 & 0 \\ 0.4 & 0.4 & 1.2 & 0 & 0 & 0 \\ 0 & 0 & 0 & 0.4 & 0 & 0 \\ 0 & 0 & 0 & 0 & 0.4 & 0 \\ 0 & 0 & 0 & 0 & 0 & 0.4 \end{bmatrix} \in^T \qquad (11)$$

which is the array of elastic stiffnesses for an isotropic material. The fibres are arranged parallel to seven directions.

It is easy to verify that another simple isotropic arrangement is obtained with two sets of fibres if 4/5 of the material is aligned parallel to cube face diagonals and 1/5 parallel to cube edges. In this case the fibres are arranged parallel to nine directions. Attainable volume fractions for these two arrangements are given in the Table.

4. Attainable volume fractions

We have given these two simple examples involving more than one set of rods because, in fact, the case of fibre bundles (or rods) distributed equally in six directions in space is much less easy to deal with. The geometry is complicated [1, 2, 3]. The fibres in the simplest case, must run normal to the faces of a regular dodecahedron and a truly repeating pattern in 3-D is not possible. Parkhouse and Kelly [1] have investigated this case is in great detail and find that for a quasi regular array a volume factor of approximately one quarter can be obtained but only with bundles of fibre of four types of cross section.

The attainable volume fractions for the 3-D arrangement of fibres in six or more directions are given in the Table. The cross sections for the cases of fibres parallel to cube face diagonals and cube edges are relatively simple ones

TABLE

Rod Directions	Rod Cross Sections	E_{max}	E_{min}	Volume Fraction
Random in space	?	0	0	0
Cube face diagonals	70.5° rhombic	$E/15$	$E/27$	$1/3$
Cube face diagonals	truncated rhombic*	$E/8$	$5E/72$	$5/8$
Normals to regular Dodecahedron	4 incl.reg.pentagonal		$EV_{f/6}$	approx 0.248
60% Cube body diagonals 40% cube edge	rhombic square	$\frac{5E}{16}\left(2-\sqrt{3}\right)$	$= 0.084\,E$	$\frac{15}{8}\left(\left(2-\sqrt{3}\right)\right)$ $= 0.50_2$
80% cube face diagonals 20% cube edges	irreg. septagon irreg. pentagon	$EV_{f/6}$	$=0.092E$	0.5525

*truncated to form a hexagon so that the short sides of the hexagon are of length one quarter of the short diagonal of the rhombus

References:

[1] J. G. Parkhouse and A. Kelly. THE REGULAR PACKING OF FIBRES IN THREE DIMENSIONS. Proc. R Soc. Lond. A 1998 in press.

[2] A. Kelly and J G Parkhouse . ELASTICALLY ISOTROPIC FIBROUS COMPOSITES. Proceedings 18th Riso International Symposium on Materials Science: Polymeric Composites- Expanding the Limits. August 1997 pp 59-64 Published by Riso National Laboratory. Roskilde, Denmark.

[3] R.M.Christensen. SUFFICIENT SYMMETRY CONDITIONS FOR ISOTROPY OF THE ELASTIC MODULI TENSOR. Trans ASME Jnl. Appl. Mech. (1987) 54 772-777

Micromechanical analysis of failure in composites at phase interface

Seiichi Nomura[*] and Harry D. Edmiston[*]

[*] Department of Mechanical and Aerospace Engineering, The University of Texas at Arlington, Arlington, TX 76019–0023, U.S.A.

Abstract

A failure envelope for a metal matrix composite is derived with the assumption that composite failure takes place at the phase interface between a fiber and the matrix by yielding. The composite is modeled as an anisotropic matrix that contains a single fiber where the matrix possesses the properties of the composite (self-consistent model). The stress field at the phase interface around a fiber is estimated analytically by Eshelby's method and is substituted into an equation of failure criterion for the matrix. Failure envelopes can be thus drawn that properly reflects the effect of fiber shapes, anisotropy and fiber volume fractions.

1. Introduction

Theoretical prediction of failure criteria for composite materials has a long history in literature including a simple empirical model [1] and more detailed micromechanical approaches [2]. Different assumptions and models naturally yield different results that can be interpreted differently.

In this paper, a composite yield criterion is derived based on the yield criterion of the matrix phase at the interface between the fibers and the matrix.

A composite material is modeled as an anisotropic matrix with the elastic properties of the composite that contains a single inclusion. The effective elastic modulus of the composite is chosen by the self-consistent approximation so that the entire composite exhibits the identical response as an equivalent anisotropic and homogeneous medium [3-6]. The self-consistent model is simple in its concept yet can take the presence of multiple fibers into account.

The stress distribution at the interface between the matrix and fibers can be obtained by Eshelby's method [7] and Mura [3] as a function of the applied stress at a far field, the properties of the constituents, the fiber volume fraction and the fiber geometry. By substituting the stress field in the immediate neighborhood of the fiber into a failure criterion of the matrix (von Mises condition), the failure criterion can be expressed in terms of the far field stress.

2. Modeling

The stress field, σ^I, inside a single ellipsoidal inclusion in an infinite matrix subject to a far field strain, $\bar{\epsilon}$, is uniform and can be expressed by the Eshelby's method [7] symbolically as

$$\sigma^f = C^f \left(I + S(C^m - C^f)S - C^m \right)^{-1} (C^f - C^m)\bar{\epsilon} \tag{1}$$

where C^f and C^m are the fiber elastic modulus and the matrix modulus, respectively, and S is the Eshelby tensor [7] whose explicit form is found in the literature [3] and is a function of the aspect ratio and the matrix elastic properties. In equation (1), all the quantities except for $\bar{\epsilon}$ and σ^f are symmetrical fourth rank tensors and the product of tensors is defined as $A_{ijkl}B_{klmn}$ if both A and B are fourth rank tensors and as $A_{ijkl}x_{kl}$ if A is a fourth rank tensor and x is a second rank tensor. The identity tensor, I, is defined as $I_{ijkl} = \frac{1}{2}(\delta_{ik}\delta_{jl} + \delta_{il}\delta_{jk})$.

When there are multiple fibers distributed in an infinite matrix in a statistically homogeneous way, the effective elastic modulus, C^*, that represents the entire composite can be approximated by the self-consistent approximation [8,9] by

$$C^* = C^m + v_f(C^f - C^m) \left(I + S(C^* - C^f)S - C^* \right)^{-1} (C^f - C^*) \tag{2}$$

where v_f is the volume fraction of the fiber.

Equation (2) is a set of non-linear simultaneous equations for the effective moduli, C^*, and can be solved numerically. The self-consistent approximation can take interaction among fibers into account. Its limitation is low volume fractions and the symmetry of the matrix and fibers. Once the effective moduli of the composite are derived, the stress field inside the fiber can be approximated by using the Eshelby's approach.

At the boundary between the matrix and the fiber, the displacements and tractions must be continuous. However, the strains across the boundary need not be continuous but may be described in terms of a "jump" parameter, λ_i, as [3]

$$\epsilon_{ij}^{out} - \epsilon_{ij}^{in} = \lambda_i n_j \tag{3}$$

where ϵ_{ij}^{out} is the strain outside the fiber, ϵ_{ij}^{in} is the strain inside the fiber and n_i is the outward unit normal vector to the surface of the fiber.

Equating the tractions at the boundary results in

$$C_{ijkl}^m \epsilon_{kl}^{out} n_j = C_{ijkl}^f \epsilon_{kl}^{in} n_j \tag{4}$$

Substituting equation (3) into equation (4) results in a set of simultaneous equations for the parameters, λ_i, as

$$C_{ijkl}^m \lambda_k n_l n_j = (C_{ijkl}^f - C_{ijkl}^m)\epsilon_{kl}^{in} n_j \tag{5}$$

Solving these equations for λ_i allows the calculation of ϵ_{ij}^{out} in terms of ϵ_{ij}^{in}.

The stresses outside the fiber may then be calculated from the applied far field stress, $\bar{\sigma}$, the elastic moduli of the fiber and the matrix, C^f and C^m, respectively, and the fiber volume fractions, v_f, as

$$\begin{aligned} \sigma_{ij}^{out} &= C_{ijkl}^m \epsilon_{kl}^{out} \\ &= G_{ijkl}(C^f, C^m, v_f)\bar{\sigma}_{kl} \end{aligned} \tag{6}$$

where $G_{ijkl}(C^f, C^m, v_f)$ is the proportionality factor to the far stress field. Equation (6) shows that σ_{ij}^{out} can be expressed as a function of the applied far field stresses.

The interface stresses, σ_{ij}^{out}, are now substituted into an appropriate failure criterion for the matrix phase as

$$F(\sigma_{ij}^{out}) = 0 \tag{7}$$

This is translated into a failure criterion in terms of $\bar{\sigma}_{ij}$ using equation (6) as

$$F(\bar{\sigma}_{ij}) = 0 \tag{8}$$

Thus, a failure criterion for the entire composite is written in terms of the applied far field stress.

For example, the von Mises failure criterion is expressed using the interface stresses as

$$\sigma_{ij}'\sigma_{ij}' = s^2 \tag{9}$$

where σ_{ij}' is the deviatoric part of the stress and s is a constant.

Substituting equation (6) into equation (9) yields the failure criterion in terms of the externally applied stress as

$$g_{ijkl}\bar{\sigma}_{kl}\bar{\sigma}_{ij} = s^2 \tag{10}$$

where g_{ijkl} is a fourth rank tensor and is a function of the composite constituents, the fiber volume fraction and the fiber shape (aspect ratio).

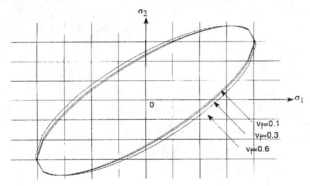

Figure 1:Typical failure envelope

3. Results and discussions

Without appropriate experimental data readily available, the developed theory is demonstrated for a fictitious composite only. Figure 1 represents the failure envelopes computed for a composite reinforced with cylindrical fibers aligned unidirectionally. The material is transversely isotropic and the envelopes represent the allowable states of axial stress which may act in the plane of isotropy. The envelopes presented represent composites with fiber volume fractions of.1,.3 and 6. In the transverse plane the reinforcement is not continuous and the strength of the material is controlled by the strength of the matrix. The failure envelopes represent this characteristic as well as demonstrate an increase in strength with increasing fiber volume fraction.

The assumption of initial failure at the phase boundary implies that this modeling technique is most applicable to composites subject to matrix dominated failure. This technique, however, is not restricted to the von Mises criterion. Any suitable failure theory can be incorporated into the algorithm to generate the failure envelopes. The work continues to address the effect of fiber shape (aspect ratios) which will be reported subsequently.

References

[1] Tsai, S.W. and Wu, E. M., 1971, "A General Theory of Strength for Anisotropic Materials," *J. Comp. Mat.*, 5, pp.58-80.

[2] Dvorak, G.J. and Bahai-El-Din, Y.A., 1979, "Elastic-Plastic Behavior of Fibrous Composites," *J. Mech. Physics Solids*, 27, p.51.

[3] Mura, T., 1987, *Micromechanics of Defects in Solids*, 2nd Ed. Martinus Nijhoff, Dordrecht.

[4] Nomura, S. and Chou, T.-W., 1984, "Bounds for Elastic Moduli of Multiphase Short-Fiber Composites," *J. Appl. Mechanics*, 51, pp.540-545.

[5] Benveniste, Y. and Aboudi, J., 1984, "A Continuum Model for Fiber Reinforced Materials with Debonding," *Int. J. Solids Structures*, 20, pp.935-951.

[6] Laws M. and McLaughlin, R, 1979, "The Effect of Fibre Length on the Overall Moduli of Composite Materials," *J. Mech. and Phys. Solids*, 27, pp.1-13.

[7] Eshelby, J. D., 1957, "The Determination of the Elastic Field of an Ellipsoidal Inclusion and Related Problems," *Proc. Roy. Soc.*, A241, pp.376-396.

[8] Hill, R., 1965, "Theory of Mechanical Properties of Fibre Strengthened Materials - III. Self-Consistent Model," *Journal of the Mechanics and Physics of Solids*, 13, pp.189-198.

[9] Chou T.-W., Nomura, S. and Taya, M., 1980, "A Self-Consistent Approach to the Elastic Stiffness of Short-Fiber Composites," *J. Composite Mat.*, 14, pp.178-188.

The influence of fibre diameter and strength on the properties of glass-fibre-reinforced polyamide 6,6

J. L. Thomason, Owens Corning Science & Technology Center, 2790
Columbus Road, Granville, OHIO 43023.1200

Abstract

We discuss the effect of fibre strength (E-glass and high strength S-2 glass®) and
diameter (17-9 microns) on the balance of mechanical properties of glass reinforced
polyamide 66. The results show that the elastic properties of short glass fibre reinforced
polyamide 6,6 are not strongly influenced by fibre diameter in the 17-10 micron range.
Surprisingly, addition of a nominally stiffer *S-2 glass* fibre did not improve composite
stiffness. The ultimate properties of these composites (strength and Izod impact) showed a
clear dependence on fibre diameter and the presence of high strength *S-2 glass* fibres.
Tensile elongation, tensile and flexural strength, and unnotched Izod impact all increased
significantly over the 17-10 micron diameter range. Notched Izod impact showed a small
but significant decrease over the same range. Addition of 20% w/w of *S-2 glass* to the 17
micron E-glass returned the composite property level to approximately the same level as
that obtained with 14 micron E-glass fibres.

Introduction

Glass fibre reinforced polyamides, such as nylon 6 and nylon 66, are excellent
composite materials in terms of their high levels of toughness, heat and oil resistance.
However, the stiffness and strength of these materials may still be too low to replace
some metal components. The mechanical properties of thermoplastic composites
containing 'short' fibres has been the subject of much attention[1-6]. These properties
result from a combination of the fibre and matrix properties and the ability to transfer
stresses across the fibre-matrix interface. The optimization of composite performance
through control of the base materials and the various steps of fibre-matrix
combination and parts production is a major technical challenge facing raw materials
suppliers to the composite industry. Variables such as the fibre content, length,
diameter, orientation and the interfacial strength are of prime importance to the final
balance of properties exhibited by injection moulded thermoplastic composites.
However, due to the complexity of the interdependence of many of the structure-process-
property variables of these systems it is extremely difficult to isolate the overall influence
of individual parameters on the final balance of composite properties. Well defined
samples where parameters have been independently varied can be difficult and expensive
to produce. We are currently engaged on a programme to further elucidate the structure-
process-property relationships in short glass fibre reinforced polyamides. In this paper we
discuss some results on the effect of fibre strength and diameter on the balance
mechanical properties of glass reinforced polyamide 66.

Experimental

The E-glass samples in this study were all produced using the Owens Corning Cratec™
process for chopped strands. Samples with nominal fibre diameters of 10, 11, 14 and 17
micron were produced. These samples were chopped to a length of 4 mm and were sized
with 123D sizing which is a chopped strand sizing for polyamide reinforcement. The *S-2
glass* sample was produced at a different facility with a nominal 9 micron diameter was
chopped to 1.6 mm and was sized with 933 size which was developed for reinforcement
of high performance thermoplastics. Nominal properties of E and *S-2 glass* fibres are
given in Table 1. The polyamide 6,6 (PA6,6) used was Zytel 101.

Glass	E	*S-2*
Tensile Strength (MPa) at 23°C	3445	4890
Tensile Modulus (GPa) at 23°C	72.3	86.9
Tensile Elongation at 23°C (%)	4.8	5.7

Table 1. Nominal Glass Tensile Properties

The glass bundles and pre-dried PA6,6 pellets were dry blended to 30% w/w glass content
and compounded on a single screw extruder (2.5 inch, 3.75:1, 24:1 L/D screw). The
compounds were moulded into test bars on a 200-ton Cincinnati Milacron moulding
machine. Melt temperatures 288-293°C for compounding and 293-299°C for
moulding, at a mould temperature of 93°C. All mechanical property testing was
performed at 23°C and at a relative humidity of 50%. The test specimens were
allowed to equilibrate for 24 hours under these conditions before testing. Tensile
properties were measured in accordance with the procedures in ASTM D-638,
flexural properties following ASTM D-790, and Izod impact properties ASTM D-
256. Fibre length and diameters were determined by image analysis and optical
microscopy on fibre samples removed from the moulded bars after high temperature
ashing.

Results

The weight averaged fibre lengths and fibre diameters determined on fibres removed
from injection moulded tensile bars are given in Table 2. Unfortunately reduction of
any distribution to a single average always results in a loss of information. Number
averages are weighted to the short side of the fibre length distribution and are useful
for examining the level of fibre damage during processing. However, the weight
average reflects the volume fraction at that length and is therefore more useful in
matching to the composite mechanical properties which are mainly volume fraction
driven. Examination of the data on fibre length reveals a clear trend for increasing
fibre length retention with increasing average fibre diameter. On adding 5% *S-2 glass*
there does not appear to be any significant change in the average fibre length,
however the addition of 20% *S-2 glass* to the 17 micron E-glass appears to have
resulted in a reduction in the average fibre length in the moulded composites.

Most composite mechanical properties are strongly influenced by the fibre length in the moulded part. Consequently, understanding the mechanisms of fibre length degradation during processing is an important area of study. There has been, and continues to be, much activity in this area[1-14]. One of the most commonly discussed mechanisms is the breaking of fibres in bending mode due to the high levels of shear in the compounding and moulding process[9-14]. If this is correct then the flexibility, or ability to survive bending, of glass fibres is important to the composite final properties. The ability of a fibre to survive bending depends on its minimum bending radius[11] defined as $R_b = r . E_f / S_f$ (r = fibre radius, E_f = fibre modulus). All other factors being equal, residual fibre length after compounding and moulding should be inversely proportional to the fibres' minimum bending radius. Consequently the above equation implies that residual fibre length should be inversely proportional to fibre diameter. However, comparing the data for aspect ratio and minimum bending radius in Table 2, we do not find this inverse proportionality. In fact the data show a direct relationship between these two measurement. It may well be that the increase in apparent viscosity of molten fibre-thermoplastic mixtures when fibre diameter is decreased may have resulted in a higher level of fibre attrition. The probability of fibre-fibre contact during processing, another possible mechanism of fibre length attrition, is also increased when the fibre diameter is decreased. This will require further detailed investigation.

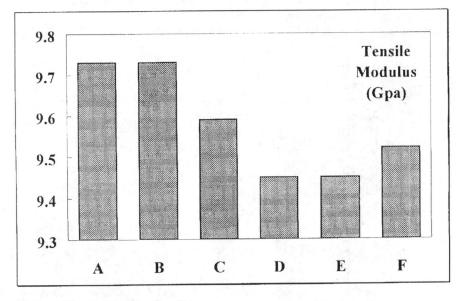

Figure 1. Tensile Modulus Results

The composite mechanical property data is summarised in Figures 1-5 and a full overview of all results is presented in Table 2. The tensile modulus of samples A-D shows a slight trend for decreasing stiffness with increasing fibre diameter. The stiffness of samples E and F does not appear to have been improved significantly by the addition of the stiffer *S-2 glass* fibres. The same general trends were also found in the flexural modulus data. The main factors affecting the apparent modulus of injection moulded test bars are the fibre content, stiffness, and orientation, and the matrix stiffness[9-16] . To a lesser extent the aspect ratio, in the normal range found in these samples, also plays a role. The slight differences in measured glass content do not explain the observed differences in composite stiffness. The fibre orientation and aspect ratio are somewhat more complex quantities which, unfortunately, are not independent variables. It has been shown in many studies that injection moulded composites have a complex layered structure with very different average fibre orientation in the different layers[1,5,6] . Furthermore the thickness of these layers and the fibre orientation within the layers has been shown to vary with fibre length. In general, longer residual fibre length in the moulded composite leads to a more random average orientation and a lower "orientation factor" in the equations commonly used to calculate composite stiffness. After taking differences in average aspect ratio into account, the reduction in stiffness between sample A and D can be theoretically calculated to be given by either a reduction in aspect ratio of 32% or a reduction in orientation factor by 3%. The data in Table 2 do not indicate a large change in aspect ratio between samples A and D, but there is an increase in average fibre length which, following the argument presented above, may have lead to a lowering of the orientation factor. Comparing the results for samples D and F we note that addition of 20% of stiffer *S-2 glass* fibres has had no significant effect on the composite stiffness. Furthermore the average fibre length in sample F is lower so it seems unlikely that the length dependent orientation factor discussion above applies to this unexpected result.

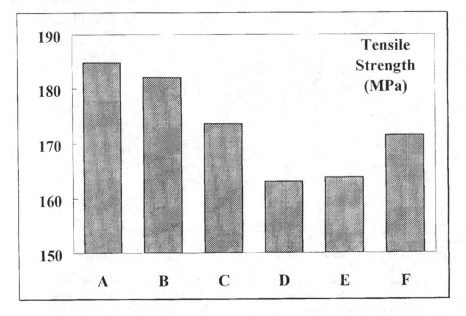

Figure 2. Tensile Strength Results

Tensile strength and elongation results are presented in Figures 2 and 3. The data for tensile and flexural strength from these samples also gave identical trends. The results for samples A-D show that the composite strength is reduced as the average fibre diameter is increased. Addition of small (5%) amounts of high strength *S-2 glass* does not apparently improved the composite strength, although addition of a larger fraction (20%) did result in a significant improvement. An identical trend was seen in the tensile elongation. A similar trend was also observed with unnotched Izod data (Figure 4) although the magnitude of the changes was larger in this case. It is interesting to note that the results for the notched Izod Impact show the opposite trend to all the preceding properties. Figure 5 shows that the notched Izod increased with increasing fibre diameter and then decreased with addition of chopped *S-2 glass*. This trend is in line with some recently published semi-empirical equations for energy absorption in random inplane composites based on notched Charpy impact results[17].

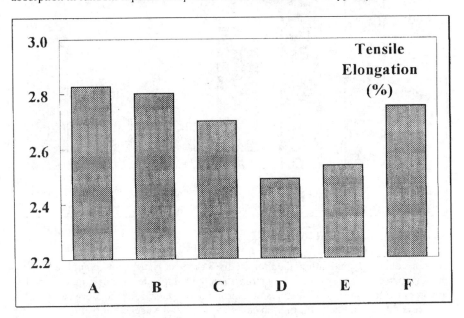

Figure 3. Tensile Elongation Results

There are many more factors which influence the ultimate properties of composites in comparison to the elastic properties. In particular, the balance of fibre aspect ratio, fibre strength, and interfacial shear strength (IFSS), is 'critical' in determining strength and impact properties[18]. This concept is embodied in the widely used definition of critical fibre length $L_c = S_f.D/2.Tau$ (S_f = fibre strength, D = fibre diameter, Tau = IFSS) as used in the well known Kelly-Tyson equation for strength prediction in discontinuous fibre reinforced composites[19]. There are considerable difficulties in obtaining data which reflect the true values of these variables such as S_f at the short lengths common in these composites. Accurate modeling of the composite properties requires more realistic values of the residual fibre strength and an independent determination of the length distributions of the two fibre types in the mixed fibre samples. Furthermore, there is a considerable spread in fibre diameter in most of the commercially available glass fibre products. It is one of the goals of our research programme to obtain more reliable data in this area. In terms of IFSS, a recent study on glass fibre reinforced polypropylene showed that IFSS as determined by the fibre pullout test was appropriate for Tau values in the Kelly-Tyson equation[18].

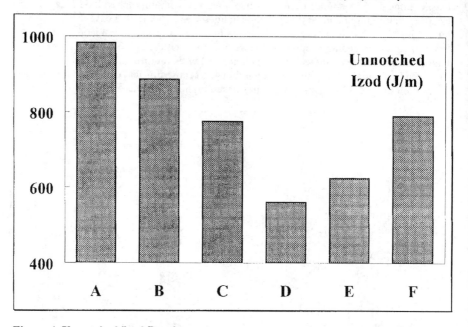

Figure 4. Unnotched Izod Results

We have obtained some initial data for the IFSS in these systems and combined with more realistic values for actual fibre strength in the composite[20] we calculate critical aspect ratios for the E-glass samples of approximately 35 and for the *S-2 glass* samples of approximately 41. In both cases we see that the aspect ratios in Table 2 all exceed these values and therefore we are operating in the regime in Kelly-Tyson where some of the fibres should be broken during the composite fracture process. Consequently the absolute value of fibre strength should have a direct influence on the measured composite strength. Using a simple ratio of the strength values in Table 1 indicates that replacing 20% E by *S-2 glass* should add approximately 8% to the fibre contribution of the composite tensile strength. If we remove a fraction of the values in Table 2 proportional to the matrix contribution (approximately 0.15x70 MPa) then we find that the fibre contribution of sample F is approximately 8% higher than sample D. Of course this simple calculation does not take differences in fibre diameter and IFSS into account. However, it does indicate that the gain in composite strength through addition of *S-2 glass* is found to be in the region of the expected value.

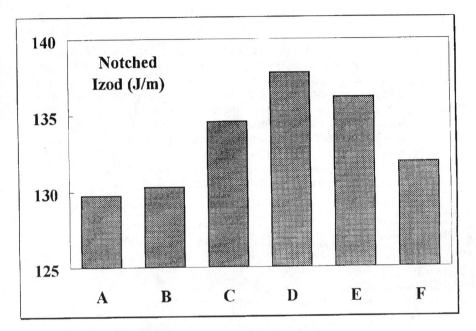

Figure 5. Notched Izod Results

Sample		A	B	C	D	E	F
Glass		E10	E11	E14	E17	E17/S-2	E17/S-2
Ratio	(% w/w)	100	100	100	100	95/5	80/20
Tensile Strength	(MPa)	184.5	181.9	173.7	163.4	163.7	171.2
Youngs Modulus	(GPa)	9.75	9.70	9.56	9.45	9.42	9.53
Tensile Elongation	(%)	2.83	2.80	2.70	2.49	2.54	2.75
Izod Unnotched	(J/m)	983	885	775	564	628	790
Izod Notched	(J/m)	130	129	135	138	136	132
Glass Content	(% w/w)	30.1	29.7	30.0	29.8	30.0	29.5
Fibre Diameter	(Micron)	9.8	10.9	13.7	17.0	17.0/9.0	17.0/9.0
Weight Av. Length	(Micron)	470	650	850	790	850	690
Fibre Minimum Bending Radius	(Micron)	103	114	144	178	178/87	178/87
Aspect Ratio		48.0	59.6	62.0	46.5	?	?

Table 2. Mechanical Properties and Fibre Data

Conclusions

Our results show that the elastic properties of 30% w/w short glass fibre reinforced polyamide 6,6 are not strongly influenced by fibre diameter in the 10-17 micron range. Surprisingly, addition of a nominally stiffer *S-2 glass* fibre did not improve composite stiffness. The ultimate properties of these composites (strength and Izod impact) showed a clear dependence on nominal fibre diameter and the presence of high strength *S-2 glass* fibres. Tensile elongation, tensile and flexural strength, and unnotched Izod impact all decreased significantly over the 10-17 micron diameter range. Notched Izod impact showed a small but significant increase over the same range. Addition of 20% w/w of *S-2 glass* to the 17 micron E-glass returned the composite property level to approximately the same level as that obtained with 14 micron E-glass fibres

References

1. S.Toll and P.O.Andersson, Polym.Composites., **14**, pp 116-125 (1993)
2. N. Takeda, D.Y. Song and K. Nakatal, Adv. Compos. Mater., **5**, pp 201-212 (1996)
3. T. Moriwaki, Composites, **27A**, pp 379-384 (1996)
4. N. Sato, T. Kurauchi, S. Sato and O. Kamigaito, J. Mater. Sci., **19**, pp 1145-1152 (1984)
5. J. J. Horst and J. L. Spoormaker, J. Mater. Sci., **32**, pp 3641-3651 (1997)
6. M. Akay and D. Barkley, J. Mater. Sci., **26**, pp. 2731-2742 (1991)
7. M. Arroyo and F.Avlos, Polym. Compos., **10**, pp 117-121 (1989)
8. R. A. Schweizer, Plast. Compound., **4**(6) 58-62 (1981)
9. J. B. Shortall and D. Pennington, Plastics Rubber Proc. Appl, **2**, 33-40 (1982)
10. B. Fisa, Polym.Compos., **6**, pp 232-239 (1985)
11. B. Franzen, C. Klason, J. Kubat and T. Kitano, Composites **20**, pp 65-76 (1989)
12. R. S. Bailey and H. Kraft, Intern. Polymer Processing, **2**, pp 94-101 (1987)
13. T. Vu-Khanh, J. Denault, P. Habib, and A. Low, Compos. Sci. Technol., **40**, pp 423-435 (1991)
14. R. S. Bailey, M. Davies and D. R. Moore, Composites, **20**, pp 453-460 (1989)
15. M .J. Folkes in "Short Fibre Reinforced Thermoplastics", Research Studies Press, Chichester (1985)
16. J.L.Thomason and M.A.Vlug, Composites, **27A**, pp 477-484 (1996)
17. J. L. Thomason and M.A.Vlug, Composites, **28A**, pp 277-288 (1997)
18. J. L. Thomason, M.A.Vlug, G. Schipper and H .G. L. T. Krikor, Composites, **27A**, pp 1075-1084 (1996)
19. A. Kelly and W. R. Tyson, J. Mech. Phys. Solids, **13**, pp 329-350 (1965)
20. J. L. Thomason, unpublished results

LIMIT ANALYSIS OF PERIODIC COMPOSITES BY A KINEMATIC APPROACH

V. Carvelli, G. Maier, A. Taliercio

Department of Structural Engineering – Technical University (Politecnico) of Milan
Piazza Leonardo da Vinci 32, 20133 Milan-Italy

Abstract

Efficient evaluation of strength domains for metal-matrix composite (MMC) materials with unidirectional periodic fibers is the objective pursued in this paper. The analyses presented are based on the kinematic limit theorem and on the finite element method in its displacement formulation. The resulting nonlinear mathematical programming problems are solved by an iterative procedure. The limit analysis method developed and implemented for MMCs takes the strain-periodicity of the microscopic displacement field into account and directly leads to an upper bound of the collapse multiplier and to a collapse mechanism. The numerical tests performed so far exhibit fast convergence and show excellent agreement with experimental and numerical results arising from different approaches.

1. Introduction

In view of their peculiar mechanical properties, metal-matrix composites (MMC) are widely used in space and aeronautical industry. They are particularly suited to technological applications where high specific stiffness, ultimate strength and ductility are required. In particular, aluminium pipes reinforced by continuous boron fibers are adopted for space shuttles, composite materials with aluminium matrix and carbon fibers are frequently employed in aircraft and helicopters.

Their importance in the above mentioned and other fields confers practical interest to the cost-effective evaluation of the strength domain for MMC materials with unidirectional periodic fibers. This is the objective pursued in this paper.

Many kinds of MMCs are produced with a regular arrangement of fibers. This circumstance allows us to interpret the composite material as a periodic multiphase system. In this context the homogenization theory of periodic composite materials can be applied.

As shown in [1], the determination of the macroscopic strength domain of a ductile periodic heterogeneous material is traceable to the solution of a limit analysis problem defined only over a Representative Volume (RV), i.e. over the smallest volume (or "unit cell") that contains all information necessary to describe the microstructure. Therefore the macroscopic strength domain is evaluated by analysing a RV subject to a state of "macroscopic" (or average) stresses.

The analyses presented herein are based on the kinematic limit theorem of classical plasticity and on the finite element method in its displacement formulation. The mathematical programming problem thus generated (see e.g. [2]) is solved by an iterative procedure. The algorithm adopted is patterned according to the one applied in [3, 4, 5] to homogeneous materials.

The original feature of the present work is the application of this approach and algorithm to the *RV* of periodic media using periodicity boundary conditions. The limit analysis method developed and implemented here for MMCs requires a suitable partition of the nodal variables to take the strain-periodicity of the displacement field at the microscale into account, and directly leads to an upper bound on the collapse multiplier and to a collapse mechanism. The numerical tests performed so far show rapid convergence and excellent agreement with experimental results [6, 7], with other numerical results based on piecewise linearization of the yield surface and linear programming [7, 8], and with evolutive analyses [9]. From a computational point of view, the present approach turns out to be definitely less expensive than evolutive analyses [9] and approaches based on piecewise linear yield domain [7, 8].

2. Fundamentals of homogenization theory for periodic media

Let a composite material consist of an array of parallel fibers with circular cross-section embedded in a bonding matrix (Fig. 1a). If the fibers form a regular array, the composite can be regarded as heterogeneous material with periodic structure. In this case a representative volume (*RV*) can be singled out which contains all information needed to describe the structure completely. The *RV* will be described in the orthogonal reference frame ($O\ x_1\ x_2\ x_3$), x_3 being in the fiber direction (Fig. 1b). Matrix notation is adopted throughout, with underlined symbols for matrices (and column-vector) and superscript T for transposition.

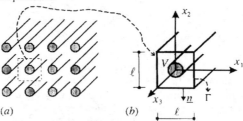

(a)   (b)

Figure 1. (a) periodic fiber reinforced composite, (b) representative volume (*RV*)

The analysis which follows will be developed in a cross section perpendicular to x_3-axis (x_1-x_2 plane), the length of the *RV* along x_3-axis being therefore immaterial [8].

As shown e.g. in [1], the determination of the macroscopic strength domain of a periodic heterogeneous material reduces to solving a limit analysis problem defined over the *RV*. Any *RV* in the composite medium is associated with a point in a fictitious homogenized medium subject to "macroscopic" stresses $\underline{\Sigma}$ and strains $\underline{E}$. These variables are defined as average:

$$\underline{\Sigma} \equiv \langle \underline{\sigma} \rangle = \frac{1}{V}\int_V \underline{\sigma}\, dV \qquad \underline{E} \equiv \langle \underline{\varepsilon} \rangle = \frac{1}{V}\int_V \underline{\varepsilon}\, dV \qquad (1)$$

where V is the volume of the *RV* and $\underline{\sigma}$, $\underline{\varepsilon}$ are the "microscopic" stress and strain fields, respectively. For a *RV* located at a sufficiently large distance from the boundary of the heterogeneous body, the microscopic strain and stress fields conform to the periodicity of the geometry [1]. This means that:

$$\tilde{\underline{u}} = \underline{u} - [x]\underline{E} \qquad \text{periodic on } \Gamma$$

$$[\underline{n}]\underline{\sigma} \qquad \text{anti - periodic on } \Gamma \qquad (2)$$

where $\underline{u}$ is the microscopic displacement field, $[\underline{x}]$ is the matrix which defines its linear addend and $[\underline{n}]$ is the matrix of Cauchy equilibrium conditions formed by the outward normal to the boundary Γ of V (Fig. 1b). Eqns. (2) impose the continuity of the displacement $\tilde{\underline{u}}$ and the traction vector $[\underline{n}]\underline{\sigma}$ between two adjacent RVs. The relation (2a) entails splitting the local strain field into its fluctuating and uniform average part

$$\underline{\varepsilon} = \tilde{\underline{\varepsilon}} + \underline{E}, \qquad \tilde{\underline{\varepsilon}} \text{ periodic on } \Gamma \quad (\langle \tilde{\underline{\varepsilon}} \rangle = 0) \tag{3}$$

Homogenization is the procedure that relates $\underline{\Sigma}$ to $\underline{E}$ by means of (1) and the local constitutive laws. If $\underline{\sigma}$ and $\underline{u}$ fulfil eqns. (2), it can be shown [1] that the average of the "microscopic" work over any RV is equal to the "macroscopic" work:

$$\langle \underline{\sigma}^T \underline{\varepsilon}(\underline{u}) \rangle = \underline{\Sigma}^T \underline{E} \tag{4}$$

Eqn. (4) expresses the virtual work principle for a RV subjected to "loads" $\underline{\Sigma}$; it is also called Hill's macrohomogeneity equality and plays an important role in the homogenization theory.

3. Finite element formulation of kinematic limit analysis for a RV

The present purpose is to evaluate the macroscopic strength domain of a periodic fiber-reinforced material under the hypothesis that both constituents are isotropic, rigid-perfectly-plastic and stable in Drucker's sense, i.e. with associated flow rules and convex yield domain. Interpreting the average stresses $\underline{\Sigma}$ as assigned loads on the RV, let eqn. (4) be written as the balance between internal and external power associated to a plastic mechanism $\underline{u}$. It is implicitly understood that the velocity field $\underline{u}$ defines the strain rate field $\underline{\dot{\varepsilon}}$ through the compatibility operator, and $\underline{\dot{\varepsilon}}$ in turn defines the dissipated power $\underline{\sigma}' \underline{\dot{\varepsilon}}$ through Hill's extremum principle. Thus, account taken of eqn. (3), the kinematic theorem of limit analysis can be formulated as the following minimization problem:

$$\begin{cases} s = \min_{\underline{u}} \dfrac{1}{V} \displaystyle\int_V \underline{\sigma}' \, \underline{\dot{\varepsilon}}(\underline{u}) \, dV, \text{ subject to:} \\[2ex] \underline{\Sigma}^T \left(\dfrac{1}{V} \displaystyle\int_V \underline{\dot{\varepsilon}}(\underline{u}) \, dV \right) = 1 \end{cases} \tag{5}$$

where s is the limit multiplier of the assigned macroscopic stresses $\underline{\Sigma}$ (i.e. the safety factor with respect to plastic collapse) and (5b) is the normalization condition for the macroscopic work rate. Two differences are worth noting between a classical limit analysis problem and the formulation (5): the periodicity condition has to be satisfied by the addend $\tilde{\underline{u}}$ of the velocity field; the loads applied to the RV are not accounted for through boundary conditions, but through an average condition (eqn. (1a)). If Mises yield criterion is assumed for both phases, eqns. (5) can be rewritten as:

$$\begin{cases} s = \min_{\underline{u}} \sqrt{\dfrac{2}{3}} \dfrac{1}{V} \displaystyle\sum_r \int_{V_r} \sigma_0^r \sqrt{\underline{\dot{\varepsilon}}^T \underline{D}\underline{\dot{\varepsilon}}} \, dV, \text{ subject to:} \\[2ex] \underline{\Sigma}^T \underline{\dot{E}} = 1 \\[1ex] \dot{\varepsilon}_V(\underline{u}) = 0 \qquad \forall \underline{x} \in V \end{cases} \tag{6}$$

where σ_0^r denotes the yield limit of each constituent (r=matrix or fiber), $\underline{D}$ is the usual symmetric matrix of constants emerging from Mises yield function and $\dot{\varepsilon}_v$ the volumetric strain rate.

By finite element (FE) modelling the periodic velocity field $\underline{\tilde{u}}$ over the RV and by adopting Gauss integration to approximate the integrals over elements, one arrives at the following algebraic equality-constrained minimization with respect to the vector $\underline{\tilde{U}}$ of nodal periodic velocities and to the vector $\underline{\dot{E}}$ of average strains over RV:

$$\begin{cases} s = \min_{\underline{\tilde{U}},\underline{\dot{E}}} \sqrt{\tfrac{2}{3}} \tfrac{1}{V} \sum_{i \in I} \sigma_0^{r(i)} \rho_i |J|_i \sqrt{\underline{\dot{E}}^T \underline{D}\,\underline{\dot{E}} + 2\underline{\dot{E}}^T \underline{D}\,\underline{B}_i \underline{\tilde{U}} + \underline{\tilde{U}}^T \underline{R}_i \underline{\tilde{U}}}, \text{ subject to :} \\[6pt] \underline{\Sigma}^T \underline{\dot{E}} = 1 \\[6pt] \underline{Y}^T (\underline{\dot{E}} + \underline{B}_i \underline{\tilde{U}}) = 0 \quad i \in I \\[6pt] \underline{\tilde{U}} \qquad \text{periodic on } \Gamma \end{cases} \qquad (7)$$

The transition from problem (6) to its discrete version (7) is performed by the following provisions: the strain rate at the Gauss point i ($i \in I$, I being the set of Gauss points) is expressed as $\underline{\dot{\varepsilon}}_i = \underline{B}_i \underline{\tilde{U}} + \underline{\dot{E}}$ using eqn. (3) and the compatibility matrix $\underline{B}_i$, vector $\underline{\tilde{U}}$ gathering all nodal d.o.f.s of the assembled aggregate of FEs; it has been set $\underline{R}_i = \underline{B}_i^T \underline{D} \underline{B}_i$; ρ_i and $|J|_i$ denote the Gauss weight and the jacobian of the mapping, respectively, computed at point i; the normalization (5b) is recovered; the plastic incompressibility constraint (6c) is enforced at each Gauss point by eqn. (7c), $\underline{Y}$ being the vector of constants which transforms a strain vector into volumetric strain.

With reference to Fig. 1b, for a square RV the periodic velocity field $\underline{\tilde{u}}$ on the boundary Γ of the RV must satisfy the following conditions:

$$\begin{aligned} \underline{\tilde{u}}(-\ell/2, x_2) &= \underline{\tilde{u}}(\ell/2, x_2) \\ \underline{\tilde{u}}(x_1, -\ell/2) &= \underline{\tilde{u}}(x_1, \ell/2) \end{aligned} \qquad (8)$$

Similar periodicity conditions have to be imposed on different (e.g. hexagonal) RVs.

These constraints reduce number of independent nodal velocities and, hence, lead to a reduced vector $\underline{\tilde{U}}^*$ of d.o.f.s and to accordingly reduced matrices $\underline{B}^*$ and $\underline{R}^*$. Therefore problem (7) becomes:

$$\begin{cases} s = \min_{\underline{\tilde{U}}^*,\underline{\dot{E}}} \sqrt{\tfrac{2}{3}} \tfrac{1}{V} \sum_{i \in I} \sigma_0^{r(i)} \rho_i |J|_i \sqrt{\underline{\dot{E}}^T \underline{D}\,\underline{\dot{E}} + 2\underline{\dot{E}}^T \underline{D}\,\underline{B}_i^* \underline{\tilde{U}}^* + \underline{\tilde{U}}^{*T} \underline{R}_i^* \underline{\tilde{U}}^*}, \text{ subject to :} \\[6pt] \underline{\Sigma}^T \underline{\dot{E}} = 1 \\[6pt] \underline{Y}^T (\underline{\dot{E}} + \underline{B}_i^* \underline{\tilde{U}}^*) = 0 \quad i \in I \end{cases} \qquad (9)$$

4. Iterative solution of the constrained minimization

In the finite element formulation (9), based on displacement modelling, the incompressibility condition (9c) is known to generate "locking" phenomena. In order to avoid this spurious effect, a penalty method is adopted here, similarly to [3, 4, 5].

Enforcing the normalization constraint (9b) by the Lagrange method and the incompressibility constraint (9c) by a penalty procedure, eqns. (9) give:

$$s = \min_{\underline{\dot{E}},\underline{\dot{E}},\lambda} \left\{ \sqrt{\tfrac{2}{3}} \tfrac{1}{V} \sum_{i \in I} \sigma_0^{r(i)} \rho_i |J|_i \sqrt{\underline{\dot{E}}^T \underline{D}\,\underline{\dot{E}} + 2\underline{\dot{E}}^T \underline{D}\,\underline{B}_i^* \underline{\tilde{\dot{U}}}^* + \underline{\tilde{\dot{U}}}^{*T} \underline{R}_i^* \underline{\tilde{\dot{U}}}^*} + \lambda(1 - \underline{\Sigma}^T \underline{\dot{E}}) \right.$$

$$\left. + \alpha \sum_{i \in I} \rho_i |J|_i (\underline{\dot{E}} + \underline{B}_i^* \underline{\tilde{\dot{U}}}^*)^T \underline{C}(\underline{\dot{E}} + \underline{B}_i^* \underline{\tilde{\dot{U}}}^*) \right\} \tag{10}$$

where λ is the Lagrange multiplier, α is the penalty factor and $\underline{C} = \underline{Y}\,\underline{Y}^T$.

The optimality conditions of problem (10) lead to a set of nonlinear equations. They are solved herein through an iterative scheme according to the algorithm adopted for the limit analysis of bodies in plane strain in [3], of structures subject to constant and proportional loading in [4] and of three-dimensional continua in [5]. In the above quoted works, classical (i.e. non-periodic) boundary conditions were employed. This algorithm reduces to solving, at the iteration $k+1$, the following set of linear equations:

$$\begin{cases} \sqrt{\tfrac{2}{3}} \tfrac{1}{V} \sum_{i \in I} \sigma_0^{r(i)} \rho_i |J|_i \dfrac{\underline{R}_i^* \underline{\tilde{\dot{U}}}_{k+1}^* + \underline{B}_i^{*T} \underline{D}\underline{\dot{E}}_{k+1}}{\sqrt{H_k^i}} + \alpha \sum_{i \in I} \rho_i |J|_i (\underline{B}_i^{*T} \underline{C}\underline{B}_i^* \underline{\tilde{\dot{U}}}_{k+1}^* + \underline{B}_i^{*T} \underline{C}\underline{\dot{E}}_{k+1}) = \underline{0} \\[2mm]
\sqrt{\tfrac{2}{3}} \tfrac{1}{V} \sum_{i \in I} \sigma_0^{r(i)} \rho_i |J|_i \dfrac{\underline{D}\underline{B}_i^* \underline{\tilde{\dot{U}}}_{k-1}^* + \underline{D}\underline{\dot{E}}_{k+1}}{\sqrt{H_k^i}} + \alpha \sum_{i \in I} \rho_i |J|_i (\underline{C}\underline{B}_i^* \underline{\tilde{\dot{U}}}_{k+1}^* + \underline{C}\underline{\dot{E}}_{k+1}) = \lambda_{k+1} \underline{\Sigma} \\[2mm]
\underline{\Sigma}^T \underline{\dot{E}}_{k+1} = 1 \end{cases} \tag{11}$$

where:
$$H_k^i = \underline{\dot{E}}_k^T \underline{D}\underline{\dot{E}}_k + 2\underline{\dot{E}}_k^T \underline{D}\underline{B}_i^* \underline{\tilde{\dot{U}}}_k^* + \underline{\tilde{\dot{U}}}_k^{*T} \underline{R}_i^* \underline{\tilde{\dot{U}}}_k^* \tag{12}$$

At each iteration the rigid and plastic zones are distinguished. Namely, before proceeding with the $(k+1)th$ iteration, eqn. (12) is computed at every integration point and the set I of Gauss points is subdivided into the rigid zone subset R_{k+1} and the plastic zone subset P_{k+1}, i.e.

$$P_{k+1} = \{ i \in I \text{ such that } H_k^i \neq 0 \} \qquad R_{k+1} = \{ i \in I \text{ such that } H_k^i = 0 \} \tag{13}$$

The constraint in (13b) is enforced in weak form by the penalty method setting:

$$H_k^i = \begin{cases} \underline{\dot{E}}_k^T \underline{D}\underline{\dot{E}}_k + 2\underline{\dot{E}}_k^T \underline{D}\underline{B}_i^* \underline{\tilde{\dot{U}}}_k^* + \underline{\tilde{\dot{U}}}_k^{*T} \underline{R}_i^* \underline{\tilde{\dot{U}}}_k^* & \forall i \in P_{k+1} \\ \beta \ll 1 & \forall i \in R_{k+1} \end{cases} \tag{14}$$

The iterative procedure is initialized, at $k=0$, by assuming $H_0^i = 1 \ \forall i \in I$.

The $k+1$ solution $\{ \underline{\tilde{\dot{U}}}_{k+1}, \underline{\dot{E}}_{k+1} \}$ yields an upper bound on the limit factor of the macroscopic stresses as:

$$s_{k-1} = \sqrt{\tfrac{2}{3}} \tfrac{1}{V} \sum_{i \in I} \sigma_0^{r(i)} \rho_i |J|_i \sqrt{\underline{\dot{E}}_{k+1}^T \underline{D}\,\underline{\dot{E}}_{k+1} + 2\underline{\dot{E}}_{k+1}^T \underline{D}\,\underline{B}_i^* \underline{\tilde{\dot{U}}}_{k+1}^* + \underline{\tilde{\dot{U}}}_{k+1}^{*T} \underline{R}_i^* \underline{\tilde{\dot{U}}}_{k+1}^*} \tag{15}$$

Numerical experiences show that the above iterative process leads to the limit load multiplier s and to a collapse mechanism through a convergent sequence with monotonically decreasing s_k.

5. Applications

The limit analysis method presented in what precedes can be applied to any RV apt to define a periodic medium. It is tested below by evaluating the macroscopic strength of two kind of heterogeneous media. The first one has a periodic array of rectangular or circular holes. The second one is a unidirectional periodic fiber-reinforced composite.

The first set of numerical tests concerns a flat aluminium specimen periodically perforated by rectangular holes as specified by its RV in Fig. 2a (setting $V=1$ i.e. $\ell=1$) and thin enough to make very accurate the plane-stress idealization. In Fig. 2b experimental data presented in [6] are compared with results of the present approach, an evolutive analysis [9] and a limit analysis based on piecewise linearization of the yield domain and linear programming [7].

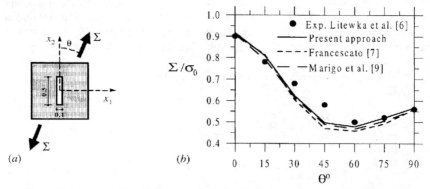

Figure 2. (a) square RV with rectangular hole, (b) limit macroscopic stress vs. angle θ

The comparison of the results obtained with the different numerical methods shows a good agreement, as can be seen in Fig. 2b. The present approach leads to the least maximum difference (of ~14% for $\theta=45°$) respect to the experimental results.

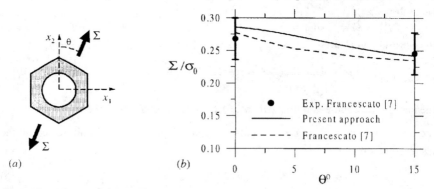

Figure 3. (a) hexagonal RV with circular hole ($v_h=0.5$), (b) limit macroscopic stress vs. angle θ

Fig. 3a shows the hexagonal RV for a periodic array of circular holes ($V=1$). Experiments on a copper perforated sheet with this geometry and a hole volume fraction $v_h=0.5$ are described in [7]. Fig. 3b illustrates the macroscopic yield tensile stresses obtained by the present upper bound approach, by the numerical method based on piecewise linearization of the copper yield domain and linear programming, and by the experiments [7]. It can be seen, Fig. 3b, that the present approach gives slightly higher values than the other numerical method. The maximum difference between the present approach and the test data is less than 6% over the whole range of θ considered.

The anisotropic behaviour of the material having a periodic array of circular holes can be appreciated from Fig. 4, where the macroscopic yield domains of a hexagonal RV (Fig. 4a) and a square RV (Fig. 4b) are depicted for two orientations θ of the principal stresses.

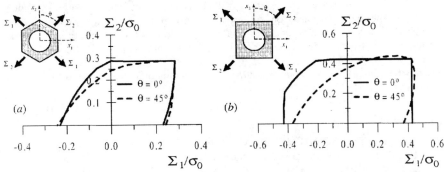

Figure 4. Macroscopic strength domains: (a) hexagonal RV (v_h=0.5), (b) square RV (v_h=0.3)

The second application deals with a periodic fiber-reinforced composite with a square RV (Fig. 5) submitted to in-plane transverse macroscopic tension, under plane-strain conditions, for a fiber volume fraction v_f=0.5 and phase yield limit $\sigma_0^f / \sigma_0^m = 5$ (setting V=1).

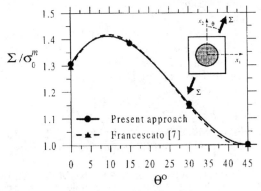

Figure 5. Square RV (v_f=0.5, $\sigma_0^f / \sigma_0^m = 5$): limit macroscopic stress vs. angle θ

Fig. 5 shows the limit macroscopic tension for different values of θ. Fully satisfactory agreement is found between the present results and those presented in [7] obtained by a numerical method based on linearization of the local yield domain and linear programming.

6. Conclusions

A method for limit analysis of periodic metal-matrix composites (MMC) and other ductile heterogeneous materials has been developed based on the kinematic limit theorem and iterative solution of the constrained minimization problem. The method leads to an upper bound on the collapse multiplier of the macroscopic stress and to a relevant collapse mechanism. The numerical tests performed show a good agreement with available experimental data and with other numerical approaches. The numerical methods compared turn out to exhibit nearly the same accuracy with equivalent finite element setting. However, the present iterative procedure greatly reduces the computing time with respect to the evolutive analysis and involves a number of variables much lower than the approach based on the linearization of the constituents yield domains. Therefore, from the computational point of view, the present approach turns out to be definitely cost-effective for the evaluation of the limit strength of periodic media.

Acknowledgement: This paper contains results obtained in the frame of EU-HCM (contract ERBC HRX-CT94-0629).

References

[1] P. Suquet, ELEMENTS OF HOMOGENIZATION FOR INELASTIC SOLID MECHANICS - Homogenization techniques for composite media, CISM Lectures, Springer-Verlag, 1985

[2] M. Z. Cohn, G. Maier, ENGINEERING PLASTICITY BY MATHEMATICAL PROGRAMMING - Pergamon Press, New York, 1979

[3] Y. G. Zhang, P. Zhang, M. W. Lu, COMPUTATIONAL LIMIT ANALYSIS OF RIGID-PLASTIC BODIES IN PLANE STRAIN - Acta Mech. Sol. Sinica, Vol. 6, 341-348, 1993

[4] Y. G. Zhang, P. Zhang, W. M. Xue, LIMIT ANALYSIS CONSIDERING INITIAL CONSTANT LOADINGS AND PROPORTIONAL LOADINGS - Computational Mech., Vol. 14, 229-234, 1994

[5] Y. H. Liu, Z. Z. Cen, B. Y. Xu, A NUMERICAL METHOD FOR PLASTIC LIMIT ANALYSIS OF 3-D STRUCTURES - Int. J. Sol. Struct., Vol. 32, 1645-1658, 1995

[6] A. Litewka, A. Sawczuk, J. Stanislawka, SIMULATION OF ORIENTED CONTINUOUS DAMAGED EVOLUTION - J. Theor. Appl. Mech., Vol 3, 675-688, 1984

[7] P. Francescato, PRÉVISION DU COMPORTEMENT PLASTIQUE DES MATÉRIAUX HÉTÉROGÈNES À CONSTITUANTS MÉTALLIQUES. APPLICATION AUX COMPOSITES À MATRICE MÉTALLIQUE À FIBRES CONTINUES ET AUX PLAQUES PERFORÉES - PhD thesis, University Joseph Fourier - Grenoble I, 1994

[8] P. Francescato, J. Pastor, LOWER AND UPPER NUMERICAL BOUNDS TO THE OFF-AXIS STRENGTH OF UNIDIRECTIONAL FIBER-REINFORCED COMPOSITES BY LIMIT ANALYSIS METHODS - Eur. J. Mech. /A Solids, Vol. 16, 213-234, 1997

[9] J. J. Marigo, P. Mialon, J. C. Michel, P. Suquet, PLASTICITÉ ET HOMOGÉNÉISATION: UN EXEMPLE DE PRÉVISION DES CHARGES LIMITES D'UNE STRUCTURE HÉTÉROGÈNE PÉRIODIQUE - J. Theor. Appl. Mech., Vol. 6, 47-75, 1987

Solution of 2-D Inhomogeneity Problems in Composite Structures of Finite Dimension

J. H. Andreasen[*] , J. Wang[*], B. L. Karihaloo[†]

[*]*Institute of Mechanical Engineering, Aalborg University, Pon 101, DK-9220 Aalborg East, Denmark*
[†]*Division of Civil Engineering, Cardiff School of Engineering, University of Wales Cardiff, Queen's Buildings, P.O. Box 686, Newport Road, Cardiff CF2 3TB, U.K.*

Abstract

In this paper, a general method of solution is presented for determining the elastic field in a finite isotropic region containing an isotropic circular inhomogeneity whose elastic constants differ from the rest of the region. The method is based on a combination of Muskhelishvili complex potentials and boundary collocation.

The use of the general method is demonstrated on two examples commonly encountered in design of composite materials. The first example concerns the residual stiffness of a damaged area in fibre-reinforced quasi-isotropic composite laminates that have suffered a low-velocity impact over their central area resulting in a circular damage area. The second example concerns the prediction of transverse moduli and stress field in composites reinforced by continuous circular fibres distributed in a square or hexagonal pattern. The theoretically predicted moduli using the general solution method are shown to be in very good agreement with experimental results.

1. Introduction

The determination of elastic fields in engineering composite materials often leads to the solution of inhomogeneity problems, in the sense defined by Eshelby (1957) who proposed an elegant method for determining the elastic field in an infinite body containing a finite inhomogeneity whose elastic constants differed from the rest of the body. He showed how the field inside and outside the inhomogeneity could be constructed using the notion of an equivalent inclusion whose elastic constants were the same as those of the rest of the body but which was subjected to unknown eigenstrains. Eshelby (1961) also suggested a method for constructing the elastic field if the finite inhomogeneity is contained in a body of finite dimensions using the notion of image fields that cancel tractions on the traction-free surfaces of the body. The construction of these image fields is a formidable task even in two-dimensions. In two-dimensions, the Eshelby formalism can of course be expressed equally elegantly in terms of Muskhelishvili complex potentials (e.g. Karihaloo & Andreasen, 1996). We shall exploit this duality between the Eshelby formalism and complex potentials on the one hand and on the other, take recourse to the boundary collocation method (e.g. Isida & Nemat-Nasser, 1987) that is known to solve two-dimensional elastostatic problems for finite domains with a high degree of accuracy.

A judicious choice of the complex potentials will allow us to satisfy the traction and displacement continuity conditions on the interface between the inhomogeneity and surrounding body, simultaneously with the prescribed traction and displacement conditions on the outer boundary of the body.

We shall illustrate the use of the general method on two examples commonly encountered in design of composite materials. The first example concerns the residual stiffness of a damaged area in fibre-reinforced quasi-isotropic composite laminates. The results of this example are useful for predicting the stress field and the compressive strength of a fibre-reinforced quasi-isotropic composite laminate that has suffered a low-velocity impact over its central area resulting in a circular damage area. It is of great interest to know how this damaged area affects the stress state in the composite laminate as it is barely visible on the surface, thus posing a hidden menace to the in-plane compressive strength of the laminate (Wang *et al.*; 1996). Repair of damaged laminates with adhesively-bonded or co-cured patches also results in the patched area being an "inhomogeneity" in the sense defined above. The second example concerns the prediction of transverse moduli in composites reinforced by continuous circular fibres distributed in a square or hexagonal pattern. The theoretically predicted moduli using the general solution method are shown to be in very good agreement with experimental results.

2. Solution of an Inhomogeneity in a Finite Region

In this section, we will outline the essence of the solution for an inhomogeneity in a finite plane region. The details of the solution procedure can be found in the recent work of Wang *et al.* (1998). Consider a finite rectangular plane region containing a circular inhomogeneity at its centre, as shown in Figure 1. The Young modulus and Poisson ratio of the region outside of the inhomogeneity and those of the inhomogeneity are E and ν, and $\overline{E}$ and $\overline{\nu}$, respectively.

For the plane problem shown in Figure 1, the stresses and displacements can be expressed in terms of two complex potentials (Muskhelishvili, 1954).

$$\sigma_{xx} + \sigma_{yy} = 2\{\phi'(z) + \overline{\phi'(z)}\},$$
$$\sigma_{yy} - \sigma_{xx} + 2i\sigma_{xy} = 2\{\overline{z}\phi''(z) + \psi'(z)\}, \tag{1}$$

where a prime denotes differentiation with respect to $z = x + iy$ and an overbar denotes the complex conjugate.

The displacements u and v in the x- and y-directions are obtained through

$$2\mu(u + iv) = \kappa\,\phi(z) - z\,\overline{\phi'(z)} - \overline{\psi(z)}, \tag{2}$$

where μ is the shear modulus and κ is the Kolosov constant related to Poisson's ratio ν through $\kappa = 3 - 4\nu$ for plane strain and $\kappa = (3 - \nu)/(1 + \nu)$ for plane stress. The complex potentials $\phi(z)$ and $\psi(z)$, like μ and κ, will be different inside and outside the inhomogeneity.

The complex potentials for the region outside the inhomogeneity can be expressed in Laurent series for any general loading on the finite outer boundary

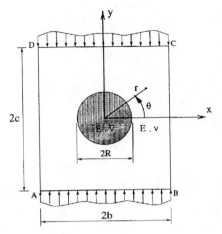

Figure 1: A finite rectangular region containing a circular inhomogeneity of radius R under arbitrary loading.

$$\phi_1(z) = \sum_{n=1} \left[a_n \left(\frac{R}{z} \right)^n + A_n \left(\frac{z}{R} \right)^n \right],$$

$$\psi_1(z) = \sum_{n=1} \left[b_n \left(\frac{R}{z} \right)^n + B_n \left(\frac{z}{R} \right)^n \right] + B_0, \tag{3}$$

whereas for the region occupied by the inhomogeneity, the potentials can be expressed in Taylor series

$$\phi_2(z) = \sum_{n=1} c_n \left(\frac{z}{R} \right)^n,$$

$$\psi_2(z) = \sum_{n=0} d_n \left(\frac{z}{R} \right)^n. \tag{4}$$

In the above series representations, the four groups of complex coefficients a_n, b_n, c_n, and d_n are eliminated analytically by satisfying the continuity conditions of displacement and traction across the boundary of the inhomogeneity. This leaves only the coefficients A_n and B_n to be determined from the boundary conditions on the outer boundary that can be of arbitrary geometry and be subjected to arbitrary loading. Solving the continuity conditions of displacement and traction across the boundary of the inhomogeneity for a_n, b_n, c_n, and c_n in terms of A_n and B_n, and substituting the resulting expressions into (3) and (4) gives the following potentials for the region outside the inhomogeneity

$$\phi_1(z) = -2\overline{A}_2\lambda_2 - \overline{A}_1\lambda_2\frac{z}{R} + \sum_{n=1}^{M+2} \left[n\overline{A}_n\lambda_2 \left(\frac{R}{z} \right)^{n-2} + A_n \left(\frac{z}{R} \right)^n \right]$$

$$+ \sum_{n=1}^{M} \overline{B}_n\lambda_2 \left(\frac{R}{z} \right)^n,$$

$$\psi_1(z) = -\left((A_1 + \overline{A}_1)(1 + \lambda_4) + \overline{A}_1(\lambda_1 - \lambda_2)\right)\frac{R}{z}$$
$$+ \sum_{n=1}^{M+2} \overline{A}_n(\lambda_1 + n(n-2)\lambda_2)\left(\frac{R}{z}\right)^n$$
$$+ \sum_{n=0}^{M} \left[B_n\left(\frac{z}{R}\right)^n + n\overline{B}_n\lambda_2\left(\frac{R}{z}\right)^{(n+2)}\right], \tag{5}$$

and for the inhomogeneity itself

$$\phi_2(z) = -\frac{1}{2}\left((A_1 + \overline{A}_1)\lambda_4 - (A_1 - \overline{A}_1)\lambda_3\right)\frac{z}{R} + \sum_{n=2}^{M+2} A_n(1 + \lambda_1)\left(\frac{z}{R}\right)^n,$$
$$\psi_2(z) = -2A_2(1 + \lambda_2) + (2A_2 + B_0)\Gamma$$
$$- \sum_{n=2}^{M+2} nA_n(\lambda_1 - \lambda_2)\left(\frac{z}{R}\right)^{n-2} + \sum_{n=1}^{M} B_n(1 + \lambda_2)\left(\frac{z}{R}\right)^n. \tag{6}$$

The constants $\lambda_{1,\ldots,4}$ are given by

$$\lambda_1 = \frac{\alpha + \beta}{1 - \beta} \qquad \lambda_2 = \frac{\alpha - \beta}{1 + \beta}$$
$$\lambda_3 = \frac{1 + \alpha}{1 - \alpha} \qquad \lambda_4 = \frac{1 + \alpha}{2\beta - \alpha - 1} \tag{7}$$

where α and β are the Dundurs' parameters (Dundurs, 1969) given by the ratio between the shear moduli $\Gamma = \mu_2/\mu_1$ and the Kolosov constants κ_1 and κ_2 through

$$\alpha = \frac{\Gamma(\kappa_1 + 1) - (\kappa_2 + 1)}{\Gamma(\kappa_1 + 1) + (\kappa_2 + 1)} \qquad \beta = \frac{\Gamma(\kappa_1 - 1) - (\kappa_2 - 1)}{\Gamma(\kappa_1 + 1) + (\kappa_2 + 1)} \tag{8}$$

The coefficients A_n and B_n are to be determined from the boundary conditions on the outer boundary using the boundary collocation procedure developed by Isida & Nemat-Nasser (1987). The boundary collocation procedure needs the expressions for the displacements and force resultants outside the inhomogeneity. They are given in the work by Wang *et al.* (1998). The choice of the number of terms M is dictated by the desired accuracy. It should be noted that the potentials (5) and (6) are valid for any external loading and for any topology of the outer boundary, so long as the stress fields are non-singular everywhere.

3. Prediction of Residual Stiffness of a Damaged Area in Composite Laminates

It is widely known that the damage in fibre-reinforced composite laminates caused by low-velocity impact can considerably reduce their in-plane compressive strength. Low-velocity impact usually causes a damaged area of limited size in a composite laminate. It is of great interest to know how this damaged area affects the stress state in the composite laminate as the damage is barely visible on the surface, thus posing a hidden menace to the in-plane compressive strength of the laminate. For a quasi-isotropic laminate, for instance, $(\pm 45^0/90^0/0^0)_n$, the experimental observations (Davies & Zhang, 1995; Wang *et al.*, 1996) showed that the internal damage almost assumes a cylindrical shape.

A typical C-scan image of an impacted carbon/epoxy quasi-isotropic composite laminate is shown in Figure 2. This image was obtained by measuring the time-of-flight of an echoed ultrasonic beam. Therefore, the different colours (degrees of grey in the black and white picture) in the damaged area reflect the depths of the delaminations. It is seen that the envelope of the damaged area is almost a perfect circle.

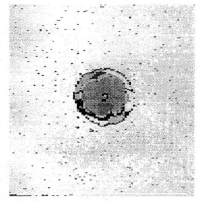

Figure 2: A time-of-flight C-scan image of an impacted quasi-isotropic laminate of thickness 4 mm. The degrees of grey in the central area reflected the depths of the delaminations.

For a quasi-isotropic laminate, it is known that the low-velocity impact causes matrix cracks and peanut-shaped delaminations at interfaces within the laminate. These peanut-shaped delaminations are oriented in the respective directions of fibres in the constituent laminae such that their envelope assumes a circular shape (Davies & Zhang, 1995; Wang *et al.*, 1996). As the fibres in a quasi-isotropic laminate are distributed such that the whole laminate assumes a quasi-isotropic nature in its own plane and the distribution of the impact-induced damage follows the directions of the fibres in the laminate, we assume that the damaged area is also a quasi-isotropic structure in the plane of the laminate. As the damaged area contains matrix cracks and delaminations, its elastic properties are different from those of the undamaged material.

Due to the complexity of the impact-induced damage, it is very hard to measure the residual stiffness of the damaged area directly. Instead, we propose to evaluate the residual stiffness of the damaged area using the experimentally measured overall "effective" modulus of the damaged specimens.

The predicted residual Young modulus of the damaged area in 5 impacted quasi-isotropic carbon/epoxy composite laminates subjected to different levels of impact energy is shown in Table 1. It is seen that the low-velocity impact has created quite a soft area in the laminates. Knowing the residual Young modulus of the damaged area, the stress field in a laminate subjected to unidirectional compression can be obtained using the general method (Wang *et al.*, 1998). This may shed light on the post-impact compressive strength of laminates which have been subjected to low-velocity impact.

Tab. 1: Experimental and theoretical results for laminates of 4 mm thickness

Specimen No.	Impact energy (J)	Damaged damaged area (mm²)	Residual modulus of of damaged area (GPa)
4-1	4.86	719.5	5.46
4-2	5.45	789.5	5.97
4-3	7.13	912.3	11.26
4-4	9.00	1013.8	3.28
4-5	9.72	863.3	6.33

4. Transverse Young moduli

As the second illustrative example, we shall use the general method to predict the transverse moduli in a unidirectional fibre-reinforced composite by assuming that the circular continuous fibres are doubly periodically distributed in the matrix. Two distributions of fibres are considered – the square and the hexagonal configurations. Figures 3 show the normal cross-sections of the composite for these configurations of fibre.

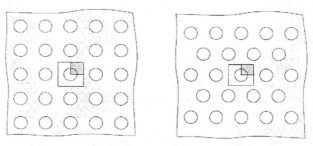

Figure 3: Square and hexagonal configuration of fibres in a composite.

For the prediction of the transverse Young modulus, it is assumed that the composite is subjected to a stress in one direction with an average value of, say σ_{yy}^0, whereas for the determination of its shear modulus, it is assumed to be subjected to a shear stress with an average value of, say σ_{xy}^0. These average measures of stress along with similar average measures of displacements determines the effective moduli. Due to periodicity, only the two quarter unit cells shown in Figure 3 by dark shadings need to be studied.

For a glass/epoxy unidirectional fibre-reinforced composite studied by Shan & Chou (1995), the variation of the transverse Young modulus E_c with the volume fraction of fibre V_f is shown in Figure 4 for the two configurations. The Young moduli and the Poisson ratios of the fibre and the matrix are 73.1 GPa and 3.45 GPa, and 0.22 and 0.35, respectively. It is found that for the same volume fraction of fibre, the hexagonal configuration of fibres gives a lower value of E_c than does the square configuration. Also shown in Figure 4 are the experimental results given by Tsai & Hahn (1980). It is seen that the experimental data fall by and large around an area bounded by the two theoretical curves. The upper bound of this area is the curve predicted using the square configuration, and the lower bound that predicted using the hexagonal configuration.

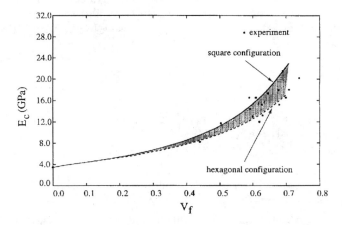

Figure 4: Variation of the transverse Young modulus with the volume fraction of fibre for a glass/epoxy composite.

5. Discussion and Conclusions

In this paper, a general method for the solution of two-dimensional elastostatic problems for finite domains containing a circular inhomogeneity was presented. The general method was developed using complex potentials, which are in the form of series expansions, and boundary collocation. Although the method was developed for a circular inhomogeneity, it can be easily extended to any inhomogeneity which can be mapped on to a circle. The loading on, and geometry of, the outer boundary can be arbitrary. The accuracy of the solution can be controlled by an appropriate choice of the number of terms in the series expansions, i.e. the number of boundary collocation points.

The general method of solution was illustrated on two examples borrowed from the field of fibre-reinforced composite materials. In the first example, an attempt was made to predict the residual stiffness of the damaged area in fibre-reinforced quasi-isotropic composite laminates which have been subjected to low-velocity impact. The results may shed light on the post-impact compressive strength of this type of laminates. In the second example, the general method was used to predict the transverse Young moduli of unidirectional fibre-reinforced composite materials. Good agreement with experimental results was found for the transverse Young modulus.

Acknowledgements

The experimental work cited in this paper was carried out by J. Wang in the Department of Aeronautics at Imperial College under support from British Aerospace and the supervision of Professor G. A. O. Davies. J. Wang is grateful for the support of British Aerospace and the supervision of Professor G. A. O. Davies. The award of a visiting post-doctoral fellowship (DANVIS) to J. Wang by the Danish Research Academy is warmly acknowledged.

References

Davies, G. A. O. & Zhang, X. 1995 Impact damage prediction in carbon composite structures. *Int. J. Impact Engng* **16**, 149–170.

Dundurs, J. 1969 Edge-bonded dissimilar orthogonal elastic wedges. *J. Appl. Mech.* **36**, 650–652.

Eshelby, J. D. 1957 The determination of the elastic field of an ellipsoidal inclusion, and related problems. *Proc. Roy. Soc. London* **A241**, 376–396.

Eshelby, J. D. 1961 Elastic inclusions and inhomogeneities, in *Progress in Solid Mechanics 2*, eds. I. N. Sneddon and R. Hill, pp. 89–140, Amsterdam: North-Holland.

Isida, M. & Nemat-Nasser, S. 1987 A unified analysis of various problems relating to circular holes with edge cracks. *Engineering Fracture Mechanics* **27**, 571–591.

Karihaloo, B. L. & Andreasen, J. H. 1996 *Mechanics of Transformation Toughening and Related Topics.* Amsterdam: Elsevier.

Muskhelishvili, N. I. 1954 *Some Basic Problems of the Mathematical Theory of Elasticity.* (English Edition, translated by J. R. M. Radok), Noordhoff International Publishing, Leyden.

Shan, H. Z. & Chou, T-W. 1995 Transverse elastic moduli of unidirectional fibre composites with fibre/matrix interfacial debonding. *Composites Science and Technology* **53**, 383–391.

Tsai, S. W. & Hahn, H. T. 1980 *Introduction to Composite Materials.* Westport: Technomic Publishing Co., Inc.

Wang, J, Davies, G. A. O. & Hitchings, D. 1996 Impact and compression-after-impact tests of T300/914 carbon/epoxy laminates. Technical report, Department of Aeronautics, Imperial College.

Wang, J., Andreasen, J. H. & Karihaloo, B. L. 1998 A general method of solution for an inhomogeneity in a finite plane region. Work in progress.

A micromechanical approach to the strength of unidirectional periodic composites with elastic-brittle constituents

Q.-C. He and J. Botsis

Laboratory of Applied Mechanics and Reliability Analysis, Department of Mechanical Engineering
Swiss Federal Institute of Technology, CH-1015 Lausanne, Switzerland

Abstract

The macroscopic first cracking strength (or damage-free) domain of an elastic-brittle unidirectional periodic composite is formulated in terms of its phase elastic and strength properties. The stress localization tensor is shown to be the key to finding this domain which is then proved to be convex provided so are the phase counterparts. In the idealized case of two-dimensional unidirectional periodic composites, the stress localization tensor is explicitly determined for the fully anisotropic phases.

1. Introduction

The importance of the problem of predicting the strength of composite materials is underscored by the considerable efforts which have been directed to it over the past three decades. Essentially, two approaches have been developed:

- The first one is phenomenological, which consists in experimentally determining failure (or yielding) threshold when only one-dimensional (1D) stress component is active, and postulating failure criteria for combined stress in terms of 1D threshold stresses. In this regard, we can cite the criterion of Tsai and Wu (1971) and that of Hashin (1980). The phenomenological approach has the main drawback of being unable to explicitly relate the strength of a composite to the strength properties of its constituents (fibers, matrix and interface) and to its structural morphology (fiber shape, fiber orientation, fiber spacing, ...). However, such relations are of prime interest when manufacturing composites or designing composite structures. In addition, the strength of several composites is scale-dependent, thus there is the necessity of taking their microstructural geometry into account in their strength prediction.

- The second one is micromechanical, which consists in deriving the failure criterion of a composite from the knowledge of the relevant mechanical properties and geometrical characteristics of its components. This approach has substantially been developed in the past two decades, leading to a number of results of practical and theoretical importance. However, up to now this approach has mainly been developed for predicting the ultimate strength of composites with *elastic-perfectly plastic* constituents (see e.g. Suquet, 1983; de Buhan and Taliercio, 1991; Ponte Castaneda and Debotton, 1992) and little has been done for the case of *elastic-brittle* constituents. The reasons for this are: (i) in the elastic-perfectly plastic case, the ultimate strength problem becomes a problem of limit analysis, so that the difficulties in finding the stress distribution in an inhomogeneous material can be avoided; (ii) on the contrary, in the elastic-brittle case, the problem cannot be solved without knowing the distribution of stress, the theory of limit analysis leading to a serious under- or over-estimate for strength.

Based on the existing literature, we believe that a rational micromechanical approach to evaluating the strength of composites with *elastic-brittle constituents* is lacking. Since a great deal of fiber-reinforced composites of technological interest belong to such a class, in this work we address and formulate the problem of strength of such materials from a continuum micromechanics standpoint and present some preliminary results from studying idealized two-dimensional unidirectional periodic composites.

2. General Formulation

The composite under investigation consists of aligned *continuous* elastic-brittle fibers which are *periodically* arranged in their transverse plane and embedded in an elastic-brittle matrix. Both the matrix and fiber phases are assumed to be homogenous but not necessarily isotropic. Before formulating the strength problem for these unidirectional periodic composites, let us note several facts which will turn out to play a prime role. Firstly, a characteristic property of unidirectional composites is that *its local (or microscopic) behavior is invariant under any translation along the fiber direction.* This invariance will be referred to as the *unidirectional homogeneity.* Secondly, a unidirectional periodic composite can be considered as generated by repeating a unit cell in the three directions of an Euclidean space, so that the *representative volume element* in the sense of micromechanics corresponds to a *unit cell.* Thirdly, the unit cell for a given unidirectional periodic composite is not unique. In particular, the unidirectional homogeneity makes it possible to arbitrarily fix the length and location of a unit cell along the fiber direction. Finally, the non-uniqueness in choosing the unit cell for a given periodic composite must have no effects on its overall (or macroscopic) behavior obtained by means of homogenization techniques.

Formulating our strength problem amounts to properly defining the strength of a unidirectional periodic composite in terms of the elastic and strength properties of its constituents. Let $\bar{\Omega}$ be the closed domain occupied by a unit cell with Ω and $\partial\Omega$ as its interior domain and boundary. If $\bar{\Omega}_1$ and $\bar{\Omega}_2$ denote the respective sub-domains occupied by the fiber and matrix phases, then $\Gamma = \partial\bar{\Omega}_1 \cap \partial\bar{\Omega}_2$ corresponds to the interface between them. By hypothesis, the matrix and fibers are linearly elastic, so that the stress-strain relation is given by

$$S(x) = \mathbb{K}(x)E(x) \quad \text{or} \quad E(x) = \mathbb{C}(x)S(x), \tag{2.1}$$

where S represents the (Cauchy) stress tensor, E the (infinitesimal) strain tensor, $\mathbb{K}$ the elastic stiffness tensor, and $\mathbb{C}$ ($= \mathbb{K}^{-1}$) the elastic compliance tensor. By hypothesis, the function $\mathbb{K}(x)$ or $\mathbb{C}(x)$ is constant in each phase. The strength (or damage-free) domains of the fibers, matrix and interface are characterized by the local failure criteria $f_i(S) \leq 0$ as follows:

$$S_i = \{\mathbf{S} \mid f_i(\mathbf{S}) \leq 0\}, \tag{2.2}$$

with i = 1 for the fibers, i = 2 for the matrix, and i = 3 for the fiber-matrix interface. Each of S_i will be assumed to a *convex* domain of the stress tensor space.

As usual in micromechanics, the macroscopic stress and strain tensors, designated by $\overline{\mathbf{S}}$ and $\overline{\mathbf{E}}$, are defined as the volume averages of the corresponding microscopic fields:

$$\overline{\mathbf{S}} = \langle\mathbf{S}\rangle = \frac{1}{\text{vol}(\Omega)} \int_\Omega \mathbf{S}(\mathbf{x}) \, dv, \qquad \overline{\mathbf{E}} = \langle\mathbf{E}\rangle = \frac{1}{\text{vol}(\Omega)} \int_\Omega \mathbf{E}(\mathbf{x}) \, dv. \tag{2.3}$$

Hereafter, a letter with (without) an over bar refers to a macroscopic (microscopic) quantity and $\langle\cdot\rangle$ stands for the volume average over Ω. Our strength problem can then be formulated in a physically meaningful and mathematically consistent way.

– *Strength Problem* (P): find all macroscopic stress tensor $\overline{\mathbf{S}}$ such that

$$\overline{\mathbf{S}} = \langle\mathbf{S}\rangle, \tag{2.4}$$

$$\mathbf{S}(\mathbf{x}) = \mathbb{K}(\mathbf{x})[\overline{\mathbf{E}} + \mathbf{E}^*(\mathbf{x})] \quad \text{for} \quad \mathbf{x} \in \overline{\Omega}\backslash\Gamma, \tag{2.5}$$

$$\mathbf{E}^*(\mathbf{x}) = [\nabla\mathbf{u}^*(\mathbf{x}) + (\nabla\mathbf{u}^*)^T(\mathbf{x})]/2 \quad \text{for} \quad \mathbf{x} \in \overline{\Omega}\backslash\Gamma, \tag{2.6}$$

$$\text{div}\mathbf{S}(\mathbf{x}) = \mathbf{0} \quad \text{for} \quad \mathbf{x} \in \Omega\backslash\Gamma, \tag{2.7}$$

$$\mathbf{Sn} \text{ is anti-periodic and } \mathbf{u}^* \text{ is periodic on } \partial\Omega, \tag{2.8}$$

$$[\mathbf{S}]\mathbf{n} = \mathbf{0} \text{ and } (\mathbf{I} - \mathbf{n}\otimes\mathbf{n})[\mathbf{E}^*](\mathbf{I} - \mathbf{n}\otimes\mathbf{n}) = \mathbf{0} \text{ across } \Gamma, \tag{2.9}$$

$$f_i[\mathbf{S}(\mathbf{x})] \leq 0 \text{ with } i = 1 \text{ for } \mathbf{x} \in \overline{\Omega}_1\backslash\Gamma, \ i = 2 \text{ for } \mathbf{x} \in \overline{\Omega}_2\backslash\Gamma, \ i = 3 \text{ for } \mathbf{x} \in \Gamma. \tag{2.10}$$

Above, $\mathbf{u}^*$ denotes the periodic part of the displacement field $\mathbf{u}$ over $\overline{\Omega}$, $\mathbf{n}$ the unit normal vector field over $\partial\Omega$ or Γ, $[\cdot]$ the jump operator across Γ, and $\otimes$ the usual tensor product. The boundary conditions imposed by periodicity are given by (2.8) while the stress and displacement continuity conditions resulting from the perfect interface hypothesis are contained in (2.9).

All macroscopic stress tensors satisfying the conditions (2.4)-(2.10) form the largest domain of the macroscopic stress tensor space within which no damage or cracking is produced in the composite. It will be convenient to call this characteristic domain the *damage-free or first cracking strength domain* of the composite. Identification of this domain is crucial for the applications for which damage of any kind is undesirable (for example, when it leads to stiffness reduction or exposes the microstructure to corrosive environment).

The foregoing strength problem is a non-classical boundary value problem with inequality constraints. To better understand the structure of this problem, we reformulate it in the following way. Let the macroscopic stress tensor $\overline{\mathbf{S}}$ in (2.4) be given. Then, equations (2.4)-(2.9) define a boundary value problem of linear elasticity. If the classical assumptions are made on the symmetry, coercivity and boundedness of $\mathbb{K}(\mathbf{x})$, such a problem has a unique solution for the microscopic stress field $\mathbf{S}(\mathbf{x})$ (Suquet, 1987). Owing to the linearity of the problem, this solution can be expressed in the form

$$S(x) = L(x)\overline{S}, \tag{2.11}$$

where the fourth-order tensor L, called the *stress localization tensor*, generally has no major symmetry and no inverse and satisfies the condition that $\langle L \rangle = I$ with I being the fourth-order identity tensor on the second-order symmetric tensor space. Next, we define *the macroscopic damage-free or first cracking strength domains for the fibers, matrix and interface*, respectively, as

$$\overline{S}_1 = \{\overline{S} \mid \overline{f}_1(x, \overline{S}) = f_1[L(x)\overline{S}] \leq 0 \text{ for all } x \in \overline{\Omega}_1\backslash\Gamma\}, \tag{2.12a}$$

$$\overline{S}_2 = \{\overline{S} \mid \overline{f}_2(x, \overline{S}) = f_2[L(x)\overline{S}] \leq 0 \text{ for all } x \in \overline{\Omega}_2\backslash\Gamma\}, \tag{2.12b}$$

$$\overline{S}_3 = \{\overline{S} \mid \overline{f}_3(x, \overline{S}) = f_3[L(x)\overline{S}] \leq 0 \text{ for all } x \in \Gamma\}. \tag{2.12c}$$

Then, *the macroscopic damage-free or first cracking strength domain $\overline{S}$ of the composite* is given by

$$\overline{S} = \overline{S}_1 \cap \overline{S}_2 \cap \overline{S}_3. \tag{2.13}$$

The assumption that the three domains defined by (2.2) are all convex implies that $\overline{S}$ is *convex*. To justify this simple but important result, we first note that (2.12a)-(2.12c) are equivalent to

$$\overline{S}_1 = \bigcap_{x \in \overline{\Omega}_1\backslash\Gamma} L^{-1}(x)S_1, \quad \overline{S}_2 = \bigcap_{x \in \overline{\Omega}_2\backslash\Gamma} L^{-1}(x)S_2, \quad \overline{S}_3 = \bigcap_{x \in \Gamma} L^{-1}(x)S_3, \tag{2.12'}$$

where $L^{-1}(x)S_i$ are understood in the sense that $L^{-1}(x)S_i = \{\overline{S} \mid L(x)\overline{S} \in S_i\}$. Since the convexity of a set is conserved under any linear transformation, $L^{-1}(x)S_i$ are convex domains. Further, as the intersection of convex sets remains a convex set, it follows from (2.12') and (2.13) that each of $\overline{S}_1, \overline{S}_2, \overline{S}_3$ and $\overline{S}$ is convex.

Now, it becomes natural to say that the failure or first cracking is *fiber-dominated* if $\overline{f}_1(x, \overline{S}) = 0$ is firstly reached for some $x \in \overline{\Omega}_1\backslash\Gamma$, *matrix-dominated* if $\overline{f}_2(x, \overline{S}) = 0$ is firstly reached for some $x \in \overline{\Omega}_2\backslash\Gamma$, and *interface-dominated* if $\overline{f}_2(x, \overline{S}) = 0$ is firstly reached for some $x \in \Gamma$. This trimodal classification of failure modes may be viewed as an extension of Hashin's bimodal one (Hashin, 1980).

3. Two-Dimensional Unidirectional Periodic Composites

In the preceding section, we have seen that, once the local failure criteria $f_i(S) \leq 0$ are specified, finding the first cracking strength domain of a composite resides mainly in determining the stress localization tensor field $L(x)$. Generally speaking, numerical methods are needed for obtaining $L(x)$. However, for a two-dimensional (2D) unidirectional periodic composite (Fig. 1), which may be viewed as an idealized material and will be used for a preliminary study, it is possible to get a simple analytical expression for $L(x)$.

Consider the 2D periodic composite as shown in Fig. 1. The axes parallel and normal to the fibers are defined by unit vectors e_1 and e_2. As in section 2, the fiber phase is referred to as phase 1 and the matrix one as phase 2. If the fiber width and spacing are d and s, respectively, then the fiber volume (or area) fraction c_1 and matrix volume (or area) fraction c_2 are given by

$$c_1 = d/s, \qquad c_2 = 1 - d/s = 1 - c_1. \tag{3.1}$$

A simple but useful formula is: $\langle\cdot\rangle = c_1\langle\cdot\rangle_1 + c_2\langle\cdot\rangle_2$ with $\langle\cdot\rangle_1$ and $\langle\cdot\rangle_2$ representing the fiber and matrix phase averages.

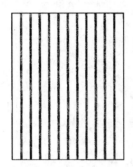 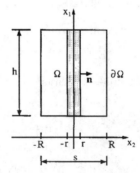

Fig. 1. A 2D unidirectional periodic composite and an associated unit cell

When the composite is subjected to plane stresses, it is convenient to set $S_1 = S_{11}$, $S_2 = S_{22}$, $S_6 = \sqrt{2}S_{12} = \sqrt{2}S_{21}$, $E_1 = E_{11}$, $E_2 = E_{22}$, $E_6 = \sqrt{2}E_{12} = \sqrt{2}E_{21}$, so that the stress-strain relation (2.1) can be written in the following alternative compact matrix forms

$$\begin{bmatrix} S_1 \\ S_2 \\ S_6 \end{bmatrix} = \begin{bmatrix} K_{11} & K_{12} & K_{16} \\ K_{21} & K_{22} & K_{26} \\ K_{61} & K_{62} & K_{66} \end{bmatrix} \begin{bmatrix} E_1 \\ E_2 \\ E_6 \end{bmatrix}, \tag{3.2}$$

$$\begin{bmatrix} E_1 \\ E_2 \\ E_6 \end{bmatrix} = \begin{bmatrix} C_{11} & C_{12} & C_{16} \\ C_{21} & C_{22} & C_{26} \\ C_{61} & C_{62} & C_{66} \end{bmatrix} \begin{bmatrix} S_1 \\ S_2 \\ S_6 \end{bmatrix}, \tag{3.3}$$

where $K_{11} = K_{1111}$, $K_{12} = K_{21} = K_{1122}$, $K_{16} = K_{61} = \sqrt{2}K_{1112}$, $K_{22} = K_{2222}$, $K_{26} = K_{62} = \sqrt{2}K_{2212}$, $K_{66} = 2K_{1212}$, and similar relations exist between C_{ij} and C_{ijkl}.

As has been pointed out at the beginning of section 2, the unidirectional composites have the characteristic that the local properties are invariant along the fiber direction. For a 2D unidirectional periodic composite, this characteristic and the periodicity conditions expressed by (2.8) have the consequence that the stress and strain are constant both in the fiber and matrix phases. Further, the continuity conditions (2.9) resulting from the perfect interface assumption require that the normal strain and shear stress along the fiber direction and the normal stress perpendicular to the fiber direction be constant:

$$E_1 = \langle E_1 \rangle = \overline{E}_1, \quad S_1 = \langle S_1 \rangle = \overline{S}_1, \quad S_6 = \langle S_6 \rangle = \overline{S}_6. \tag{3.4}$$

It is due to this fact that a simple analytic expression can be obtained for $\mathbb{L}(\mathbf{x})$. Introducing two orthogonal projection operators $\mathbb{P}$ and $\mathbb{P}^{\perp}$ for any 2D symmetric second-order tensor, say $\mathbf{S}$, such that

$$\mathbb{P}\mathbf{S} \triangleq \begin{bmatrix} 1 & 0 & 0 \\ 0 & 0 & 0 \\ 0 & 0 & 0 \end{bmatrix} \begin{bmatrix} S_1 \\ S_2 \\ S_6 \end{bmatrix} = \begin{bmatrix} S_1 \\ 0 \\ 0 \end{bmatrix}, \quad \mathbb{P}^{\perp}\mathbf{S} \triangleq \begin{bmatrix} 0 & 0 & 0 \\ 0 & 1 & 0 \\ 0 & 0 & 1 \end{bmatrix} \begin{bmatrix} S_1 \\ S_2 \\ S_6 \end{bmatrix} = \begin{bmatrix} 0 \\ S_2 \\ S_6 \end{bmatrix}, \quad (3.5)$$

then (3.4) is equivalent to writing

$$\mathbb{P}\mathbf{E} = \mathbb{P}\overline{\mathbf{E}}, \qquad \mathbb{P}^{\perp}\mathbf{S} = \mathbb{P}^{\perp}\overline{\mathbf{S}}. \tag{3.4'}$$

Define the fourth-order tensors $\mathbb{A}$ and $\mathbb{B}$ by

$$\mathbb{A} = (\mathbb{P}^{\perp}\mathbb{K}\mathbb{P}^{\perp})^{-1} \triangleq \begin{bmatrix} 0 & 0 & 0 \\ 0 & K_{22} & K_{26} \\ 0 & K_{62} & K_{66} \end{bmatrix}^{-1} = \frac{1}{K_{22}K_{66} - K_{26}^2} \begin{bmatrix} 0 & 0 & 0 \\ 0 & K_{66} & -K_{26} \\ 0 & -K_{62} & K_{22} \end{bmatrix}, \tag{3.6}$$

$$\mathbb{B} = (\mathbb{P}\mathbb{C}\mathbb{P})^{-1} \triangleq \begin{bmatrix} C_{11} & 0 & 0 \\ 0 & 0 & 0 \\ 0 & 0 & 0 \end{bmatrix}^{-1} = \begin{bmatrix} 1/C_{11} & 0 & 0 \\ 0 & 0 & 0 \\ 0 & 0 & 0 \end{bmatrix}, \tag{3.7}$$

where $(\mathbb{P}^{\perp}\mathbb{K}\mathbb{P}^{\perp})^{-1}$ and $(\mathbb{P}\mathbb{C}\mathbb{P})^{-1}$ are understood in the sense that $(\mathbb{P}^{\perp}\mathbb{K}\mathbb{P}^{\perp})^{-1}(\mathbb{P}^{\perp}\mathbb{K}\mathbb{P}^{\perp}) = \mathbb{P}^{\perp}$ and $(\mathbb{P}\mathbb{C}\mathbb{P})^{-1}(\mathbb{P}\mathbb{C}\mathbb{P}) = \mathbb{P}$. Then, using (3.4'), (3.2) and (3.3), it can be deduced that the stress localization tensor is given by

$$\mathbb{L} = \mathbb{K}\mathbb{A} + \mathbb{B}\langle\mathbb{B}\rangle^{-1}\langle\mathbb{B}\mathbb{C}\rangle$$

$$\triangleq \frac{1}{K_{22}K_{66} - K_{26}^2} \begin{bmatrix} 0 & K_{12}K_{66} - K_{16}K_{26} & -K_{12}K_{26} + K_{16}K_{22} \\ 0 & K_{22}K_{66} - K_{26}^2 & 0 \\ 0 & 0 & K_{22}K_{66} - K_{26}^2 \end{bmatrix}$$

$$+ \frac{1}{\langle C_{11}^{-1}\rangle C_{11}^2} \begin{bmatrix} C_{11} & C_{12} & C_{16} \\ 0 & 0 & 0 \\ 0 & 0 & 0 \end{bmatrix}. \tag{3.8}$$

In other words, the microscopic stress components S_1, S_2, S_6 are related to the macroscopic stress components $\overline{S}_1$, $\overline{S}_2$, $\overline{S}_6$ by

$$S_1 = \frac{1}{\langle C_{11}^{-1}\rangle C_{11}} \overline{S}_1 + \left(\frac{K_{12}K_{66} - K_{16}K_{26}}{K_{22}K_{66} - K_{26}^2} + \frac{C_{12}}{\langle C_{11}^{-1}\rangle C_{11}^2}\right)\overline{S}_2$$

$$+ \left(\frac{-K_{12}K_{26} + K_{16}K_{22}}{K_{22}K_{66} - K_{26}^2} + \frac{C_{16}}{\langle C_{11}^{-1}\rangle C_{11}^2}\right)\overline{S}_6, \tag{3.9a}$$

$$S_2 = \overline{S}_2, \qquad S_6 = \overline{S}_6. \tag{3.9b}$$

We see from (3.9a) that in the general case, either the macroscopic transverse normal stress $\overline{S}_2$ or longitudinal shear stress $\overline{S}_6$ may induce a microscopic longitudinal normal stress field even though the corresponding volume average $\overline{S}_1$ equals zero.

Assuming that both the fiber and matrix phases are *orthotropic* relative to the fiber direction e_1, then (3.3) reduces to

$$
\begin{bmatrix} E_1^{(i)} \\[4pt] E_2^{(i)} \\[4pt] E_6^{(i)} \end{bmatrix} = \begin{bmatrix} 1/E_L^{(i)} & -\nu_T^{(i)}/E_T^{(i)} & 0 \\[4pt] -\nu_L^{(i)}/E_L^{(i)} & 1/E_T^{(i)} & 0 \\[4pt] 0 & 0 & 1/G_L^{(i)} \end{bmatrix} \begin{bmatrix} S_1^{(i)} \\[4pt] S_2^{(i)} \\[4pt] S_6^{(i)} \end{bmatrix}, \qquad (3.10)
$$

where $E_L^{(i)}$, $E_T^{(i)}$, $\nu_L^{(i)}$ and $\nu_T^{(i)}$ are the longitudinal and transverse Young's moduli and Poisson's ratios satisfying the relation $\nu_T^{(i)}/E_T^{(i)} = \nu_L^{(i)}/E_L^{(i)}$, and $G_L^{(i)}$ is the longitudinal shear modulus ($i = 1$ for the fiber phase and $i = 2$ for the matrix phase). Correspondingly, (3.9) is simplified into

$$
S_1^{(1)} = \frac{1}{c_1\alpha + c_2}\,(\alpha\overline{S}_1 + c_2\eta\overline{S}_2), \quad S_1^{(2)} = \frac{1}{c_1\alpha + c_2}\,(\overline{S}_1 - c_1\eta\overline{S}_2), \quad (3.11a)
$$

$$
S_2^{(1)} = S_2^{(2)} = \overline{S}_2, \quad S_6^{(1)} = S_6^{(2)} = \overline{S}_6, \qquad (3.11b)
$$

where

$$
\alpha = E_L^{(1)}/E_L^{(2)}, \qquad \eta = \nu_L^{(1)} - \alpha\nu_L^{(2)}. \qquad (3.12)
$$

It is remarkable that the microscopic stress field depends on only two coefficients, α and η, in place of the eight independent elastic constants appearing in (3.10). In the simplest isotropic case, (3.11) also holds provided (3.12) is replaced by

$$
\alpha = E_f/E_m, \qquad \eta = \nu_f - \alpha\nu_m, \qquad (3.13)
$$

where E_f and E_m are the fiber and matrix Young's moduli, and ν_f and ν_m the fiber and matrix Poisson's ratios.

Now, if (2.2) is specified, we can obtain the first cracking strength domain of 2D unidirectional periodic composites by substituting (3.8) into (2.12) together with (2.13). As an example of application, consider glass/epoxy unidirectional composites. For an E-glass/epoxy composite, the elastic properties are: $E_f = 76$ GPa, $\nu_f = 0.22$, $E_m = 4$ GPa, $\nu_m = 0.39$ (Hull and Clyne, 1996). The numerical values of the coefficients defined by (3.13) are: $\alpha = 19$, $\eta = -7.19$. The failure criteria for the fiber and matrix phases under consideration are assumed to take the form

$$
\sqrt{\{\varepsilon_1^{(i)}\}_+^2 + \{\varepsilon_2^{(i)}\}_+^2 + \{\varepsilon_3^{(i)}\}_+^2} - \varepsilon_\sigma^{(i)} \leq 0. \qquad (3.14)
$$

Here, $\varepsilon_1^{(i)}$, $\varepsilon_2^{(i)}$ and $\varepsilon_3^{(i)}$ are the eigenvalues of the fiber or matrix strain tensor $\mathbf{E}^{(i)}$; $\{x\}_+$ is equal to x if $x > 0$ and to zero if $x \leq 0$; $\varepsilon_\sigma^{(i)}$ are the tensile failure strains, whose numerical values for the E-glass fiber and epoxy resin are: $\varepsilon_\sigma^{(1)} = 2.6\%$, $\varepsilon_\sigma^{(2)} = 4\%$ (Hull and Clyne, 1996). The criterion (3.14), due to Mazars (1985), is suitable for a wide class of isotropic elastic-brittle materials. Assuming that the fiber/matrix interface is perfect, the macroscopic first cracking strength domain of the composite under the bi-axial loading along and normal to the fiber direction is numerically determined for $c_1 = 0.3$ (Fig. 2). We observe that the failure is matrix-dominated or fiber-dominated when the ratio of $\overline{S}_1$ to $\overline{S}_2$ varies.

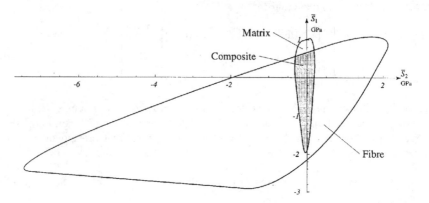

Fig. 2. The first cracking strength domain of a 2D unidirectional periodic E-glass/epoxy composite ($c_1 = 0.3$ and $c_2 = 0.7$) under the bi-axial loading along and normal to the fiber direction

4. Concluding Remarks

The strength problem formulated in section 2 can only be solved numerically in the general three-dimensional case. The analytical results presented in section 3 allow us gain qualitative insight into the problem and are of practical use to layered composites. The evolution of the strength domain of a composite with its microscopic damage state should be taken into account if its load carrying capacity before ultimate rupture is employed.

References

de Buhan, P., and Taliercio, A. (1991). A homogenization approach to the yield strength of composite materials. *Eur. J. Mech., A/Solids* **10**, 129-154.

Hashin, Z. (1980). Failure criteria for unidirectional fiber composites. *J. Appl. Mech.* **47**, 329-334.

Hull, D., and Clyne, T.W. (1996) *Introduction to Composite Materials.* Cambridge University Press.

Mazars, J. (1984) Application de la mécanique de l'endommagement au comportement non linéaire et à la rupture du béton de structure. Thèse de doctorat d'état, Université Paris-VI.

Ponte Castaneda, P., and Debotton, G. (1992). On the homogenized yield strength of two-phase composites. *Proc. R. Soc. Lond. A* **438**, 419-431.

Suquet, P. (1983). Analyse limite et homogénéisation. *C. R. Acad. Sci. Paris Sér. II* **296**, 1355-1358.

Suquet, P. (1987). Elements of homogenization for inelastic solid mechanics. in *Homogenization Techniques for Composite Media.* Lecture Notes in Physics 272 (ed. by E. Sanchez-Palencia & A. Zaoui), 193-278. Springer, Berlin.

Tsai, S.W., and Wu, E.M. (1971). A general theory of strength for anisotropic materials. *J. Composite Materials* **5**, 58-81.

Kinetics of adhesive contact formation between thermoplastic polymers and reinforcing fibres

K. Schneider, E. Pisanova, V. Dutschk, S. Zhandarov

Institute of Polymer Research Dresden, Hohe Str. 6, 01069 Dresden, Germany

Abstract

Formation of an adhesive contact between a polymer melt (or solution) and reinforcing fibres is considered from a viewpoint of kinetics. A two-stage model of this process has been proposed, and an expression for the interfacial bond strength as a function of time and temperature has been derived. Experimental data on bond strength in adhesive joints between thermoplastic polymers and reinforcing fibres formed under various conditions have been obtained, and the concept of activation energy has been used to analyze them. Since the process is controlled by the stage having the larger activation energy, the adhesive contact formation between fibres and polymer solutions is governed by the rate of adhesive bonding, whereas that between fibres and polymer melts — by the rate of the melt spreading.

1. Thermodynamic and kinetic factors in polymer composites processing

Adhesion between polymer matrices and reinforcing fibres is one of the important parameters in controlling the composite performance. Adhesive interaction occurs through molecular forces acting when two dissimilar media (the fibre and the matrix, in the case of fibre-reinforced composites) are brought into close contact. The degree of this interaction can be measured in terms of the thermodynamical work of adhesion, W_A ("fundamental" adhesion), and the interfacial bond strength, τ ("practical" adhesion) [1, 2]. According to modern knowledge, the work of adhesion can be represented as consisting of two main components, one of which (W_A^d) is corresponding to the van-der-Waals (mainly dispersive) forces, and the other (W_A^{ab}) — to the acid-base interaction [3]:

$$W_A = W_A^d + W_A^{ab}.$$ (1)

The van-der-Waals interaction is universal and non-local. It means that it always exists between any two molecules, its intensity being a function of the distance between them. As a consequence, the dispersive component of adhesion is present in adhesive contact between any two bodies and is weakly dependent on their physical and chemical nature. On the contrary, the acid-base component is due to local bonds between acid and base centers situated at the interacting surfaces. These bonds are similar to chemical ones, in particular, through their ability to saturation, though their strength is, as a rule, much lesser [2]. The value of W_A^{ab} is proportional to the number of formed acid-base bonds. An important example of these bonds are hydrogen bonds [2].

Strictly speaking, eqn (1) was derived for contacting bodies being in thermodynamical equilibrium. In equilibrium, both work of adhesion and interfacial bond strength reach their maximum (ultimate) values determined by thermodynamic properties of the interacting materials. However, these maxima are practically never achieved in real composite materials manufacturing processes, which is due to the following two physical mechanisms. First, the rate of bond formation between the acid and base centers of the fibre and the matrix is finite. Second, when a composite material is formed of polymer melt, high melt viscosity results in non-perfect contact between the reinforcing elements and the polymer. Thus, kinetics, together with thermodynamics, is of extreme importance in composite materials production [2, 4]. It is this kinetic factor that is responsible for the fact that the strength of adhesive bonding substantially depends on such processing conditions as temperature, pressure, and time of formation. Process engineers have empirically designed manufacturing regimes to obtain composites having "optimal" properties. However, the kinetics of adhesive contact formation has been studied only qualitatively. At the same time, appearance of a great variety of new thermoplastic polymers, and especially polymer blends, calls for a model approach to composite design which should include kinetic parameters as well.

2. Stages of adhesive contact formation and activation energies

Formation of an adhesive contact is a complicated physicochemical interaction and can include several stages. It is very important to know what stage controls the whole process for each particular polymer/reinforcement system. For instance, in the case of a polymer melt contact with a solid surface, kinetics of spreading is believed to determine the system behaviour [2]. However, spreading (and wetting) is only one of the stages of the adhesive contact formation. As the polymer and the reinforcement come into more and more intimate contact, forces of physical and chemical nature which are responsible for the adhesion at the atomic and molecular level begin to act. The manifestation of these forces for each particular case, i.e. for two surfaces having specific physicochemical structures at a given temperature, is also characterized by certain kinetic regularities. Thus, the formation of adhesive contact between a thermoplastic polymer and a fibre surface can be represented as consisting of two successive stages:

$$
\begin{array}{ccccc}
& \text{Spreading} & & \text{Physical and} & \\
& & & \text{chemical interactions} & \\
\text{Polymer} & & & & \\
+ & \underset{\xleftarrow{k_2}}{\xrightarrow{k_1}} & \text{Wetted} & \underset{\xleftarrow{k_4}}{\xrightarrow{k_3}} & \text{Adhesive} \\
\text{fibre surface} & & \text{surface} & & \text{contact}
\end{array}
\tag{2}
$$

In the above scheme, $k_1 ... k_4$ are rate constants for corresponding elementary reactions. These constants depend strongly on temperature. One of possible approaches to analyse the kinetics of the processes is the concept of activation energy. According to this approach, rate constants can be described by Arrhenius-type equations

$$k_i = k_{0i} \exp\left(-\frac{W_i}{kT}\right),\qquad(3)$$

where k is the Boltzmann's constant, T is the temperature, k_{0i} are pre-exponents (weakly dependent on temperature), and W_i is the activation energy for ith elementary process. Each elementary stage is characterized by its own activation energy value.

The process of an adhesive contact formation can be studied quantitatively by means of the system of equations which describe the variations of fractions of unwetted (u), wetted (w), and adhesively bonded (b) fibre surface area, proceeding from the scheme of the process (2):

$$\begin{cases} \dfrac{du}{dt} = -k_1 u + k_2 w; \\[4pt] \dfrac{dw}{dt} = k_1 u - k_2 w - k_3 w + k_4 b; \\[4pt] \dfrac{db}{dt} = k_3 w - k_4 b; \\[4pt] u + w + b = 1. \end{cases}\qquad(4)$$

Having system (4) solved for b under initial conditions $u=1$, $w=b=0$, we obtain the variation of the interfacial bond strength with time, and taking equations (3) into account — also with temperature. However, the constants $k_1 ... k_4$ appear in the solution in a very complex manner, and they are, as a rule, *a priori* unknown. Fortunately, the rates of the two stages of process (2), spreading and physicochemical interactions, usually differ considerably, and the rate of the process as a whole is controlled by the slowest stage, depending on the nature of the components and the conditions of the adhesive contact formation. In this case, the solution of system (4) becomes more simple. For instance, if the contact formation is controlled by physicochemical interactions, the interfacial bond strength as a function of time can be written as

$$\tau = \frac{k_3}{k_3 + k_4}\tau_0 \left\{1 - \exp\left[-(k_3 + k_4)t\right]\right\},\qquad(5)$$

the ultimate τ value at $t \to \infty$ being a function of temperature:

$$\tau_\infty = \frac{k_3(T)}{k_3(T) + k_4(T)}\tau_0.\qquad(6)$$

Within the frames of the activation energy approach and in view of eqns (3), eqn (6) takes the form

$$\tau_\infty = \tau_0 \left\{1 + \frac{k_{04}}{k_{03}}\exp\left[-\left(\frac{W_4 - W_3}{kT}\right)\right]\right\}^{-1}$$

and at $\dfrac{k_{04}}{k_{03}}\exp\left[-\left(\dfrac{W_4 - W_3}{kT}\right)\right] \gg 1$

$$\tau_{\infty} = \tau_0 \frac{k_{03}}{k_{04}} \exp\left[-\left(\frac{W_3 - W_4}{kT}\right)\right] = \tau_0 \frac{k_{03}}{k_{04}} \exp\left(-\frac{\Delta W}{kT}\right). \qquad (7)$$

In eqns (5)–(7) τ_0 is the absolute limit of the bond strength, corresponding to 100% adhesive bonds between the matrix and the fibre ($b = 1$). And if the process is controlled by the melt spreading, subscripts 3 and 4 in (5)–(7) should be replaced with 1 and 2, respectively.

The value of ΔW, the **apparent activation energy** for the adhesive contact formation, can be measured experimentally, and from it one can suggest what mechanism controls this process.

3. Results and discussion

To estimate activation energies separately for each stage, an investigation of the adhesive contact formation under different conditions is necessary. In this paper, adhesive contact between thermoplastic polymers (polycarbonate, polyethylene, polystyrene) and reinforcing fibres (carbon and glass) was formed from melts and solutions at different temperatures and for different consolidation times. The bond strength was characterized using the single-fibre-composite (SFC) technique (the fragmentation test) [5] and the microbond test [6]. The apparent activation energy was determined by plotting experimental data on the $\ln \tau$ — $1/T$ coordinates. As follows from eqn (7),

$$\ln \tau_{\infty} = \ln\left(\tau_0 \frac{k_{0i}}{k_{0j}}\right) - \frac{\Delta W}{k} \cdot \frac{1}{T}; \qquad (8)$$

thus, the plot of $\ln \tau_{\infty}$ against $1/T$ yields a straight line whose slope is proportional to

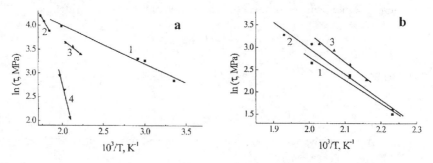

Fig. 1. Interfacial bond strength (on the $\ln \tau$ — $1/T$ coordinates) for thermoplastic polymer — reinforcing fibre (or flat solid substrate) systems as a function of the temperature of joint formation. Activation energies: 0.07 eV for polycarbonate solution — carbon fibre (*a*, 1); 0.24 eV for polycarbonate melt — carbon fibre (*a*, 2); 0.11 eV for penton coating (from the melt) on steel (*a*, 3) [11]; 0.62 eV for polystyrene melt — glass fibre (*a*, 4). Fig. 1*b* for polyethylene melt with carbon fibres ($\Delta W = 0.45$ eV, 1); glass fibres ($\Delta W = 0.50$ eV, 2) and polyethylene coating (from the melt) on steel ($\Delta W = 0.54$ eV, 3).

Table 1

Bond strength in thermoplastic polymer/reinforcing fibre systems

Matrix	Fibre	Specimen formation conditions		Bond strength (MPa)	Ref.
		Temperature/time (^{0}C/min.)	Method of combining		
Polycarbonate	Carbon (AS4)	25/1440	Solution	17.1	[7]
		60/1440	Solution	26.2	[7]
		70/1440	Solution	27.3	[7]
		230/15	Solution ($P = 475$ kPa)	53.7	[7]
Polycarbonate	Carbon (UKN)	25/1440	Solution	17.5	[5]
		270/15	Melt	48.7	[5]
		25/1440+270/15	Solution + melt	51.2	[5]
		305/15	Melt	66.7	[10]
		270/180	Melt	83.0	[5]
		290/15	Melt	59.4	[10]
		290/15	Melt + electric charge*	78.4	[10]
		305/15	Melt + electric charge*	121.7	[10]
Polyethylene	Carbon (UKN)	150/15	Melt	4.5	This paper
		175/15	Melt	10.8	—//—
		200/15	Melt	14.3	—//—
		150/15	Melt + electric charge*	6.1	—//—
		175/15	Melt + electric charge*	17.4	—//—
		200/15	Melt + electric charge*	21.0	—//—
Polyethylene	E-glass	150/15	Melt	5.0	—//—
		175/15	Melt	10.3	—//—
		200/15	Melt	21.8	—//—
		220/15	Melt	26.5	—//—
		220/15	Melt + electric charge*	31.4	—//—
Polystyrene	E-glass	205/15	Melt	7.9	[6]
		220/15	Melt	14.1	[6]
		235/15	Melt	19.1	[6]

* Electrodeposition of charged polymer particles from a fluidized bed with subsequent thermal treatment [10].

the apparent activation energy ΔW. Moreover, eqn (8) holds true for any technique of the bond strength measurement, including adhesion of films, coatings, particles etc.

The adhesive contact formation from a solution is distinguished by fast proceeding of the first stage (spreading and wetting) and, as a rule, slow second stage, because fibre-matrix combining in a solution is usually carried on at moderate temperatures. As a result, the process is controlled by the rate of the physicochemical interaction between the components. The activation energy value we have calculated for the polycarbonate solution — carbon fibre system in the processing temperature range from 25^0C to 230^0C (using experimental data from [7]) was about $\Delta W_s = 0.07$ eV (see Fig. 1), which is between typical energies of the van-der-Waals interaction and acid-base (hydrogen) bonds [2]. Indeed, as was shown, e.g. in [8], both of these interactions contribute to the adhesion of polycarbonate to carbon fibres. On the contrary, our analysis of the time and temperature dependences of the bond strength between carbon fibres and polycarbonate melt yielded the activation energy $\Delta W_m \approx 0.24$ eV. This value corresponds to the activation energy of viscous flow in polycarbonate [9]. Since ΔW_m, corresponding to the first stage of the contact formation in scheme (2), is three times as large as ΔW_s, the activation energy for the second stage, the contact formation is controlled by the first stage (spreading+wetting). The kinetics of melt spreading can be influenced by varying technological parameters; indeed, the measured bond strength for this system strongly depends on the temperature and time of consolidation [5, 7] (Table 1). Spreading of a polymer melt can also be facilitated by application of external pressure or electric field [10]. In this paper, the electric field at the interface was produced by the polymer particles, which were electrically charged in a fluidized bed and then deposited onto the fibre surface, with subsequent thermal treatment. As was shown in [10], in an electric field induced intensification of molecular forces and macromolecules orientation in the melt are possible, which results in further increase in the bond strength (Fig. 2; see also Table 1).

The activation energy value estimated for the polyethylene melt/glass fibre pair appeared to be 0.50 eV (Fig. 1b, curve 2). Moreover, very close values have been obtained for adhesive joints of polyethylene with carbon fibres (curve 1) and for polyethylene coatings on steel (curve 3, calculated using experimental data from [11]). Such behaviour is attributed to the fact that the activation energy for the second stage (physicochemical interactions) is considerably less than 0.5 eV. For this system, the bond strength dependence on the processing temperature and time was also observed but

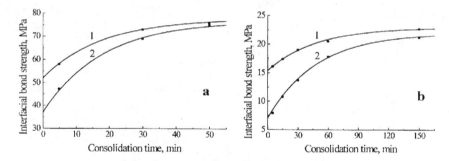

Fig. 2. Time dependences of the interfacial bond strength in polycarbonate melt (275^0C) — carbon fibre (a) and polyethylene melt (175^0C) — carbon fibre (b). Curves 1 correspond to charged polymer particles electrodeposited from a fluidized bed; curves 2 — to uncharged polymer matrices.

it was much less pronounced as compared with polycarbonate (Table 1, Fig. 2). Non-polar polyethylene molecules obviously do not form local bonds with the surface of the solid reinforcement, and the interaction at the second stage is determined by van-der-Waals forces only. In other words, the thermodynamic limit of adhesive bond strength is easily achieved for this system, and technological parameters have only marginal effect on interfacial adhesion.

In contrast to polyolefines, most polar polymers are capable of acid-base interaction (i.e. form local bonds) with reinforcing fibres, and the bond strength in their joints is sensitive to the processing conditions. Information about the kinetics of polymer interaction with the fibre surface is necessary for target-oriented control of the properties of polymer composites under design. The kinetic parameters of interactions between the components (in particular, activation energies of viscous flow and local bond formation) can be substituted into models which simulate the adhesive contact formation at the molecular level. Such molecular modelling should be very useful to predict interfacial bond strength values or even (at the macro level) mechanical properties of composites under different modes of loading.

Acknowledgement

The authors would like to acknowledge the German Research Foundation (DFG) and the Saxon Ministry of Science and Culture (SMWK) for financial support.

References

[1] *K. L. Mittal* (Ed.), in ADHESION MEASUREMENT OF FILMS AND COATINGS. VSP, Utrecht, The Netherlands, 1995, pp. 1–13.

[2] *A. J. Kinloch*, ADHESION AND ADHESIVES: SCIENCE AND TECHNOLOGY. Chapman & Hall, London, 1994.

[3] *F. M. Fowkes*, in PHYSICOCHEMICAL ASPECTS OF POLYMER SURFACES, vol. 2. K. L. Mittal (Ed.), Plenum Press, New York, 1983, pp. 583–603.

[4] *Yu. S. Lipatov*, PHYSICAL CHEMISTRY OF FILLED POLYMERS. Khimia, Moscow, 1977.

[5] *E. V. Pisanova, S. F. Zhandarov, V. A. Dovgyalo*, INTERFACIAL ADHESION AND FAILURE MODES IN SINGLE FILAMENT THERMOPLASTIC COMPOSITES. Polymer Composites, 1994. Vol. 15, No. 2, pp. 147–155.

[6] *E. Pisanova, V. Dutschk, B. Lauke*, WORK OF ADHESION AND LOCAL BOND STRENGTH IN GLASS FIBRE — THERMOPLASTIC POLYMER SYSTEMS. J. Adhesion Sci. Technol., 1998 (in press).

[7] *M. C. Waterbury, L. T. Drzal*, in CONTROLLED INTERFACES IN COMPOSITE MATERIALS. H. Ishida (Ed.), Elsevier Science, New York, 1990, pp. 731–739.

[8] *M. Nardin, J. Schultz*, RELATIONSHIP BETWEEN FIBRE-MATRIX ADHESION AND THE INTERFACIAL SHEAR STRENGTH IN POLYMER-BASED COMPOSITES. Composite Interfaces, 1993. Vol. 1, No. 2, pp. 172–192.

[9] *V. P. Privalko*, MOLECULAR STRUCTURE AND PROPERTIES OF POLYMERS. Khimia, Leningrad, 1986.

[10] *S. F. Zhandarov, E. V. Pisanova, V. A. Dovgyalo*, THE EFFECT OF ELECTRIC FIELD ON THE BOND STRENGTH BETWEEN THERMOPLASTIC POLYMERS AND CARBON FIBERS. J. Adhesion Sci. Technol., 1994. Vol. 8, No. 9, pp. 995–1005.

[11] *V. A. Belyi, N. I. Egorenkov, Yu. M. Pleskachevskii*, POLYMER ADHESION TO METALS. Navuka i Technika, Minsk, 1971.

The effect of transcrystalline interface on the mechanical properties of composite materials

N. Klein, H. Nuriel and G. Marom

Casali Institute of Applied Chemistry, Graduate School of Applied Science, The Hebrew University of Jerusalem, 91904 Jerusalem, Israel

Abstract

The existence of a transcrystalline layer is expected to influence the fibre dominated longitudinal properties of the composite material to an extent which cannot be accounted for by a simple 'rule-of-mixtures'. The results of a study of the tensile properties of unidirectional composites with low fibre contents are reported in this paper, where composites with and without transcrystallinity are compared. It is shown that in this aramid fibre reinforced nylon 66 the contribution of the transcrystalline layer is insignificant. The proposed explanation is based on a previous observation that in these low fibre content composites a relatively thick transcrystalline layer develops, in which - due to the sheafing growth mechanism - the polymer chain axis (c-axis) is mostly oriented at right angle to the fibre direction. Hence, the transcrystalline layer is less effective compared with systems where the polymer chain axis is parallel to the fibre.

1. Introduction

The interface between fibres and polymers has been an important subject of investigation since the first days of the new science of composite materials. The interface has been referred to as the "heart of the composite" because of the wide range of properties which are dependent on its type and quality. Whereas traditionally, the interface has been viewed simply as a stress transfer region from the matrix to the fibres, recently its significance as a third phase has been recognized. As a result many research groups are currently engaged in investigating the interfacial layer or interphase, and layer thicknesses of the order of the fibre diameter are considered. Of special interest is the interface in 'single-polymer composites' whose unique properties result from their composition, wherein the reinforcement and the matrix are made of the same polymer. In polyethylene-based single polymer composites, for example, a sizable crystalline interface (transcrystalline layer) can develop on the fibre surface, thereby adding yet a third phase to those of the extended chain fibres and bulk crystalline matrix.

The existence of the transcrystalline layer influences the fibre dominated longitudinal properties of the composite material to an extent which cannot be accounted for by a simple 'rule-of-mixtures'. The suggested explanation, based on a series of studies aiming to link the mechanical performance to the microstructure, attributes the effect of the transcrystalline layer to a preferred crystallite orientation relative to the fibre, thereby conferring to the matrix in the fibre direction higher rigidity and reduced thermal expansion, which in turn lower the residual thermal stresses.

The properties of the transcrystalline layer, when compared with those of the matrix, reflect a higher degree of order, which results from a more compact crystal packing and possibly from a preferred crystalline alignment. The crystalline alignment may be characterized by a particular distribution in orientation of the c-axes of the crystallites. On the molecular level, since the c-axis is parallel to the chain axis of the polymer, the physical and mechanical properties in this direction reflect the covalent nature of the polymer chain, while the properties in the perpendicular directions, along the a- and b-axes, reflect weaker intermolecular interactions (van-der-Waals and hydrogen-bonding). For both reasons, it is expected that the orientation distribution of the polymer chains in the transcrystalline layer will determine the nature and extent of its effect on the properties of the composite material.

X-ray diffraction analysis performed on various polyethylene and polyamide transcrystalline layers in microcomposites revealed preferential orientation of the polymer chain with respect to the fibre axis, so that their c-axes are inclined at specific angles relative to the fibre axis. The specific orientation in each case is a weighted average determined by the crystal growth mechanism, which results in an orientation distribution, and geometrical factors, such as the thickness. Consequently, the effect on the mechanical properties will depend on the thickness and thereby on the fibre volume fraction. The largest effects are expected in high fibre contents where the c-axes of the epitaxially nucleated transcrystalline layer is mostly aligned in the fibre direction.

The objective of this work, whose results are reported below, was to measure the effect of the transcrystalline layer on the longitudinal mechanical properties of aramid fibre reinforced nylon 66 microcomposites.

2. Experimental

Pellets of nylon-66, obtained from Nilit, Israel, were used for the matrix. The polymer was characterized by the following weight and number average molecular weight values: $\overline{M_w}$ =32500 and $\overline{M_n}$= 16800 , calculated from the intrinsic viscosity in formic acid and from end-group analysis, respectively. Single filaments of aramid Kevlar 29 (Du Pont) were used as reinforcement. The filaments were retrieved from 1000 filament, 1500 denier yarns with a proprietary rope processing finish (1500-1000-R 80-961). Two sets of samples were prepared for tensile testing, namely, quenched and isothermally crystallized unidirectional microcomposites. Additional information is given in [1,2].

3. Results and discussion

Figure 1 presents the tensile stress-strain behaviour of quenched and crystallized neat nylon 66 matrix specimens. It shows that the crystallization process results in increasing both the strength and the modulus and in decreasing the ultimate strain. The actual values (in MPa) of the strength and modulus, respectively, are 38±7 and 989±144 for the quenched and 51±8 and 1649±424 for the crystallized matrix. It is seen that the tensile strength and modulus of the crystallized matrix are higher by 34% and 67%, respectively, than those of the quenched matrix, and thus, the relevant question with regard to the properties of the composites is whether or not the contributions of the transcrystalline layer exceeds these differences.

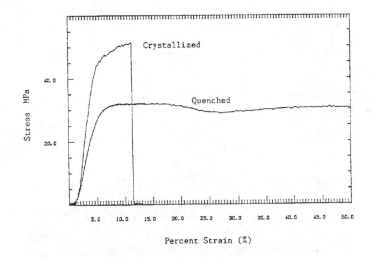

Figure 1. Stress-strain traces of quenched and crystallized nylon 66 matrix

The first difference to be observed between the quenched and crystallized composites was their respective modes of failure. Whereas the isothermally crystallized (treated) composites, which included a significantly thick transcrystalline layer [3], failed by longitudinal splitting, the quenched (untreated) composites failed in a much more brittle mode by sharp transverse fracture. This is demonstrated in Figure 2 by showing a picture of failed specimens of the two types: The top specimens exhibit a typical sharp fracture resulting from a brittle failure, and the bottom ones exhibit the longitudinal splitting. Naturally, those differences were also apparent in the stress-strain traces (not shown here).

However, the picture is not as clear with regard to the actual values of the strength and modulus. Figure 3 presents the tensile modulus and ultimate strength of the composites as a function of the fibre volume fraction and the contribution of crystallization appears to be insignificant. Both the modulus and the strength increase in a linear fashion with the fibre content, the longitudinal properties of the composites are dominated by the fibre properties with no apparent contribution of the transcrystallinity.

The proposed explanation is based on observations by two recent studies. In one study of aramid and carbon fibre-reinforced nylon 6,6 [4] it was concluded that in the nucleation and initial growth stages the first chain folds were oriented so that the chain axis was aligned in the fibre direction, and in the crystal growth that followed a typical sheaf structure was formed, leading gradually to spherulite formation, as in bulk crystallization. In another study of polyethylene fibre-reinforced high density polyethylene matrix, the observed preferential orientation of the transcrystalline layer reflected lamellar twisting as it grew in the radial direction [5]. The two studies have pointed out a direct relationship between preferential crystallite orientation and the radial distance from the fibre surface. Consequently, the effect of the transcrystallinity on the mechanical properties will depend on its thickness and thereby on the fibre volume fraction. The largest mechanical effects are expected in high fibre contents where the c-axis of the epitaxially nucleated transcrystalline layer is mostly aligned in the fibre direction.

In these low fibre content composites a relatively thick transcrystalline layer develops, in which - due to the sheafing growth mechanism - the polymer chain axis (c-axis) is mostly oriented at right angle to the fibre direction. Hence, the transcrystalline layer is less effective compared with systems where the polymer chain axis is parallel to the fibre.

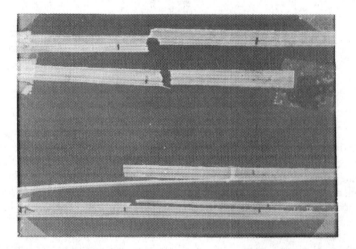

Figure 2. Failed microcomposite specimens: Quenched (top) and crystallized (bottom)

4. Conclusion

It is concluded that the results of the nylon composites (Figure 3) reflect a fibre volume fraction effect, wherein, only for high volume fractions which are accompanied by thin transcrystalline layers, are the c-axis of the crystallites aligned mostly in the fibre direction, yielding higher longitudinal properties.

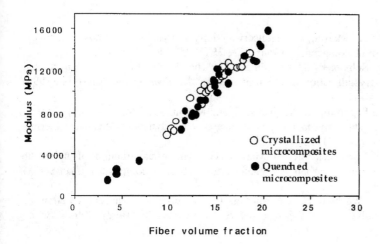

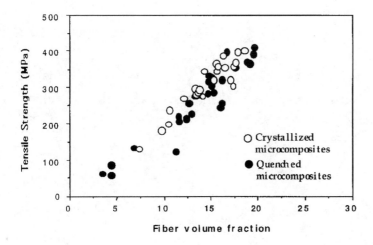

Figure 3. The tensile modulus and strength as a function of the fibre volume fraction: A comparison of Quenched and crystallized materials

References

[1]   Klein, N. and Marom, G. "Transcrystallinity in nylon 66 composites and its influence on thermal expansivity " *Composites* **25** (1995) 706-710.
[2]   Klein, N., Selivansky, D. and Marom, G. "The effects of a nucleating agent and of fibers on the crystallization of nylon 66 matrices" *Polymer Composites* **16** (1995) 189-198.
[3]   Klein, N., Pegoretti, A., Migliaresi C. and Marom, G. "Determining the role of interfacial transcrystallinity in composite materials by dynamic mechanical thermal analysis" *Composites* **26** (1995) 707-712.
[4]   Klein, N., Wachtel, E. and Marom, G. "The microstructure of nylon 66 transcrystalline layers in carbon and aramid fibre reinforced composites" *Polymer* **37** (1996) 5493-5498.
[5]   Stern, T., Wachtel, E. and Marom, G. "Epitaxy and lamellar twisting in transcrystalline polyethylene" *J. Polym. Sci. B: Polym. Phys.* **35** (1997) 2429-2433.

Fibre/Matrix Adhesion and Residual Strength of Notched Composite Laminates

L. Ye, A. Afaghi-Khatibi, G. Lawcock and Y.-W. Mai

Centre for Advanced Materials and Technology
Department of Mechanical and Mechatronic Engineering
University of Sydney, NSW 2006, Australia

Abstract

Effects of fibre/matrix adhesion on residual strength of notched polymer matrix composite laminates (PMCLs) and fibre reinforced metal laminates (FRMLs) were investigated. Two different levels of adhesion between fibre and matrix was achieved by using the same carbon fibres with or without surface treatments. Residual strength tests were performed for PMCLs and FRMLs containing a circular hole/or a sharp notch for the two composite systems. It was found that laminates with poor interfacial adhesion between fibre and matrix exhibit higher residual strength than those with strong fibre/matrix adhesion. Major failure mechanisms and modes in two composite systems were studied using SEM fractography.

1. Introduction

Interfacial behaviour between fibres and polymer matrices has long been recognised as a key factor influencing the overall properties of composite materials [1-4]. Fibre/matrix interfacial phenomena control stress transfer between fibre and matrix, stress redistribution as well as mechanisms of damage accumulation and propagation. It has been qualitatively known that poor fibre/matrix adhesion produces composite materials with 'poor' properties [5]. Numerical analyses illustrated that fibre/matrix adhesion has a strong effect on the composite transverse properties [7], but the fibre-dominated longitudinal tensile properties only change marginally [6]. Generally, composite materials with weak interfaces have relatively low strength and stiffness but high resistance to fracture if the crack grows perpendicular to the fibres, whereas materials with strong interfaces have high strength and stiffness but are somewhat brittle. The main objective of this study was to investigate the effect of interfacial adhesion between fibre and matrix on residual strength of notched polymer matrix composite laminates (PMCLs) and fibre reinforced metal laminates (FRMLs). Experimental studies on the residual strength of notched PMCLs and FRMLs with two different levels of interfacial adhesion were conducted. Then effects of different parameters such as notch size, specimen geometry, and constituent properties on residual strength of notched PMCLs and FRMLs with different fibre/matrix adhesion were evaluated using an effective crack growth model (ECGM).

2. Materials and Experimental Procedure

The composite materials were two carbon fibre/epoxy systems based on surface-treated and -untreated carbon fibres, respectively, in a rubber modified, low temperature cure epoxy resin. The carbon fibres were G34-700U fibres without any surface treatment after production and G34-700T fibres treated with an electrochemical oxidation step, which optimises the adhesion to epoxy matrices, and then coated with a thin layer of epoxy (0.4% by mass). The epoxy resin was RIGIDITE® 5228 that is a commercially available system and used in the construction of structural composite prepreg.

PMCLs were produced from 300 mm by 300 mm sheets of prepreg made from the two types of carbon fibres and the epoxy resin. The laminate stacking configurations selected for this study were $[0]_{12}$ unidirectional and $[0/90]_{4S}$ cross-ply. The nominal thickness of the unidirectional and cross-ply laminates was 2.0 mm and 2.3 mm, respectively. FRMLs were produced with a 2/1 lay-up consisting of two layers of 2024-T3 aluminium alloy sheets and one layer of CF/Epoxy unidirectional composite. The composite layer was composed of two plies of either G34-700U or G34-700T prepreg. The thickness of the aluminium alloy sheets and the composite layer was 0.406 mm and 0.26 mm, respectively. The composite layer is sandwiched between the two aluminium sheets with the fibres aligned in the aluminium rolling direction.

The completed laminates were cut into different coupon sizes, using a diamond saw or abrasive water-jet. The specimen geometry for various tests is shown in Figure 1. Diamond drills were used to make circular holes. Aluminium or fibre glass/vinyl ester loading tabs were bonded onto both ends of PMCL specimens using a high strength Araldite® epoxy adhesive. The specimens made from the surface-treated and -untreated carbon fibres for PMCLs were designated as T-PMCL, U-PMCL respectively, and for those FRMLs as T-FRML and U-FRML, respectively.

All tests were performed on universal testing machines such as Instron 5567/1195, Shimadzu 500 kN or MTS 250 kN with a constant crosshead rate of 1 mm/min at ambient temperature. The load and displacement data were recorded through a computer data acquisition system. At least three specimens were tested for each individual case. To characterise failure mechanisms of the composite systems with treated and untreated fibres, fractography was conducted using a JEOL 35C scanning electron microscope (SEM) with a beam voltage of 20 kV on samples coated with gold.

3. Results and Discussion

3.1 Transverse Tensile Strength and ILSS

The transverse tensile properties show a strong sensitivity to the fibre/matrix adhesion with a 23% drop in the transverse tensile strength for the untreated carbon-fibre composites. The reduction in the transverse tensile strength reveals a reduced interfacial adhesion between the untreated fibres and the epoxy resin. Especially, the untreated carbon-fibre composites show an 18% drop in interlaminar shear strength (ILSS) due to the shear loading on the fibre/matrix interface.

For the FRMLs, the transverse tensile strength was almost constant although a significant drop was observed for that of the PMCLs. This is due to the fact that for FRMLs the transverse properties are dominated by the properties of the metallic layer. Hence, the fibre/matrix adhesion has a negligible effect on the transverse strength of FRMLs. However, the reduction in ILSS of fibre reinforced metal laminates was obtained from three-point bend tests, with the untreated fibre/metal laminate showing approximately 7.5% drop.

3.2 Unnotched Strength
As expected, the unnotched strength, σ_o, of T-PMCL with treated fibres is about 6% higher than that of U-PMCL containing the untreated fibres. This is in agreement with previous studies of Madhukar and Drzal [4] and Ahlstrom and Gerard [8]. However, the results for T-FRML and U-FRML show that the unnotched strength of these laminates is not remarkably affected by the fibre/matrix adhesion. This is due to the fact that for fibre/metal laminates the tensile properties in both longitudinal and transverse directions are dominated by the reinforcing fibres or the metallic layer, respectively.

3.3 Notched Strength
The effect of hole diameter on the normalised strength of two groups of [0/90]₄ₛ polymer matrix composite laminates (U-PMCL and T-PMCL) is shown in Figure 2, where the width of laminate was 40 mm and the diameter was increased from 4mm to 20mm. It can be clearly seen that the strength decreases with increasing the diameter for both laminate systems. Although the unnotched strength of T-PMCL is slightly higher than that U-PMCL, the notched strength of U-PMCL with untreated fibres is up to 32% higher than that of T-PMCL with treated fibres. This means that the dependence of the notched and unnotched strength on the fibre/matrix adhesion is attributed to different failure modes and failure processes. Photographs of typical failure sections of notched [0/90]₄ₛ U-PMCL and T-PMCL are shown in Figure 3. The primary damage modes in U-PMCL, having a poor fibre/matrix adhesion, were fibre-matrix splitting and fibre bundle pullout that extended to the end-tabs. The notch sensitivity of a material is attributed to the response of the material at the tip of a notch to the high local stresses. If microcracking processes occur without the notch size increasing, the stress concentration is reduced because the tip of the notch is blunted and the high local stress is redistributed over a larger volume of the material. In contrast, if no damage occurs, the stress concentration dominates the process of crack growth and brittle fracture takes place at the tip of the notch.

In notched U-PMCL specimens with weak fibre/matrix adhesion the stress concentration leads to fibre debonding when the stress is well below that required to cause fibre fracture. The debonding spreads along fibres and results in stress relaxation due to unloading of fibres, which causes the notch tip to open up and become blunt, reducing the stress concentration. However, in T-PMCL where the fibre/matrix adhesion is optimised, the crack went across the specimen without clear fibre-matrix splitting (Figure 3b). The strong interfacial adhesion between fibre and matrix in these specimens is able to transfer stress effectively, and the stress relaxation can only occur by some local mechanisms such as resin shear or fibre fracture. Thus, the crack blunting mechanism is not extensive in this case, and the notch sensitivity can not be significantly reduced, leading to a brittle fracture with low residual strength. Figure 4 shows typical SEM micrographs of the fracture surfaces of 90-deg plies in U-PMCL and T-PMCL specimens, respectively. In U-PMCL specimen (Figure 4a), fibre surfaces are almost completely devoid of matrix material, indicating extensive interfacial failure. In addition, there is little matrix material between fibres, ie. fibres are loosely held by the matrix material after failure. On the other hand, in T-PMCL (Figure 4b), the failure occurred in both matrix failure and fibre breakage, and matrix failure is more apparent in this case. The broken matrix pieces still adhered to fibres, and the fibres are held together by the matrix material.

The normalised notched residual strength versus notch size of FRMLs with a sharp notch and/or a circular hole is shown in Figures 5 and 6, where the width of specimens is W = 90.2 mm and notch sizes varies from 2a (or D) = 10 mm to 40 mm. An increase of up to 14 % or 20% in strength is evident for the specimens with the untreated fibre containing a sharp notch or a circular hole, respectively. The experimental results also showed that the applied load required to initiate crack growth was lower for the laminates with treated fibres. At 95% of the peak load, the treated fibre specimen showed crack growth in both aluminium and composite layers, while the untreated fibre specimen showed no crack growth in the aluminium layer and only a small amount of damage in the composite layer. However, the untreated fibre specimen showed a longer delamination zone length in the fibre direction. This is caused by fibre/matrix splitting which induces retardation of the crack growth in the composite layer.

4. Evaluation of Notched Residual Strength

An effective crack growth model (ECGM) has been developed in the previous studies [9-10] to simulate residual strength of notched composite laminates. Using mechanical properties of different composite systems, the residual strength of PMCLs and FRMLs with treated and untreated fibres is evaluated using the ECGM and compared with the evaluations from the other models, ie. Point Stress Criterion (PSC) [11] and Damage Zone Criterion (DZC) [12]. The characteristic distance, d_o, for PSC, the critical damage zone length, d_l^*, for DZC and the apparent fracture energy, G_c^*, for the ECGM were determined for the PMCLs and FRMLs [9-10], listed in Table 1.

For T-PMCL (Figure 2a), all three models provide predictions with good accuracy. For U-PMCL with the same hole size and specimen width, somewhat less accurate results were produced by the PSC and DZC. But the predictions from the ECGM are well in agreement with the experimental data, shown in Figure 2b. The critical effective crack length (CECL) and the critical damage zone length (CDZL) of both T-PMCL and U-PMCL, were evaluated from the ECGM. The CDZL is defined as the corresponding effective crack length where the crack opening displacement, v, is equal or greater than the critical crack opening displacement, v_c. For both composite systems, long CECL and CDZL are predicted for large hole sizes. However, the CECL and CDZL in U-PMCL are almost 50% longer than those in T-PMCL, which is consistent with experimental observations.

For T-FRML and U-FRML, the simulations from the ECGM are also in good agreement with experimental results. It can be identified that the specimens with treated fibres exhibited a high notch-sensitivity than those with untreated fibres. For both fibre reinforced metal laminate systems with a sharp notch, the crack tip opening displacement (CTOD) in the CF/epoxy layer exceeds its critical value, ie. transition from the fictitious crack to a real crack occurs. However, the transition of crack growth in the aluminium layer occurs much later. In U-FRML specimen with a low level of fibre/matrix adhesion (Figure 7a), the real crack in the CF/epoxy layer initiates when the applied load gets about 92% of the residual strength. However, in T-FRML specimen with the same geometry (Figure 7b) this occurs at a low level of the applied load (about

80% of the residual strength). All these predictions are in agreement with experimental observations.

5. Conclusion

Residual strength tests were conducted for the notched cross-ply CF/Epoxy composite laminates and fibre reinforced metal laminates with different fibre/matrix adhesion. It was found that the laminates with weak interfacial adhesion between fibre and matrix show higher residual strength than the laminates with strong fibre/matrix adhesion. Increases in notched residual strength up to 20% and 32% were achieved for the untreated FRMLs and PMCLs, respectively, over the laminates containing the treated carbon fibres. The reduced interfacial adhesion promoted fibre/matrix splitting and increased fibre pull-out lengths which delayed crack growth in the composite layer. Conversely, the laminates containing the composite layer with the high level of fibre/matrix adhesion showed a fracture surface indicating rapid crack growth through the composite layer. The ECGM was used to simulate the residual strength of notched polymer matrix composite laminates and fibre reinforced metal laminates with different fibre/matrix adhesion. It was demonstrated that the ECGM produces results which are in agreement with failure behaviour and experimental data of these materials.

References

1. *Zhou, L.-M., J.-K. Kim and Y.-W. Mai.* Composite Science and Technology, 48 (1993) 227-236.
2. *Subramanian, S., J. S. Elmore, W. W. Stinchcomb, and, K. L. Reifsnider.* in Composite Materials: Testing and Design, Volume Twelve), ASTM STP 1274, (1996) 69-87.
3   *Piggott M. R.* Polymer Composites, 8 (1987) 291-297.
4. *Madhukar M. S. and L. T. Drzal.* Journal of Composite Materials, 25 (1991) 932-957.
5. *Mullin, J. V.* in Analysis of Test Methods for High Modulus Fibres and Composites, ASTM STP 521, (1973) 349-366.
6. *Jasiuk, I. and Y. Tong.* in Mechanics of Composite Materials and Structures, ASME, (1989) 49-54.
7. *Shih, G. C. and L. J. Elbert.* Journal of Composite Materials, 21(1987) 207-224.
8. *Ahlstrom, C. and J. F. Gerard.* Polymer Composites, 16 (1995) 305-312.
9. *Afaghi-Khatibi, A., L. Ye and Y.-W. Mai.* Journal of Composite Materials, 30 (1996) 142-163.
10. *Afaghi-Khatibi, A., L. Ye and Y.-W. Mai.* Composites Science and Technology, 56 (1996) 1079-1088.
11. *Whitney, J. M. and R. J. Nuismer.* Journal of Composite Materials, 8 (1974) 253-265.
12. *Eriksson I. and C. G. Aronsson.* Journal of Composite Materials, 24 (1990) 456-482.

Table 1 Characteristic parameters and apparent fracture energy.

Laminate	d_o [mm]	d^*_1 [mm]	G^*_c [kJ/m^2]
Untreated Laminates	0.8	0.85	85
Treated Laminates	0.6	0.6	55

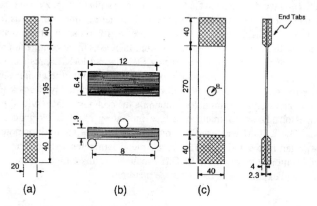

Figure 1 Geometry and dimensions of (a) tension, (b) short beam and (c) notched residual strength test specimens. All dimensions are in mm.

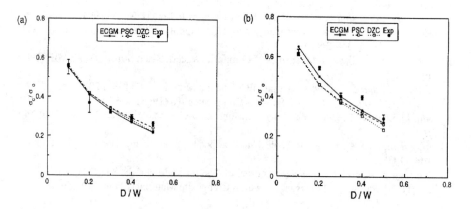

Figure 2 Notched residual strength of [0/90]$_{4S}$ laminates for (a) U-PMCL and (b) T-PMCL.

Figure 3 Typical fracture patterns of [0/90]4S notched laminates (a) U-PMCL and (b) T-PMCL

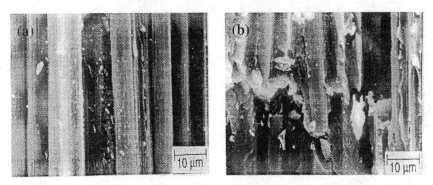

Figure 4 SEM micrographs of fracture surfaces, (a) U-PMCL and (b) T-PMCL.

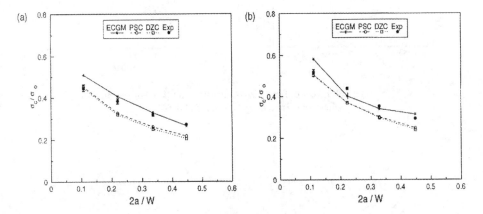

Figure 5 Notched residual strength of FRMLs with a sharp notch (a) T-FRML and (b) U-FRML.

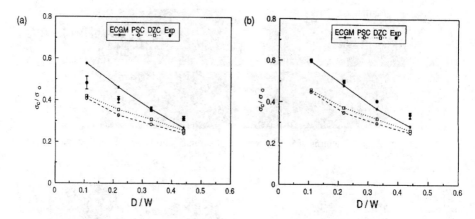

Figure 6 Notched residual strength of FRMLs with a circular hole (a) T-FRML and (b) U-FRML.

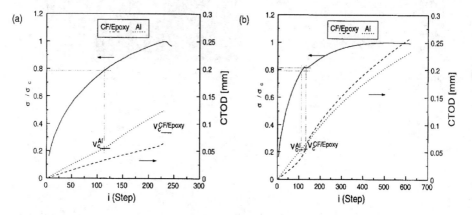

Figure 7 Effect of fibre/matrix adhesion on CTOD of FRMLs
with a sharp notch (a)U-FRML and (b)T-FRML.

Study on interface of composites reinforced with high strength polyethylene fiber: (II) The effect of fiber surface treatment on the interface shear stress

Muhuo Yu, Cuiqing Teng and Lixia Gu

Department of Materials, China Textile University,1882 West Yan-an Road, Shanghai 200051, P. R. CHINA.   E-mail: ymh@ctu.edu.cn

Abstract

In previous paper ,we have reported on the surface chemical and physical properties of high strength polyethylene(HSPE) fiber treated by controlling oxidation in chromic acid. In this work, the interface shear stress(IFSS) between treated HSPE fibers and epoxy resin was measured by Microbound Test . The results show that the IFSS is related to the surface morphology, the wettability and the functional group contents on the fiber surface, the interface adhesion between HSPE fibers and epoxy resin can be effectively improved, and the loss of tensile strength of fiber can be controlled by controlling the oxidation conditions..

1. Introduction

High strength, light weight structural composites made from fibers, such as aramid, carbon fiber, and ultra hijh strength polyethylene(HSPE), are beginning to be used routinely in place of more convention metals and metallic alloys. In these cases weight savings are an important consideration. Among most high performance fibers, UHSPE fibers show high strength, comparable with other fibers, but have the lowest density. For its high molecular weight(up to Mn= 200000) and its high orientation in crystalline regions[1], they have the highest specific strength and modulus. They also exhibit higher impact toughness, abrasion resistance, and chemical resistance than the aramid fibers. As a result, the HSPE fibers are now replacing aramid fibers for such uses as antiballistic applications, marine ropes sails, surgical gloves, etc. The HSPE fibers also have the lowest dielectric constant and loss tangent, and the highest transmission coefficient to radar wave, among the reinforce fibers,   therefore,   combing with its light weight, the composites reinforced with HSPE fibers are excellent materials for radome[2]. The HSPE fiber composites have unusually higher impact and flexural toughness than carbon and glass fiber composites. Their impact energy absorption can be up to six times that of aramid fiber composites. The combination of impact and flexural toughness of HSPE fiber with high stiffness of carbon fibers or high strength of glass fibers to tailor the properties of hybrid composites, can be used to meet any requirements for most low temperature applications.

The HSPE fiber, however, have the poorest adhesion to matrices, the lowest melting point, and the highest creep among all the reinforcing fibers. This undesirable properties limit the fibers'use in advanced composites. The fundamental cause of poor adhesion of HSPE fibers to matrices lies in their chemical composition, consisting sleekly of methylene groups. The nonpolar nature makes them difficult to wet and impossible to be chemically bonded to matrices. Besides, HSPE fibers also have smooth surfaces, which exclude mechanical interlocking, and their relatively larger diameters(40 um) reduce the specific contact area with matrices.

The interface between fiber and matrix plays a major role in the physical and mechanical properties of composite materials. Mechanically, the interface is responsible for the transfer of load between fiber and matrix, thus controlling the process of the delamination and therefore, the etching, have belong term stability of the composites. The condition at the interface between fiber and matrix prevents the propagation of cracks along the fiber length. In addition, the strong interface prevents moisture and other chemicals from diffusion along fibers, thus improving the mechanical performance and the structural stability of the composites. It is important to realize that since the reinforcing fibers range in diameter from 5 to 40 um, at a minimum fiber volume of 50%, the interface constitutes a large area within the composite. Weakness within this interfacial area can be easily magnifies to the detriment of composite properties, particularly when the

composite is under load.

Several fiber surface treatment methods[3], including plasma and chemical etching, are investigated to improve the adhesion between HSPE fiber and matrix. Among these methods, the chemical etching is a potential candidate for its easy practice in industry. However, as reported in literature [4,5.], these chemical etching require a prolonged immersion of the fiber in acids in order to have a significant improvement in interfacial bond strength, and they are accompanied by an undesirable loss in fiber strength. The motivation for enhancing adhesion through etching is to improve composite properties through improved stress transfer. The deterioration in fiber failure properties will affect composite properties and limit the advantage of any improvement of adhesion . Therefore, a balance must be struck between improved stress transfer and poorer fiber properties to achieve optimal composite properties. In previous papers[6], we have reported on the control oxidation of HSPE fiber in chromic acid by controlling the temperature and times of oxidation. In present study, the interface adhesion between epoxy resin and the HSPE fiber treated by controlling oxidation is investigated.

2. Experimental

a. Materials:

The HSPE fiber is kindly supplied by Professor Zhaofong Liu at Chemical Fiber Insitutte, China Textile University, Which was manufactured by gel spinning process with tensile strength 20g/dtex.

The E51 epoxy resin and a room temperature hardener, 593, are kindly supplied by Shanghai Resin Factory.

The oxidation of HSPE fiber in Chromic acid , the measurement of wetability, M%, and the active hydrogen contents , Eg/g, the measurement of fiber tensile strength were performed as details in previous paper[6].

b. The Microbound Test

The procedure involves the deposition of a small amount of resin onto the surface of a fiber in the form of one or more discrete microdroplets. The droplets form concentrically around the fiber in the shape of ellipsoids and retain their shape after appropriate curing.. Once cured, the microdroplet dimensions and the fiber diameter are measured with the aid of an optical microscope. The embedded length is fixed by the diameter of the microdroplet along the fiber axis, which is dependent on the amount of resin deposited on the fiber. The fiber specimens are pulled out of the microdroplets at a rate of 1 mm/min using an Instron tensile tester. The top end of the fibber is attached to a load-sensing device, and the microdroplet is contacted by load points affixed to the crosshead of a load frame. When the load points are made to move downward, the interface experiences a shear stress that ultimately causes debonding of the fiber from the microdroplet. The typical microbond test samples are shown in figure 1.

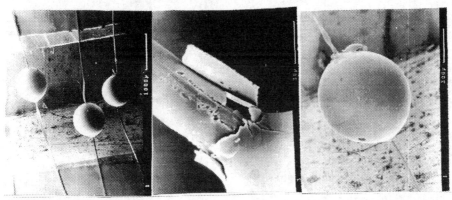

a. Before Micodebound   b. Initial Microdebound   c. After Micodebound

Figure 1. SEM Photograph of Microbond test Samples

3. Result and Discussion

A. The Microbound Test

In the microbound test, the debonding force F was taken as the maximum force preceding partial debonding. The interface shear stress can be calculated by follow equation:

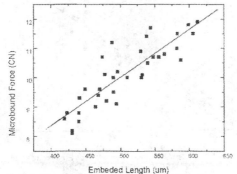

Embeded Length (um)
Figure 2. :Plot of Microbound Force versus EmbedeLength for
As received HSPE fiber

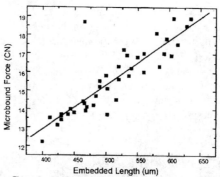

Figure 3   Plot of Microbound Force versus Embeded Length
for HSPE fiber oxidation at 50°C/5min.

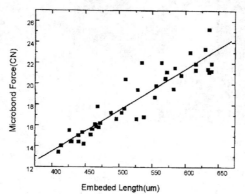

Figure 4.   Plot of Microbound Force versus Embeded length for
HSPE fiber oxidation at 60°C/5min.

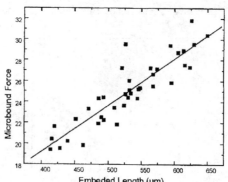

Figure 5   Plot of Microbound Force versus Embeded Length
for HSPE fiber oxidation at 70°C/5min.

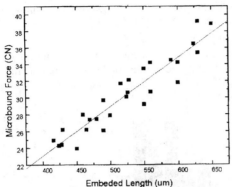

Figure 6 Plot of Microbound Force versus Embeded Lengt
for HSPE fiber oxidation at 80°C/2min

$$F = \pi d. l \tau$$

where τ is interface shear stress(IFSS), l is embedded length and d is the diameter of fiber.
In order to minimize the data scatter, a series of samples (typically 25 to 35 samples) with
various embedded length were prepared, therefore, we can get IFSS value from the plots of
debound force F versus embedded length. Figure 2 to 6 show this plot results, which
clearly reveals that the debond force F linearly increase with the embedded length
increasing and the data scatter is good than the literature reported[7].   With the linear
regression, the τ and the relative coefficient are obtained and are listed in table I. From
table I, we can see that the regression relative coefficient R is rather far from one which
means the datum are still rather scatter.   This scatter may be result from the random
distribution of the adhesion point at fiber surface which can be proven by the fact that the R
is increases as the degree of surface treatment increasing, for example, the R value for as
received fiber is the lowest, 0.87, and the R value for deep oxidation fiber is high up to 0.94.
The second reason for this datum scatter may be from the scatter of fiber diameter,   for in
figure 2 to 6, the fiber diameter was taken as the average value , 40 um.. The histogram of
HSPE fiber diameters is presented in figure 7 for untreated fibers. The histogram has been
constructed from average diameter values measured on at least 150 different fiber fragment
of 1 cm length taken from different parts of the fiber tow. The spinning process obviously
involves fiber section irregularities. The distribution is roughly centered around a value of
40 um and spread between   35 and 46 um.

Table I   The Interface Shear Stress(IFSS) and its Regression
Relative Coefficient(R)

Oxidation Temp.(C)	control	50	50	60	70	80
Oxidation Times(min)	control	5	10	5	5	2
IFSS (Mpa)	1.31	1.91	3.73	3.12	3.54	4.69
R	0.87	0.86	0.94	0.92	0.90	0.94

As the improvement method, the fiber diameter, di , embedded length , li, and the debond force Fi were measured for every microbond test, e.g.

$$Fi \quad = \quad \pi \, \tau \, di \, li$$

then, the interface shear stress τ can be calculated from equation(2). The results were presented in figure 8. The reduce of the scatter can be seen from the figure 8

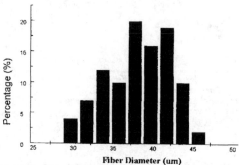

Figure 7   The histogram of HSPE fiber diameter distribution

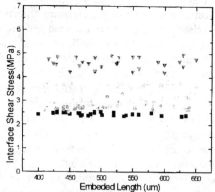

Figure 8   Plot of Interface Shear Stress versus Embeded Length for
HSPE fibers Oxidation : ■ As received fiber, ◆ 50 C/5min
, ◈ 60 C/5min,   ▲ 70 C/5min and   ▼ 80 C/2min

B. The effect of surface treatment on the interfacial shear stress

After chromic acid etching , the surface physical and chemical properties of HSPE fibers have been modified. The surface group contents   (Eg/g) and the wettability(M%) increase as the oxidation degree increasing. Table II is the datum collection of the interface shear stress, τ , the wetability , M%, and the active hydrogen contents, Eg/g. From this table , we can find that the surface oxidation of HSPE fibers can greatly improve the interface shear stress of the HSPE/epoxy composites which can be result from three facts: a) the wetability was greatly improved which can reduce the void contents in composites, b) the active hydrogen group, e.g. -COOH, and -OH group can be as the chemical bond point. The Eg/g value are changed very little from the slightly oxidation(50℃/5min) to deep oxidation (80℃/2min), however, the interface shear stress

Table II. The physical and chemical properties of HSPE fiber surface after oxidation at various conditions

Oxidation condition(C/min)	controlled	50/5	50/10	60/5	70/5	80/2
wetbility (M%)	19	49	52	55	58	68
group contents(Eg/g)	0.227		0.466	0.469	0.462	0.456
τ (Mpa)	1.31	1.91	3.73	3.12	3.54	4.69
Tensile Strength σ (cN)	130.2	14.6	106.7	111.8	108	107.5
σ/σ_o	1	0.88	0.82	0.859	0.829	0.826
τ/τ_o	1	1.40	2.85	2.38	2.70	3.58
$\tau/\tau_o \cdot \sigma/\sigma_o$	1	1.23	2.34	2.04	2.24	2.96

a. As received fiber   b. 50℃/5min   c. 80℃/2min
Figure 9. The SEM photography of broken interface of composites reinforced with HSPE fibers oxidation at various conditions

was greatly improved(from 1.91 to 4.69 Mpa). This may be due to the different ratio of -COOH/-OH, at the strong oxidation condition, the oxidation products may be mainly -COOH groups which are active to react with epoxy resin to form the chemical bond on the interface of composites; and at the mild oxidation condition, the oxidation products may be mainly -OH groups which are rather inactive to react with epoxy resin, and   c) the surface roughness of HSPE are increased[6] which can increase the mechanical interlocking at the interface of composites.

In order to balance between the fiber tensile strength loss and the improvement of interface adhesion, we take the value of $\tau/\tau_o * \sigma/\sigma_c$ as the combing properties of composites, which value are also listed in table II. From these values, we can conclude that the fiber oxidized at 80℃ for 2 minutes is the optimum surface treatment condition for epoxy resin matrix. The SEM photograph of broken interface of the composites reinforced with this HSPE fibers are shown in figure 9 which clearly reveals the improvement of the interfacial adhesion and their difference. For fiber oxidized at 80 ℃/2min, the fibrillation on the broken interface can be clearly seen. This phenominum suggests that the interfacial adhesion strength has approached its limit and the fiber splitting is an important factor which affect or govern the interface shear strength for HSPE fiber composites.

4. Conclusion

By controlling oxidation of HSPE fibers in chromic acid solution, the loss of tensile strength of fibers can be controlled and the wetability, roughness, and the functional group contents on the fiber surface can be improved. This improvement of surface properties has result in enhancing the interfacial adhesion which has exhibited in the high interface shear stress.

5. Reference

1. Pennings,A.J. and Smook,j. J. Mater. Sci., 19,3444(1984)
2. Prevorsek D.C., TRIP , 3, 4(1995)
3. Brennan, A.B., TRIP., 3, 12(1995)
4. Nardin,M and Ward, I.M, Mater. Sci. Technol., 3, 814(1987)
5. Taboudoucht,A., Opalko,R. and Ishida,H., Polym. Compos., 13(2), 81(1992)
6. Muhuo Yu, Cuiqing Teng and Lixia Gu, to be publicated.
7. Bernard ,M., Pierre M. and Ludwig R., Compos. Sci. Technol., 32, 18(1987).

Fracture mechanical analysis of fragmentation and pull-out tests

C. Marotzke, L. Qiao

Federal Institute for Materials Research and Testing (BAM), Division VI.2,
Unter den Eichen 87, 12205 Berlin, Germany

Abstract

A fracture mechanical analysis of pull-out and fragmentation tests by means of the finite element method is performed. The energy release rate arising in the pull-out test is calculated for glass and carbon fibers embedded in a thermoplastic matrix. The influence of the thermal stresses as well as of the fiber length is shown. Since a mixed mode failure arises in the pull-out test, the discrimination between the failure modes is done using the virtual crack closure method. It comes out that the begin of the interface crack is dominated by mode I while further crack propagation is governed by mode II. Furthermore, the influence of thermal stresses and interfacial friction on the crack propagation in the fragmentation test is studied. The analysis shows that a large amount of energy is released during fiber breakage resulting in an unstable propagation either of an interface or of a matrix crack. During the following phase of stable crack propagation the interfacial friction leads to a strong reduction of the energy release rate while the thermal stresses are of less importance.

1. Introduction

In the past, the determination of the bond strength of fiber reinforced polymers with pull-out and fragmentation tests was commonly performed by using a stress based analysis [1-3]. However, around the interface a complex, threedimensional stress state is encountered. Within the linear theory of elasticity, singularities arise at the fiber break in the fragmentation test and at the fiber end as well as at the fiber entry in the pull-out test, leading to severe stress concentrations at these points in a real specimen. A stress based evaluation of the test data, accordingly, is not reasonable. Moreover, in addition to the interfacial shear stresses also radial stresses are active, which are tensile in a small zone near the fiber entry or at the crack tip causing an opening of the crack [4, 5]. This is, a mixed mode failure takes place. Out of this zone, the radial stresses are compressive, resulting in a closure of the crack with interfacial friction in the debonded interface. The ratio of the two stress components strongly depends on the material properties of fiber and matrix. In the past, substantial simplifications concerning the stress field around the interface were used for the evaluation of the test data, the constant shear stress model (Kelly-Tyson) or the shear lag model [1-3]. Both models suffer, among others, from the neglection of the radial stresses, which promote the failure in the pull-out specimen and cause frictional stresses in the debonded interface in either tests. In case of the Kelly-Tyson model, the inherent assumption of an unlimited yielding of the matrix would give rise to unrealistic strains.

In pull-out and fragmentation tests, the interfacial failure process develops in a completely different manner. In the pull-out test, in addition to the initial stresses due to the thermal mismatch of fiber and matrix or due to curing, the interfacial stresses build up continuously with the increase of the external force until the interface crack starts to grow. Experimental studies with transparent matrices have shown that after crack initiation, a phase of stable crack propagation follows [6, 7]. If the crack approaches the fiber end, it starts to propagate unstably, resulting in an abrupt decay of the force at the force maximum. For a long time, the sudden drop of the force was regarded as the total debonding of the interface. Accordingly, the force maximum instead of the force at the onset of interfacial failure was used for the calculation of the interface strength. The moment of crack initiation can be detected as a kink in the force-displacement curve if the experiments are performed in a special device with high stiffness using a very short free fiber length of only a few fiber diameters [7].

In the fragmentation test, on the other hand, the loading of the interface starts abruptly with the breakage of the fiber, releasing a large amount of stored energy. As a result, the first phase of the interface crack is characterised by an unstable crack propagation. In the subsequent phase, the crack propagates stably, in general offering the opportunity for a fracture mechanical analysis. A further difference to the pull-out test is the existence of strong radial compressive stresses along the whole interface. Accordingly, a pure mode II failure with high frictional stresses in the debonded interface takes place in the fragmentation test. It is a serious drawback of the fragmentation test that, contrary to the pull-out test, the frictional stresses cannot be measured directly with the test.

As a result of the experimental findings together with the theoretical analyses, the characterisation of the interface by a fracture mechanical analysis leading to a fracture toughness or an energy release rate is preferred by many researchers meanwhile [4-6, 8]. A fracture mechanical analysis is much more adequate to the stress state arising in the interface. However, in some studies, a simple energy balance is calculated based on simplified stress distributions, which brings out unsufficient results. Alternatively, the finite element method can be employed resulting in a more realistic modelling and, moreover, allowing a separation of the failure modes in the pull-out test. In the present paper, a fracture mechanical analysis of pull-out and fragmentation tests is performed by means of the finite element method. As a first step, linear elastic behaviour of fiber and matrix is presumed.

2. Finite element model

The pull-out test model consists of a fiber of 10 micrometers in diameter embedded perpendicular into a halfsphere of matrix up to a length of 150 micrometers. The fragmentation test model of 400 microns in length corresponds to one half of a fragment, applying symmetry conditions at the edges. For both specimens, axisymmetrical finite element meshes are used (figs. 1a, 2a). The meshes are strongly refined around the interface with respect to the high stress gradients arising there (figs. 1b, 2b). In the interface, contact elements are defined together with links between the two anticipated crack faces, which can be removed in order to simulate the crack propagation. In the debonded interface, Coulomb friction can be activated. In the pull-out test, the total debonding of the fiber is simulated in 160 increments, i.e., one crack increment is about 1/10 of the fiber diameter. In the fragmentation test, a crack of 20 micrometers in length analysed using crack increments of 1/60 of the fiber diameter. The aim of the present analysis is to study the general influence of various parameters on the energy release rate, such as the residual thermal stresses, the fiber length and the fiber type. To this end, the propagation of the interface crack is simulated under fixed load conditions in order to get a clear separation of the parameters under consideration. In a forthcoming paper, real experiments will be simulated, using actual force versus crack length curves. Since a mixed mode failure takes place in the pull-out test, the energy release rate is calculated by the virtual crack closure method, in order to separate the failure modes. In the fragmentation test, the external work, the strain energy and the work of friction are calculated directly.

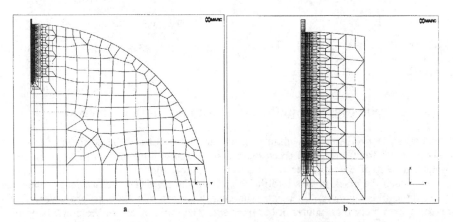

Figure 1. Finite element mesh pull-out test   a) total system b) surrounding of fiber

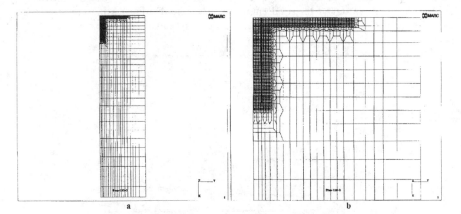

Figure 2. Finite element mesh fragmentation test
a) total system b) surrounding of fiber

The analyses are performed for glass and carbon fibers embedded in a thermoplastic matrix (polycarbonate). While the glass fibers behave isotropic, the carbon fibers are highly anisotropic (table 1). The high axial E-modulus of the carbon fiber leads to an axial stiffness ratio of fiber to matrix of 176. In case of the glass fiber the ratio is 32. The length to diameter ratio of the fibers is 15.

Table 1 Elastic and thermal constants of fibers and matrix
(a = axial direction, t = transversal direction)

matrix (polycarbonate) isotropic	E-glass fiber isotropic	HM-carbon fiber transversely isotropic
$E = 2300$ N/mm²	$E = 73500$ N/mm²	$E_a = 405000$ N/mm²
$v = 0.38$	$v = 0.18$	$E_t = 8600$ N/mm²
		$G_{at} = 13700$ N/mm²
		$v_{at} = 0.35$
		$v_{tt} = 0.53$
$\alpha_T = 70$ E⁻⁶ K	$\alpha_T = 4.5$ E⁻⁶ K	$\alpha_{Ta} = -1.4$E⁻⁶ K
		$\alpha_{Tt} = 12.0$E⁻⁶ K

3. Energy release rate of pull-out test without thermal stresses

The total energy release rate without thermal stresses as well as the portions of mode I and mode II during interfacial crack extension are shown in figure 3a for a glass fiber system. The total energy release rate increases rather slowly over 80% of the fiber length. When the crack comes into the vicinity of the fiber end, this is, into a zone of about three fiber diameters in length, the energy release rate starts to grow very fast, reaching extremely high values at the fiber end. The initial phase of the crack is dominated by mode I (fig. 3b), which, however, decreases very rapidly within 5% of the total crack length. Simultaneously, the mode II part increases, although slightly slower, resulting in the minimum of the total energy release rate. During further crack propagation, the mode II part dominates the fracture process by far.

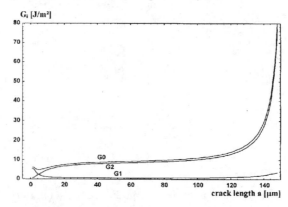

Figure 3a. Energy release rate in pull-out test (glass fiber, no thermal stresses)

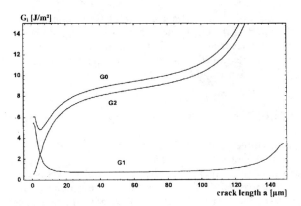

Figure 3b. Energy release rate in pull-out test - modified scale
(glass fiber, no thermal stresses)

4. Influence of thermal stresses and fiber length on the energy release rate in pull-out test

In a thermoplastic matrix, thermal residual stresses are induced during the cooling process due to the thermal mismatch between fiber and matrix. As a result of the lower expansion coefficient, the fiber suffers compressive stresses, causing positive interfacial shear stresses on one end while negative on the other end [5]. This is, the shear stresses are increased at the fiber entry while diminished at the fiber end. Apart from the fiber entry, the thermal radial stresses are compressive. However, analogous to the external force, likewise tensile stresses occur at the fiber entry, enlarging tensile stresses acting there. The residual stresses lead to an initial strain energy, which is released during interfacial crack propagation increasing the total energy release rate significantly (figs. 3a, 4 (l/d=15)). The comparison of the fiber lengths shows that during the first phase of the crack the fiber length has no significant influence on the energy release rate.

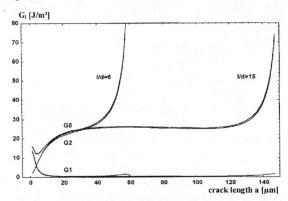

Figure 4. Energy release rate in pull-out test for different fiber lengths (l/d=15, l/d=6)
(glass fiber, thermal stresses)

However, while in case of the long fiber, a long phase of constant energy release rate exists in the middle part of the crack, the initial phase is followed directly by the steep increase of the energy release rate near the fiber end in case of the short fiber. As a result, the experimental determination of the critical energy release rate becomes more complicated for shorter fiber lengths.

5. Influence of fiber stiffness on the energy release rate in pull-out test

In order to study the influence of the fiber type, the energy release rate is calculated for a carbon fiber (table 1) with and without thermal stresses (fig. 5). Besides the anisotropy of the elastic constants, also the thermal expansion coefficients are highly anisotropic. In axial direction, the thermal expansion coefficient is even negative. The course of the energy release rate in the first phase of the crack with thermal stresses (DT=130) is similar to that encountered in the glass fiber system (fig.4), although somewhat lower. Instead of the plateau in the middle part of the crack, the energy release rate decreases slightly in case of the carbon fiber. In comparison to the glass fiber system, the contribution of the external loading is significantly lower. This is due to the higher stiffness of the carbon fiber, which releases much less energy during crack propagation. The analyses of the glass as well as of the carbon fiber system reveal that an evaluation of the experimental data without taking into account the residual stresses is not reasonable.

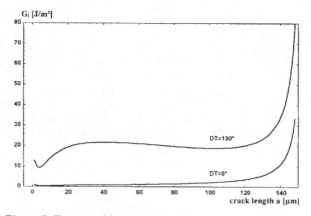

Figure 5. Energy release rate in pull-out test for carbon fiber

6. Released energy and work of friction in fragmentation test

The course of the energy release rate during a fragmentation test is completely different from that of a pull-out test. With the breakage of the fiber, a large amount of stored energy is released, which is much higher than that dissipated by the fiber break. The unused energy causes an unstable interface or matrix crack. In general, both cracks may develop simultaneously. The complex process occurring directly after fiber break requires an analysis including dynamical effects and is not subject of this paper. The present analysis is confined to the phase of static crack propagation after the dynamic phase.

First, the released strain energy of both components, i.e., fiber and matrix as well as the work of friction are studied during interfacial crack propagation. In order to estimate the maximum influence of friction, a friction coefficient of one is chosen. The released energy is calculated as the difference between the strain energy stored before fiber break and the energy stored at corresponding interface crack lengths. The energy released due to fiber fracture, accordingly, is plotted at zero interfacial crack length. By the fracture of the fiber, the fiber is unloaded, releasing a large amount of strain energy (fig. 6). At the same time, the stresses in the fiber near the break are partially transferred to the matrix, enlarging the stress state of the matrix. As a result, the matrix consumes a large amount of the energy released by the fiber. If the crack propagates along the interface, friction develops in the debonded zone, dissipating energy. The accumulated energy released by the fiber increases almost linearly during further crack propagation. The work of friction also grows linearly while the part of the matrix remains nearly constant.

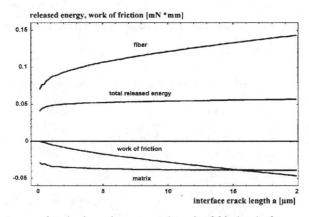

Figure 6. Accumulated released energy and work of friction in fragmentation test

7. Energy release rate in fragmentation test

The influence of thermal stresses and friction on the energy release rate of an interface crack developing in a fragmentation test is shown in figure 7. In addition, the results of a matrix crack extending perpendicular to the fiber without thermal stresses are included. Within a crack length of about one fiber radius, the influence of the fiber fracture has leveled off. All traces show a negative slope, indicating stable crack propagation. If friction is neglected, the influence of the thermal stresses is insignificant. The frictional stresses, however, lead to a very strong reduction of the energy release rate. The effect is much more pronounced if the thermal residual stresses are taken into account. The energy release rate of the pure matrix crack is in between the corresponding values of the interface crack. It depends on the fracture toughness of the interface and the pure matrix, which crack will develop in an actual test. Although the results for perfect friction ($\mu=1$) represent a limiting case, the analysis reveals the strong influence of friction on the failure process arising in a fragmentation test. This shows that the evaluation of test data requires the knowledge of the work of friction.

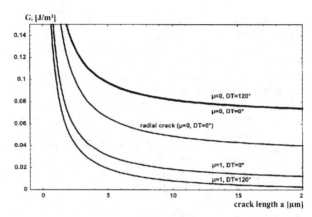

Figure 7: Energy release rate arising in fragmentation test with and without friction

Conclusions

The analyses of pull-out and fragmentation tests have shown that very different fracture processes take place in the respective tests. In pull-out tests, a mixed mode failure arises. The crack initiation is dominated by mode I, while the further crack propagation is governed by mode II. In fragmentation tests, on the other hand, a pure mode II fracture is encountered with high frictional stresses arising in the debonded zone. In order to get reasonable results, the evaluation of the experimental data of either test requires an analysis taking into account residual stresses as well as interfacial friction.

References

[1] A. Kelly, W. R. Tyson, TENSILE PROPERTIES OF FIBRE REINFORCED METALS, J. Mech. Phys. Solids, 1965, Vol. 13

[2] M. R. Piggott, LOAD BEARING FIBRE COMPOSITES, Pergamon Press, 1980

[3] C. H.Hsueh, ELASTIC LOAD TRANSFER FROM PARTIALLY EMBEDDED AXIALLY LOADED FIBRE TO MATRIX, J. Mater. Sci. Lett. , 1980, Vol.7

[4] C. Atkinson, J. Avila, E. Betz, R. E. Smelser, THE ROD PULL_OUT PROBLEM: THEORY AND EXPERIMRENT, J. Mech. Phys. Solids , 1982, Vol. 30

[5] C. Marotzke, A. Hampe, FRACTURE MECHANICAL ANALYSIS OF PULL-OUT AND FRAGMENTATION TESTS, Proceedings of the 11th Conference on Composite Materials (ICCM 11), Gold Coast , 1997, Ed. M. L. Scott, Woodhead Publishing Ltd.

[6] M. R. Piggott, Y. J. Xiong, VISUALISATION OF DEBONDING OF FULLY AND PARTIALLY EMBEDDED GLASS FIBRES IN EPOXY RESINS, Comp. Sci. Tech., 1995, Vol. 52

[7] A. Hampe, C. Marotzke, THE ENERGY RELEASE RATE OF THE FIBER/POLYMER MATRIX INTERFACE: MEASUREMENT AND THEORETICAL ANALYSIS, Journal of Reinforced Plastics and Composites, 1997, Vol. 16, No. 4

[8] H. D. Wagner, J. A. Nairn, M. Detassis, TOUGHNESS OF INTERFACES FROM INITIAL FIBER-MATRIX DEBONDING IN A SINGLE FIBER COMPOSITE FRAGMENTATION TEST, Applied Composite Materials, 1995, Vol. 2

Friction measurements on fibre reinforced polymer composites with using the indentation technique

Kalinka, G.; Leistner, A*.; Schulz, E.
Federal Institute for Materials Research and Testing,
Unter den Eichen 87, D 12205 Berlin, Germany
*Technical University Berlin,
Englische Str. 20, D 10587 Berlin, Germany

Abstract

In this paper we report some experiences of the application of 'push-out' type indentation tests applied on polymeric composites. Using a home made stiffness optimized test equipment the begin of the debonding, the delamination process and the friction after debonding can be monitored. In contrast to fragmentation tests and pull-out tests, fibre in realistic composites can be tested. The friction level after debonding can be used as an indicator for radial clamping stresses caused by the shrinkage of the polymer during curing and/or cooling down. It was found that the radial stresses are different in multifibre composites from single fibre composites as prepared for fragmentation tests or pull-out tests.

1. Introduction

Since the fibre matrix interface influence the composite properties strongly the determination of its characteristics is important for the proper use of fibre reinforced composites. Among other micromechanical methods [1] such as the single fibre pull-out test, the droplet strip-off and the fragmentation test, the interface behaviour of composites can be characterised with the indentation test with using the fibre 'push-in' or 'push-out' variation (Fig. 1).

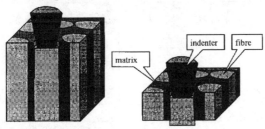

Fig. 1
Left: 'Push-in' type indentation test (cross section). The thickness of the sample is large, so that the interfacial crack does not reach the opposite side of the sample.
Right: 'Push-out' type indentation. In this case, a thin slice is prepared and the crack can reach the opposite side of the sample.

The indentation technique is the only micromechanical test which allows the in-situ characterisation of interfaces in both, custom made model samples and in realistic multifibre composite materials.

2. Push-in and push-out test type

The basic idea of the indentation test is to press an embedded single fibre in its axial direction into the matrix until the interface fails under the arising shear stress. From the load displacement curves information of failure processes of the composite such as debonding begin, frictional stress and debonding energy can be determined. In the case of push-in test type, with using the model of Marshall and Oliver [2-4] the debonding energy and the frictional stress in the debonded area of the interface can be calculated from the stiffness reduction which appears during the crack propagation under the increasing load. However, this model was developed for ceramic and glass matrix composites, where the energy dissipated by the matrix material can be neglected. In the case of polymer based composites, the plastic and/or viscoelastic deformation of the matrix near the interface crack has a significant contribution to the entire dissipated energy. Since the length of the debonding crack is unknown with the push-in test type, it is a problem to use the model of Marshall and Oliver for separating the contributions of friction and debonding energies.

For polymeric composites, it seems to be more convenient to perform the 'push-out' test type. Here, a thin slice of composite material (100-200μm) has to be prepared carefully and a single fibre segment is loaded until it is completely debonded. Then the fibre segment is pushed against the frictional stress through the slice. So the average frictional stress can be easily determined. Beside surface properties the friction level is an indicator for radial stresses on fibres in a composite caused by shrinkage and cooling down in a composite during the manufacturing process. The remaining part of the energy is the debonding energy, which is the sum of the surface energy, the plastic deformation near the surface, and the viscoelastic deformation of the matrix (Fig. 2).

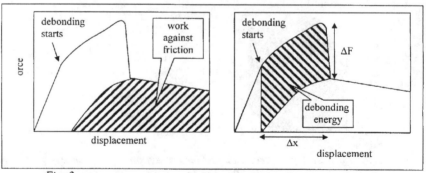

Fig. 2
Method for the estimation of the average interfacial friction and the debonding energy (the sum of surface energy, plastic and viscoelastic deformation attending the interface separation).

3. Experimental

The main requirements for a indentation test system are a high resolution of the axial displacement (better than 0.1 μm) and the in-plane positioning (better than 1 μm) as well as a precise force measurement system with a resolution better than 1 mN. In addition, a stiff design is necessary for monitoring the stiffness reduction caused by the partial debonding of a single fibre. With our apparatus, a stiff steel frame is used and highly stiff components for displacement generations and force measurement based on piezo ceramics. Function and design of the apparatus is described in Ref 6.

For a carefully preparation of the thin slices, a 'microtome saw' with diamond powder covered blades and a normal polishing devices were used. As indenters, we use custom made hard-metal needles prepared with using a home made polishing device in a flat-cone geometry (Fig. 3). The shape of the indenters match to the fibres which should be tested. The front surface of the indenter should cover about 70% of the fibre surface and the cone angle should be small enough to avoid any contact with the matrix material.

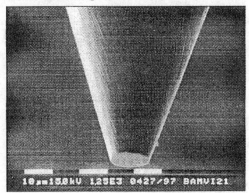

Fig. 3
Example for an indenter used for glass
fibres of 18-22μm diameter. The diameter of
the front area is abut 16 μm.

3. Results

A small block of glass fibre reinforced composite was made and a thin slice was cutted out of this block. In this slice, two groups of fibres were tested: (a) fibres which are closely surrounded with neighbour fibres and (b) fibres having at least one fibre diameter distance to any other fibre. Fibres of group a are under conditions of a realistic multifibre composite, whereas fibres of group (b) are in a situation of artificial made samples for e.g. pull-out tests of fragmentation tests. The obtained frictional stresses indicate, that the radial stresses on a single fibre in a mutifibre composite are smaller than on 'insulted' fibres. The debonding energy is smaller, too. This could be explained by the reduced contribution of plastic matrix material.

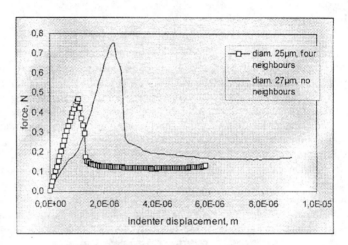

Fig. 4
Examples for push-out test data for fibers with and
without neighbour fibres.

References

1 Hampe, A., Kalinka, G., Schulz, E., in "Composites Testing and Standardisation, ECCM-CTS 2"; Ed. Hogg, P.J. et al, Woodhead Publ. Ltd., 1994 Cambridge, p 59-67
2 Marshall, D.B., *J. Ceram. Am. Soc.* 67 (1984) C259
3 Marshall, D.B., *J. Ceram. Am. Soc.* 68 (1985) 225
4 Marshall, D.B., Oliver, W. C., *J. Am. Ceram. Soc.* 70 (1987) 542
5 Kalinka, G., Hampe, A. *ICPM 1995* Eindhoven
6 Kalinka, G.; Leistner, A.; Hampe A: *Comp. Sci. Techn.* 57 (1997) 845-851

MIXED MODE ANALYSIS OF EDGE DELAMINATION
IN COMPOSITE LAMINATES

M. RUGGIU*, P. PRIOLO
DIPARTIMENTO DI INGEGNERIA MECCANICA,UNIVERSITA' DI CAGLIARI
PIAZZA d'ARMI,1 - 09123 CAGLIARI, ITALY

C.T.SUN
SCHOOL OF AERONAUTICS AND ASTRONAUTICS, PURDUE UNIVERSITY
WEST LAFAYETTE, INDIANA 47907 U.S.A.

ABSTRACT

A method to separate fracture modes in interface delamination in composite laminates is proposed. Owing to the oscillatory behavior of stresses and displacements close to the crack tip, individual strain release rates no longer exist. However, strain energy release rates for finite crack extensions can be calculated. The stress intensity factors are calculated by using an efficient method based on the crack surface displacement ratios. Explicit relationships between the finite energy release rates calculated by the modified crack closure scheme and the stress intensity factors are used to validate the method. Several edge delamination tests give the critical total strain energy release rate as well as the individual finite strain energy release rates for the interface modeled.

Keywords : interlaminar crack, strain - energy release rate, stress intensity factor, edge delamination.

INTRODUCTION

Even in very simple loading conditions, such as simple tension, composite laminates show interlaminar stresses at the free edges [1]. This stress field, which because of the Poisson ratio mismatch is not accounted for by classical laminate theory, can cause interfacial delamination failure. Unlike crack problems for isotropic materials, delamination in composite materials may involve all three modes of fracture. Moreover, the stress field shows an oscillatory behavior near the delamination crack tip.

Interface cracks in isotropic and anisotropic materials have been studied by many authors, such as Williams [2], who first studied the crack problem between two dissimilar isotropic solids, England [3], Wang and Choi [4], Wu [5], Hwu [6-7], Sun and Manoharan [8], Qian and Sun [9-10]. Many of them adopted the *small scale contact zone* proposed by Rice [11] and assumed that the oscillatory zone of the near tip is confined to a very small area compared to the crack length. However, because of this oscillatory stress field behavior, the energy release rates G_I^*, G_{II}^* and G_{III}^* for the three fracture modes oscillate as well. In fact, these individual strain energy release rates no longer exist.

In this work, the energy release rates for finite crack extension, G_I, G_{II}, G_{III}, are extended to include the interface between two monoclinic materials (one elastic symmetry plane). The finite element analysis together with the modified crack closure scheme [12] is used to obtain the G_I, G_{II}, G_{III}, and the total G values. Determination of

---

*Author to whom correspondence is to be addressed.

the stress intensity ratios K_{II}/K_I, K_{III}/K_I is based on the near tip crack surface solution. These ratios and the total energy release rate G are used to extract the stress intensity factors. The exact nonlinear relationships between the individual finite extension energy release rates and the stress intensity factors [9] are recalculated and used to validate the analytical procedure applied to evaluate the exact displacement and stress field.

The finite extension energy release rates are scaled to the critical load measured by several edge delamination tests (E.D.T.) to obtain the critical values for two different layups. The individual mode ratios among the G_i's and among the K_i's can be determined as well. Also, edge delamination tests were performed to obtain the mode I critical energy release rate without involving the mixed mode problem.

ANALYTICAL PROCEDURE

Consider an interface crack lying between two arbitrarily oriented composite laminae loaded by a constant strain ε_0 in the x_3 direction. This represents the generalized plane strain deformation commonly used to describe the free edge delamination problem in composite laminates. For this situation, the relations between the stress intensity factors and the stress field ahead of the crack tip are expressed by Hwu [6] as

$$\begin{Bmatrix} \sigma_{12} \\ \sigma_{22} \\ \sigma_{23} \end{Bmatrix} = \frac{1}{\sqrt{2\pi r}} \Lambda \left\langle\!\left\langle (\frac{r}{2a})^{i\varepsilon_\alpha} \right\rangle\!\right\rangle \Lambda^{-1} \begin{Bmatrix} K_{II} \\ K_I \\ K_{III} \end{Bmatrix} \tag{1}$$

where the angular brackets stand for a 3 x 3 diagonal matrix, r is the distance from the crack tip, $2a$ is the crack length, Λ is the complex 3 x 3 eigenvector matrix and ε_α are the oscillating indices.

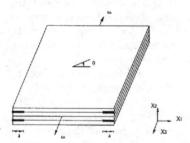

FIG. 1 Delamination geometry and loading condition

The relations between the stress intensity factors and the crack opening surface displacements behind the crack tip are

$$\begin{Bmatrix} \Delta u_1 \\ \Delta u_2 \\ \Delta u_3 \end{Bmatrix} = \sqrt{\frac{2r}{\pi}} (\bar{\Lambda}^T)^{-1} \left\langle\!\left\langle \frac{(r/2a)^{i\varepsilon_\alpha}}{(1+2i\varepsilon_\alpha)\cosh(\pi\varepsilon_\alpha)} \right\rangle\!\right\rangle \Lambda^{-1} \begin{Bmatrix} K_{II} \\ K_I \\ K_{III} \end{Bmatrix} \tag{2}$$

in which the polar coordinate (r,θ) is in the x_1 - x_2 plane with the origin at the crack tip. The Λ and the indices ε_α are obtained by solving the following eigenvalue problem:

$$(M + e^{2i\pi\delta}\,\overline{M}\,)\Lambda = 0 \tag{3}$$

where the overbar stands for the conjugate of a complex matrix and the bimaterial complex matrix M is defined as $M = D - iW$. The eigenvalue as given by Ting [13] is

$$\delta_\alpha = -\frac{1}{2} + i\varepsilon_\alpha$$

The D and W matrices are related to the Barnett and Lothe tensors [14], and the oscillating indices are calculated by applying the sextic formalism of Stroh [15]. By using the explicit expressions of ε_α and Λ derived by Qian and Sun [9], equations (1) and (2) can be written in terms of the material properties contained in the D and W matrices.

By using Irwin's integrals, the energy release rates for a finite crack extension Δa can be written as

$$G_I = \frac{1}{2\Delta a}\int_0^{\Delta a}\sigma_{22}(r,0)\Delta u_2(\Delta a - r,\pi)dr$$

$$G_{II} = \frac{1}{2\Delta a}\int_0^{\Delta a}\sigma_{12}(r,0)\Delta u_1(\Delta a - r,\pi)dr \tag{4}$$

$$G_{III} = \frac{1}{2\Delta a}\int_0^{\Delta a}\sigma_{23}(r,0)\Delta u_3(\Delta a - r,\pi)dr$$

Using the explicit formulas obtained for the crack tip stresses, equation (1), and the relative crack surface displacements, equation (2), the individual finite extension energy release rates of equation (4) are evaluted as functions of the stress intensity factors. It has been shown by Qian and Sun [9] that these individual strain energy release rates contain terms like $(\Delta a/4a)^{-2i\varepsilon}$ and thus do not converge as $\Delta a \rightarrow 0$. However, the total $G = G_I + G_{II} + G_{III}$ is still well defined. We have

$$G = \frac{1}{4}\left[\frac{D_{22}}{\cosh^2 \pi\varepsilon}K_I^2 + (D_{11} - \frac{W_{21}^2}{D_{22}}t)K_{II}^2 + (D_{33} - \frac{W_{32}^2}{D_{22}}t)K_{III}^2 + 2(D_{13} + \frac{W_{21}W_{32}}{D_{22}}t)\right]$$

$$t = \frac{1}{\beta^2}(1 - \frac{1}{\cosh^2 \pi\varepsilon}) \tag{5}$$

$$\beta = \left[-\frac{1}{2}tr(WD^{-1})^2\right]^{\frac{1}{2}}$$

where t is close to one and was neglected in [9].

FRACTURE PARAMETER CALCULATIONS

The equations given in the preceding section represent the tools for calculating the stress intensity factors. The general procedure is as follows :

1. W and D matrices and ε_α indices are calculated by a MATLAB code based on the solution from the Stroh formalism.
2. A finite element code developed for generalized plane strain deformation (pseudo 3D) is used to obtain the displacement ratios at the crack surface nodes and to obtain the energy release rates for finite crack extension and the total G by using the modified crack closure scheme.
3. The stress intensity factors are calculated by implementing the displacement ratio method [9] in the F.E. code.
4. A MATLAB code that is able to express the relations between those of the individual G_i's and K_i's is written to verify the analytical approach.

When performing the modified crack closure scheme, using the pseudo 3D finite element, it is necessary to subtract the effective body force resulting from the loading strain (applied in the x_3-direction) from the total crack tip nodal force.

Regarding the K_i calculations, Matos et al. [16] showed that for isotropic bimaterials the K_I's obtained directly from equation (2) are not accurate. However, the displacement ratio method [9] allows one to derive the stress intensity ratios K_{II}/K_I and K_{III}/K_I accurately by means of the relative displacement ratios $\Delta u_1/\Delta u_2$ and $\Delta u_3/\Delta u_2$ calculated by the F.E. analysis. The stress intensity factors are obtained by solving equation (5) in conjunction with the ratios K_{II}/K_I and K_{III}/K_I.

NUMERICAL APPLICATIONS

An F.E. modeling of the delamination at the free edge was performed. The model was loaded by a longitudinal strain $\varepsilon_0 = 10^{-6}$. A four-node isoparametric finite element, the third order Gauss quadrature, was used. The K_i's were evaluated at the first couple of nodes from the crack tip.

Eight-ply quasi-isotropic $[45/-45/0/90]_s$ and eighteen-ply $[(25/-25)_4/90]_s$ laminates of IM6/3501-6 (Hercules) carbon-epoxy composite were used. The material properties are listed in Table 1:

TABLE 1   Material Properties

E_{11} (GPa)	188.5
E_{22} (GPa)	10.0
E_{33} (GPa)	10.0
G_{12} (GPa)	6.20
v_{12}	0.29
t_0 (mm) ply thickness	76.2 x10^{-3}

The delamination was modeled between each of the 0/90 interfaces in the quasi-isotropic (Q.I.) laminate and between each of the −25/90 interfaces in the other laminate (T.N.). Because of symmetry, only a quarter of the laminate cross-section was modeled.

The crack was taken to be 6 mm long according to the real precrack design. In addition, the single mode I delamination crack between the 90/90 interfaces for both laminates was studied.

Tables 2 through 7 show the finite extension energy release rates obtained by the modified crack closure scheme (M.C.C.S) and equation (4), in conjunction with the displacement ratio method and the exact near tip solution based on the Stroh formalism (D.R.M), for various F.E. meshes.

TABLE 2 $\Delta a = 10^{-3}a$, $h_e = 2\Delta a$
(Δa is the distance from the crack tip node, h_e is the element height
M.C.C.S. is the finite element solution , D.R.M. is the numerical-analytical solution)

Q.I. 0/90	G_I/t_0 (N/m^2)	G_{II}/t_0 (N/m^2)	G_{III}/t_0 (N/m^2)	G/t_0 (N/m^2)
M.C.C.S.	1.990 x10^{-2}	9.240 x10^{-4}	1.244 x10^{-5}	1.995 x10^{-2}
D.R.M.	1.918 x10^{-2}	7.605 x10^{-4}	1.318 x10^{-5}	

TABLE 3 $\Delta a = t_0/6$, $h_e = \Delta a$

Q.I. 0/90	G_I/t_0 (N /m^2)	G_{II}/t_0 (N /m^2)	G_{III}/t_0 (N /m^2)	G/t_0 (N /m^2)
M.C.C.S.	1.868 x10-^{2}	1.259 x10^{-3}	1.248 x10^{-5}	1.996 x10^{-2}
D.R.M.	1.880 x10^{-2}	1.130 x10^{-3}	1.370 x10^{-5}	

TABLE 4 $\Delta a = t_0/8$, $h_e = \Delta a$

Q.I. 0/90	G_I/t_0 (N /m^2)	G_{II}/t_0 (N /m^2)	G_{III}/t_0 (N /m^2)	G/t_0 (N /m^2)
M.C.C.S.	1.885 x10^{-2}	1.091 x10^{-3}	1.242 x10^{-5}	1.995 x10^{-2}
D.R.M.	1.894 x10^{-2}	1.002 x10^{-3}	1.360 x10^{-5}	

TABLE 5 $\Delta a = 10^{-3}a$, $h_e = 2\Delta a$
(Δa is the distance from the crack tip node, he is the element height
M.C.C.S. is the finite element solution - D.R.M. is the numerical-analytical solution)

T.N. –25/90	G_I/t_0 (N /m^2)	G_{II}/t_0 (N /m^2)	G_{III}/t_0 (N /m^2)	G/t_0 (N /m^2)
M.C.C.S.	6.941 x10^{-2}	2.536 x10^{-2}	3.618 x10^{-4}	9.514 x10^{-2}
D.R.M.	7.570 x10^{-2}	1.888 x10^{-2}	5.920 x10^{-4}	

TABLE 6 $\Delta a = t_0/6$, $h_e = \Delta a$

T.N. –25/90	G_I/t_0 (N /m^2)	G_{II}/t_0 (N /m^2)	G_{III}/t_0 (N /m^2)	G/t_0 (N /m^2)
M.C.C.S.	6.669 x10^{-2}	2.808 x10^{-2}	3.409 x10^{-4}	9.512 x10^{-2}
D.R.M.	7.040 x10^{-2}	2.410 x10^{-2}	6.070 x10^{-4}	

TABLE 7 $\Delta a = t_0/8$, $h_e = \Delta a$

T.N. –25/90	G_I/t_0 (N /m^2)	G_{II}/t_0 (N /m^2)	G_{III}/t_0 (N /m^2)	G/t_0 (N /m^2)
M.C.C.S.	6.806 x10^{-2}	2.667 x10^{-2}	3.256 x10^{-4}	9.507 x10^{-2}
D.R.M.	7.170 x10^{-2}	2.270 x10^{-2}	6.460 x10^{-4}	

From the results we note that the differences between the two methods are reduced by using smaller regular-shape elements around the crack tip ($\Delta a = h_e$). This trend is basically true for the dominant mode, mode I, for both stacking sequences, but not always for the smallest mode. In the quasi-isotropic laminate, even though mode III is three orders smaller than mode I, the difference between these two methods is small because for the 0/90 interface, mode III is uncoupled from mode I and mode II, and G_{III} does not oscillate [9]. In contrast, it is seen that for the -25/90 interface in the $[(25/-25)_4/90]_s$ laminate the smallest fracture mode is not so well predicted.

EXPERIMENTAL PROGRAM

In order to obtain the critical energy release rates for the interface cracks in composite laminates, edge delamination tests [17-18] must be performed. For delamination between 0/90, -25/90, and 90/90 interfaces, the specimens were prepared by placing 6 mm wide teflon strips at these interfaces in both specimen edges during the layup procedure. The geometry of the specimen is shown in Figure 5.

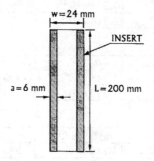

FIG.5 Geometry of the specimen

Ten specimens for each interface were tested in tension on an MTS servo-hydraulic testing system, with load control (0.05 kN/s for the Q.I. specimen and 0.1 kN/s for the T.N. specimen). Strain was measured by means of strain gages in the Q.I. laminate and by an extensometer (50 mm base length) for the T.N. Figures 6 and 7 show the delamination onset strains for the quasi-isotropic laminate, and figures 8 and 9 show those for the $[(25/-25)_4/90]$ laminate. It is noted that the 90/90 delamination exhibits stable crack growth while delaminations along the 0/90 and -25/90 show the characteristic plateau of the curve. In the 90/90 case, the onset of delamination was detected by the change in stress-strain curve.

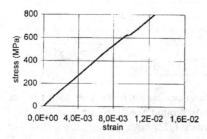

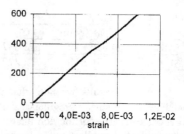

FIG.6 0/90 interface in Q.I. laminate   FIG.7 90/90 interface in Q.I. laminate

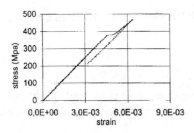

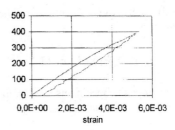

FIG.8 –25/90 interface in T.N. laminate   FIG.9 90/90 interface in T.N. laminate

Tables 8 and 9 resume the experimental data.

TABLE 8 Mixed Mode Fracture Toughness

Interface	$\varepsilon_c \times 10^{-3}$	G_c (J/m²)	G_I (J/m²)	G_{II} (J/m²)	G_{III} (J/m²)
0/90	9.6	140	132	7.67	0.08
-25/90	4.5	147	105	41.2	0.50

TABLE 9   Mode I Fracture Toughness along 90/90 Interface

Lay-up	Q.I.	T.N.
ε_c (x10⁻³)	5.6	3.0
G_{Ic} (J/m²)	90	87

The critical energy release rate G_c and the individual G_I, G_{II}, G_{III} shown in Tables 8 and 9 are obtained by scaling the numerical results to the delamination onset strain. For delaminations along the 0/90 interface (in the Q.I. laminate) and the –25/90 interface (in the T.N. laminate), all three fracture modes are present. However, G_{III} is observed to be negligibly small, and G_{II} is much smaller than G_I in the case of the 0/90 interface. From the results listed in Table 8, it is seen that the critical energy release rate G_c for the –25/90 interface is greater than that for the 0/90 interface. It can be concluded that larger mode mixity tends to increase the interfacial fracture toughness G_c.

In the case of delamination along the 90/90 interface, both laminates give about the same mode I critical energy release rate.

CONCLUSION

The finite element-based method has been shown to be accurate for calculating stress intensity factors and finite-extension energy release rates. Mode mixity can be taken into account using either the finite strain energy release rates or the ratios of the stress intensity factors. This method was demonstrated to be very efficient in extracting the critical energy release rate and mode mixities for edge delamination cracks in composite laminates. Further delamination tests using other types of specimen are necessary to verify the use of G_c and mode mixities as a delamination fracture criterion for composite laminates.

ACKNOWLEDGEMENT

The composite specimens used in this study were fabricated at the McDonnell Douglas Composite Materials Laboratory during the first author's visit to Purdue University in 1997.

REFERENCES

1. Pipes, R.B. and Pagano, N.J., "Interlaminar stresses in composite laminates under uniaxial extension", *J. Compos. Mater.*, Vol. **4**, 1970, pp. 538-548
2. Williams, M.L., "The stresses around a fault or crack in dissimilar media", *Bull. Seismol. Soc. Am.*, 1959, **49**, pp. 199-204
3. England, A.H., "A crack between dissimilar media", *J. Appl. Mech.*, 1965, **32**, pp. 400-402
4. Wang, S.S. and Choi, I., "The interface crack between dissimilar anisotropic composites under mixed mode loading.", *J. Appl. Mech.*, 1983, **50**, pp. 179-183
5. Wu, K.C., "Stress intensity factor and energy release rate for interfacial cracks between dissimilar anisotropic materials", *J. Appl. Mech.*, 1990, **57**, pp. 882-886
6. Hwu, C., "Fracture parameters for the orthotropic bimaterial interface cracks", *Eng. Fract. Mech.*, 1993, **45**, pp. 89-97
7. Hwu, C., "Explicit solutions for collinear interface crack problems", *Int. J. Solids Struct.*, 1993, **3**, pp. 301-312
8. Sun, C.T. and Manoharan, M.G. "Strain energy release rate of an interfacial crack between two orthotropic solids.", *J. Compos. Mater.*, 1989, **23**, pp. 460-478
9. Qian, W. and Sun, C.T. "Calculation of stress intensity factors for interlaminar cracks in composite laminate", *Comp. Science and Techn.*, 1997, **57**, pp. 637-650
10. Qian, W. and Sun, C.T., "Method for calculating stress intensity factors for interfacial cracks between two orthotropic solids", to appear in the *Int. J. Solids and Struct.*
11. Rice, J.R., "Elastic fracture mechanics concepts for interfacial cracks.", *J.Appl. Mech.*, 1988, **55**, pp. 98–103
12. Rybicki, E.F., Schmueser, D.W. and Fox , " An Energy release Rate Approach for stable crack growth in the free–edge delamination problem", *J. Compos. Mat.*, Vol. **11**, 1977, pp. 470-487
13. Ting, T.C.T., "Explicit solution and invariance of the singularites at an interface crack in anisotropic composites", *International Journal of Solids and Structure*, 1986, **22**, pp. 965-983
14. Barnett, D.M. & Lothe, J., "Synthesis of the sextic and integral fomalism for dislocation, Green's function and surface waves in anistropic elastic solids." *Phys. Norv.*, **7**, 1973, pp. 13–19
15. Stroh, A. N., "Dislocations and cracks in anistropic elesticity.", Phil. Mag.,1958, **7**, pp. 625–646
16. Matos, R.M., McMeeking, R.M., Charalambides, P.G. and Drory, D., "A method for calculating stress intensities in bimaterial fracture", *Int. J. Fract.*, 1989, **40**, pp. 235-254
17. O'Brien, T.K., Johnson, N.J., Raju, I.S., Morris, D.H. and Simonds, R.A., "Comparisons of various configurations of the edge delamination test for interlaminar fracture toughness", *Toughened Composites, ASTM STP 937*, Norman J. Johnson Ed., American Society for Testing and Materials, PA, 1987, pp. 199-221
18. Whtney, J.M., & Knight, M., "A modified free-edge delamination specimen", *Delamination and Debonding of Materials, ASTM STP 876*, W.S. Johnson Ed., American Society for Testing and Materials, PA, 1985, pp. 298-314

Rate and Temperature Effects on Interlaminar Fracture Toughness in Interleaved PEEK/CF Composites

R. Fracasso*, M. Rink*, A. Pavan*, R. Frassine*

* Dipartimento di Chimica Industriale e Ingegneria Chimica "G. Natta"
  Politecnico di Milano - P.zza Leonardo da Vinci 32 - 20133 Milano (Italy)

Abstract

The interlaminar fracture behaviour of unidirectional and woven fabric fibre composites based upon continuos carbon fibres (CF) and polyetheretherketone (PEEK) has been investigated over a wide range of temperatures and crack speeds. The fracture data obtained from mode I and mode II tests have been analysed using the time-temperature superposition approach, and the results are discussed in terms of fracture micromechanisms. Finally, by increasing the thickness of the resin rich interlaminar region a significant increase in delamination resistance was found.

1. Introduction

Polymer-matrix/fibre composites are being increasingly incorporated into structures which are subjected to various types of loading, either static or dynamic, in a wide range of temperatures. Delamination of composite parts is usually associated with failure of the matrix and/or the fibre/matrix interface, without a significant amount of fibre fracture. This particular type of failure requires a relatively low energy to occur and is often a limiting factor in the use of those structures. Some authors [1,2] have shown that delamination resistance may be improved by increasing the thickness of the resin-rich layer between laminae, and called it "interleaving". This method consists in sandwiching thin layers of ductile polymer film between the composite plies. Because the resin layers have lower stiffness and strength, their application has to be optimised in order not to alter the overall composite performance.

Few papers, however have dealt with the effect of the viscoelastic behaviour of the matrix on the delamination resistance of the composites. In this paper, an experimental investigation on the interlaminar fracture behaviour of unidirectional and woven carbon fibre/ polyetheretherketone (PEEK) composite laminates in mode I and mode II loading at varying testing temperature and rate is presented. Results obtained on interleaved unidirectional and woven fabric laminates are also presented.

2. Experimental procedure

2.1 Materials

The material studied was kindly supplied by EniChem S.p.A. as tows of unidirectional continuous carbon-fibres pre-impregnated with a PEEK matrix. The manufacturing technique, denominated "Fibre Impregnated with Thermoplastic" (FIT), allows a fine dispersion of the PEEK powder within each roving of the tow while providing a continuous protective coating of the same PEEK matrix. The product is a flexible continuous thread, having a diameter of about 0,2 mm, of PEEK matrix with embedded carbon fibres. The fibres were T-300 type and their volume fraction was about 58%. A 8HS woven fabric made of the same thread was also made available by EniChem.

2.2 Test specimens

The 200 x 280 mm woven fabric plies were obtained by cutting the fabric with scissors. Unidirectional plies having the same in-plane dimensions were obtained by filament winding on a flat plate. The aligned tows were welded with a soldering iron along several different transverse lines to prevent them from slacking and moving off, and then the layer was removed by cutting along the edges, so obtaining two unidirectional plies.

10-ply woven fabric and 14-ply unidirectional laminates were compression moulded at 380°C and 2 MPa according to manufacturer specifications, yielding to laminates about 2,3 mm and 3,2 mm thick, respectively. In all laminates a 50mm wide and 50μm thick release-agent coated foil (Thermalimide by Air Tech) was inserted between the central plies to produce an initial edge delamination. The interleaved laminates were prepared placing two additional 200 x 280 mm PEEK layers on either side of the starter film. Two different nominal interleaving thicknesses of 50 and 100 μm were adopted.

Specimens 20 x 210mm were obtained by saw-cutting of the moulded plates. For mode I testing 20 x 15 x 20 mm aluminium alloy end-blocks were bonded on each side of the specimen using a modified epoxy adhesive (DP 490 by 3M). The curing cycle was 80°C for 90 min. The lateral side of the specimen was painted white with typewriter correction fluid, and marked at 1 mm intervals so that the position of the crack front could be monitored during the fracture tests using a high-resolution video camera connected to a video-recorder.

2.3 Fracture test methods

Interlaminar fracture testing was conducted at constant displacement rates ranging from 0.05 to 200 mm/min and test temperatures from 23° to 160°C. At least two replicas for each testing condition were conducted. Fracture tests were carried out under mode I (double cantilever beam test, DCB) and mode II (end notched flexure test, ENF). Load, displacement and crack length were determined at any time during the tests. Interlaminar fracture energies were calculated as suggested by the European Structural Integrity Society (ESIS) protocol [3] by using the beam theory for DCB and the direct beam theory for ENF. Large displacements, stiffening due to end-blocks and transverse shear effects were taken into account, as appropriate. Before mode II tests, the specimens were precracked under mode I loading, so as to propagate the initial edge delamination about 10 mm. Then, the starter film was removed and a small coupon of PTFE film, 50 μm thick, was inserted into the crack to reduce friction.

The experimental fracture data obtained at different temperatures were reduced according to the time-temperature superposition principle. This was achieved by simply multiplying the crack speed ($\dot{a}$=da/dt) at a certain test temperature by the correspondent polymer matrix shift factor ($a_T^{T_o}$), obtained in a previous work [4]. In this way, the interlaminar fracture energy values obtained at a given rate and temperature were shifted along the crack speed axis to produce a master curve. The reference temperature, T_o, was 23°C.

3. Results and discussion

3.1 Uninterleaved laminates

3.1.1 Unidirectional composite

Interlaminar fracture tests in mode I and mode II were performed on the composite prepared from the unidirectional carbon fibre/PEEK composite. Crack propagation was always stable and took place at a crack speed ($\dot{a}$) approximately constant during the test. No evidence of a crack resistance curve was noticed. The master curves obtained from the average values of the interlaminar fracture toughness as a function of the reduced crack speed are shown in Fig. 1.

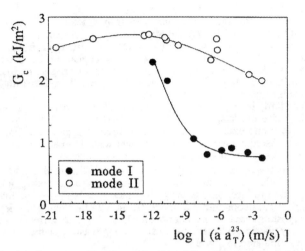

Figure 1- *Interlaminar fracture energy, G_C, as a function of the reduced crack speed, for the unidirectional carbon fibre/PEEK composite (T_o=23°C)*

Results show a regular decreasing trend with increasing crack speed in mode I, while for mode II the trend is non-monotonic, with a slight maximum for crack speeds of about 2 x 10^{-13} m/s. Mode II values appear to be significantly larger than those in mode I, especially at high crack speeds. The fracture surfaces were examined using a scanning electron microscope, for two different testing conditions (Fig. 2).

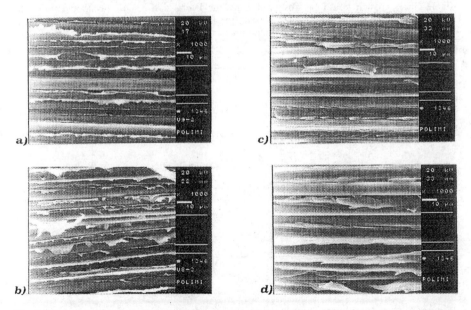

Figure 2- *Scanning electron micrographs of fracture surfaces from the unidirectional carbon fibre/PEEK composite at two different testing conditions, (a)-(b): mode I; (c)-(d): mode II.*

Micrographs (a) and (b) in Fig. 2 were obtained from fracture surfaces of specimens tested in mode I at high (3,3 x 10^{-4}m/s) and low (3,3 x 10^{-11}m/s) crack speeds, which are associated with low (0,82 kJ/m²) and high (1,98 kJ/m²) interlaminar fracture toughness, respectively (see Fig. 1). In both cases, the fibres appear to be clearly exposed, suggesting that a low degree of adhesion exists between fibres and matrix. The larger plastic deformation of the matrix observed at the lower speed (Fig. 2(b)) may explain the larger toughness value obtained.

Micrographs (c) and (d) in Fig. 2 were obtained from specimens tested in mode II at approximately the same crack speeds as for (a) and (b). The fibres now appear well coated with the matrix and extensive plastic deformation is apparent in both cases. The similarity of the two fracture surfaces is in agreement with the relatively small difference observed between G_{IIc} values at varying crack speed (see Fig. 1).

It may therefore be concluded that delamination in unidirectional carbon fibre/PEEK composites is controlled by a competition between matrix cracking and interfacial debonding. For mode I failure, only at very low crack speeds may some matrix yielding occur before matrix/fibre debonding. For mode II, on the other hand, no debonding was observed thus suggesting that fracture is dominated by plastic deformation of the matrix. The greater extent of plastic deformation may also be the reason for the higher toughness values obtained as compared to mode I.

3.1.2 Woven fabric composite

Only mode I fracture tests were performed for the woven fabric composite. The material showed typical unstable ("stick-slip") crack growth with series of rapid crack propagations. Results obtained in a previous work [4] show that some slow crack growth may occur before each crack jump. Such crack growth may, in the first approximation, be assumed to take place at a speed equal to the applied displacement rate, $\dot{x}$. Although no stable crack growth could be observed from the lateral surface of the specimen in the present work, we made here the same assuption ($\dot{a}=\dot{x}$). The interlaminar fracture energy, G_{Ic}, as a function of the nominal reduced crack speed is shown in Fig. 3.

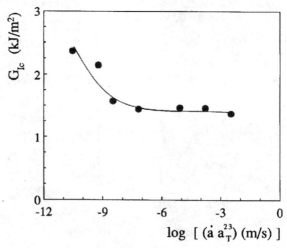

Figure 3- *Mode I interlaminar fracture energy, G_{Ic}, as a function of the reduced crack speed for the woven fabric carbon fibre/PEEK composite ($T_o=23\,°C$)*

The trend is similar to that of the unidirectional composite (see Fig. 1): fairly regular decreasing values of G_{Ic} are observed with increasing crack speed. The energy values for the woven composite are, however, far higher than those for the unidirectional composite. This is most probably due to the geometrical and spatial differences between the unidirectional and the woven fabric materials. Micrographs (a) and (b) in Fig. 4 were obtained from fracture surfaces of specimens tested at high $(1,7 \times 10^{-4} m/s)$ and low $(3,0 \times 10^{-11} m/s)$ crack speeds which are associated with low $(1,46 \ kJ/m^2)$ and high $(2,36 \ kJ/m^2)$ interlaminar fracture toughness, respectively (see Fig. 3). The transverse carbon fibres act as a barrier to the propagating crack, and delamination sometimes changes its plane of propagation, involving fibre breakage. Therefore more energy is dissipated than in the unidirectional composite, leading to higher fracture toughness. Another mechanism of energy dissipation could be the plastic deformation of localized PEEK-rich regions, which are not present in the unidirectional composite. The latter mechanism, however, could not be assessed from the micrographs.

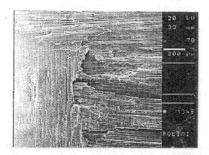 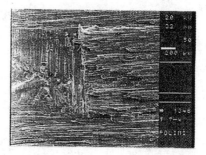

Figure 4- Scanning electron micrographs of fracture surfaces of the woven fabric carbon fibre/PEEK composite at two different testing conditions (mode I)

3.2 Interleaved laminates

In laminated composites, interlaminar cracks are essentially confined in a narrow resin-rich region between plies, in which the plastic deformation process at the crack tip is severely restricted by the adjacent rigid plies. The interleaving method relies on the possibility for the crack tip plastic zone to expand perpendicular to the crack plane and to allow for more energy dissipation during crack extension.

3.2.1 Unidirectional composite

The effect of the interleaving on the fracture behaviour of the unidirectional composites in mode I and mode II is shown in Fig. 5. Two different nominal thicknesses of the interleaving film were adopted: 50 and 100 µm. The fracture toughness of the interleaved laminates is always greater than that of the uninterleaved ones, and increases regularly with the interleaving thickness. The trend of G_C values with crack speed is essentially the same for all materials examined.

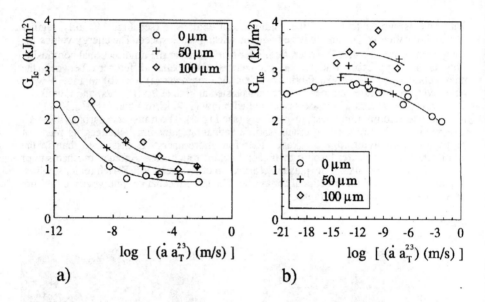

Figure 5- *Intelaminar fracture toughness, G_C, as a function of crack speed for unidirectional composites with different nominal interleaving thicknesses: (a) mode I and (b) mode II (T_o=23 °C)*

The analysis of the fracture surfaces in mode I shows that most of the fibres are now coated with the matrix, suggesting that delamination in the interleaved material propagates essentially through the PEEK layer. Therefore, adhesion between fibres and matrix does not seem to be involved any more.

Furthermore, the fracture surface of the specimen with 100 μm thick interleaving shows extensive plasticity that may explain the higher values of G_{Ic} for this material.

The interleaved fracture surfaces in mode II were similar to those of the uninterleaved material. This makes it difficult to explain the higher values of the fracture toughness for the interleaved material in terms of fracture surface features.

3.2.2 Woven fabric composite

The effect of interleaving on mode I fracture behaviour of woven fabric composites is shown in Fig. 6. Delamination toughness is enhanced by increasing interleaving thickness.

The micromechanism of failure of the interleaved woven fabric could not be clearly identified from the SEM micrographs of the fracture surfaces.

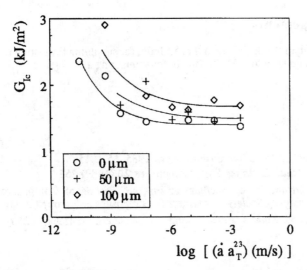

Figure 6- *Mode I interlaminar fracture energy, G_{Ic}, as a function of crack speed for the woven fabric composite at varying interleaving thickness (T_o=23°C)*

4. Conclusions

The effects of temperature and test rate on the interlaminar fracture toughness of unidirectional and woven fabric carbon fibre composites based upon a thermoplastic polyetheretherketone (PEEK) matrix and prepared with FIT technology, have been studied. In order to improve delamination toughness, the interleaving method has also been tested. Finally, some failure micromecanisms have been postulated. The following conclusions can be drawn from the results obtained.

The temperature and crack propagation rate effects upon fracture toughness may readily be appreciated by constructing a master curve, i.e. by plotting the interlaminar fracture toughness as a function of a reduced crack propagation rate. For the unidirectional composite tested in mode I, toughness values were strongly dependent on the crack speed, whilst in mode II only a moderate effect was observed. The fracture toughness in mode II turned out to be significantly greater than that in mode I, especially for high crack speeds. Fracture surfaces analysis by SEM indicates that failure in mode II is associated with significant shear plastic deformation of the matrix while failure in mode I occurs essentially by fibre/matrix debonding.

The mode I toughness of the woven fabric composite shows the same trend with crack speed as the unidirectional composite, but with higher energy values, probably due to additional dissipative failure mechanisms.

Interleaving with the same matrix resin has proved to be a viable technique of toughening. The introduction of resin central layers of 50 and 100 μm produced an increase in delamination resistance for both unidirectional and fabric composites at all loading conditions examined.

Acknowledgements

The authors wish to thank EniChem S.p.A., Italy, for supplying the testing materials. R. Fracasso is also grateful to CAPES, Brazil, for sponsoring a PhD grant.

References

[1] *R.B. Krieger*, SAMPE Symposium, April 1985, 1570-1585
[2] *N. Sela, O. Ishai, L. Banks-Sills*, Composites **20** (1989) 257
[3] *P. Davies*, Protocols for Interlaminar Fracture Testing of Composites, *European Structural Integrity Society* - Polymers & Composites task group (1993)
[4] *R. Frassine, M. Rink, A. Pavan*, Comp. Sci. Technol. **56** (1996) 1253

Delamination growth in epoxy-matrix composites under cyclic loading: implications for design and certification

Sunil Singh* and Emile Greenhalgh*

*Structural Materials Centre, DERA Farnborough, Hants GU14 0LX, United Kingdom

Abstract

This paper discusses quantitative results from fatigue delamination tests (MMB) at unidirectional and cross-ply ply interfaces in carbon-fibre/epoxy-matrix laminates at three different mixed-mode loading conditions. The implications for material characterisation, structural design and certification are addressed. The results indicate the critical design parameter for delamination in thermoset-matrix composites under cyclic loading is the strain energy release rate threshold, $G_{threshold}$. The results suggest that a design optimised for static loading with an appropriate safety factor would be approximately optimised for fatigue loading. This would significantly reduce the number of expensive time-consuming fatigue tests required for certification. The cost of fatigue testing may also be reduced by using accelerated test programmes at appropriate low-load cut-off levels, providing careful consideration is given to creep and environmental effects. There is scope for further optimisation by taking a stochastic approach to represent the local variations in toughness in delamination modelling.

1. Introduction

The full commercial benefits of using high performance fibre reinforced plastics have not yet been realised. This is in part due to insufficient understanding of the conditions which may provoke failure, necessitating the use of high safety factors in design and extensive test matrices for characterisation and certification. One area of concern is the effect on structural performance of delaminations, which arise due to manufacturing defects or impact damage. It is well known that cracks in metals can grow in fatigue at loads well below those which would provoke static failure; the fatigue performance of delaminated composites is yet to be fully characterised and understood. The loading on a delamination crack may be peel (mode I) or shear (mode II); in reality the loading conditions are usually a combination of both peel and shear (mixed-mode loading).

Previous studies of interlaminar fatigue have focused on unidirectional ply interfaces. In real structures, containing multidirectional lay-ups, delamination growth is not observed between plies at the same fibre orientation. Delaminations most frequently occur between orthogonally-oriented plies, i.e. at +45°/-45° and 0°/90° ply interfaces. Delamination propagation is essentially the same for both cases since delaminations locally grow parallel to the fibres.

This paper discusses quantitative results from fatigue delamination resistance tests at unidirectional (0°/0°) and cross-ply (0°/90°) interfaces in carbon-fibre/epoxy-matrix

© British Crown Copyright 1998/DERA.
Published with the permission of the Controller of Her Britannic Majesty's Stationary Office

laminates at three different mixed-mode loading conditions. The Implications for material characterisation, structural design and certification are addressed. An earlier paper reported results from a fractographic study of the micromechanisms of delamination growth in the same specimens [1].

2. Experimental details

The material under investigation here was a commercial carbon-fibre/epoxy-matrix pre-preg system (Hexcel T800/924), cured and post-cured according to manufacturer's recommended procedure. Standard 24-ply double cantilever beam (DCB) coupons containing a 10μm PTFE insert film artificial delamination were used to characterise unidirectional ply interfaces, while 32-ply quasi-isotropic specimens with edge inserts [2] were used for cross-ply interfaces. For the the latter a stacking sequence was chosen with a 0°/90° ply interface at the mid-plane which maximised specimen axial and transverse bending stiffnesses whilst minimising residual thermal stresses and bend-twist coupling terms of the stiffness matrix.

Delamination toughness testing was carried out using the mixed-mode bending (MMB) rig [3]. A short 75% mode I pre-crack was generated quasi-statically before testing at a frequency of 5Hz (to a maximum of 1.8×10^6 cycles) at one of three selected mixed-mode conditions: 25%, 50% and 75% mode I. The amplitude of the cross-head displacement was kept constant during the tests. The maximum displacement was chosen such that the initial peak applied strain energy release rate was in the range of 40% - 80% of the strain energy release rate for static failure. The specimen was almost entirely unloaded during each cycle.

After testing, typical specimens were chosen, sputter-coated and examined using a scanning electron microscope (SEM) at magnifications between 50x and 5 000x. Further experimental details may be found in the earlier paper [1].

3. Analysis

The fatigue results were compared with a theory for crack growth in composites, similar to that proposed by Paris [4]. The Paris theory suggests that for crack propagation in metals under cyclic loading there is a threshold of stress intensity amplitude $\Delta K_{threshold}$, below which a crack will not grow. Crack propagation rates da/dN above this threshold are proportional to a power-law function of ΔK (equation 1).

In fibre reinforced composites the stress-field at the crack tip is highly distorted by the fibres, making a representative evaluation of K problematic. A more applicable power-law model in terms of the amplitude of applied strain energy release rate, ΔG, has been used for over thirty years (equation 2) [5]. However, there is some ambiguity in the definition of ΔG for reversed cyclic loading, where the maximum and minimum loads are of opposing sign. The stress intensity factors corresponding to maximum and minimum load are then of opposing sign and $\Delta K > K_{max}$. The strain energy release rates, however, are of the same sign. Three different expressions of ΔG have been used in articles concerning the fatigue behaviour of composites, making direct comparison of results from different sources difficult [6-9].

There is no physical reason for fatigue crack propagation rates in thermoset-matrix composites to depend on the amplitude of strain energy release rates, and some authors [10, 11] have chosen to replace ΔG in the power-law with peak strain energy release rate G_{max}, avoiding the above ambiguity (equation 3). This was the approach taken here.

Values for G_{max} were calculated at peak loads using corrected beam-theory expressions [12, 13] and propagation rates da/dN were determined by differentiating polynomial fits of crack length a versus number of cycles N. The parameters in the power law were then obtained from the intercept and gradient of the modified Paris plots.

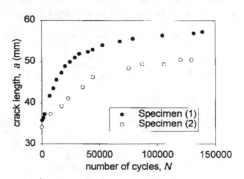

Figure 1. a versus N.

$$\frac{da}{dN} = A(\Delta K)^p \qquad (1)$$

$$\frac{da}{dN} = B(\Delta G)^q \qquad (2)$$

$$\frac{da}{dN} = C(G_{max})^r \qquad (3)$$

$(A, p, B, q, C, r$ are fitting constants)

4. Results

Figure (1) shows curves of a versus N for two specimens, of which the curve for specimen (1) was more typical.

Figure (2) shows curves of G_{max} versus a for the same two specimens. The fatigue tests were carried out under constant amplitude of displacement. As the crack length increased the specimen usually became more compliant. It was therefore usual for the applied strain energy release rate to decrease as the test progressed, as in the case of specimen (1). However, as demonstrated by the results for specimen (2), this was not always the case.

The data in Figures (1) and (2) were combined to produce Paris plots of the type shown in Figure (3), showing da/dN versus G_{max} on log-log scales.

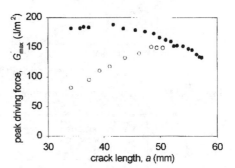

Figure 2. R Curves.

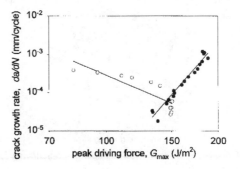

Figure 3. Paris plots.

The behaviour of specimen (1) was described reasonably well by equation (3). However for some specimens it was observed that da/dN was falling whilst G_{max} was constant or even increasing (as in the case of specimen (2)). When these results are included there is significant scatter in the Paris plot. This phenomenon has not been previously reported and such results may in the past have been dismissed as being anomalous. It should be noted that this scatter was not significantly greater than that observed in static tests carried out as part of the same programme.

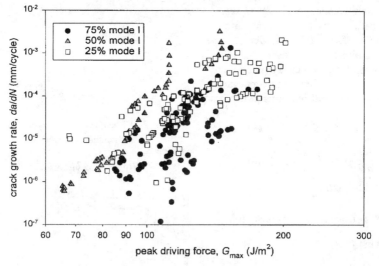

Figure 4. Modified Paris plots for delaminations at 0°/0° ply interfaces in T800/924.

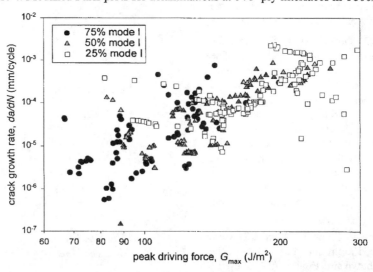

Figure 5. Modified Paris plots for delaminations at 0°/90° ply interfaces in T800/924.

Fatigue test results for unidirectional ply interfaces tested at 75%, 50% and 25% mode I loading are shown in Figure 4. Fatigue crack propagation did not occur below the relatively high threshold of $G_{max}/G_{static} \approx 30\%$. The tendency for strain energy release rate to increase with increasing proportion of mode II loading [14] was less apparent in the fatigue data (at a given propagation rate) than in comparable static data. There was too much scatter to draw any firm conclusions other than that, in general, da/dN tended to decrease with decreasing strain energy release rate.

Fatigue results for cross-ply interfaces are shown in Figure 5. These results may indicate a slightly enhanced toughness and higher degree of scatter at 0°/90 ply interfaces when compared with 0°/0° ply interfaces.

Fracture surfaces generated under fatigue at magnifications up to 2000x were not easily distinguishable from surfaces generated quasi-statically at the same mixed mode conditions (Figure 6). At higher magnifications features characteristic of cyclic loading, striations and matrix rollers, have been observed [8]. Further details of surface morphology observed in this study are given elsewhere [1].

Figure 6. Micrographs of static and fatigue fracture surfaces generated at 0°/0° ply interface at 50% mode I (50% mode II) loading. (x1500, 30° tilt).

5. Discussion

The scatter in the critical strain energy release rates of laminated composites was attributed to variations in the local distribution of fibres near the delaminating ply interface [1].

These results indicate that thermoset-matrix composites do not suffer from fatigue in the same manner as metals. Fatigue crack growth in metals relies on the migration of dislocations. Thermosets are amorphous and therefore do not contain dislocations. Furthermore they are cross-linked, severely hindering plastic flow. Crack propagation can only occur when strong covalent bonds are permanently broken. Similar mechanisms and energies are involved, whether the material fails statically or in fatigue; this is reflected in the similarity of static and fatigue fracture surfaces. The domain in which crack rates can be described by a Paris relationship (as used for metals) is therefore narrow. The dependence of propagation rate on applied strain energy release rate is weak, which is reflected in a very high gradient in Paris plots.

As a result of the steepness of the Paris plot, the narrowness of its domain and the significant local variations in toughness exhibited by fibre composites, the Paris propagation model is of severely limited value in predicting component lifetime. Others [11] have argued that, for a material perfectly obeying the Paris law with an exponent of 9, a 10% uncertainty in the loads would lead to a 61% uncertainty in crack propagation rate. If crack growth in composites is governed by a power-law function of (G_{max}/G_{static}) and G_{static} varies locally by 10% then over short distances a similar uncertainty in propagation rates would result, even if the loads were known perfectly.

In thermoset-matrix composite components where delamination growth over short distances could prove critical, cyclic loads should be kept below their threshold levels (*i.e.* a "no growth" criterion should be used). This is consistent with current design philosophy. Contrary to the design of metal components, this does not imply a huge safety factor relative to the static case. Figure 7 shows crack propagation rates in terms of the proportion of the static critical strain energy release rate applied from all the laminate tests carried out and, for comparison, published results for 2024 T3 aluminium [15]. An aluminium component containing a crack, loaded cyclically at 10% of its static critical strain energy release rate, will fail relatively quickly (1mm crack growth in 10^4 cycles). A composite component containing a delamination, loaded in the same manner, has infinite life.

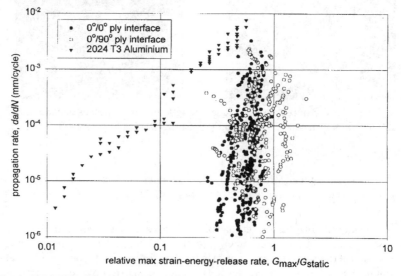

Figure 7. Modified Paris plots for delaminations at 0°/0° and 0°/90° ply interfaces in T800/924. Peak driving force normalised against value for quasi-static failure.

The suggested 'safe' proportion of G_{static} of 10% for this material corresponds to about 30% of P_{static}. Some aircraft manufacturers and operators are currently ignoring the parts of the loading spectrum which fall below 30% of peak load, in order to accelerate certification test programmes. In one example 5 000 flights were simulated in 23 000 cycles, compared with 650 000 cycles for the equivalent metal structure. These results indicate that the approach is probably appropriate for thermoset-matrix composites;

however more work is required to confirm how this behaviour observed in coupons may be extrapolated to structures. There must also be some concern over the practise of simulating component life by cycling at frequencies as high as 5-10Hz since creep and environmental effects may not be correctly reflected.

6. Conclusions

The critical design parameter for delamination in thermoset-matrix composites under cyclic loading is the strain energy release rate threshold, $G_{threshold}$. This may be determined over a long time-scale by using a small number of specimens and incrementing the load until failure is observed, or over a shorter time-scale by using a larger number of specimens to focus rapidly on the relevant part of the S-N curve.

If, as may be indicated in these results, $G_{threshold}/G_{static}$ in epoxy-matrix systems does not depend strongly on the mixed-mode loading conditions, then a design optimised for static loading with an appropriate safety factor would be "approximately optimised" for fatigue loading. The design process could be based principally on static tests, with expensive, time-consuming fatigue testing being reserved for verification. The cost of this final phase may also be reduced by using accelerated test programmes at appropriate low-load cut-off levels, providing careful consideration is given to creep and environmental effects.

There is scope for further optimisation by taking a stochastic approach [16] to represent the local variations in toughness and model delamination behaviour. This would further reduce the magnitude of the safety factors. Demonstration of improvements in the understanding of the failure of composites under fatigue loading will lead to reductions in the size of test matrices required for certification.

Acknowledgements

This work was carried out as part of Technology Group 4 (Materials and Structures) of the MOD Corporate Research Programme, and as part of the DTI CARAD Programme.

References

[1]   S. Singh, E.S. Greenhalgh, MICROMECHANISMS OF INTERLAMINAR FRACTURE IN CARBON-EPOXY COMPOSITES AT MULTIDIRECTIONAL PLY INTERFACES, Proceedings of the Fourth Conference on Deformation and Fracture of Composites (DFC-4), Manchester, UK, 1997

[2]   S. Foster, P. Robinson, J. Hodgkinson, AN INVESTIGATION OF INTERLAMINAR FRACTURE TOUGHNESS AT 0°/THETA INTERFACES IN CARBON-EPOXY LAMINATES, Proceedings of the Fourth Conference on Deformation and Fracture of Composites (DFC-4), Manchester, UK, 1997

[3]   J.R. Reeder, J.H. Crews, NON-LINEAR ANALYSIS AND REDESIGN OF THE MIXED-MODE BENDING DELAMINATION TEST, NASA Technical Memorandum 102777, 1991

[4]   P. Paris, FATIGUE - AN INTERDISCIPLINARY APPROACH, Proceedings of the Tenth Sagamore Conference, New York, USA, 1964

[5]   S. Mostovoy, E. Ripling, FRACTURE TOUGHNESS OF AN EPOXY
      SYSTEM, Journal of Applied Polymer Science, Vol. 10, 1966

[6]   M. Hojo, C. Gustafson, K. Tanaka, R. Hayashi, NEAR-THRESHOLD
      PROPAGATION OF DELAMINATION FATIGUE CRACKS IN
      UNIDIRECTIONAL CFRP IN AIR AND IN WATER, Proceedings of
      Composites 86: Recent advances in Japan and in the United States, Tokyo,
      Japan, 1986

[7]   A. Russell, K. Street, THE EFFECT OF MATRIX TOUGHNESS ON
      DELAMINATION: STATIC AND FATIGUE FRACTURE UNDER MODE II
      SHEAR LOADING OF GRAPHITE FIBER COMPOSITES, American Society
      for Testing and Materials (ASTM) Special Technical Publication 937, 1987

[8]   M. Hiley, P. Curtis, MODE II DAMAGE DEVELOPMENT IN CARBON
      FIBRE REINFORCED PLASTICS, 74th AGARD Structures & Materials
      Meeting, Patras, Greece, AGARD, 1992

[9]   Y. Prel, P. Davies, M. Benzeggagh, F. de Charentenay, MODE I AND MODE
      II DELAMINATION OF THERMOSETTING AND THERMOPLASTIC
      COMPOSITES, American Society for Testing and Materials (ASTM) Special
      Technical Publication 1012, 1989

[10] D. Wilkins, J. Eisenmann, R. Camin, W. Margolis, R. Benson,
      CHARACTERIZING DELAMINATION GROWTH IN GRAPHITE-EPOXY,
      American Society for Testing and Material (ASTM) Special Technical
      Publication 775, 1980

[11] R. Martin, G. Murri, CHARACTERIZATION OF MODE I AND MODE II
      DELAMINATION GROWTH IN GRAPHITE-EPOXY, NASA Technical
      Memorandum 100577, 1988

[12] A. Kinloch, Y. Wang, J. Williams, P. Yayla, THE MIXED-MODE
      DELAMINATION OF FIBRE COMPOSITE MATERIALS, Composites
      Science & Technology, Vol. 47, 1993

[13] Y. Wang, J. Williams, CORRECTIONS FOR MODE II FRACTURE
      TOUGHNESS SPECIMENS OF COMPOSITE MATERIALS, Composites
      Science & Technology, Vol. 43, 1992

[14] E.S. Greenhalgh, CHARACTERISATION OF MIXED-MODE
      DELAMINATION GROWTH IN CARBON-FIBRE COMPOSITES, PhD
      Thesis, Imperial College of Science, Technology & Medicine, London, UK, to
      be published 1998

[15] R. Hertzberg, P. Paris, Proceedings of the First International Fracture
      Conference, Sendai, Japan, 1965

[16] R. Ganesan, S.V. Hoa, S. Zhang, M. El-Karmalawy, A STOCHASTIC
      CUMULATIVE DAMAGE MODEL FOR THE FATIGUE RESPONSE OF
      LAMINATED COMPOSITES, Proceedings of the eleventh International
      Conference on Composite Materials (ICCM-11), Gold Coast, Australia, 1997

Delamination Growth under Cyclic Loading

M. König, R. Krüger

Institute for Statics and Dynamics of Aerospace Structures, University of Stuttgart
Pfaffenwaldring 27, D-70550 Stuttgart, Germany

Abstract

Cyclic loading has been applied to CFRP (Carbon Fibre Reinforced Plastic) laminates, containing artificial delaminations, as well as ply cuts at various interfaces. For the experimentally detected delamination contours local energy release rates have been computed. Plots of measured delamination progression per load cycle versus computed energy release rates have been included in a Paris Law diagram, as obtained experimentally using simple specimens for material characterization. Results obtained for delamination growth in a 0°/0° interface suggest that growth prediction, based on Paris Law, is possible. Additional results for specimens where delamination growth occurs in a 0°/45° interface indicate that Paris Law parameters depend on the orientations of the adjacent plies. It appears that – at least in the case of the shear mode – the slope of the Paris Law is much lower than in the case of a 0°/0° interface. This implies, that the possibility of a delamination growth approach should be considered in the design of CFRP structures.

1. Introduction

Delamination, the disbond of two adjacent fibre reinforced layers of a laminate is a prevalent damage mode for CFRP structures. A survey on problems concerning composite parts of civil aircraft shows that delamination, mainly caused by impact, presents 60% of all damage observed [1]. Up to now failure caused by delamination is prevented by using empirically determined design criteria – based on maximum allowable strains (typically 0.3%) – during layout and construction of components made of fibre reinforced materials [2]. However, if one consideres that these materials are capable of withstanding about 1% strain in fibre direction, it is evident that this is not an optimal strategy. Application of the concept of damage tolerance would improve the situation. This, however, requires the possibility to predict delamination growth.

In the present paper, criteria based on fracture mechanics are used to describe the delamination failure. Delamination propagation therefore is to be expected, when a function of the mixed mode energy release rates G_I, G_{II}, G_{III} along the delamination front locally exceeds a certain value G_c. This value can be regarded as a property of the interface and depends on the material and on the ply orientations of the layers adjacent to the plane of delamination. Looking at the fracture toughness data for untoughened carbon/epoxy materials we notice that the mode II values are about three to eight times higher than the corresponding mode I values. It is therefore necessary to consider the individual mode contributions of the energy release rate along a delamination contour. The goal of the investigation presented is to obtain information on the dependence of delamination growth on the mixed mode energy release rates for cyclic loading, using a combined *experimental-numerical procedure*.

2. Computation of Energy Release Rates

A significant step in the current investigation is the computation of energy release rates along arbitrarily shaped delamination fronts. Computational methods based on three-dimensional finite element modelling are a meaningful and efficient tool, by which the energy release rate at delamination growth along the entire delamination front can be evaluated. A *layered element with eight nodes*, formulated according to a continuum based three-dimensional shell theory [3] has been used for the three-dimensional models of the specimens. The virtual crack closure method was found to be most favourable for three-dimensional finite element analysis, as the separation of the total energy release rate into the contributing modes is inherent to the method and only one complete finite element analysis is necessary [4, 5]. Interpenetration of the layers is prevented by using a contact processor that utilizes a contactor target concept applying the penalty method [6].

For quasi-static loading the concept has been verified by analyzing delamination contours of double cantilever beam (DCB) and end notched flexure (ENF) specimens [7]. For ENF specimens in which the delamination propagates at an interface between layers of dissimilar orientations, curved fronts are observed. This is due to an increased mode III contribution [8]. Computing energy release rates along these fronts yields a constant value for the total energy release rate G_T, rather than for the mode II component [9]. This induced the authors to postulate that the total energy release rate controls the shape of a growing front in the case of pure shear and that mode separation may be restricted to separate the total energy release rate G_T into the normal opening mode G_I and the shear mode $G_{shear} = G_{II} + G_{III}$.

3. Experimental-Numerical Approach

Following a *global/local testing and analysis approach* as described in [10], failure criteria need to be verified and improved at each stage, i.e. from the level of material characterization via coupon tests up to the sub-structure and full structure level. Therefore the applicability of the assumptions above was investigated using structure related specimens, subjected to tension-tension (R=0.1) and tension-compression (R=−1) fatigue loading. The tests carried out, focused on the critical areas of the structure and the local analyses incorporated fracture mechanics for damage prediction in these critical areas. The specimens were made of prepreg material T300/914C and were designed to simulate specific different states of damage. The A- and B-type specimens (see figures 1 through 3) with a stacking sequence of $[0_2/+45/0_2/-45/0/90]_s$, containing an artificial 10×10 mm square delamination at interface 3/4 for the A-type and at interface 5/6 for the B-type specimen, were designed to simulate a structural component subjected to tension-tension loading. An additionally introduced 10 mm cut through ply 4 for the A-type (plies 4 and 5 for the B-type) specimens served to simulate fibre fracture. The C-type specimen (see figures 4 and 5) with $[\pm5/+45/\pm5/-45/0/\pm85/0/-45/\mp5/+45/\mp5]$ layup and an artificial circular delamination of 10 mm diameter at interface 2/3 was designed to simulate a structural component subjected to tension-compression loading, which due to an impact has been damaged near the surface.

During the experiments delamination growth was monitored using C-scan, grid reflexion, X-ray photography and Moiré technique, yielding the delamination contours which were then used as input to the numerical models. In the nonlinear finite element analysis delaminations with averaged and smoothed delamination contours were introduced as discrete discontinuities. Observed growth was compared with Paris Law lines as obtained via simple DCB, ENF and MMB specimens. This led to the conclusions drawn at the end of the paper.

4. Delamination Growth in 0°/0° and in 0°/45° Interfaces

Specimen A was subjected to R=0.1 fatigue loading with a maximum stress level at 30% of the ultimate tensile stress of the laminate. Using C-scan, delamination growth was observed at ply interface 4/5, while the deliberate delamination at interface 3/4 remained unaffected. Energy release rates were computed along averaged measured fronts s_1, s_2, s_3, s_4 and s_7 (figure 2) which correspond to 1000, 2000, 3000, 4000 and 7000 load cycles. Plots of measured delamination progression per load cycle (da/dN-values) versus computed energy release rates G have been included in a Paris Law diagram as obtained experimentally using simple specimens to characterize mode I and II failure [11]. As shown in figure 6 these results lie well between the experimentally determined straight lines for mode I and mode II failure, with the majority of data points located around the Paris Law line for $G_I/G_{II} = 0.25$, which agrees with the computed mixed mode ratio [12].

The B-type specimen was also subjected to R=0.1 fatigue loading with a maximum stress level at approximately 30% of the ultimate tensile stress of the laminate. Using C-scan, delamination growth was observed at ply interface 3/4 while the deliberate delamination at interface 5/6 remained unaffected. Energy release rates were computed along averaged measured fronts s_1, s_2, s_4, s_5, s_7 and s_8 (figure 3) which correspond to 12,500, 15,000, 20,000, 30,000, 50,000 and 100,000 load cycles. A comparison of the delamination growth in this specimen, which now occurs in a 0°/45° interface, with that for the 0°/0° interface is presented in figure 7. It is obvious that the delamination growth in the 0°/45° interface is much slower than in the 0°/0° interface. However, it is not possible to fit a Paris Law to the data points for specimen B alone.

Specimen C was subjected to tension-compression fatigue loading (R=-1) with stress maxima varying between 220 and 240 N/mm^2. These stress maxima have shown to yield stable delamination growth. The out-of-plane (i.e. buckling) deformation was monitored by Moiré technique. Using numerical post-processing procedures, the size and shape of the delaminated sublaminate were determined from this information, yielding the delamination contours. Obtained fronts are in satisfactory agreement with those extracted from corresponding X-ray photographies. The smoothed fronts ($s_1 - s_6$) for 100,000, 200,000,

300,000, 400,000, 500,000 and 1 million load cycles (figure 5) have been used as input to the finite element model [13].

Although the external loading varies with R=-1, the energy release rates vary with R=0. They are completely zero in the tension phase of the load cycle. Delamination growth occurs in a -5°/45° interface, which is assumed to behave very similar to the 0°/45° interface. Hence, the measured delamination progression (da/dN) for specimen C, plotted over the computed energy release rates, has been added to the diagram in figure 7. It should be mentioned here, that only those data points which have a mixed mode ratio in the interval $0 \leq G_I/(G_{II} + G_{III}) \leq 0.11$ have been included in figure 7. By combining the results for the specimens B and C, a Paris Law line can be obtained by a least squares fit. This line is presented in figure 7.

5. Conclusions and Outlook

In structural configurations delamination usually occurs between plies of different orientations. From the results presented, it appears that in this situation (at least in the case of very small mode I contribution and for the material considered) the slope of the Paris Law is much smaller than that obtained for a 0°/0° interface. This means that delamination growth prediction would be less sensitive to errors in the applied load and, hence, for the application of the damage tolerance concept a delamination growth approach could be realized. This is in contrast to the common recommendation, to apply a *no growth* criterion. Moreover, it appears from figure 7 that in the present case threshold values for delamination growth in the angle ply interface are much lower than those for the 0°/0° interface, or even perhaps not existing. This makes the application of a *no growth* criterion doubtful.

It should be emphasized, however, that the above statement is based on experiments and analysis of rather complex specimens. Investigation of the behaviour of angle ply interfaces in fatigue, employing simple specimens for material characterization, is still missing and should be undertaken to validate the above conclusions.

Acknowledgement

The authors gratefully acknowledge the support by the *Deutsche Forschungsgemeinschaft (DFG)*, Az.: Kr 668/17.

References

[1] A.G. Miller, D.T. Lovell, and J.C. Seferis. The evolution of an aerospace material: Influence of design, manufacturing and in-service performance. *Composite Structures*, 27:193–206, 1994.

[2] H. Eggers, H.C. Goetting, and H. Bäumel. Synergism between layer cracking and delaminations in MD-laminates of CFRE. In *AGARD Structures and Material Panel*, Patras, Greece, 1992.

[3] H. Parisch. A continuum-based shell theory for non-linear applications. *Int. J. Num. Meth. Eng.*, 38:1855–1883, 1995.

[4] E.F. Rybicki and M.F. Kanninen. A finite element calculation of stress intensity factors by a modified crack closure integral. *Eng. Fracture Mech.*, 9:931–938, 1977.

[5] K.N. Shivakumar, P.W. Tan, and J.C. Newman Jr. A virtual crack-closure technique for calculating stress intensity factors for cracked three dimensional bodies. *Int. J. Fracture*, 36:R43–R50, 1988.

[6] H. Parisch. A consistent tangent stiffness matrix for three-dimensional non-linear contact analysis. *Int. J. Num. Meth. Eng.*, 28:1803–1812, 1989.

[7] R. Krüger, M. König, and T. Schneider. Computation of local energy release rates along straight and curved delamination fronts of unidirectionally laminated DCB- and ENF-specimens, AIAA-93-1457-CP. In *The 34th SDM Conference*, pages 1332–1342, 1993.

[8] B.D. Davidson, R. Krüger, and M. König. Three Dimensional Analysis and Resulting Design Recommendations for Unidirectional and Multidirectional End-Notched Flexure Tests. *Journal of Composite Materials*, 29(16):2108–2133, 1995.

[9] R. Krüger. Three Dimensional Finite Element Analysis of Multidirectional Composite DCB, SLB and ENF Specimens. ISD-Report No. 94/2, Institute for Statics and Dynamics of Aerospace Structures, University of Stuttgart, 1994.

[10] R.H. Martin. Local fracture mechanics analysis of stringer pull-off and delamination in a post-buckled compression panel. In A. Poursartip and K. Street, editors, *The Tenth International Conference on Composite Materials, Vol. I*, pages 253–260. Woodhead Publishing Ltd., 1995. ISBN 1-85573-222-1.

[11] M. König, R. Krüger, K. Kussmaul, M. von Alberti and G. Gädke. "Characterizing Static and Fatigue Interlaminar Fracture Behavior of a First Generation Graphite/Epoxy Composite," *13th Composite Materials: Testing and Design, Thirteenth Volume, ASTM STP 1242*, S.J. Hooper Ed., American Society for Testing and Materials, 1997, pp. 60-81.

[12] M. König, R. Krüger, and M. Gädke. Prediction of Delamination Growth Under Cyclic Loading Using Fracture Mechanics. In A.H. Cardon, H. Fukuda, and K. Reifsnider, editors, *Progress in Durability Analysis of Composite Systems, Proceedings of the international conference DURACOSYS 95, Brussels*, pages 45–52. A.A. Balkema Publishers, 1996. ISBN 90-5410-809-6.

[13] R. Krüger, S. Rinderknecht, C. Hänsel, and M. König. Computational Structural Analysis and Testing: An Approach to Understand Delamination Growth. In E. A. Armanios, editor, *Interlaminar Fracture of Composites*, pages 181–202. Key Eng. Mater., Vols. 120-121, Transtec Publications Ltd., 1996. ISSN 1013-9826.

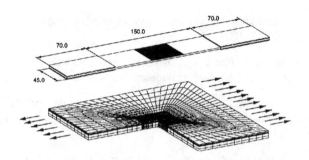

Figure 1: A-type and B-type specimen and section modelled.

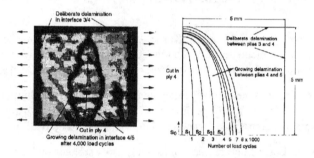

Figure 2: C-scan of delaminated area and resulting fronts after smoothing for A-type specimen.

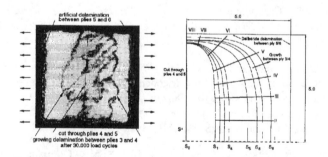

Figure 3: C-scan of delaminated area and resulting fronts after smoothing for B-type specimen.

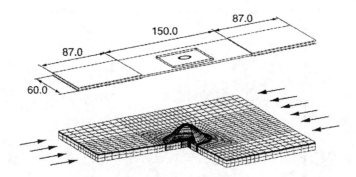

Figure 4: C-type specimen, section modelled and deformed geometry.

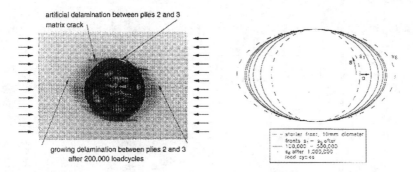

Figure 5: X-ray photography of delaminated area and resulting fronts after smoothing for C-type specimen.

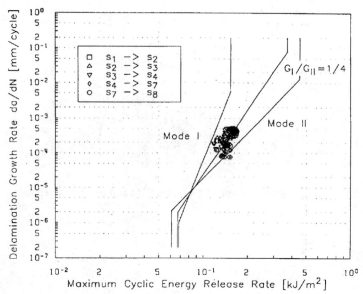

Figure 6: Results obtained for A-type specimen in comparison to Paris Law from material tests ($0°/0°$ interface).

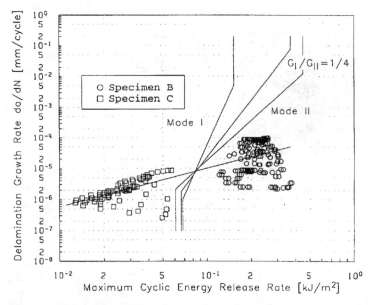

Figure 7: Results obtained for B-type and C-type specimens ($0 \leq G_I/(G_{II}+G_{III}) \leq 0.11$) in comparison to Paris Law for the $0°/0°$ interface.

Analysis and Prediction of Delamination Growth from Embedded Defects in Composite Materials

S J Lord*, E S Greenhalgh*

*Structural Materials Centre, DERA, Farnborough, Hants, United Kingdom
Email: slord@taz.dera.gov.uk, esgreenhalgh@taz.dera.gov.uk

Abstract

Experimental and modelling studies were carried out on CFRP panels to study damage growth from embedded defects under compressive load. In the experiments growth was monitored using Moiré interferometry, and the damage mechanisms were determined using optical and electron microscopy. The computations used the ABAQUS finite element code in conjunction with a virtual crack closure FORTRAN routine, in an iterative manner, to grow the delamination; ply splitting in the delaminated ply was also monitored. The results showed excellent agreement for the shape of the delamination blister and the location of ply splitting, but were disappointing for the prediction of strain for delamination initiation and ply splitting The problems with the modelling were attributed to the fact that the migration of the delamination to other planes, via ply splitting, was not modelled; here the experiments showed that this was an important mechanism and has implications for tailoring lay-ups to minimise delamination growth.

1. Introduction

The use of carbon-fibre reinforced plastic composites is increasing because of their high specific strength and stiffness. However, the full potential of composite structures is not being realised because delamination and its effects on failure are not fully understood. A delamination may be the result of defective manufacture or assembly. If a component has been in-service for sometime, the delamination may have grown from a stress concentration or from impact damage. The compressive performance of the structure would be critical since buckling of the delaminated material can lead to significant reductions in strength [1]. An assessment as to the severity of the damage would have to be made to determine whether repair was necessary.

The aim of this work was to understand and predict delamination growth under compressive loading using two complimentary approaches; experiments and finite element models. There have been numerous investigations into delamination growth in composite materials which are reviewed elsewhere [2]. Although delamination growth from embedded defects has been extensively characterised, the detailed propagation mechanisms are not understood. Currently, most finite element models assume that delamination growth remains within the defect plane [2,3,4]; this work has shown that

© British Crown Copyright 1998/DERA
Published with the permission of the Controller of Her Britannic Majesty's Stationary Office.

this is not the case. In this work the detailed mechanisms were investigated using a novel approach; fractographic analysis.

2. Experimental and Modelling Details

In the experimental studies, sandwich stabilised panels were used to study damage growth from embedded defects under compressive load. The panels were manufactured from Hexcel T800/924 tape using a lay-up of $[(+45°/-45°/0°/90°)_3]_S$ and the core was Aeroweb 4.5-1/8-10(5052) honeycomb. Embedded defects of various shapes and sizes were positioned at different depths through the skin; in this work only circular defects (10μm PTFE film) of diameter 50mm were considered. Each panel had a defect at one of two ply interfaces; +45°/-45° (5/6 ply interface) or 0°/90° (3/4 ply interface). The tests were conducted in quasi-static compression and the growth was monitored using Moiré interferometry. After testing, the detailed damage mechanisms were determined using optical and electron microscopy.

In the finite element studies, the delaminations were modelled using the ABAQUS finite element code in conjunction with a virtual crack closure (VCC) routine. The elements were eight-noded shell elements (S8R) with reduced integration [5]. The whole model was 200mm x 200mm in area and one element deep, the thickness varying with stacking sequence; each ply was 0.125mm thick. There was a central region, 50mm in diameter, which modelled an embedded defect within the laminate. The entire mesh was flat except the delamination which was a sinusoidal half-wave perturbation of peak height 0.1mm; this was to model the effect of the embedded material and negated the need for an instability calculation. One edge of the model was clamped and an in-plane displacement, parallel to the 0° ply, was applied to the opposite edge. The outer area of the model was constrained in the z-plane and rotations about the x- and y- axes, whilst the delamination was free to displace and rotate in all directions.

Located 0.01mm beneath the model was a rigid surface to simulate the sublaminate beneath the delaminated plies and was linked to the mesh using contact elements (IRS1) [5]. The displacement was applied in increments, using a non-linear algorithm (Riks Method). The output from the simulation was put through a FORTRAN routine to calculate the mode I and mode II strain energy release rates (G_I and G_{II}) at each point on the delamination boundary. The values of G_I and G_{II} were then input into a failure criterion to identify the strain and location of delamination growth. A range of criteria were used to predict initiation [2] but the General Interaction Criterion [6] was chosen since earlier studies [2] had indicated that this gave the best fit to the experimental toughness data (Equation 1). The criterion parameters were derived from characterising the toughness of the 0°/90° ply interface for dry T800/924 tested at room temperature and were; G_{IC}=184.95J/m², G_{IIC}=537.10J/m², β=-1.3939 and Φ=0.00747. After modifying the mesh to incorporate the growth, the model was rerun. This process was repeated six times to simulate a significant amount of growth from the initial defect.

$$\left[\frac{G_I}{G_{IC}}-1\right]\left[\frac{G_{II}}{G_{IIC}}-1\right]+\left[\beta+\Phi\frac{G_I}{(G_I+G_{II})}\right]\frac{G_I G_{II}}{G_{IC}G_{IIC}}\geq 0 \qquad (1)$$

In addition to delamination, a second damage mechanism was identified; ply splitting in the delaminated plies and to monitor this failure mode, the transverse strain (ε_{22}) in each ply was also output. Splitting was deemed to have occurred when this strain exceeded the ultimate transverse tensile strain of the ply.

Before running the models a convergence test was conducted to determine the optimum mesh density. If the mesh was too coarse, the results would be incorrect whilst, if the mesh was too dense, the model would be unmanageable. The optimum mesh density, which was within 10% of the extrapolated continuum result, had 768 elements.

3. Experimental and Modelling Results

The experimental results showed that the blister shape and orientation were dependant on the stacking sequence of the delaminated plies (Figure 1; the loading direction for each image was vertical). The predicted blister shape, orientation and height showed excellent agreement with the experimental results. As the load increased the originally circular blister flattened at the longitudinal extent, forming an ellipse. For the defect at the +45°/-45° ply interface, the blister also rotated such that, at initiation of delamination growth, the major axis of the ellipse was at 15° to the horizontal. The mode I and II components were greatest at the major and minor axes of the elliptical blister respectively; at initiation of delamination, the mode I component dominated.

| Ply Interface | Delamination Initiation Strain | | | Splitting |
	Experiment	Average Prediction	General Interaction [6]	Prediction
+45°/-45°	2900με	3693με	3700με	3070με
0°/90°	2500με	2857με	2810με	950με

Table 1: Comparison between predicted and experimental initiation strains

The experimental and predicted strains at which delamination initiated are shown in Table 1. The predicted strains varied with criterion, although they were all conservative (*i.e.* greater than the experimental results). For the defect at the +45°/-45° ply interface the difference between the experimental and predicted strains was 27%, with a 34% spread between different criteria. For the defects at the 0°/90° ply interface the difference between the experimental and predicted strains was 14% with a 22% spread between the criteria. All the criteria indicated that growth had initiated at the boundary of the major axis of the blister, as was observed experimentally.

The correlation between the experiments and predictions for ply splitting was good (Figure 2). For the defect at the +45°/-45° ply interface, the splitting initiated at a location 80% from the centre of the defect. For the defect at the 0°/90° ply interface, the splitting developed prior to the delamination growth at the 90° and 270° sites. Although the initiation sites for splitting appeared to be realistic, for both models the predicted development of the splitting during the subsequent growth steps was incorrect.

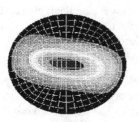 

(a) +45°/-45° ply interface

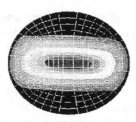 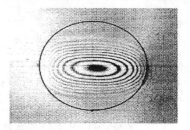

(b) 0°/90° ply interface

Figure 1 Experiments and models of delamination blister shape

(a) +45°/-45° ply interface

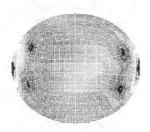 

(b) 0°/90° ply interface

Figure 2 Comparison between experiments and predicted for ply splitting

The prediction of the subsequent delamination growth was disappointing (Figure 3). For the defect at the +45°/-45° ply interface, the blister grew stably along the major axis of the blister. Fractographic evidence showed that the delamination growth was initially at the defect plane, but then migrated through +45° ply splits into the adjacent 90°/+45° interface. The delamination then extended within this interface, whilst simultaneous growth occurred within the adjacent 0°/90° ply interface, via 90° ply splits. However, although the model predicted a similar delamination shape, the growth was unstable and the correct mechanisms were not reproduced.

For the defect at the 0°/90° ply interface (Figure 3) the predicted and observed blister shapes differed. The model predicted unstable growth transverse to the loading direction whilst, in reality, the delamination had grown stably at 45° and 135° to the loading direction, developing into a star shaped blister. Fractographic analysis showed that the delamination had migrated into the adjacent -45°/90° ply interface and grow parallel to the -45° ply. In the later stages, splitting developed in the -45° ply, leading to migration of the delamination into the +45°/-45° ply interface.

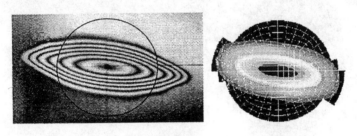

(a) +45°/-45° ply interface (5000μɛ)

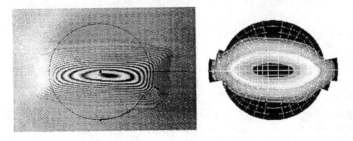

(b) 0°/90° ply interface (7000μɛ)

Figure 3 Experiments and models of delamination growth from embedded defects

4. Discussion

A key finding of this work was that damage growth from a single plane defect develops on several planes (Figure 4). This Figure shows the complex mechanisms which occur during delamination growth from a single plane defect at a 0°/90° ply interface. Delaminations migrate through the plies towards the upper surface, until an interface was reached in which the upper ply and growth directions were approximately coincident; the delamination then propagates within this interface. Other damage mechanisms occur such as ply splitting and fibre fracture. To realistically predict damage growth, modelling of splitting and migration of the delamination plane is essential. In the models reported here, there was no scope to allow the delamination to migrate. Consequently, the models could predict the blister shape and splitting but not the delamination initiation and growth. To model delamination migration would be very difficult and time-consuming, requiring sophisticated interface and three dimensional brick elements. However, without such a facility, realistic modelling of delamination growth from an arbitrary interface.

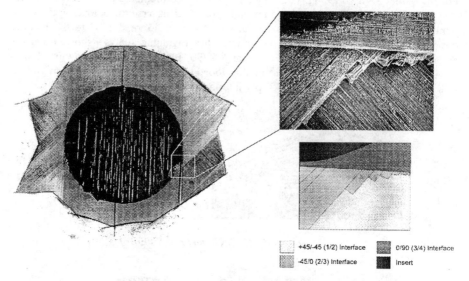

+45/-45 (1/2) Interface   0/90 (3/4) Interface
-45/0 (2/3) Interface   Insert

Figure 4 Fractographic analysis of the damage growth from an embedded defect

Modelling of delamination growth proved to be very demanding on resources due to the non-linear nature of the models and the incorporation of contact between the delaminated plies and the sublaminate. A number of simplifications were used to reduce the processing time; the most significant was to model the sublaminate beneath the defect plane as a rigid surface. However, this simplification would affected the mixed-mode conditions at the delamination boundary [4]. Attempts were made to model the sublaminate but this led to problems with model size and processing time.

Firstly, the shape of the delamination blister and the mixed-mode conditions around the boundary were basically controlled by two mechanisms; bending moment and stiffness coupling effects. At the longitudinal extents of the blister the crack faces were closed, promoting mode II whilst, at the lateral extents, the loading promoted mode I forces, leading to the elliptical blister shape. In addition, the stacking sequence of the delaminated plies led to rotation of the blister and the mode I and II maxima. These effects are controlled by the stiffness of the delaminated plies and, on these aspects, the model exhibited excellent agreement with the experimental results.

The prediction of initiation of growth was poor; for both models the strains were greater than those observed experimentally. This was attributed to ply splitting generated by the high local curvatures of the delaminated plies close to the defect boundary. A wide range of failure criteria were used to predict initiation of delamination growth. The results of this work indicated that, for a given loading condition, the predicted delamination initiation strains were significantly affected by the choice of criterion.

The experimental studies showed that the subsequent damage growth processes were complex, involving multi-plane delamination, ply splitting and fibre fracture. The models performed poorly at simulating the subsequent damage growth which was attributed to the incorrect modelling of the physics of the problem.

Most models developed to predict delamination growth assume the fracture toughness, at a given loading condition, is a constant. This means that identical defects under identical loading should behave in an identical manner. However, coupon tests typically exhibit up to 10% variability[2] which has been attributed to the inhomogeneous nature of composites and the sensitivity of the toughness to these inhomogenieties [7]. To develop fully realistic models of delaminations in composite structures, a stochastic approach to modelling the toughness should be developed. This is particularly pertinent for modelling components under cyclic loading or when predicting the stability of delamination growth.

5. Conclusions

Experimental and numerical (finite element) studies were conducted on embedded defects in carbon-fibre/epoxy composites, with detailed fractographic analysis used to ascertain the detailed damage growth mechanisms. From this study, the following conclusions were drawn.

1. The delamination blister shape and deflection is controlled by the loading conditions and the stacking sequence of the delaminated plies.

2. Delamination initiation and growth are controlled by the peel (mode I) and shear (mode II) components at the defect boundary; the blister shape of the delaminated material dictated the position of the maxima of the components. The mode I maximum, which was the main driving force for delamination growth, was approximately transverse to the applied loading.

3. A key finding was that single plane defects do not imply single plane delamination growth. Migration of the delamination to planes other than that of the original defect, via ply splits, was an important mechanism and has implications for tailoring lay-ups to minimise delamination growth. Realistic modelling of delamination growth in composite structures will not be achieved without simulating this mechanism.

4. The models accurately predicted the shape of the delamination blister, the location of ply splitting and, in one case, the blister shape after initiation of delamination growth. However, the predicted delamination initiation strains were significantly

above those observed experimentally and the subsequent growth mechanisms were not correctly simulated.

6. Acknowledgements

The authors acknowledge the support of DTI CARAD and the MOD Applied Research Programme.

The contributions of Mr R Dadey, for helping with hardware difficulties, and Dr L Iannucci, for his general advice, and HKS staff for their advice, are also acknowledged.

The contributions of Mr P J Mobbs and Mrs A Dewar, for manufacturing the panels, Mr F Matthews and Dr S M Bishop, for advice on the testing and damage mechanisms and Mr B Sluce for specimen preparation. Mr S Singh is also acknowledged for his contributions in understanding the damage mechanisms.

7. References

[1]   A. Rubin, 'EVALUATION OF MULTILEVEL DELAMINATIONS INDUCED DURING AIRCRAFT COMPOSITE STRUCTURES ASSEMBLY', Proceedings of the 8th International Conference on Composite Materials (ICCM8), Honolulu, Hawaii, pp36A1-36A15, 1991.

[2]   E. Greenhalgh, 'CHARACTERISATION OF MIXED-MODE DELAMINATION GROWTH IN CARBON-FIBRE COMPOSITES', PhD Thesis, Imperial College, London, (to be published), 1998.

[3]   T. Ireman, J. Thesken, E. Greenhalgh, R. Sharp, M. Gädke, S. Maison, Y. Ousset, F. Roudolff & A. La Barbera, 'DAMAGE PROPAGATION IN COMPOSITE STRUCTURAL ELEMENTS - COUPON EXPERIMENTS AND ANALYSES', Composite Structures, Vol 36, pp209-220, 1997.

[4]   K. Nilsson, L. Asp & J. Alpman, 'DELAMINATION BUCKLING AND GROWTH AT GLOBAL BUCKLING', FFA TN-1997-39, 1997.

[5]   Hibbitt, Karlsson & Sorensen Inc., 'ABAQUS STANDARD USERS MANUAL (VOL 1)', Version 5.5, 1995.

[6]   S. Hashemi, A. Kinloch & J. Williams, 'MIXED MODE FRACTURE IN FIBRE POLYMER COMPOSITE LAMINATES', In 'Composite Materials; Fatigue And Fracture (3rd Conference)', Ed T. O'Brien, ASTM STP 1110, Philadelphia, pp143-168, 1991.

[7]   S. Singh & E. Greenhalgh, 'MICROMECHANISMS OF INTERLAMINAR FRACTURE IN CARBON-EPOXY COMPOSITES AT MULTIDIRECTIONAL PLY INTERFACES', Proceedings of the 4th Conference on Deformation and Fracture of Composites (DFC4), UMIST, Manchester, (1997).

Delamination resistance due to fibre cross-over bridging: The bridging law approach

Torben K. Jacobsen[*] and Bent F. Sørensen[**]

[*]LM Glasfiber A/S, R&D Dept., Rolles Møllevej 1, DK-6640 Lunderskov, Denmark
[**]Materials Research Department, Risø National Laboratory, DK-4000 Roskilde, Denmark

Abstract

Delamination is a major problem in laminated fibre composites. Fibre cross-over bridging can enhance the crack growth resistance during crack growth. An experimental study was undertaken to illuminate this effect in a unidirectional C/Epoxy fibre composite. Double cantilever beam specimens, loaded with pure bending moments, were used. For this configuration the global energy release rate is independent of the details of the bridging law. Also, utilizing the path independent J integral the bridging law can be obtained. Specimens with different heights were tested. In contrast to conventional R-curves the bridging law was found to be independent of the specimen geometry. Thus, the bridging law may be regarded as a material property under large scale bridging conditions.

1. Introduction

Studies of crack growth in fiber composites show that the fracture resistance G_R can increase substantially during crack growth [1,2]. This increase is attributed to fibre cross-over bridging. The increase in G_R is often described by R-curves, i.e. G_R as a function of the crack extension Δa. However, it is questionable whether R-curves are material properties under Large Scale Bridging conditions (LSB).

As long as the process zone associated with crack growth is small, linear elastic fracture mechanics concepts can be applied, and an R-curve can be regarded as a material property. This is in analogy with small scale yielding at the crack tip in metals. In fibre composites the fibre cross-over bridging zone can be of the order of several cm's. Thus, characterization of such composites under Small Scale Bridging would require very large specimens (dimensions of the order of meters). This is not feasible from a fracture mechanics testing viewpoint.

Instead, one has to accept that LSB occurs in most cases during fracture mechanics tests of fibre composites and in structural applications. Having recognized that, two issues stand clear: (1) Under LSB R-curves are no longer material properties as they depend on the specimen geometry [1,2]. (2) Under LSB the energy release rate cannot be computed from standard equations for fracture mechanics test specimens without bridging [3].

The two problems are overcome by (i) using the concept of bridging laws, and (ii) by using one of the few test configurations for which the global energy release rate can be computed without the knowledge of the bridging law.

2. Application of the J integral to crack bridging

A prerequisite for the following approach is that a specimen configuration must be used for which the *J integral taken along the external edges is independent of the details of the bridging law*, such as the symmetric Double Cantilever Beam (DCB) specimen loaded with pure bending moments M (Fig. 1), [3].

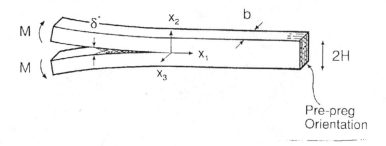

Figure 1. The geometry and loading of the DCB-specimen loaded with pure bending moments.

Consider a crack with bridging fibres across the crack faces near the crack tip. The closure stress σ is assumed to depend only on the local crack opening δ. Since fibres will fail when being loaded sufficiently, it is assumed that a maximum crack opening δ_0 exists, beyond which the closure traction vanishes. Shrinking the path of the J integral to the crack faces and around the crack tip gives

$$\mathcal{G} = \int_0^{\delta^*} \sigma(\delta)d\delta + \mathcal{G}_0 \tag{1}$$

where $\mathcal{G}_0$ is the energy release rate at the crack tip, the integral is the energy dissipation in the bridging zone and δ^* is the end-opening of the bridging zone. When δ^* attains δ_0 the R-curve reaches its steady-state value $\mathcal{G}_{ss}$. During subsequent crack growth the bridging zone translates along the beam with the crack tip. Note, that such a self-similar crack growth requires the use of a steady-state specimen, such as the DCB specimen loaded with pure bending moments.

By measuring δ^* together with the energy release rate $\mathcal{G}_R$, the bridging law can be determined by [3]

$$\sigma(\delta^*) = \frac{\partial \mathcal{G}_R}{\partial \delta^*} \tag{2}$$

3. Practical implementation

The material examined in the present study is a unidirectional carbon-fibre/epoxy composite processed in-house from pre-pregs. Specimens of different heights (ranging from $H = 4$ to 10 mm) were cut from a plate, such that the notch was parallel with the fibre direction and perpendicular to the plane of the plate. The subsequent crack growth was therefore intralaminar, see Fig. 1.

Extensometers were mounted at each face of the specimens just at the end of the cut notch to record δ^* as a function of M, see Fig. 2. They were mounted at pins so that they did not constrain the rotation of the beams. In order to minimize the influence of the pins on the stress state in the beams the pins were mounted by glue in holes drilled through the neutral axis of the beams (Fig. 2). The pure bending moments were applied to the specimens using a special fixture [4] mounted at a screw-driven tensile testing machine. The specimens were tested at a constant displacement rate.

4. Experimental results

The $\mathcal{G}_R$-δ^* data recorded on two nominally identical specimens ($H = 6$ mm) loaded continuously are shown in Fig. 3a. $\mathcal{G}_{ss}$ is reached at $\delta^* = \delta_0 = 4$ mm where at steady-state bridging length zone of approximately $10H$ is observed from replicas. Similar results were found for $H = 4$ mm and $H = 10$ mm. This clearly demonstrates that C/Epoxy experiences

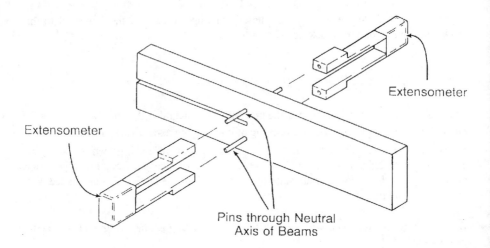

Figure 2. Schematic drawing showing how extensometers, for the measurement of δ^*, were mounted at pins through the neutral axes of the beams.

LSB. Furthermore, the fibre bundles bridging the crack were clearly visible by eye when steady-state was reached. The experimental data were fitted with the following function

$$G_R(\delta^*) = G_0 + \Delta G_{ss}\left(\frac{\delta^*}{\delta_0}\right)^{\frac{1}{2}} \qquad\qquad 0 < \delta^* < \delta_0 \qquad\qquad (3)$$

where $\Delta G_{ss} = G_{ss} - G_0$. The fitting parameters are given in Fig. 3a. Fig. 3b shows the G_R-δ^* data recorded from loading/unloading experiments for $H = 4$, 6 and 10 mm. The data points for all specimen geometries follow (3) closely, indicating that the bridging law $\sigma(\delta)$ is readily independent of specimen geometry and thus a material property. The bridging law is found by (2) and (3)

$$\sigma(\delta) = \frac{\Delta G_{ss}}{2\sqrt{\delta_0\delta}} \qquad\qquad 0 < \delta < \delta_0 \qquad\qquad (4)$$

The experimentally observed bridging law (4) has the same form as a micromechanical model for fibre cross-over bridging proposed by Spearing and Evans [1].

5. Conclusion

Crack growth resistance due to fibre cross-over bridging of a unidirectional carbon fibre epoxy composite was investigated by loading DCB-specimens with pure bending moments. The bridging law was determined experimentally simply by the measurement of the end-opening of the bridging zone. The bridging law was found to be independent of specimen geometry, i.e. a material property.

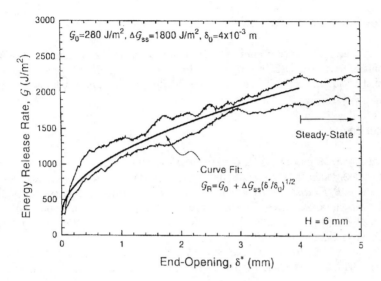

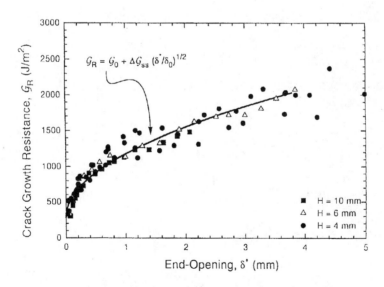

Figure 3. The crack growth resistance as a function of end-opening. Top graph shows data from two identical specimens loaded without interruption, and bottom graph shows data for specimens with different H's .

Acknowledgement

The Engineering Science Centre for Structural Characterization and Modelling of Materials at Risø, The Danish Programme for Wind-Energy EFP95, and Akademiet for de tekniske Videnskaber (ATV grant EPD 015/RISØ) are gratefully acknowledged.

References

[1] *Spearing, S. M. and Evans, A. G.*, THE ROLE OF FIBER BRIDGING IN THE DELAMINATION RESISTANCE OF FIBER-REINFORCED COMPOSITES, Acta Metall. Mater., 1992, **40**, 2191.
[2] *Sørensen, B. F., and Jacobsen, T. K.*, LARGE SCALE BRIDGING IN COMPOSITES: R-CURVES AND BRIDGING LAWS, Composites Part A, accepted for publication 1998.
[3] *Suo, Z., Bao, G. and Fan, B.*, DELAMINATION R-CURVE PHENOMENA DUE TO DAMAGE, J. Mech. Phys. Solids., 1992, **40**, 1.
[4] *Sørensen, B. F., Brethe, P. and Skov-Hansen, P.*, CONTROLLED CRACK GROWTH IN CERAMICS: THE DCB-SPECIMEN LOADED WITH PURE MOMENTS, J. Euro. Ceram. Soc., 1996, **16**, 1021.

Interlaminar Non-Linear Fracture Energy Measurements of Thermoplastic Composites Using The DCB Specimen Configuration

G. C. Christopoulos, Y. P. Markopoulos, V. Kostopoulos
Applied Mechanics Laboratory, University of Patras, University Campus, GR, 265 00 Patras, Greece

Abstract

The present work deals with the determination of the interlaminar mode I crack growth resistance curve (R-curve) of a unidirectional carbon fibre reinforced polyamide 12 (CF-PA12), using the standard DCB configuration. The loading condition applied was not a monotonic one but a cycle loading-unloading pattern. The determination of the R-curve was based on a compliance calibration technique. The behaviour of the material was modeled as a sum of two contributions; a linear elastic one that fails according to the principles of linear elastic fracture mechanics (LEFM) and an independent damage mechanism which absorbs energy but does not alter the ultimate fracture load.

1. Introduction

The use of advanced materials for high performance applications requires an insight understanding of their mechanical behaviour. It is therefore necessary to have accurate material properties information in order to perform damage and failure analysis.
On the other hand it is well recognized that the microstucture, such as crystallinity and morphology, of semicrystalline thermoplastic composites highly depends on the processing conditions employed during the manufacturing of laminates [1].
There are a number of reports in the literature, concerned with the measurement of mode I interlaminar fracture toughness in composite materials [2-7]. Several different approaches and specimen geometries have been used. Among them Double Cantilever Beam, (DCB), configuration, is one of the most common specimen geometries applied.
In the present paper cyclic loading-unloading conditions were applied to the test specimens, for the determination of the R-curve based on a compliance calibration method.

2. Experimental Procedure

2.1. Manufacturing of the Material

The material used in this study was a Carbon Fibre Polyamide 12, CF-PA12, tape 30 mm in width and 0.55 mm in thickness supplied by Baycomp Co., Canada. Laminates were manufactured using a steel mold in a hot press. Each laminate contained eight layers of CF/PA12, and the final thickness of the plates was 3.5 mm. A thin steel foil, 30µm thick, coated with release agent, was inserted in the middle of the laminate during the moulding operation in order to obtain a starter crack.

Specimens were cut out from the manufactured plates by using a diamond wheel. Their dimensions of 150 mm long by 20 mm wide.

2.2. Interlaminar Fracture Tests

The most commonly used test for interlaminar fracture toughness under mode I is the DCB test, Fig.1. Specimens were tested in order to reveal the irreversible damage mechanisms occurring in the vicinity of the crack tip, subjected to loading/unloading test patterns in mode I, at room temperature (20°C), with a crosshead speed of 1mm/min for both loading and unloading cycles. Each test specimen was unloaded up to 50% of each previous displacement, Fig. 2, so that the slope of the unloading line (e.g. the inverse of compliance) could be determined. The specimen was then reloaded. It was possible to repeat this loading/unloading pattern five to six times until crack advances about 50 mm from its initial value. Crack opening displacement δ, load P, and crack length α, were monitored in real time acquisition mode. For the accurate detection of crack length a traveling microscope with a magnification of twenty times was used. All tests performed under position control at a closed loop servohydraulic universal testing machine.

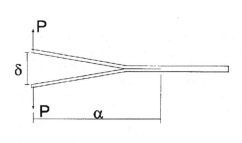

Figure 1. Double Cantilever Beam test specimen

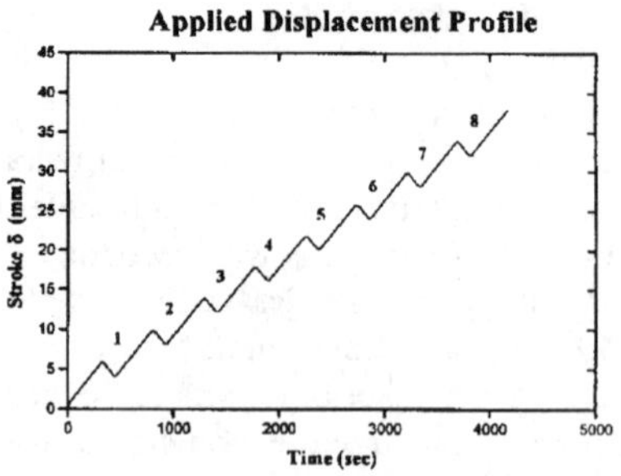

Figure 2: Applied Displacement Profile (Loading/Unloading)

The present analysis aims to determining the crack growth resistance R, as the sum of two energy rate contributions:

- the non-linear strain energy release rate G^*_R, which can be directly correlated to G_{IC} fracture parameter, calculated by LEFM;
- the plastic energy dissipation rate Φ_{ir} which is the energy consumed due to the process of the formation of the damage zone which surrounds the crack tip region.

The crack growth resistance R, the nonlinear strain energy release rate G^*_R and the plastic energy dissipation rate Φ_{ir} were calculated using the loading/unloading procedure applied for the DCB specimen configuration. The establishment of this nonlinear semi-empirical approach, based on the form of the P-δ curve at each stage of crack increment, is a necessary requirement in order to understand the fracture behaviour, track down the inherent irreversible mechanisms and calculate the consumed energy rates needed for the formation of the crack area.

In Fig.3 a representative curve in case of the linear elastic fracture behaviour is shown. At the instant a critical load is reached, a crack increment occurs. As a result the curve exhibits a non-linear form. Successive unloading at different stages of crack propagation, shows that no plastic phenomena are present. In this case the characteristic parameters are the load P, the crack opening displacement δ, and the compliance from the origin, C=δ/P. Therefore the effective crack length may be easily calculated. Additionally it is obvious that the non linearity of the P-δ curve is only attributable to the irreversible loss of energy consumed to create a new crack surface.

Figure 4, shows a schematic representation for the case of the linear non-elastic material behaviour. The crack propagates when a critical load is exceeded. In this case, successive unloading at different crack increments conclude to plastic deformation as a result of the development of various additional fracture mechanisms such as fibre bridging and pull-out, fibre-matrix debonding, debris effects, and extensive matrix microcracking.

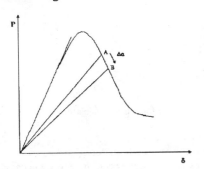

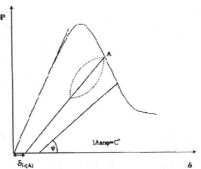

Fig.3: Linear Elastic Fracture Behaviour   Fig. 4: Linear Non-Elastic Material Behaviour

The consumed total energy now is consisted by the part necessary for the formation of new crack surface and the energy necessary for supplying all the irreversible damage mechanisms appeared.

The characteristic parameters now are the load P, the COD δ, the remaining displacement at each unloading step $δ_{ir}$ as well as the modified compliance

$$C^* = \frac{δ - δ_{ir}}{P} \qquad (1)$$

In case that the loops of the loading cycles are not linear, a non-linear, non-elastic behaviour is assigned. Using the medians to the loops one may apply the same calculation procedure which is presented in the following.

The main reason of using the loading-unloading procedure is that the irreversible mechanisms are traced via the registration of the residual displacements $δ_{ir}$. The variation of compliance C^* between two successive cycles, n-1 and n, permits an estimation of the crack increment using an experimentally well established formula [8,16]:

$$\alpha_n = \alpha_{n-1} + \frac{b_{n-1}}{2} \frac{C_n^* - C_{n-1}^*}{C_n^*}$$

(2)

where:

α_n, α_{n-1} : crack length at the two successive cycles n, n-1 respectively

C_n^*, C_{n-1}^* : the corresponding modified compliance during the cycles n and n-1

$b_n = W - \alpha_n$: uncracked ligament of the specimen after n-cycles.

Thus at each unloading stage one may assign an effective crack length a_{eff}. In general the total energy consumed for crack propagation from α to $\alpha + \Delta\alpha$ consists of two contributions:

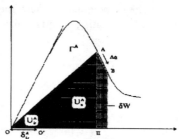

Fig. 5: Determination of the energy values in a P- δ curve for a linear non elastic material

- the energy to create a new crack surface ΔA, Γ
- the energy dissipated to the irreversible mechanisms due to material structure, U_{ir}.

The energy contribution U_{ir} under some circumstances may exceed the energy necessary for the creation of the crack surface, Γ. Thus its magnitude belongs among the important parameters which must be determined for the characterization of the fracture behaviour of a macroscopically non-elastic material. In Figure 5, which is a (P- δ) curve of such a material, the definition of all the energy parameters involved in the procedure of fracture characterization are given. The assumption made is that the crack propagates in a slow and stable way due to the external energy supply and the kinetic energy of the system is negligible.

The total mechanical energy, W, given to the system for crack growth has the following form:

$$W = U_e + U_p + \Gamma$$

(3)

where:

W: the total work given to the system externally

U_e: the elastic energy

U_{ir}: the energy loss due to irreversible phenomena

Γ: the energy for the formation of a new crack surface.

For a quasi-static increment of crack surface $\delta A = B\delta\alpha$, G_R^* is the elastic energy release rate, given by the rate of change of the area OAO'. As a result G_R^* is the rate of change of crack advance energy Γ. For the elastic deformation, the strain energy component is given by:

$$U_e = \frac{1}{2} P(\delta - \delta_{ir}) = \frac{1}{2} P^2 (\frac{\delta - \delta_{ir}}{P}) = \frac{1}{2} P^2 C^* \tag{4}$$

The elastic energy release rate is therefore:

$$\frac{dU_e}{d\alpha} = PC^* \frac{dP}{d\alpha} + \frac{P^2}{2} \frac{dC^*}{d\alpha} \tag{5}$$

According to Mai [14] and Figure 5, a fraction λ (%) of the total mechanical energy input W corresponds to the irreversible mechanisms energy U_{ir} at each crack growth stage, and it is given as the ratio of the areas, OAO' (U_{ir}) and OAEO (W):

$$\lambda_i = \frac{U_{ir}}{W} \Big|_{\alpha_{eff}} \tag{6}$$

Differentiating eq. (1) and taking into account eq. (4) the modified potential energy release rate G_R^* is given by:

$$G_R^* = \frac{(1-\lambda)}{B} \frac{Pd\delta}{da} - \frac{PC^*}{B} \frac{dP}{da} - \frac{P^2}{2B} \frac{dC^*}{da} \tag{7}$$

It should by noted, that the component dP/da cannot be neglected since irreversible mechanisms are presented, thus this quantity should be calculated for each loading-unloading cycle. Furthermore neglecting the quantity dP/da and assuming that λ ratio tends to zero LEFM approach is obtained as a special case of eq. (7).

The energy rate Φ_{ir} coming from the non-elastic energy part, which is associated with the irreversible mechanisms due to development of the damage zone in the vicinity of the crack tip is given by the relation:

$$\Phi_{ir} = \lambda \frac{Pd\delta}{Bda} \tag{8}$$

According to the previous analysis, the experimental evaluation of the energy release rate values G_R^* and Φ_{ir} is based on the following steps:

- Monitoring of the $(P - \delta)$ curve during loading-unloading cycles.
- Calculation of δ, δ_{ir}, P and C^* for each cycle.
- Evaluation of effective crack increment using eq. (2) for n, n-1 loading-unloading cycles.
- Determination of the functions $P(a_{eff})$ and $\delta(a_{eff})$.
- Numerical integration for the evaluation of the external work offering to the system, W, and the non-elastic energy U_{ir} for the calculation of energy ratio λ.

3. Results and Discussion

A typical loading/unloading curve from the tests performed so far for CF/PA12 is presented in Fig. 6. Then a detailed analysis is conducted in order to calculate the load P, the crack opening displacement δ, the irreversible remaining displacement δ_{ir} at each unloading stage and the modified compliance C*. Based on that, the effective crack length, α_{eff}, at the nth loading unloading cycle is evaluated using the recurrence relation of eq.(2). From the data displayed in Fig. 7 is very important to point out the very good agreement between the experimental measured crack length during the tests and the effective crack length obtained from the analysis described above.

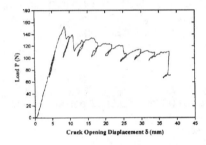

Figure 6: Load versus crack opening displacement

Figure 7: Experimental and effective crack length versus applied load

Furthermore, based on (P-δ) curve the plastic dissipated energy ratio λ can be easily defined according to eq. (6). The obtained results from the present analysis are given in Fig. 8.

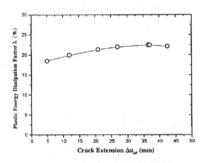
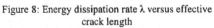
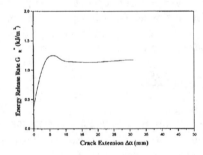

Figure 8: Energy dissipation rate λ versus effective crack length

Figure 9: Energy release rate versus crack extension

It is evident that at the early steps of crack propagation the ratio of irreversible plastic energy dissipated to the total mechanical energy input is low, namely 10-15%, and this becomes clear from the form of (P-δ) curve together with the small value of the irreversible displacements δ_{ir} for each loading-unloading loop. When stable crack propagation initiates the ratio λ increases and reaches a plateau value. This is due to the complete development of the fracture process zone which is mainly consisted by cracked matrix, bridged by intact and/or failed fibers, which debond, slip and pull out. Crack bridging enhances material crack resistance by partially shielding the crack tip from the applied load. Figure 9, presents the elastic energy release rate G_R^*, versus the effective crack extension $\Delta\alpha_{eff}$. Initially the energy increases as the crack propagates for $\Delta\alpha_{eff} \approx 5$ mm. Then the energy remains almost constant. This plateau value is, $G_R^* = 1.2$ kJ/m^2.

In Figure 10, the plot of the irreversible energy rate Φ_{ir} is given versus the effective crack extension $\Delta\alpha_{eff}$. It is shown that again there is a crack propagation period where the energy increases. Figure 11 shows the crack resistance $R(=G_R^* + \Phi_{ir})$ versus the effective crack extension, $\Delta\alpha_{eff}$.

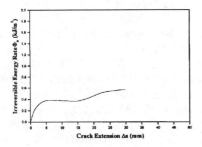

Figure 10. Plastic energy dissipation rate versus effective crack length.

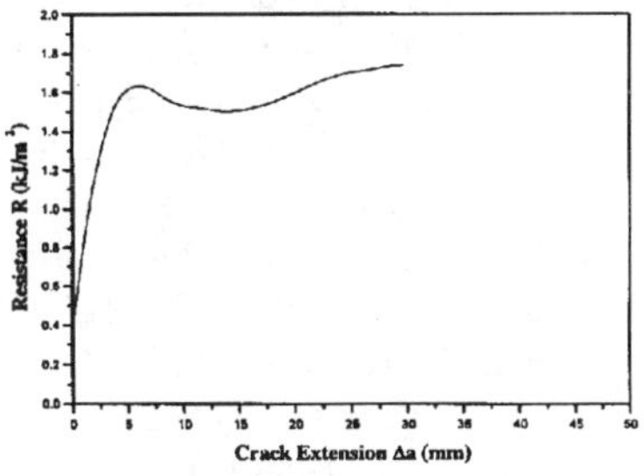

Figure 11. Resistance, R, versus effective crack length.

4. Conclusions

An effort has been made to adopt a different testing technique for the characterization of the fracture parameters of CF\PA12 composite through the evaluation of R-curve behaviour. Based on the use of DCB test geometry, cyclic loading/unloading pattern have been applied in test coupons.

A stress shielding mechanism at the crack tip was detected, which is mainly consisted on the formation of fibre bridging between the upper and the lower parts of the DCB test specimen, together with the presence of resin rich areas.

Furthermore it was found that the loading/unloading testing procedure leads to a more realistic and self consistent calculations where the contribution of the crack tip phenomena and the stress shielding mechanisms, which enhance the material crack resistance to the material fracture behaviour are calculated separately.

This approach appear to provide consistent prediction of the fracture resistance of CF\PA12 composite material.

5. Acknowledgements

The supply of the prepreg tape by Bay Comp, Canada, as well as the financial support by Advisory Group for Aerospace Research and Development (*AGARD*), are gratefully acknowledged.

References

[1] J. M. Whitney, ELASTICITY ANALYSIS OF ORTHOTROPIC BEAMS UNDER CONCENTRATED LOADS - Composites Science and Technology, Vol. 22, Pages 167-184.
[2] W. D. Bascom, G. W. Bullman, D. L. Hunston, R. M. Jensen THE WIDTH-TAPERED DOUBLE CANTILEVER BEAM FOR INTERLAMINAR FRACTURE TESTING - Proceedings of the 29th National SAMPE Symposium, April 3-5, 1984.
[3] F. X. De Charentenay, M. Benzeggagh FRACTURE MECHANICS OF MODE I DELAMINATION COMPOSITE MATERIALS - Proceedings of the 3rd International Conference on Composite Materials (ICMM 3), Paris, 1980.
[4] D. F. Devitt, R. A. Schapery, W. L. Bradley A METHOD FOR DETERMINING THE MODE I DELAMINATION FRACTURE TOUGHNESS ON ELASTIC AND VISCOELASTIC COMPOSITE MATERIALS - Composite Materials, Vol. 14, 1980.
[5] K. Friedrich, APPLICATION OF FRACTURE MECHANICS TO COMPOSITE MATERIALS - Elsevier Scien. Publ., Amsterdam, 1989.
[6] N. Sela, O. Ishai, INTERLAMINAR FRACTURE TOUGHNESS AND TOUGHENING OF LAMINATED COMPOSITE MATERIALS: A REVIEW - Composites, Vol. 20, 1989.
[7] P. Davies, INTERLAMINAR FRACTURE TESTING OF COMPOSITES - Protocol for Joint Round Robin, ESIS, 1990.
[8] K. Kageyama, T. Kobayashi, T. W. Chou ANALYTICAL COMPLIANCE METHOD FOR MODE I INTERLAMINAR FACTURE TOUGHNESS TESTING OF COMPOSITES - Composites, Vol. 18, 1987.
[12] E. H. Lunz, R-CURVE AND COMPLIANCE CHANGE UPON RENOTCHING - American Ceramic Society, Vol. 76, 1993.
[13] B. N. Cox, D. B. Marshall, CONCEPTS FOR BRIDGED CRACKS IN FRACTURE AND FATIGUE - Acta Metall. Mater., Vol. 42, 1994.
[14] Y. W. Mai, M. I. Hakeem, SLOW CRACK GROWTH IN CELLULOSE FIBRE CEMENTS - Material Science, Vol. 19, 1984.
[15] S. Mototsugu, K. Urashima, M. Inagaki, ENERGY PRINCIPLES OF ELASTIC-PLASTIC FRACTURE AND ITS APPLICATION TO THE FRACTURE MECHANICS OF A POLYCRYSTALLINE GRAPHITE - American Ceramic Society, Vol. 66, 1982.
[16] J. W. Cao, M. Sakai, THE CRACK FACE BRIDGING OF BRITTLE MATRIX COMPOSITES - Proceedings of the 6th International Conference on Fracture Mechanics of Ceramics, Stuttgart, 1995.

Effect of hybrid yarn structure on the delamination behaviour of thermoplastic composites

B. Lauke, U. Bunzel

Institute of Polymer Research Dresden, 01069 Dresden, Hohe Str. 6

Abstract

Continuous glass-fibre reinforced polyamide matrix composites have been manufactured on the basis of hybrid yarns with different arrangements of the reinforcing glass and polyamide matrix fibres. In addition glass fibres with different sizings were used. To characterize delamination behaviour of composites made with these semi-finished products, mechanical and fracture mechanical tests have been performed, as: longitudinal and transverse tensile test, compression shear, DCB (double cantilever beam) test, SEN (single edge notched) test, and ENF (end notch flexure) test. The results demonstrate the significant influence of different hybrid yarn structures and glass fibre coatings on fracture toughness. The best mechanical and fracture mechanical properties have been obtained for composites manufactured with air textured commingled yarns on the basis of glass fibres only with aminosilane as coupling agent.

1. Introduction

Hybrid yarns consisting of reinforcing and matrix fibres are one kind of basic materials (semi-finished products) to construct continuous-fibre-reinforced thermoplastic composites[1,2]. The composite properties are mainly influenced by the arrangement of the reinforcing fibres and the homogeneity of the fibre distribution in the composite as well as the impregnation of the glass fibres with polymer matrix. Hybrid yarns are processed usually into thermoplastic composites by hand laying[3], filament winding[4,5] or as recently done by the pultrusion process[6].

In this study it is aimed to consider the influence of the different fibre arrangements of the semi-finished materials on the interlaminar and intralaminar crack propagation behaviour of thermoplastic unidirectional composites made by compression moulding. The delamination resistance of the material is characterized by the critical energy release rate. For the commingled yarn, glass-fibres of four different sizings embedded in a polyamide fibre matrix are investigated.

2. Materials

Hybrid Yarns
The hybrid yarns were manufactured on the basis of glass and polyamide fibres whose properties are summarized in Table 1. The fineness of fibres are given in tex (tex = g/1000m).

		SBS	KEM	COM	FS	SCH
Glass						
filament diameter	[µm]	10	10	10	9	10
fineness of filament yarn	[tex]	40	40	40	136	-
Polyamid 6						
fiber fineness	[dtex]	-	-	-	1.3	-
fineness of filament yarn	[tex]	40	40	40	-	-
Hybrid Yarn						
glass:polyamid	[mass-%]	60:40	66:34	66:34	67:33	72:28
fineness	[tex]	200	1440	720	200	588
producer		IPF	IPF	IPF	ITA *	Schappe°

* Inst. für Textiltechnik der RWTH Aachen
° prospectus Schappe Techniques

Table 1:   Basic materials for hybrid yarn manufacturing

Using these filaments, hybrid yarns have been produced by different technologies, resulting in different structures of the semi-finished product:

• parallel arrangement of glass- and polyamide fibres (side-by-side, SBS)
• parallel arrangement of matrix fibres surrounded by parallel glass fibres in the core, sheathed by matrix fibres in the skin („Kemafil"-technology, KEM)
• commingled glass- and polyamide fibres made by air texturing (commingled yarn, COM)
• parallel arrangement of glass fibres in the core and spun matrix fibres in the skin (friction spinning, FS)
• mixture of glass- and matrix fibres, both discontinuous, surrounded by a continuous matrix filament (Schappe-technology, SCH)

Three material arrangements (SBS, KEM, and COM) have been produced in our laboratory. The FS-material was provided by the RWTH Aachen. The Schappe-hybrid yarn, produced by the Schappe-techniques, is a commercial type.
As SEM observations show, the homogeneity of the glass fibre distribution within the matrix is strongly dependent on the hybrid yarn structure. The SCH- and COM-composites show the best degree of mixing of reinforcing and matrix fibres. This is a merely qualitative description of the yarns microstructure, which is influencing, however, more profound quantities that determine the consolidation quality of the composite. The impregnation of the reinforcing fibres with matrix material is mainly determined by the average flow distance of the polymer, a number also difficult to express quantitatively but a parameter depending on the degree of mixing. The yarns are ranked according to the increasing flow distance: SCH, COM, KEM, SBS and FS. A second parameter is the possibility of fibre flow, i.e. the fibres themselves can move together with the matrix. In compression moulding of a flat plate, this movement is negligible for continuous yarns, that can considered fixed. The Schappe yarn contains discontinuous fibres and thus such movements are possible. If all these parameters are compared, the Schappe and COM yarns come out best, which are comparable, followed by the SBS, the KEM is worse, and the worst material is the FS.

Glass Fibre Sizings

For the manufacturing of air textured commingled yarns four different glass fibre sizings (B, C, E, G) have been used (see Table 2).

size	A	without sizing
size	B	γ-aminopropyltrietoxysilane (γ-APS)
size	C	epoxy dispersion
size	E	γ-APS + epoxy dispersion
size	G	modified aminosilane and polyurethan dispersion

Table 2. Glass fibre suface treatments (sizings)

The size modifies the wetting of the glass fibres and consequently it is one of the factors that influence the impregnation process of the glass fibres with the polyamide matrix and the adhesion strength of the fibre/matrix interface. The coating material (size) of the fibres involves different agents. It contains in general a film-forming agent for protection, lubricants to reduce wear during handling, and coupling agents. For our examinations we have used different film-forming and coupling agents. The size A contains no coupling agent and no film-former, only water. The size B contains no film-former but aminosilane as the coupling agent. Size C contains only a film-forming epoxy dispersion and size E is a combination of epoxy-dispersion and aminosilane. Size G is a coating that contains both the film-forming agent: polyurethane dispersion and a coupling agent: modified aminosilane.

Composite Manufacturing

The composites consisting of 15 unidirectional laminae have been manufactured by compression moulding. Hybrid yarns were wound with a filament winding device on a plate core and consolidated under 3 MPa pressure at a temperature of 245 °C into plates. These plates were cut into the different sample geometries necessary for the delamination and tension tests. Though it was tried to achieve composites with the same glass fibre volume fractions by using the same mass ratio between glass and polyamide fibres (see Table 1) for each yarn, the outcoming composites show a weak variation in the glass fibre volume fraction as shown in Table 3.

YARN	FIBRE VOLUME FRACTION [%]	SIZE	FIBRE VOLUME FRACTION [%]
SBS	45.5	A	48.6
KEM	52.9	B	48.5
COM	47.5	C	51.4
FS	59.1	E	47.5
SCH	56.0	G	57.1

Table 3. Fibre volume fractions of the composites for the hybrid yarn structures and of the COM-composites for different sizings

3. Experiments

Longitudinal and Transverse Tension
The tension properties of laminates were determined by loading appropriately cut samples by a Zwick Material Testing Machine with a rate of 1mm/min.

Compression Shear Test
To assess the bonding quality between the different layers of composites it is necessary to choose a useful shear test. We have developed a compression shear test[7,8] that allows the shear loading of a certain pre-determined plane.

Stress Intensity and Energy Release Rate
Within the limits of linear-elastic fracture mechanics and for orthotropic materials the two alternative characteristics stress intensity factor, K and energy release rate, G are related by the following equation, derived by Sih et. al.[9]:

$$G_I = \frac{K^2}{E_0} \quad \text{with} \quad \frac{1}{E_0} = (\frac{a_{11}a_{22}}{2})^{1/2}\left\{(\frac{a_{22}}{a_{11}})^{1/2} + (\frac{2a_{12}+a_{66}}{2a_{11}})\right\}^{1/2} \tag{1}$$

where the a_{ij} are the components of the compliance tensor.

For an isotropic material the relations for the mode I stress intensity factor $K_{I(iso)} = \sigma_A \cdot Y$ (crack length, geometry) are available for many test geometries. However, for orthotropic materials the calibration factor Y and thus K_I becomes a function of a_{ij} and is known only for some special cases. For example, Sweeney[10] derived the Y-factor for an orthotropic material numerically for a SEN-test geometry by a finite element analysis.

To apply the isotropic stress intensity relations for the indirect determination of the energy release rate, we have used the following equation

$$G_I = \frac{K^2_{I(iso)}}{E_{ij}} \tag{2}$$

With this relation the energy release rate of an orthotropic material can be calculated approximately with the stress intensity of a corresponding sample geometry of an isotropic material.

For the DCB and ENF test geometries the data reduction scheme based on the compliance calibration as given in the ESIS protocol, reviewed in ref 11, and applied recently in refs 12 and 13 has been used. In these tests the crack is propagating parallel to the laminae (interlaminar) while for the SEN-test the crack plane is parallel to fibre direction but perpendicular to the laminae (intralaminar). For the determination of the critical energy release rate G_{Ic}, i.e. the point of crack growth initiation, we have used the maximum load in the load-displacement curve (P_{max}, δ_{max}, a).

4. Results and Discussion

Longitudinal and Transverse Tension

The longitudinal tensile strength of unidirectional fibre reinforced polymer composites can usually be increased only in an insignificant manner by improving the interface shear strength i. e. to enhance the ability to transfer stresses between possibly broken fibres and matrix[14,15]. On the other hand, orientation of fibres out of the loading direction causes a decrease of composite strength[16], which becomes now sensitive to the fibre-matrix interface properties. For the hybrid yarns manufactured into composites, the structural situation is more complex than in unidirectional composites. Though the yarns are placed parallel, the composite is only "quasi-unidirectional" because for some structures the fibres are not aligned but waved, and in the case of SCH the composite contains discontinuous fibres. Consequently also the longitudinal tensile strength and elastic modulus should be dependent on the yarn type and fibre sizing. The results for the different yarns are summarized in Table 4. Table 5 shows the variation of longitudinal strength and modulus with fibre sizing for the COM composites.

	n		SBS	KEM	COM	FS	SCH
Longitudinal tensile strength	5	[MPa]	697.1 ± 21.7	583.7 ± 37.5	580.0 ± 21.7	605.4 ±21.7	489.4 ±51.1
Longitudinal modul	5	[GPa]	33.7 ± 1.3	37.6 ± 0.6	33.6 ±0.9	40.7 ± 0.7	39.4 ± 1.1
Transverse tensile strength	5	[MPa]	3.5 ± 0.5	4.1 ± 0.5	13.3 ± 2.5	2.7 ± 0.6	24.7 ± 1.7
Transverse modul	5	[GPa]	0.9 ± 0.2	1.1 ± 0.1	3.4 ± 0.5	0.9 ± 0.2	4.9 ±0.2
Interlaminar shear strength	8	[MPa]	17.2 ± 0.3	17.1 ± 1.0	37.1 ± 2.3	11.4 ± 0.5	39.3 ± 0.5
Intralaminar shear strength	8	[MPa]	19.6 ± 1.9	22.3 ± 2.6	43.0 ± 2.2	15.8 ± 1.7	46.8 ± 0.7
Critical energy release rate							
DCB-test (mode I)	4 - 5	[J/m²]	1850	2060	2300	2110	1568
SEN-test (mode I)	1 - 5	[J/m²]	305	162	387	79	911
ENF-test (mode II)	3	[J/m²]	4950	3415	6774	-	6447

Table 4.   Influence of hybrid yarn structure on tensile properties, compression shear strength and critical energy release rate, size E, fibre volume fractions see in Table 3, n: number of specimens

	n		size A	size B	size C	size E	size G
Longitudinal tensile strength	5	[MPa]	421.6 ± 39.9	639.5 ± 27.4	562.1 ± 33.0	580.0 ± 21.7	588.6 ± 12.9
Longitudinal modul	5	[GPa]	33.2 ± 1.2	41.4 ± 0.6	34.2 ± 1.8	33.6 ± 0.93	38.9 ± 1.6
Transverse tensile strength	5	[MPa]	8.1 ± 0.1	33.7 ± 1.1	9.3 ± 0.5	13.3 ± 2.5	3.1 ± 0.2
Transverse modul	5	[GPa]	1.2 ± 0.02	7.0 ± 2.1	2.5 ± 0.1	3.37 ± 0.47	0.9 ± 0.1
Interlaminar shear strength	8	[MPa]	19.9 ±0.7	58.4 ± 1.9	24.1 ± 1.0	37.09 ± 2.26	21.6 ± 2.3
Intralaminar shear strength	8	[MPa]	29.1 ± 1.2	74.5 ± 1.8	42.6 ± 1.4	42.99 ± 2.21	46.9 ± 1.7
Critical energy release rate							
DCB-test (mode I)	2 - 5	[J/m²]	-	3668	2066	2300	1679
SEN-test (mode I)	1 - 5	[J/m²]	537	1462	315	387	112
ENF-test (mode II)	3 - 4	[J/m²]	3119	6936	4633	6774	2498

Table 5.   Influence of glass fibre sizings on tensile properties, compression shear strength and critical energy release rate of COM-composites, fibre volume fractions see in Table 3, n: number of specimens

If we consider the different yarn structures and take the knowledge about strength of fibre reinforced composites into account, we would anticipate the following ranking for longitudinal strength and modulus: the SBS should provide the highest values because the reinforcing fibres (and they are responsible for the properties in fibre direction) are all arranged parallel. Within the yarns KEM and FS the fibre arrangement is similar but may be slightly influenced by the surrounding matrix filaments during consolidation, the commingled yarn is characterized by the arrangement of the fibres out of the yarn axis, and the SCH material is reinforced with discontinuous fibres.

This trend is confirmed by the values in Table 4 with the ranking: SBS, (FS, KEM), COM, SCH. If we consider also the variation of the fibre volume fraction in Table 3 it becomes clear that the SBS- composite with the lowest value of v =0.46 has a much higher strength as the SCH-composite with v=0.56. Applying only a rule of mixture type relation for strength of continuous fibre reinforced composites would provide the opposite trend with a roughly 20% higher strength of SCH- compared to SBS-composites. However, even if the reduction of mean fibre stress in discontinuous fibres is considered, such an approach neglects the peculiarities of the strengh-limiting processes determined by failure of the component that survives initial failure. The calculation of strength of hybrid yarn is still an unsolved problem and it is mainly determined by the arrangement of the reinforcing fibres in the composite.

The longitudinal modulus should not dependent on the interface quality, because at low loads the interface still is intact. The difference in the data, shown in Table 5, results mainly from the variation of the fibre volume fraction (compare Table 3). Only size B, with one of the lower fibre contents surprisingly provides a longitudinal modulus higher than that of the other composites, and this trend is observed also for the other mechanical properties.

Transverse strength and transverse modulus depend strongly on the composite microstructure, especially on the homogeneity of the distribution of the reinforcing fibres, on the directionality, on the quality of fibre/matrix adhesion and the impregnation quality that is by itself dependent on adhesion and wettability.

The SCH and COM yarn structures result in composites with the highest values for transverse strength and modulus. As discussed above these two hybrid yarns contain discontinuous fibres and misorientated fibres respectively. This fact results in the better transverse properties compared to the other yarn structures. The FS-yarn is constructed by parallel bundles of glass fibres surrounded by the matrix filament, and the flow distances are the highest one resulting in a bad impregnation quality and consequently in the worst transverse properties.

For transverse loading, normal forces have to be transferred between the constituents. An optimized sizing is necessary as fibre/matrix interfaces are extremely sensitive to normal loads (cf. ref 17). Again size B provides the highest strength and modulus (see Table 5).

The sizing B on the basis of aminosilane provides the highest longitudinal and transverse properties. Unfortunately, this product cannot be applied in technical applications because it leads to rather brittle and electrostatically charged glass fibres not appropriate for technical processing in the air texturing process. On the other hand the sizing G results in mechanical properties below all other variants. This sizing really prohibits the consolidation of the commingled yarn into a composite material, and the bad mechanical properties correspond to the bad visual appearance of these samples.

Inter- and Intralaminar Shear Strength

The shear strength values are being influenced by the yarn structure, and the interface properties. During the winding process the yarn is placed side by side, one lamina over the other. Afterwards this plate is consolidated into the laminate.

After consolidation the composite should have similar properties in the two perpendicular planes. If not, it would reveal that the whole compression moulding process is not optimal, leading to a heterogeneous structure. As the data show, the differences between inter- and intralaminar shear strength for the hybrid yarns are not significant. In analogy to the above discussions one would expect that the SCH-composites should provide the highest interlaminar strength because this material contains discontinuous fibres which are able to move into out off axis directions during compression moulding. It should be followed by the COM material because the shear planes are crossed by waved fibres.

The decrease in shear strength follows the increase in fibre alignment and is ranked: SCH, COM, (SBS, KEM), FS.

Critical Energy Release Rate

Mode I energy release rates are measured by the DCB- and SEN-test , mode II energy release rates by the ENF-test. In our case of delamination cracks, the energy dissipation mechanisms are fibre related energies as fibre bridging and fibre fracture, debonding between fibres and matrix, fibre pull-out and matrix related contributions as plastic deformations (crazing or shear band formation) or brittle matrix fracture. The highest mode I critical energy release rates as well as mode II values for crack initiation with the DCB, the SEN and the ENF tests have been obtained for the laminate made by COM-yarns.

The results of the hybrid yarns with the technically practicable size E are in the region of the G_{Ic} values for thermoplastic laminates published elsewhere[18,19]. Only the COM-composite with sizing B shows an excellent value of $G_{Ic}= 3.7$ kJ/m^2. Fractografic examinations reveal more clean fibres, i. e. crack propagation along the fibre/matrix interface for sizing E while the fracture pattern for the specimens with size B show large regions with plastic matrix deformations and crack propagation within the matrix material.

5. Conclusion

The interlaminar and intralaminar delamination of unidirectional reinforced thermoplastic composites can be strongly influenced by a slight disturbance of fibre orientation with an appropriate choice of the hybrid yarn structure. The arrangement of reinforcing and matrix fibres in the prepreg determines the impregnation quality, the distribution and orientation of the reinforcement in the final composite material. Well mixed hybrid yarns do have a positive impact on fracture mechanical properties. Yarns with reinforcing fibres not strongly directed in the yarn axis, as in the air textured commingled yarn (COM), show the highest fracture toughness for crack initiation and the highest crack resistance during propagation, especially caused by fibre bridging. The benefit of the increased fracture toughness and transverse tensile and shear strength and modulus is paid by a reduction of the longitudinal tensile strength. For example if an linear increase of strength with fibre volume fraction is supposed, than the SCH composite shows a reduction of longitudinal strength compared to the SBS material of about 43% and the COM composite of about 20%. The longitudinal modulus for a corrected identical fibre volume fraction shows a much smaller variation of about 6%, the COM material provides even a 6% higher modulus as SBS composites.

Summarizing all experimental results, the conclusion can be drawn that the air textured commingled yarn (COM) with optimized glass fibre sizing provides the highest transverse modulus and strength as well as the highest mode I and mode II crack resistance of glass fibre reinforced polyamide composites. Research on impregnation behaviour during compression moulding of these materials is necessary in the future to get a more quantitative evaluation how the degree of mixing, average flow distance, and compaction stability influence consolidation.

Literature

1   Fujita, A, Maekawa, Z., Hamada, H., Matsuda, M., and Matsuo, T., J. of Reinforced Plastics and Composites 1993, 12, 156-172

2   Mäder, E., Bunzel, U., and A. Mally *Technische Textilien* 1995, **38**, 205-208

3   Bader, M.G. „Reinforced Thermoplastics", in *Handbook of Composites* (Ed. Kelly, A., Mileiko, S.T.), Vol. IV, Elsevier, New York, 1983

4   Haupert, F., and Friedrich, K. *Advanced Composites* 1993, **2** , 14

5   Lauke, B., and Friedrich, K. *Composites Manufaturing* 1993, **4**, 93-101

6   Michaeli, W., and Jürss, D. *Composites Part A* 1996, **27A**, 3-7.

7   Lauke, B., Schneider, K., and Friedrich, K. ECCM-Conference, April, 1992, Bordeaux, Preprint p. 313-318

8   Lauke, B., Beckert, W., and Schneider, K. *Applied Composite Materials* 1994, **1**, 267-271

9   Sih, G.C., Paris, P.C., Irwin, G. R. *Int. J. Fract. Mech.* 1965, **1**, 189-203

10   Sweeney, J. *Journal of Strain Analysis* 1986, **21**, 99-107.

11   Brunner, A.J., Flüeler, P., Davies, P., Blackman, B.R.K., and Williams, J.G. "Determination of the delamination resistance of fibre-reinforced composites, current scope of test protocols and future potential", ECCM-7, 14.-16. May 1996, London, UK

12   Beckert, W., and Lauke, B. *Mat.-wiss. u. Werkstofftech.* 1996, **27**, 14-24

13   Starke, C., Beckert, W., and Lauke, B. *Mat. wiss., u. Werkstofftechn.* 1996, **27**, 80-89

14   Shih, G.C. and Ebert L.J. Journal of Composite Materials 1987, 21, 207-224

15   Hoecker, F. "Grenzflächeneffekte in Hochleistungsfaserverbundwerkstoffen mit polymerer Matrix", Fortschrittsberichte VDI, Reihe 5, Nr. 439, 1996

16   Hull, D. "An introduction to composite materials", Cambridge University Press, 1985

17   Beckert, W. and Lauke, B. *Computational Materials Science* 1996, **5**, 1-11

18   Davis, P. and Moore, D. R. *Composite Sci. and Technology* 1990, 38, 211-227

19   Ye, L. and Friedrich, K. *Composite Sci. and Technology* 1993, 46, 187-198

Novel Laminated Composites with Nanoreinforced Interfaces

Yuris A. Dzenis[1] and Darrell H. Reneker[2]

[1]Department of Engineering Mechanics, Center for Materials Research and Analysis, University of Nebraska-Lincoln, Lincoln, NE 68588-0347, U.S.A.

[2]Maurice Morton Institute of Polymer Science, Department of Polymer Science, University of Akron, Akron, OH 44325-3909, U.S.A.

Abstract

A problem of delamination in advanced polymer laminated composites is addressed. A new concept of thin continuous polymer nanofiber reinforcement of interfaces is formulated. A novel electrospinning technology is used as a method to produce continuous polymer fibers of submicron diameters. Experimental proof-of-concept is performed on a graphite/epoxy composite. It is found that the addition of 2.5 mass% of polybenzimidazole nanofibers results in 130% improvement of the Mode II critical energy release rate and 15% improvement of the Mode I critical energy release rate. Fractographic observations indicate that intralaminar fracture of the primary plies may limit improvement of interlaminar fracture toughness in peel and shear.

1. Introduction

Advanced composite materials are usually utilized in structural applications as laminates with plies of unidirectional composites. Delamination is an intrinsic and severe problem of these materials [1]. Interlaminar stresses due to mismatch of anisotropic mechanical and thermal properties of plies occur at free edges, joints, matrix cracks, and under out-of-plane loading. Delamination is often the dominating failure mode in laminates subjected to impact and fatigue loading. Idealized models of edge delamination exhibit singular concentrations of interlaminar shear and peel stresses near edges [2]. High interlaminar stresses occur in the vicinity of cracks in primary reinforcing plies. This is especially dangerous because interlaminar stresses act on an unreinforced plane. Such a plane is always present between the plies with different fiber orientations.

A number of methods to prevent delamination were developed over the years [3-4]. These include matrix toughening, optimization of stacking sequence, laminate stitching, braiding, edge cap reinforcement, critical ply termination, and replacement of a stiff ply by one that has softer regions. Most designs to reduce delamination resulted in significant cost or weight penalties. Ductile interleaving [5] is an effective method to improve delamination resistance. Layers of toughened resins (interleaves) are inserted at interfaces between the primary plies in this method. Rubber or thermoplastic particle toughening is often utilized. Substantial improvement in interlaminar fracture toughness was achieved by this method. However, the thickness of the toughened interleaves is usually comparable to the thickness of the primary reinforcing plies. The use of these interleaves at multiple interfaces may substantially increase weight or reduce in-plane properties of laminates.

In this paper, a concept of thin polymer nanofiber reinforcement of interfaces in laminates is proposed. A new manufacturing technology of electrospinning is suggested as a method to produce continuous polymer nanofibers. An experimental verification of the concept is performed on an example of a graphite/epoxy laminate.

2. Nanoreinforcement of Interfaces: A Novel Concept

Reinforcement of interfaces in laminates by fibers of conventional diameters was experimentally explored in [6]. Random mats of commercial polyaramid, polyester, and glass fibers were studied. Improvement in the Mode I critical energy release rate was reported for all in-lay materials. Best results were obtained with polyaramid fibers combined with a toughened epoxy adhesive. However, the thickness of the inner, randomly reinforced layer was equal or higher than the thickness of primary reinforcing plies. Fiber mats required preimpregnation. The toughened adhesive layer required additional curing, leading to some embrittlement of the primary plies. Structurally, thick interlayers reinforced by fibers of conventional diameters can be regarded as additional plies rather than interface modifiers. High thickness of these layers and low volume fraction of randomly distributed fibers lead to substantial reduction of in-plane properties of laminates.

A novel concept of thin, continuous fiber reinforcement of interfaces is proposed. The key in this delamination suppression concept is the small fiber diameter. Thin fibers are expected to reinforce laminate interfaces without substantial reduction of in-plane properties. Each laminate interface can be reinforced by thin fibers without substantial increase in weight. Thin fiber reinforcement may be capable of both arresting matrix crack propagation and suppressing crack induced and edge delamination. No preimpregnation of the thin fibers is needed. The thin fiber mat can be attached to one side of the conventional prepreg prior to lamination. The thin fibers are flexible and are expected to conform to the shape and distribution of the primary reinforcing fibers at the interface as the resin flows. Therefore, the resin rich zones near the primary ply surfaces will be reduced or eliminated. The delamination suppression will be achieved through both improvement of interlaminar fracture toughness and reduction of interlaminar stresses by smearing the mismatch of ply properties in multidirectional laminates. The diameter of conventional reinforcing graphite fibers used in advanced composites varies from 5-7 micrometers. Fibers of smaller diameters are needed to implement the concept described above. An order of magnitude difference is expected to be sufficient. Therefore, fibers of submicron diameters (nanofibers) are needed for interface reinforcement. Strong continuous fibers of submicron diameters are not readily available in general.

Recently, a new technology of manufacturing of thin polymer fibers by electrospinning was developed and studied. The method is based on the principle of spinning polymer solutions in a high-voltage electric field. It consists of the ejection, from the end of a capillary tube, of a charged jet of a polymer solution which is elongated and accelerated by an external field. Nanofibers have been prepared by electrospinning from over thirty different synthetic and natural polymers, including high-temperature polyimide and polyaramid fibers. Fibers with diameters as small as three nanometers have been produced by this method. The method produces random nonwoven mats of fibers. Unlike short submicron diameter microfibers or whiskers, electrospun polymer fibers are continuous. That eliminates the potential for ingestion of airborne fibers into the lungs. The electrospun polymer nanofibers possess substantial degree of macromolecular orientation which is promising for their use as reinforcement in composites. At the same time, polymer nanofibers can be expected to possess inherent ductility which is promising for their use to improve fracture toughness. Thus both reduction of interlaminar stresses and improvement of interlaminar crack propagation resistance can be expected. It is proposed here, to use non-woven mats of electrospun polymer nanofibers as reinforcement at interfaces in advanced composite laminates.

3. Experimental Verification

Non-woven fabrics of polybenzimidazole (PBI) nanofibers with diameters ranging from 300-500 nanometers were manufactured by electrospinning. This temperature resistant, linear polymer with glass transition temperature about 430°C is used in thermal-protective, fire-blocking, and other aggressive temperature and environment applications. The sheets of electrospun PBI nanofibers were washed in water, treated with 50% sulfuric acid to improve the strength of the fibers, and dried at 200°C. The areal weight of the resulting fiber mats was 6.85 g/m^2. A scanning electron microscope (SEM) picture of the non-woven PBI sheet used in this study is shown in Fig. 1a.

Laminates with and without nanoreinforcement at interfaces were manufactured from a unidirectional graphite/epoxy prepreg T2G190/F263 provided by Hexcel Corporation. No additional resin was applied in the specimens with nanofibers. Unidirectional [0$_{20}$] composite panels were manufactured. Strips of thin Teflon$^{®}$ film were placed as crack starters at the midplane interface. No additional epoxy resin was used during manufacturing of specimens with nanofibers. A specialized press-clave was designed and built for proper impregnation and consolidation of PBI nanofibers in the laminates. Several graphite/epoxy composite panels with nanofiber reinforcement of interfaces were manufactured in controlled temperature, pressure, and vacuum environments. Reasonable quality consolidation was confirmed by SEM observations of the cross-sections (Fig. 1b) and NDE.

The consolidated thickness of the nanoreinforced interlayer was around 21 micrometers. Volume content of nanofibers in the interlayer, calculated from the measured interlayer thickness, mat areal weight, and PBI density (1.4 g/cm^3), was 24%. The consolidated thickness of the primary graphite/epoxy ply in the laminate without PBI nanofibers was around 180 micrometers. Therefore, less than 12% increase in the total laminate thickness is expected if the nanofibers were used at each interlaminar interface. Estimated increase in the laminate weight calculated as a ratio of the aerial weights of the nanofiber sheet and the prepreg tape was 2.5%.

The Arcan test method [7] was used for interlaminar fracture testing. This test was selected for the experimental study due to the relatively small specimen size. The possibility of the mixed-mode interlaminar fracture testing was also taken into account. The two major fracture modes were studied: the opening, or peel mode (Mode I) and the sliding shear mode (Mode II). The 20-ply specimens were used in the comparative study. Interlaminar fracture testing was performed using MTS servohydraulic testing machine that was digitally upgraded with an Instron 8500 Series control and data acquisition system. Load-displacement curves were recorded at the constant stroke rate 0.5 mm/min. The maximum loads were extracted and used to calculate the critical stress intensity factors. The following expressions for the Mode I and Mode II stress intensity factors were used [7]:

$$K_I = \sigma_\infty \sqrt{\pi a}\, f_I(a/c)$$

$$K_{II} = \tau_\infty \sqrt{\pi a}\, f_{II}(a/c)$$

where σ_∞ and τ_∞ are the "far-field" normal and shear stresses, a is the crack length, c is the dimension of the specimen along the crack, and the correction factors f_I, f_{II} are defined as follows:

$$f_I(a/c) = 1.12 - 0.231\,(a/c) + 10.55\,(a/c)^2 - 21.27\,(a/c)^3 + 30.39\,(a/c)^4$$

$$f_{II}(a/c) = \frac{1.122 - 0.561\,(a/c) + 0.085\,(a/c)^2 + 0.180\,(a/c)^3}{[1 - (a/c)]^{1/2}}$$

The strain energy release rate was obtained from the stress intensity factors using the following correlations [7]:

$$G_I = K_I^2 \left(\frac{S_{11}S_{22}}{2}\right)^{1/2} \left[\left(\frac{S_{22}}{S_{11}}\right)^{1/2} + \frac{2S_{12} + S_{66}}{2S_{11}}\right]^{1/2}$$

$$G_{II} = K_{II}^2 \frac{S_{11}}{\sqrt{2}} \left[\left(\frac{S_{22}}{S_{11}}\right)^{1/2} + \frac{2S_{12} + S_{66}}{2S_{11}}\right]^{1/2}$$

where S_{ij} are the elements of the compliance matrix for a transversely isotropic unidirectional composite. Elastic constants of the unidirectional T2G190/F263 composite were obtained from the standard tensile quasi-static tests on $[0_8]$, $[90_{16}]$, and $[(\pm\ 45)_8]_s$ composite coupons: $E_1 = 133$ GPa, $E_2 = 9.4$ GPa, $G_{12} = 5.7$ GPa, and $\nu_{12} = 0.37$. The elements of the compliance matrix of the composite were computed using elastic relations for a transversely isotropic material.

Ten or more fracture specimens were tested for each mode. The calculated values of the critical energy release rates for laminates with and without nanofiber reinforcement of interface are presented below. Increase of 130% in the Mode II and 15% in the Mode I critical energy release rates are observed.

Critical Energy Release Rate

Material	G_{Ic}, kJ/m^2	G_{IIc}, kJ/m^2
Original Laminate	0.089 ± 0.036	1.17 ± 0.37
Laminate with Nanoreinforced Interface	0.101 ± 0.044	2.67 ± 1.07

Optical images of the Mode I fracture surfaces of the specimens with nanoreinforced interface exhibited a variety of fracture modes including rather extensive intralaminar delamination. SEM images in Fig. 2 show the transition from the fracture through the nanoreinforced interlayer to the fracture in the resin rich zone between the nanoreinforced interlayer and the primary ply. Pulled-out PBI nanofibers in Fig. 2b indicate their substantial mechanical strength. Longitudinal splitting in the primary ply, observed in Fig. 2b, may have facilitated deviation of the interlaminar crack into the primary ply. Optical and SEM images of the Mode II fracture surfaces revealed the crack path primarily through the nanoreinforced interlayer or through the interface between the interlayer and the primary ply (Fig. 3). Some intralaminar fracture of the primary ply was also observed. The extent of intralaminar fracture varied from specimen to specimen. Analysis showed that specimens with considerable intralaminar fracture demonstrated lower G_{IIc} values.

Overall, the laminate with nanoreinforcement at the interface exhibited substantial increase in interlaminar fracture resistance compared to the unmodified laminate. This improvement was obtained with the minimum increase of laminate weight. Fractographic observations indicate that intralaminar fracture of the primary plies may limit improvement of interlaminar fracture toughness in peel and shear.

Acknowledgments

This work was funded in part by NSF grants DMI-9523022 and EPS-9640815. The support of the National Science Foundation is gratefully acknowledged. The authors wish to thank Hoechst Celanese Corporation for providing PBI, Hexcell Corporation for providing graphite/epoxy prepreg, J.-S. Kim for producing PBI nanofibers, and S. Sergiyenko for laminate manufacturing and testing.

References

[1] Carg, A.C., "Delamination - A Damage Mode in Composite Structures," Engineering Fracture Mechanics, 1986, Vol. 29, pp. 557-584.

[2] Pagano, N.J., "Interlaminar Response of Composite Materials," Composite Materials Series, Vol. 5, Elsevier: Amsterdam, 1989

[3] Chan, W.C., "Design Approaches for Edge Delamination Resistance in Laminated Composites," Journal of Composite Technology and Research, 1991, Vol. 14, pp. 91-96.

[4] Chan, W.S. and Ochoa, O.O., "Edge Delamination Resistance by a Critical Ply Termination," Key Engineering Materials, 1989, Vol. 37, pp. 285-304.

[5] Masters, J.E., "Improved Impact and Delamination Resistance Through Interleafing," Key Engineering Materials, Vol. 37, Trans Tech Publ: Switzerland, 1989, pp. 317-348.

[6] Browning, C.E. and Schwartz, H.S., "Delamination Resistant Composite Concepts," Composite Materials: Testing and Design, J.M. Whitney, Ed., ASTM STP 893, American Society for Testing and Materials, Philadelphia, 1986, pp. 256-265.

[7] Carlsson, L.A. and Pipes, R. Experimental Characterization of Advanced Composite Materials, Prentice-Hall, 1987, p. 173

a

b

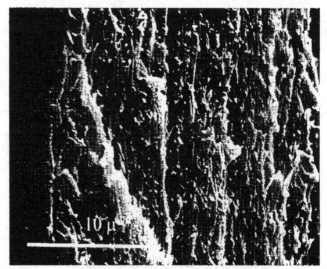

FIGURE 1. SEM images of the PBI nanofibers and the longitudinal cross-section of the laminate with nanoreinforced interface.

a

b

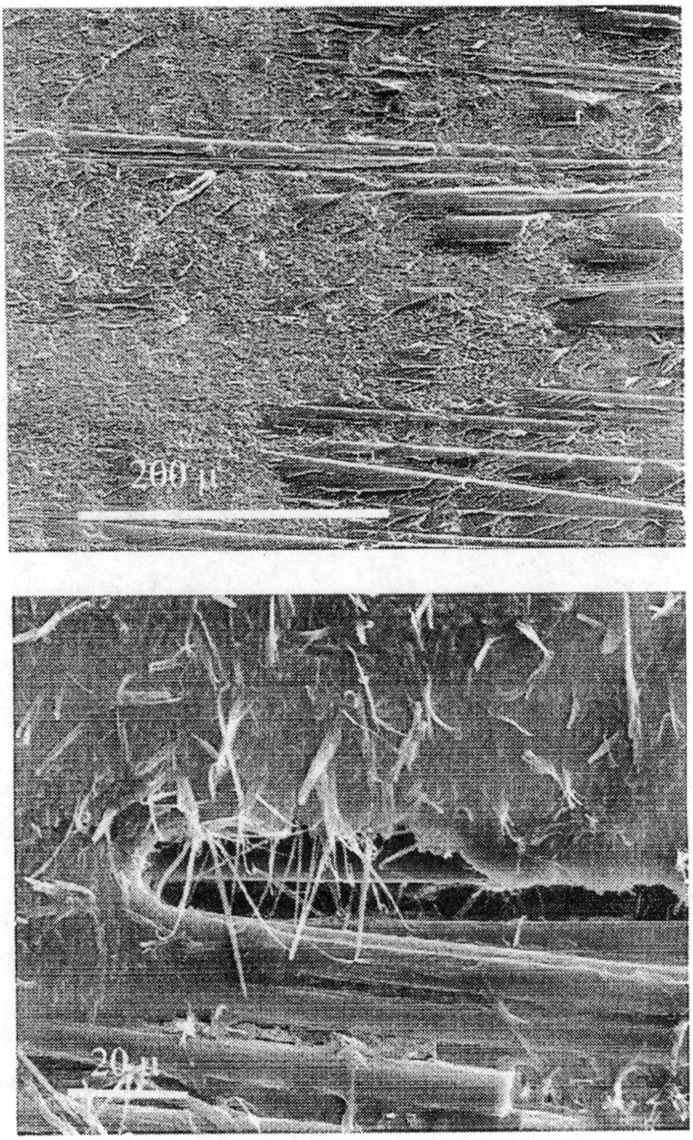

FIGURE 2. SEM images of the Mode I fracture surface of the laminate with
nanoreinforced interface.

a

b

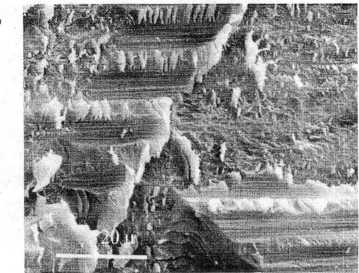

FIGURE 3. SEM images of the Mode II fracture surface of the laminate with
nanoreinforced interface.

Towards a structural identification of delamination initiation and growth

O. Allix*, P. Ladevèze*, D. Lévêque*

* Laboratoire de Mécanique et Technologie (LMT - Cachan)
ENS de Cachan / CNRS / Université Paris 6
61, Avenue du Président Wilson - 94235 Cachan cedex - France
e-mail: allix@lmt.ens-cachan.fr

Abstract

The present study, supported by CNES (French National Center for Space Studies), concerns the modeling and identification of an interface damage model devoted to the prediction of delamination. In this paper, a damage mechanics approach – including the single-layer model and the interlaminar interface model – is presented to solve delamination problems in laminated structures, such as the delamination onset and growth around holes. For the M55J/M18 (high-modulus carbon-fibres / epoxy resin) material, numerical simulations are presented and compared with experimental results, in particular, for holed plates in tension.

1. Introduction

This paper is focused on the identification of an interface damage model and on its prediction possibilities in the complex case of delamination growth around holes. The method, that allows identifying the few intrinsic characteristics of the interface model, consists of the development of reliable tests with specialised identification software.

An initial step, which has been achieved in other studies, is to define what we call a laminate mesomodel. At the mesoscale, characterized by the thickness of the ply, the laminated structure is described as a stacking sequence of homogeneous layers throughout the thickness and interlaminar interfaces. The main damage mechanisms are described as: fiber breaking, matrix micro-cracking and adjacent layers debonding, i.e. delamination [1-3]. The single-layer model includes both damage and inelasticity [4-6]. The interlaminar interface is described by a two-dimensional model which ensures traction and displacement transfer from one ply to another. Its mechanical behavior depends on the angle between the fibers of two adjacent layers [7, 8].

Two models have to be identified: the single layer model and the interface model. Each composite specimen, which contains several layers and interfaces, is computed in order to derive the material quantities intrinsic to the single layer [5, 6] or to the interlaminar interface [8, 9]. In the case of the interlaminar connection, a first identification procedure was tested in [10]. It makes use of standard tests : delamination propagation (fracture mechanics tests : DCB, ENF, MMF and CLS) and of initiation (edge delamination tension tests). The E.D.T. test are, in general, not reliable enough. In fact they are unstable and thus sensitive to defects. Therefore we develop a global identification method based on more reliable delamination tests of holed-plate specimens in tension. First simulation of such test are given.

2. Mesomodeling concept

Delamination often appears as the result of the interaction between different damage mechanisms, such as fiber breaking, transverse micro-cracking and the debonding of adjacent layers itself [1-3, 11]. Our aim herein is to build a bridge between damage mechanics and delamination by including all these damage mechanisms within the delamination analysis. To perform this, a damage meso-model, which allows us to predict both delamination initiation and growth, has been defined. Thanks to the proposed approach, both delamination and intralaminar damage prediction are included in a single model.

An initial step, which has been conducted in other studies, was to model a laminate as a stacking sequence of non-linear layers and non-linear interlaminar interfaces (figure 1). At the layer level, the inner damage mechanisms are taken into account by means of internal damage variables. These damage variables are prescribed to be uniform throughout the thickness of each ply which defines what we call a damage meso-model [4]. This plays a crucial role for a mesh-independent damage prediction. The single-layer model and its identification, including damage and inelasticity, were previously developed in other studies [5, 6].

The interlaminar interface is a two-dimensional entity which ensures traction and displacement transfer from one ply to another. Its mechanical behaviour depends on the angle between the fibers of two adjacent layers. Its primary interest is to allow the modeling of more or less progressive degradation of the interlaminar connection [7, 8].

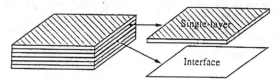

Figure 1. Laminate meso-modeling.

3. Single-layer modeling

The damaged material strain energy proposed below is written in the case of the plane stress assumption [5]. In what follows, subscripts 1, 2 and 3 designate respectively the fiber direction, the transverse direction inside the layer and the normal direction. The unilateral aspect of cracking in the transverse direction is taken into account by splitting the transverse energy into a "tension" energy and a "compression" energy. One then obtains the following energy for the damaged single-layer:

$$E_D^{cp} = \frac{1}{2(1-d_F)} \left[\frac{<\sigma_{11}>_+^2}{E_1^0} + \frac{\zeta(<-\sigma_{11}>_+)}{E_1^0} - (\frac{v_{21}^0}{E_2^0} + \frac{v_{12}^0}{E_1^0}) \sigma_{11}\sigma_{22} \right] \tag{1}$$

$$+ \frac{1}{2} \left[\frac{<\sigma_{22}>_+^2}{(1-d')E_2^0} + \frac{<-\sigma_{22}>_+^2}{E_2^0} + \frac{\sigma_{12}^2}{(1-d)G_{12}^0} \right]$$

where ζ is a material function describing the non-linear compressional behavior in the fibre direction [6]. d_F, d and d' are three scalar internal damage variables which remain constant within the thickness of each layer. d is describing the shear damage and d' the transverse cracking.

The thermodynamical forces associated with the mechanical dissipation are classically defined as:

$$Y_d = \frac{\partial}{\partial d} <<E_D^{cp}>> \Big|_{\sigma:cst} = \frac{<<\sigma_{12}^2>>}{2 G_{12}^0(1-d)^2} \tag{2}$$

$$Y_{d'} = \frac{\partial}{\partial d'} <<E_D^{cp}>> \bigg|_{\sigma:\mathrm{cst}} \frac{<< <\sigma_{22}>_+^2 >>}{2\,E_2^0(1-d')^2}$$

where $<X>_+$ is the positive part of X and $<< >>$ denotes the mean value within the thickness. For static loadings, the damage evolution laws can be formally written:

$$d\big|_t = A_d\,(Y_d\big|_\tau\, Y_{d'}\big|_\tau,\ \tau \le t) \tag{3}$$

$$d'\big|_t = A_{d'}\,(Y_d\big|_\tau\, Y_{d'}\big|_\tau,\ \tau \le t)$$

where the operators A_d and $A_{d'}$ are material functions to be identified. The evolution of d_F corresponds to the standard brittle fracture mechanism of the carbon-fibers in tension. More details, in particular for the modeling of inelastic strains as well as for the identification procedure, can be found in references [4-6].

4. Interlaminar interface modeling

4.1. Damageable behavior modeling

The effect of the deterioration of the interlaminar connection on its mechanical behaviour is taken into account by means of three internal damage variables. The energy per unit area proposed in [7] is:

$$E_d = \frac{1}{2}\left[\frac{<-\sigma_{33}>_+^2}{k_3^0} + \frac{<\sigma_{33}>_+^2}{k_3^0(1-d_3)} + \frac{\sigma_{32}^2}{k_2^0(1-d_2)} + \frac{\sigma_{31}^2}{k_1^0(1-d_1)}\right] \tag{4}$$

where k_i^0 is an interlaminar stiffness value and d_i the internal damage indicator associated with its Fracture Mechanics mode, while subscript i corresponds to an orthotropic direction of the interface, defined by the bissector of the fiber directions of the two adjacent layers (see figure 2).

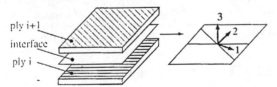

ply i+1

interface

ply i

Figure 2. Orthotropic directions of the interlaminar interface.

Classically, the damage energy release rates, associated with the dissipated energy ω by damage and by unit area, are introduced as:

$$Y_{d3} = \frac{1}{2}\frac{<\sigma_{33}>_+^2}{k_3^0(1-d_3)^2} \tag{5}$$

$$Y_{d_1} = \frac{1}{2} \frac{\sigma_{31}^2}{k_1^0 (1-d_1)^2}$$

$$Y_{d_2} = \frac{1}{2} \frac{\sigma_{32}^2}{k_2^0 (1-d_2)^2}$$

and

$$\phi = Y_{d_3}\dot{d}_3 + Y_{d_1}\dot{d}_1 + Y_{d_2}\dot{d}_2 \tag{6}$$

with $\phi \geq 0$ to satisfy the Clausius-Duheim inequality.

In what follows, an "isotropic" damage evolution law is described. In this model [8], the damage evolution law is assumed to be governed by means of an equivalent damage energy release rate of the following form:

$$\underline{Y}(t) = \sup \mid_{\tau \leq t} \left[\left((Y_{d_3})^\alpha + (\gamma_1 Y_{d_1})^\alpha + (\gamma_2 Y_{d_2})^\alpha \right)^{1/\alpha} \mid_\tau \right] \tag{7}$$

The evolution of the damage indicators is thus assumed to be strongly coupled. γ_1, γ_2 and α are material parameters.

A damage evolution law is then defined by the choice of a material function ω, such that:

$$d_3 = d_1 = d_2 = \omega(\underline{Y}) \quad \text{if } d_3 < 1 \tag{8}$$

$$d_3 = d_1 = d_2 = 1 \quad \text{otherwise}$$

One simple case, used for application purposes, is:

$$\omega(\underline{Y}) = \left[\frac{n}{n+1} \frac{<\underline{Y}-Y_0>_+}{Y_c-Y_0} \right]^n \tag{9}$$

where a critical value Y_c and a threshold value Y_0 are introduced. High values of the n case correspond to a rather brittle interface.

To summarize, the damage evolution law is defined by means of the six intrinsic material parameters Y_c, Y_0, γ_1, γ_2, α and n. As regards the onset of delamination, the significant parameters are Y_0, n and the interface stiffnesses. It will be shown in the next paragraph that Y_c, γ_1, γ_2 and α are all related to the critical energy release rates.

4.2. Links with Fracture Mechanics

A simple way to identify the parameters related to the delamination growth is to compare Damage Mechanics with Linear Elastic Fracture Mechanics. This was performed by comparing the mechanical dissipation yielded by the two approaches [8], and only the results are presented below.

In the case of pure-mode situations, when the critical energy release rate reaches its stabilised value at the propagation denoted by G_c^p, we obtain:

$$G_{cI}^p = Y_c ; \quad G_{cII}^p = \frac{Y_c}{\gamma_1} ; \quad G_{cIII}^p = \frac{Y_c}{\gamma_2} \tag{10}$$

For a mixed-mode loading situation, we simply derive a standard LEFM model:

$$\left(\frac{G_{\mathrm{I}}}{G_{c\mathrm{I}}^{p}}\right)^{\alpha} + \left(\frac{G_{\mathrm{II}}}{G_{c\mathrm{II}}^{p}}\right)^{\alpha} + \left(\frac{G_{\mathrm{III}}}{G_{c\mathrm{III}}^{p}}\right)^{\alpha} = 1 \qquad (11)$$

wherein α governs the shape of the failure locus in the mixed-mode case.

4.3. Identification methodology

Several standard fracture mechanics tests have been chosen to identify the interface damage model. The tests, conducted in an earlier work [12], are the pure-mode I DCB (Double-Cantilever Beam) Test, the pure-mode II ENF (End-Notched Flexure) test, and two mixed-mode tests: the MMF (Mixed-Mode Flexure) test and the CLS (Cracked-Lap Shear) test. Each specimen tested was a $[(+\theta/-\theta)_{4s}/(-\theta/+\theta)_{4s}]$ laminate with $\theta = 0°$, 22.5° or 45°, according to the three kinds of $\pm\theta$ interlaminar interfaces investigated. An anti-adhesive film is inserted at the mid-plane in order to initiate cracking. The critical energy release rate G_c is classically obtained by deriving the compliance of the specimen, as is usually carried out within the concept of Linear Elastic Fracture Mechanics. Nevertheless, in the case of carbon-epoxy laminates, the main assumptions of LEFM are not always satisfied even in the simple case of a DCB specimen. This is true, in particular, in the case of: non-unidirectional stacking sequences [13] and R-curve-like phenomena [14]. In the former case, inner layer damage mechanisms may be activated; they lead to an apparent energy release rate greater than the local interfacial one. In this case, it's necessary to do a non-linear re-analysis for the evaluation of the damage state inside the layers [10]. Then the critical energy release rates at propagation are corrected with the part of the energy dissipated inside the layers.

From these corrected rates and from the relationships existing between Fracture Mechanics and Damage Mechanics (10-11), we deduce the values of the critical energies Y_c, the coupling coefficient γ_1 and parameter α. The identification results are reported in table 1. The interface parameters seem to be independent of θ for all $\pm\theta$ interfaces with $\theta \neq 0°$. Let us also note that the 0°/0° interface appears to be something artificial. However, such an "artificial" interface can be introduced, for example, to describe a delamination crack in a thick layer.

Interface	Y_c (N mm⁻¹)	γ_1	α
0°/0°	0.11 ± 0.01	0.37 ± 0.15	1.59
±22.5°	0.17 ± 0.01	0.36 ± 0.17	1.12
±45°	0.19 ± 0.01	0.44 ± 0.16	1.19

Table 1. Interface model parameters.

The other parameters associated with the delamination onset, namely Y_0, n and the interface stiffnesses, are indirectly identified by comparing the tension strain at onset of delamination between experimental results of Edge Delamination Tension tests and the results given by a specialized software dedicated to the free-edge effects [15, 16]. More information on this identification methodology can be found in reference [17].

5. Delamination test of a laminated holed-plate in tension

More reliable delamination tests would be those on laminated plates with a circular hole. In fact, for such specimens, the delamination crack initiation is reproducible and the growth is often stable within a certain range. Also of interest in this test is the valuable information it provides in terms of the shape and size of the delaminated area revealed by means of X-ray photography. The difficulty herein is that, due to the complexity of the state of stresses, the interpretation of this test requires complex computations. For such tests, a specialized software DSDM (Delamination Simulation by Damage Mechanics) has been previously developed [18] with the intralaminar damage being taken into account in the analysis.

In what follows, we present an initial comparison between the experimental observations in the M55J/M18 $[0_3/\pm45_2/90]_s$ laminate loaded in tension and the computation. Figure 3 shows the evolution of the X-revealed damage map near the hole for an increasing applied load. The first damage, appearing at 55 % of rupture (figure 3a), is transverse cracking in 90°-plies near the hole, and matrix cracking in the 0°-plies tangent at the hole and in the fibre direction called "splitting". Delamination only begins at about 80 % of rupture (figure 3b). Just before the rupture (figure 3c), the delaminated area is always found to be located between the splittings and developed in the 0°-direction with about two hole diameters in length. Micrographs were performed and show (figure 4) that the damage is well-developed in several ways: splittings, transverse cracking not only in the 90°-plies but also in the ±45°-plies, multiple delamination at the 0°/+45°, ±45° (the most damaged) and -45°/90° interfaces. From the computation, the splitting can be seen as a shear damage in the 0°-layer (see figure 5a). In fact, when the first 0°-fibres near the hole crack, the local load is transferred by shear in the matrix at the adjacent fibres. The delaminated area computed in the ±45° interface is shown in figure 5b as an example (the delaminated area corresponds to $d_3 = 1$). In the same manner, the other interfaces – except for the mid-plane – are found to be less delaminated.

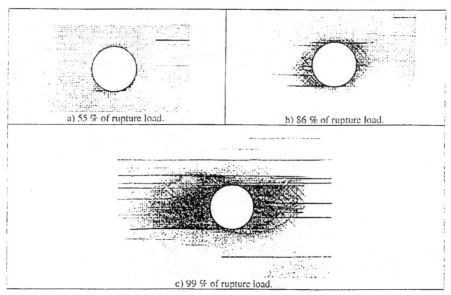

a) 55 % of rupture load.

b) 86 % of rupture load.

c) 99 % of rupture load.

Figure 3. X-ray photographs (×1.5) of a M55J/M18 $[0_3/\pm45_2/90]_s$ holed-specimen (rupture load 430 MPa).

Figure 4. Micrograph in a section tangent at the hole of a $[0_3/\pm45_2/90]_s$ specimen at 92 % of the rupture load

In order to achieve a good comparison, we have not, for the time being, used the Y_c parameter identified previously (table 1) and which have led to the prediction of no delamination. Different explanations for this feature are currently under consideration, and some progress is being hoped for shortly.

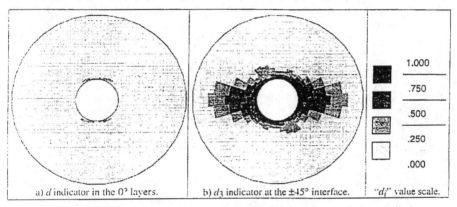

| a) d indicator in the $0°$ layers. | b) d_3 indicator at the $\pm45°$ interface. | "d_i" value scale. |

Figure 5. Damage maps computed in a $[0_3/\pm45_2/90]_s$ holed-specimen at the rupture load

6. Conclusions

This work is a first step towards a structural identification of delamination in composite laminates. In the damage mechanics approach proposed, the laminate behavior is described by means of a meso-model including the single-layer model and the interlaminar interface model.

The identification of the interface damage model is necessarily indirect, since we cannot measure interfacial characteristics from global data obtained at the laminate level. An initial set of parameters was then determined using standard initiation and propagation tests conducted on M55J/M18 material specimens. Lastly, a comparison between computation and holed-specimen tested in tension has revealed the value of using such an approach as a standard identification test. In order to perform further predictions of delamination, a global identification software, based on this latter kind of test, still has to be developed.

References

[1]   A.L. Highsmith, K.L.Reifsnider, STIFFNESS REDUCTION MECHANISM IN COMPOSITE MATERIAL - ASTM-STP 775, Damage in Composite Materials, A.S.T.M., pp. 103-117, 1982.
[2]   R. Talreja, TRANSVERSE CRACKING AND STIFFNESS REDUCTION IN COMPOSITE LAMINATES - Journal of Composite Materials, Vol. 19, pp. 355-375, 1985.

[3] *J.M. Whithney*, EXPERIMENTAL CHARACTERIZATION OF DELAMINATION FRACTURE - Interlaminar response of composite materials, Comp.Mat. Series, Vol. 5, Pagano N. J. Ed, pp. 111-239, 1989.

[4] *P. Ladevèze*, A DAMAGE COMPUTATIONAL METHOD FOR COMPOSITE STRUCTURES - J. Computer and Structure, Vol. 44 (1/2), pp. 79-87, 1992.

[5] *P. Ladevèze, E. Le Dantec*, DAMAGE MODELING OF THE ELEMENTARY PLY FOR LAMINATED COMPOSITES - Composite Science and Technology, Vol. 43, pp. 257-267, 1992.

[6] *O. Allix, P. Ladevèze, E. Vittecoq*, MODELLING AND IDENTIFICATION OF THE MECHANICAL BEHAVIOUR OF COMPOSITE LAMINATES IN COMPRESSION - Composite Science and Technology, Vol. 51, pp. 35-42, 1994.

[7] *O. Allix, P. Ladevèze*, INTERLAMINAR INTERFACE MODELLING FOR THE PREDICTION OF LAMINATE DELAMINATION - Composite Structures, Vol. 22, pp. 235-242, 1992.

[8] *O. Allix, P. Ladevèze*, DAMAGE MECHANICS OF INTERFACIAL MEDIA: BASIC ASPECTS, IDENTIFICATION AND APPLICATION TO DELAMINATION - Damage and Interfacial Debonding in Composites, Studies in applied Mechanics, Vol. 44, Eds Allen D. and Voyiadjis G., Elsevier, pp. 167-88, 1996.

[9] *O. Allix, A. Corigliano, P. Ladevèze*, DAMAGE ANALYSIS OF INTERLAMINAR FRACTURE SPECIMENS - Composite Sructures, Vol. 31, pp. 61-74, 1995.

[10] *O. Allix, P. Ladevèze, D. Lévêque, L. Perret*, IDENTIFICATION AND VALIDATION OF AN INTERFACE DAMAGE MODEL FOR DELAMINATION PREDICTION - Computational Plasticity, Eds Owen D.R.J., Oñate E. and Hinton E., Barcelone, pp. 1139-1147, 1997.

[11] *F.W. Crossman, A.S.D. Wang*, THE DEPENDENCE OF TRANSVERSE CRACKING AND DELAMINATION ON PLY THICKNESS IN GRAPHITE/EPOXY LAMINATES - Damage in Composite Materials, ASTM STP 775, K.L. Reifsnider Ed, pp. 118-139, 1982.

[12] *O. Allix, D. Lévêque, L. Perret*, INTERLAMINAR INTERFACE MODEL IDENTIFICATION AND FORECAST OF DELAMINATION IN COMPOSITE LAMINATES - Composite Science & Technology , Special Issue: 10th French National Colloquium on Composite Materials, to appear, 1998.

[13] *P. Robinson, D.Q. Song*, A MODIFIED DCB SPECIMEN FOR MODE I TESTING OF MULTIDIRECTIONAL LAMINATES - Composite Science and Technology, Vol. 26, pp. 1554-1577, 1992.

[14] *F.-X. De Charentenay, J.M. Harry, Y.J. Prel, M.L Benzeggagh*, CHARACTERIZING THE EFFECT OF DELAMINATION DEFECT BY MODE I DELAMINATION TEST - Effect of Defects in Composite Materials, ASTM STP 836, pp. 84-103, 1984.

[15] *L. Daudeville, P. Ladevèze*, A DAMAGE MECHANICS TOOL FOR LAMINATE DELAMINATION - Journal of Composite Structures, Vol. 25, pp. 547-555, 1993.

[16] *O. Allix, D. Lévêque, L. Perret*, ON THE IDENTIFICATION OF AN INTERFACE DAMAGE MODEL FOR THE PREDICTION OF DELAMINATION INITIATION AND GROWTH - First Int. Conf. on Damage and Failure of Interfaces, DFI-1, Sept. 21-24 1997, Vienna, to appear.

[17] *D. Lévêque*, ANALYSE DE LA TENUE AU DÉLAMINAGE DES COMPOSITES STRATIFIÉS: IDENTIFICATION D'UN MODÈLE D'INTERFACE INTERLAMINAIRE - PHD thesis, École Normale Supérieure de Cachan, January 5 1998.

[18] *O. Allix*, DAMAGE ANALYSIS OF DELAMINATION AROUND A HOLE - New Advances in Computational Structural Mechanics, P. Ladevèze and O. C. Zienkiewicz eds., Elsevier Science Publishers B.V., pp. 411-421, 1992.

Interaction Between Polymer-based Composite Friction Materials and Counterface

F. Abbasi[†], A. Shojaei Segherlou, G. A. Iranpoor
[†]Research Center of Iranianian Railways, Italia Crossing, Vesal St., Tehran, Iran.

and A. A. Katbab[‡]
[‡]Department of Polymer Engineering, Amirkabir University, Hafcz Avc., Tehran, Iran.

Abstract

with the purpose of improving performance of polymeric composite disc brake pad, and eliminating undesirable mechanical and thermal effects on opposing surface, we started wide investigations about friction materials for railways brake system. In this way, we firstly determined the most important characteristics of brake pads, especially compression modulus and thermal conductivity, which control interaction between the disc and pad, and we concluded that with controlling these parameters it could be avoided to undesirable effects on the disc.

Keywords: brake, disc pad, friction material, composite, thermal damage

INTRODUCTION

Polymeric composite friction materials are widely used as stator components of sliding friction couple of brake system for rail vehicles. For axle mounted discs these materials are called "brake pads", and for braking via contacting with thread of the wheels are called "brake shoes", and we are going to substitute the metallic brake shoes with polymeric composite ones.

During stopping, brake converts the kinetic energy of the moving vehicle into heat, absorbs the heat, and gradually dissipates it into the atmosphere. The brake is a sliding friction couple consisting of a rotor (disc) connected to the wheel and a stator on which the friction material is mounted (pad, lining, or block)[1].

The most widely used fillers in friction materials are reinforcing fibers such as asbestos, cotton, steel , and carbon fibers ; mineral particles such as barium sulfate, silica, and alumina ; metal particles such as copper, brass, zinc ; and solid lubricants, for example graphite and molybdenum disulfide. The matrix may be of metal, silicates, sulfides etc., or an organic polymers. The organic polymers are used in the greater proportion of materials [2]. Fillers are used to friction component as an auxiliary material for achieving required friction coefficient(μ) and wear properties. The various components of the material have to be present in the optimum proportion, for example, if there is too small a proportion of resin the material may be physically weak, if too great a proportion the friction coefficient may fall at high temperatures.

Depending on the interactions between filler particles and hard abrasive asperities, remarkably different wear features were observed for different sizes of the same filler [3]. When the filler is significantly smaller than the asperity spacing, no load supporting action is provided

provided and effects of cavitation-debonding at the filler matrix interfaces (damage zone model) become important. Increasing the adhesion at the filler-matrix interfaces significantly improves the wear resistance of elastomers owing to the suppression of filler pull out (medium particles) and debonding-cavitation at the interface (small particles)[4].

A brake system must provide high and stable friction, low wear and noise and vibration-free performance for reason of safety, comfort and durability. These performance characteristics can be influenced by the friction film (glaze) formed on the sliding surfaces of the brake rotor and pads. The used ingredients in friction material, not only has to have the specified μ and wear properties for the particular application, but also has to meet a number of other requirements, for example, it must not cause thermal damage of the opposing surface, or the wear the later unduly[5,6].

For eliminating undesirable effects on opposing surface such as thermal damage, hot spotting, and unduly wear. we started wide investigations about friction materials for railways brake system. In this way, we firstly determined the most important characteristics of brake pads, which control interaction between the disc and pad, and concluded that with controlling these parameters to undesirable effects on the disc could be prevented. In this work we studied the effect of the variation of compression modulus and thermal conductivity of composite materials on the opposing surface.

EXPERIMENTAL

For the compression modulus (E) the test specimens are tested on a Rockwell hardness machine[7] on which the ball is replaced by a cylindrical mandrel 13.3 mm in diameter. The minimum (preliminary) load is 10 kgf , and the maximum (total) load of 35 kgf. Before beginning the tests, the deflection of the test machine between minimum and maximum loads shall be recorded without the test specimen in piston. The following test sequence is then applied to each test specimen . The minimum load is applied and the dial is reset to zero (black graduated scale), with the test specimen placed in a central piston under the mandrel. The maximum load is then applied for 45 seconds, followed by a 10-second application of the minimum load. The dial is reset to zero, then the maximum load is applied again. The reading is then taken at the point when, after about 10 seconds, the needle shows a sudden deceleration. The deflection of the machine is then subtracted from this reading, and the net deflection expressed in graduations on the scale is multiplied by two to obtain the deflection Δh of the test specimen in μm.

The modulus E is calculated (average of results on the 6 test pieces) in accordance with the formula:

$$E = \frac{3.122 \times 10 \times h}{d^2 \times \Delta h} \qquad (1)$$

h=height of the specimen (mm)
d=diameter of the test specimen (mm)
Δh=deflection of test specimen (μm)

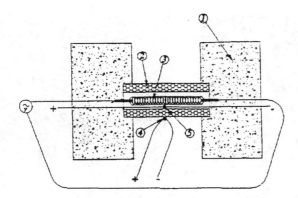

Fig.1. Schematic of the apparatus used for thermal conductivity measurement
(1) asbestos isolator; (2) sample; (3) heating element; (4),(5) thermocouples.

Thermal conductivity (K) measurement was carried out using cylindrical shells prepared from composite brake pads. Fig.1 shows the apparatus used for thermal conductivity measurement. The two ends of the cylinder were accurately isolated, so that to sure the heat flux could only be radial. Consequently, using two thermocouples, temperatures of the inside and outside of the shell were detected. Then K was calculated from the equation:

$$K = \frac{V\,I\,ln(R_o\,/\,R_i)}{2\pi L\,(T_i - T_o)} \qquad (2)$$

R_i , R_o = Inner and outer diameter of shell
T_i , T_o = Inner and outer temperature of shell
V , I = Electrical voltage and current

The structure of the different types of composite materials, particle size and distribution and qualitative elemental analysis, were studied using a scanning electron microscope (SEM Model S360 Cambridge).

The ideal means to test hot spotting is to assemble different types of composite brake pads on the axle mounted discs with complete vehicle, then performance of discs are investigated during the same periods of time.

RESULTS AND DISCUSSION

It has became obvious now, that the mechanical and tribological properties of composite fiction materials are related to the properties of their components[7], and the way they interact. The size and distribution of particles , particularly metallic particles, also determine the thermal conductivity of composite , which is the one of the most important factors of composite material. The SEM micrographs of figs. 2(a)-2(d) shows the size and distribution of metallic particles for different types of brake pads. It can be concluded that the pads made from sample No. 1 have high thermal conductivity, and those made from sample No. 4 have low thermal conductivity.

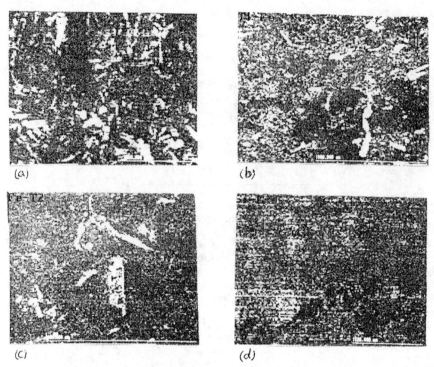

Fig. 2. Scanning electron micrographs of different samples: Fe EDS maps for (a) sample No.1; (b) sample No. 2; (c) sample No.3; (d) sample No. 4.

The results of compression modulus (obtained from the eq. 1) and thermal conductivity (obtained from eq. 2) measurmentes are shown in Table 1.

Table 1. The values of modulus and thermal conductivity of four different types of composite friction material.

Sample No.	1	2	3	4
Average E (N/mm^2)	271	4350	4115	9946
Average K (W/m^2 °C)	3.44	0.92	2.5	0.51

The typical photograph of unworked disc surface, before assembling on the axle , is shown in Fig. 3. Sample No. 1, as shown in Fig. 4, after three months (more than 70000 km) working under the service conditions ,did not have any undesirable effects on the disc. Fig. 5 is a photograph of critical hot spot sites, formed during the brake application of a few days duration with sample No. 4. The light elliptical spot is believed to have formed first [8], followed by the merged dark spot below it. Fig. 5(a) shows early stage in a martensite band during formation on a disc. With continued hot spotting, the original hot spot site expands to become a region of tempered martensite, laced with predominantly axial cracks . Such discs

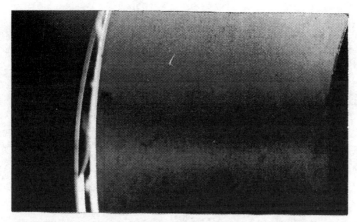

Fig. 3. The surface of unworked discs before contact with composite brake pads.

Fig. 4. Photograph of disc after three months (more than 70000 km) working, the used pads had the composition of the sample No. 1.

may be replaced well before fracture, since , as shown in Fig. 5(b) bending and hoop stresses from high brake loading cause these cracks to grow into structural cracks . In the case of the samples with medium E (samples No. 2 and 3), if the thermal conductivity of composite material were low, behavior of its opposing surface would be the same as the sample No. 4 [Fig. 6(a)]. But the behavior of the composite material with high thermal conductivity (sample No. 3) was very similar to the sample N0. 1[Fig. 6(b)].

The heat generated in the friction contact was conducted off mainly through the cast iron disc and through the elastomer composite. However, if stick-slip occurs, heat due to hysteresis will be created inside the elastomer material. This gave rise to higher temperatures in the elastomer materials [9,10]. The dissipation of the energy (frictional heat) put into the disc,

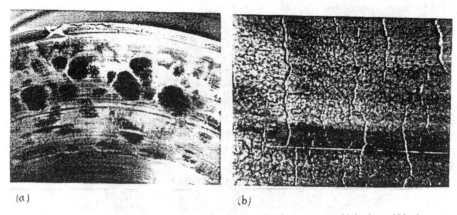

(a) (b)

Fig. 5. Photograph of disc after three days (nearly 3000 km) contact with brake pad No.4 :
(a)early stage in the formation of martensite band on the disc. cracking has started;
(b) conversion of martensite band to radial crack on disc after 7 days.

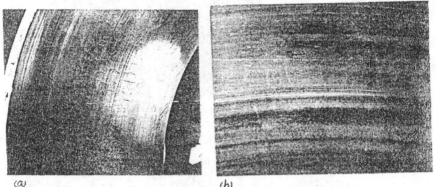

(a) (b)

Fig. 6. Photographs of disc after 30 days (nearly 30000 km) contact with
the brake pads (a) No.2; (b) No.4.

raises the disc temperature. But, especially in severe braking, this increase of temperature is very unequal in its distribution, because of the formation of hot spots in the disc. These hot spots are slightly elevated above the surrounding rolling surface. The input heat is concentrated on discs, and this results in high local temperatures, for example above 800°C. The elevated spots continually wear down and new ones form. The mechanism of the formation of hot spots are probably as follows: During heating, the compressive stress under the brake pad rises by the restrained thermal expansion up to yielding (in radial direction). For some reason this yielding is not always a general yielding, but it starts at separate points, resulting in the formation of "hills". These hills are fed by thermal expansion of the local metal and by pushing in of metal yielding from the surrounding area, which is in compression. The cause of the local (instead of general) yielding could be an unequal temperature distribution, an existing unequal stress field or another cause of instability. The high temperature of the hot spots certainly influences the thermal fatigue [11], and the formulator must therefor attempt to design the friction material so that the frictional heat generated is uniformly distributed over the working area of the friction material and so avoid local high temperatures. The real area of contact between friction material and counterface is much less than the nominal area of contact, and during braking the contact area tends to move cyclically over the latter. The true area of contact depends on the amount of elastic and plastic deformation produced on the roughness peaks of the contacting surfaces.

METAL ON METAL

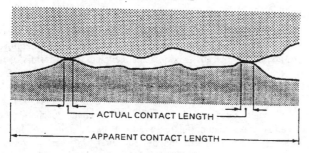

RUBBER ON METAL

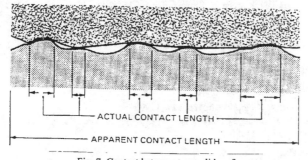

Fig. 7. Contact between two solid surfaces

The surface of the disc which is in contact with the pads will become flattened due to wear. The pads will deformed elastically and conform to some degree with the contacting surface of the disc. The less rigid the material of the pad, the greater will be the degree of conformity, as illustrated in Fig. 7.

Consequently, in order to minimize the temperature the contact area should be relatively large, (and therefore the material should be soft), and the locus of the contact area should traverse the wheel area of the pad once every revolution of the wheel and not just traverse parts of it, that is, the material should have a low modulus and be conformable to the opposing surface, since frictional heat could be dissipate mechanically and the temperature of the contact area can not increase more than a definite level. In the case of the friction material with medium modulus, high thermal conductivity helps to conduct a major part of heat generated through the pad and therefore increasing of contact temperature will be stopped.

CONCLUSION

The dissipation of the energy (frictional heat) put into the disc, raises the disc temperature. This increase of temperature is very unequal in its distribution. The input heat is concentrated on discs, and this results in high local temperatures. We must therefore attempt to design the friction material so that the frictional heat generated is uniformly distributed over the working area of the friction material and so avoid local high temperatures. In order to minimize the temperature the contact area should be relatively large, (and therefore the material should be soft), that is, the material should have a low modulus and be conformable to the opposing surface, since frictional heat could be dissipate mechanically and the temperature of the contact area can not increase more than a definite level. In the case of the friction material with modulus, more than that definite level, high thermal conductivity helps to conduct a major part of heat generated through the pad and therefore increasing of contact temperature will be stopped.

REFERENCES

[1] K. Othmer , Encyclopedia of Chem. Eng. vol.4, pp. 202-212.
[2] P. Walsh and R. T. Spurr, J. I. Mech. E. (Railway Div), C161, 137 (1979).
[3] Arnold C. M. Yang, Juan E. Ayala, Adam Bell and J. Camphell Scott, Wear, 146,349 (1991).
[4] A. J. Kinloch and R. J. Young, Fracture Behaviour of Polymers, chap. 10, p. 375 (1985).
[5] B. W. Klein, Bendix Technical Journal , 2(3), 109 (Automn 1969).
[6] S. K. Rhee, M. G. Jacko and P. H. S. Tsang, Wear, 146, 89(1991).
[7] UIC Code 541-3, OR 4th edition 1.7.1993.
[8] F.H. Kouta, A.A. Khatab and M.S. Salama, Modelling , Simulation and Control , B, AMSE Press , 14(2), 23 (1988).
[9] A. E. Ancerson and R. A. Knapp, Wear, 135, 319 (1990).
[10] A. J. Kinloch and R. J. Young, Fracture Behaviour of Polymers, chap. 10, p. 394 (1985).
[11] D. Fritzson, Wear, 139, 17 (1990).

The Micromechanics of Crack Bridging in Fibre-Reinforced Composites

Robert J. Young* and James A. Bennett*

*Materials Science Centre,UMIST/University of Manchester, Manchester M1 7HS, UK

Abstract

Raman spectroscopy has been used to study the effect of matrix cracking on the deformation micromechanics of high-performance fibres embedded in a brittle matrix. The ability of a fibre to bridge a matrix crack is considered in terms of a balance between the interfacial strength, the fibre strength and the fibre modulus. Compact tension epoxy resin specimens were prepared with HM Twaron aramid fibres embedded normal to the direction of the notch and then loaded critically such that a matrix crack ran across the fibres. The point-to-point distribution of strain was measured in the aramid fibres bridging the faces of a static crack, thus enabling the modes of interfacial failure to be monitored. The specimen was subsequently unloaded and reloaded incrementally. The behaviour of the fibres bridging the faces of the crack was observed, particularly with respect to fibre/matrix debonding and the occurrence of reverse sliding at the fibre matrix interface.

1. Introduction

The reinforcement of a brittle material with high-performance fibres can greatly improve its mechanical properties, resulting in a composite which can have high levels of stiffness and strength. Recent theoretical and experimental studies have shown that the effectiveness of this reinforcement is dependent on the efficiency of stress transfer between the matrix and the fibres [1]. It is widely acknowledged that the fibre/matrix interface influences the mechanical behaviour of fibre reinforced composites. Hence, the overall mechanical behaviour of the composite is critically dependent on, and can be controlled by varying, the properties of the fibre/matrix interface.

An important toughening mechanism in fibre-reinforced composites is crack-bridging by embedded fibres [2]. The main toughening mechanisms include interfacial debonding and frictional sliding associated with fibre pull-out. The overall toughness of the composite may be affected by the strength of the interface. In this study Raman spectroscopy [3] has been applied to the study of the deformation of high-modulus crack-bridging fibres. It has been possible to obtain fundamental information about the micromechanics of toughening by observing the debonding and frictional sliding mechanisms taking place at the fibre/matrix interface. Variations in the detailed state of deformation along the interfacial region by measuring the point-to-point variation of strain.

2. Experimental Procedure

The fibres used in this study were a development grade of high modulus (HM) Twaron aramid fibre [4], which had received no surface treatment or sizing, supplied by Akzo Nobel Research (Arnhem). The matrix material consisted of a two part, cold-curing epoxy resin consisting of 100 parts by weight of resin (LY5052) and 38 parts by weight of hardener (HY5052), both supplied by Ciba-Geigy. The resin has a Young's modulus, E_m, of about 3 GPa and a shear yield stress, τ_{my}, of approximately 43 MPa.

The CCT specimens were prepared using a window-card and a square 'picture-frame' mould [5]. Continuous Twaron fibres were mounted across the holes on the window-card; the mounting of the fibres was such that they were normal to the direction of the notch and parallel but isolated (~1 mm). The window-card was then fixed into the mould which was subsequently filled with resin. The mould was the left to cure at room temperature (22 ± 2°C) for a minimum of 7 days. The 3 mm thick resin was then machined into compact tension specimens (Figure 1(a)).

Raman spectra were obtained during deformation of the fibres in air while following the strain dependent shift of the 1610 cm⁻¹ aramid band as described elsewhere [3,4]. The CCT specimens were mounted on a 'Minimat' miniature materials straining rig (Polymer Laboratories, UK). The specimens where then placed on a straining rig and loaded until a crack propagated through the fibres such that the fibres remained intact and were observed to be bridging the crack (Figure 1(b)). The peak position of the strain-sensitive 1610 cm⁻¹ Raman band was determined for fibres embedded in the resin and such measurements were used to map the fibre strain profiles at various degrees of crack opening displacement. The straining rig was subsequently positioned on the stage of an Olympus BH-2 microscope connected to a Spex 1000M single monochromator via a spatial filter assembly. Spectra were obtained from the fibres both in air and within the resin as described elsewhere [5].

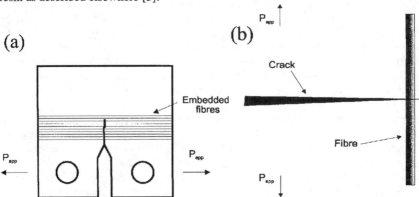

Figure 1 (a) Schematic representation of crack-bridging fibres in a CCT specimen. (b) Schematic illustration of a symmetrically-loaded matrix crack impinging on and advancing slightly beyond an embedded fibre.

3. Strength Criterion For Crack-Bridging

The tendency of a matrix crack to deflect along the fibre/matrix interface has been observed both theoretically and experimentally and has been shown to be dependent on the degree of mismatch between the elastic properties of the two materials [6]. More recently Bennett and Young [5,7] have observed fibre/crack interaction in aramid epoxy composites and noted for fibres ahead of the crack tip that the fibre strain distributions indicated interfacial failure when using untreated (weak interface) fibres in contrast to the behaviour of surface-activated (strong interface) fibres which remained fully bonded.

The general case of a semi-infinite matrix crack, normal to the direction of the fibre axis and loaded symmetrically about the crack-plane, which has impinged upon and advanced slightly beyond an embedded fibre is shown in Figure 1(b). Two specific initial outcomes are envisaged:-

a) Fibre matrix debonding occurs and a mode II crack deflects along the fibre/ matrix interface with no fibre fracture.

b) No debonding occurs and fracture of the fibre occurs when it is loaded beyond its ultimate tensile strength (σ_f^*).

Clearly, a combination of these two processes is possible when the crack has propagated sufficiently far beyond the fibre, such as initial fibre/matrix debonding followed by eventual fracture of the crack-bridging fibre.

Bennett and Young [5] have established a relationship between the strength of the fibre, σ_f^* and the interfacial shear stress, τ, such that if for a pull-out geometry,

$$\sigma_f^* \langle \frac{2\tau}{n} \tag{5}$$

then the *fibre will fracture*. The parameter n is given by [5]

$$n^2 = \frac{E_m}{E_f} \frac{1}{\ln(R/r)} \frac{1}{(1+v_m)} \tag{2}$$

where E_m is the matrix modulus, E_f is the fibre modulus, r is the fibre radius, R is the diameter of a cylinder of resin around the fibre and v_m is the matrix Poisson's ratio. Similarly if τ_{max} is the interfacial shear strength then if,

$$\tau_{max} \langle \frac{n\sigma_f}{2} \tag{6}$$

debonding of the fibre matrix interface results. This approach suggests that the outcome of a matrix crack impinging on an embedded fibre depends on the balance between the fibre strength and the shear strength of the interface. If τ_{max} is known then it is also possible to define a critical ratio of fibre strength to fibre modulus [5] to give,

$$\frac{\sigma_f^{*2}}{E_f} \rangle a\tau_{max}^2 \tag{7}$$

where a is constant and a function of the geometric and matrix of the material parameters only [5].

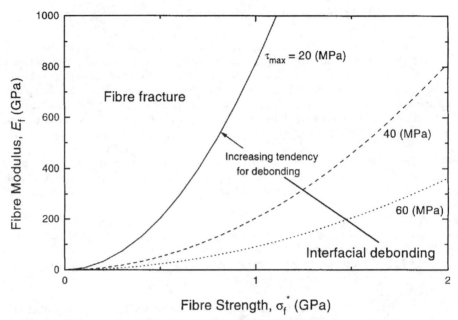

Figure 2 Dependence of fibre fracture/interfacial debonding upon interfacial shear strength, τ_{max}, fibre strength, σ_f^*, and fibre modulus E_f.

Figure 2 illustrates the correlation between τ_{max}, σ_f^* and E_f. Plotting σ_f^* against E_f for the above condition defines the boundary between interfacial debonding and fibre failure. For a fibre of fixed modulus and strength, as τ_{max} is decreased through a reduction in the strength of the interface, it can be seen that the balance between interfacial failure and fibre fracture shifts towards debonding.

For the fibre/matrix system used in this study it has been shown, using Raman measurements in conjunction with experiments on pull-out and fully-embedded geometries that for well bonded fibres the interface fails at τ_{max} = 43 MPa, which is approximately equal to the shear yield stress of the resin, τ_{my}. Figure 3 shows the plot of E_f versus σ_f^* with the boundary between fibre fracture and interfacial debonding drawn for a matrix system with a shear yield stress of 43 MPa. The data point for Twaron falls below the theoretical curve indicating when a matrix crack impinges on the fibres, the interface is likely to fail with intact fibres bridging the faces of the crack. Data points for other type of polymer fibres are also shown. It is assumed that the value of R/r is the remains constant for other fibres embedded in the same matrix. The data for PBO [8] and PE [9] fibres indicates that for a composite system with these fibres and τ_{max} = 43 MPa interfacial failure rather than fibre fracture is likely to occur. It has been found that HM Twaron [7], PBO [8] and PE [9] fibres the interface always debonds to leave a partially-debonded, intact fibre, bridging the faces of the crack.

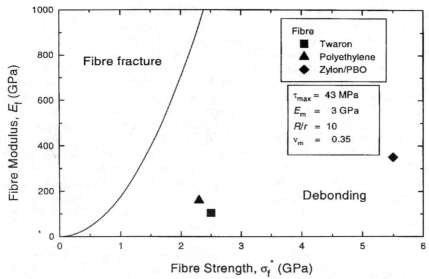

Figure 3 Dependence of fibre fracture/interfacial debonding upon interfacial shear
strength, τ_{max}, fibre strength, σ_f^*, and fibre modulus, E_f, for HM Twaron,
PE/Spectra 1000 and PBO/Zylon fibres [8,9].

4. Load/Unload/Reload in Crack-Bridging Fibres

Loading

Figure 4 shows a typical strain distribution, measured from the Raman band shift 5]
along an intact crack-bridging HM Twaron fibre. In this case the sample has been loaded
critically and a matrix-crack has been propagated beyond the fibre; the load the on
sample has been maintained such that the crack is held open at its maximum
displacement. Each point represents the axial fibre strain calculated form a single
Raman spectrum. The point $x = 0$ defines the location of the crack plane; x is positive to
the right and negative to the left of the crack. A Bannister partial-debonding model [20]
has been fitted to the experimental data and it can be seen that the data follows the
model closely. The profile (either side of the crack) can be divided clearly into two
regions; the first a higher-strain debonded region - marked by a linear profile in the
strain distribution - where the process of stress/strain transfer is being controlled by
friction and the strain decays linearly [5]. The second region, where the strain decays
exponentially, is typical of elastic stress transfer and is indicative of a bonded
fibre/matrix interface.

Unloading

Figure 5 shows the strain profiles of a crack-bridging fibre being unloaded. It can be
clearly seen that the micromechanics of unloading resemble the pattern envisaged by
Marshall [11]. Relaxation of the load on the sample allows the crack to close and

reverse sliding to take place in the debonded region. As the sample is unloaded strain on the fibre is relieved and reverse sliding appears to propagate unevenly and gradually in a 'stick-slip' fashion from the crack faces towards the tip of the debond. This is particularly evident to the left hand side of the crack where there is a 'sticking-point' at $x = -0.2$ mm. The 'sticking point' is consistently present at each level of crack closure. As unloading and reverse sliding progress, the strain at the point of debonded/bonded transition relaxes. It can be clearly seen at each level of crack closure the value of fibre strain at the transition point decreases, suggesting that a process of stress relaxation is occurring. The values of 0.92 % to the left- and 0.75 % strain to the right-hand side of the crack-plane relax to approximately 0.68 % and 0.62 % respectively, when the crack is fully closed. When the crack is fully closed there is evidently fibre compression over the distance $x = \pm 0.15$ mm either side of the crack plane, which may be due to frictional resistance towards fibre reseating of the segment of the fibre bridging the crack. In fact there must be residual compression in the region of the fibre either side of the crack plane to compensate for the residual tension around the debonded/bonded transition points.

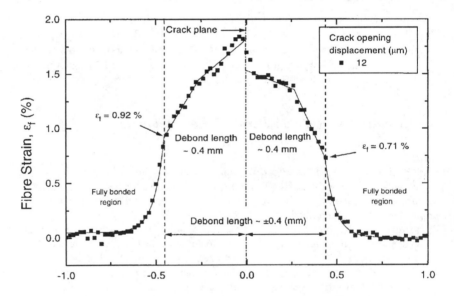

Figure 4 Distribution of fibre strain, ε_f, along an unbroken, loaded, crack-bridging HM Twaron fibre.

Reloading

The strain profile of a fibre being reloaded is shown in Figure 6; considering the data to the left hand side of the crack, it can be clearly seen that when the crack is opened to 4 μm strain builds up from the region immediately adjacent to the crack-plane. Frictional sliding takes place over a distance of ~0.2 mm, but from this point onwards the fibre

remains pinned by friction. Further opening of the crack to 8 - 10 μm permits the strain
to be transferred further along the debonded region. When the fibre is fully reloaded the
strain is taken up along the whole of the debonded region. However the debond front on
the right-hand side has advanced by ~ 0.2 mm.

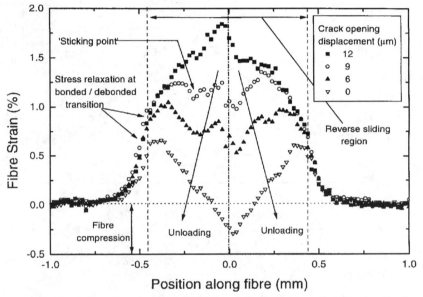

Figure 5 Strain profiles showing the relaxation of fibre strain, ε_f, during unloading/crack
closure.

5. Conclusions

Raman spectroscopy has been shown to be a powerful tool in the observation of
composite micromechanics. Predictions based on a balance between fibre strength,
interfacial shear strength and fibre modulus, indicated that an incident crack was likely
to cause failure of the fibre matrix/interface, but would not fracture HM Twaron fibres.
Experimental data confirmed the prediction showing HM Twaron fibres to bridge the
matrix crack, with interfacial debonding propagating either side of the crack plane,
resulting in a situation which could be likened to two-way pull-out of the fibre.

Crack closure resulted in reverse sliding of the fibre in the debonded region. The sliding
was shown to be uneven and followed a 'stick-slip' type pattern with sticking points
occurring consistently at each level of crack closure. On re-opening of the crack the
fibre strain built up inwards from the crack plane, at each level of crack opening the
strain was transferred along the fibre until full reloading was achieved. The
experimental data from unloading and reloading was qualitatively similar to the model
proposed by Marshall [11], and supports its validity.

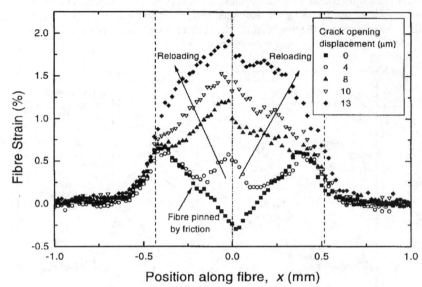

Position along fibre, x (mm)

Figure 6 Strain profiles showing build up of fibre strain, ε_f, during reloading/crack opening.

6. References

[1] Drzal, L. T., '*Advances in Polymer Science 75*-Epoxy Resins and Composites II' edited by K. Dusek, Springer, Berlin, Heidelberg; (1986) 3.

[2] Andrews, M. C., Day, R. J., Hu, X. & Young, R. J., Deformation micromechanics in high-modulus fibres and composites, *Comp. Sci. & Technol.,* **48** (1993) 255-261.

[3] Chawla, K K. *Ceramic Matrix Composites*, Chapman and Hall, London,1993.

[4] Andrews, M. C., Young, R. J. & Mahy, J., Interfacial failure mechanisms in aramid/epoxy model composites, *Composite Interfaces*, **2** (1994) 433-456.

[5] Bennett, J. A. and Young R. J. The effect of fibre-matrix adhesion upon crack bridging in fibre-reinforced composites, *Composites A*, in press.

[6] He, M. Y., and Hutchinson, J. W., Crack deflection at an interface between dissimilar elastic materials, *Int. J. Solids and Structures*, **9** (1989) 1053-1067.

[7] Bennett, J. A. and Young, R. J., Micromechanical aspects of fibre/crack interaction in an aramid/epoxy composite, *Composites Science and Technology*, **57** (1997) 945.

[8] Bennett, J. A., *Fibre/crack interactions in polymer matrix model composites*, PhD Thesis, UMIST, 1997.

[9] Gonzalez-Chi, P. I., *Deformation micromechanics in polyethylene/epoxy fibre reinforced composites*, PhD Thesis, UMIST, 1997.

[10]Bannister, D.J., Andrews, M.C., Cervenka, A.J., Young, R.J., Analysis of the single fibre pull-out test by means of Raman spectroscopy: Part II. Micromechanics of deformation for an aramid/epoxy system, *Comp. Sci. & Technol.*, **53** (1995) 411-4.

[11]Marshall, D. B., Analysis of fiber debonding and sliding experiments in brittle matrix composites, *Acta Metall. Mater.*, **40** (1992) 427-441.

Composite toughness evaluation through the concept of a fiber-bridged damage zone.

J. Lindhagen*, L.A. Berglund*

*Division of Polymer Engineering, Luleå University of Technology,
 SE-971 87 Luleå, Sweden

Abstract

The toughnesses of three different types of glass mat reinforced plastic were investigated. Sensitivity of tensile specimens to center hole notches was used as the measure of material toughness. Two specimen sizes were used with the objective of comparing notch sensitivities for different hole sizes and to compare tensile properties measured on different specimen width. All three materials exhibited high toughness. The wider specimens showed reduced scatter in measured properties. The critical opening displacement of the damage zone or crack at the onset of failure was measured using electronic speckle photography, and a toughness parameter was calculated based on fibre bridging theory. The values of the toughness parameter did not reflect the measured toughnesses, but were able correctly to predict the onset of notch sensitivity.

1. Introduction

Glass mat reinforced composites are gaining widespread use, notably in automotive applications. They are cheap, light, have reasonably good mechanical properties and are insensitive to corrosion. Manufacture of large components is readily carried out at short cycle-times by press-forming. For materials with thermoplastic matrix the possibility of recycling into new components is also an advantage.
Notch sensitivity is an important property since holes for fasteners etc. are present in most components. It also reflects the material toughness, which is important for example in designing for energy uptake at crashes. In previous studies GMT materials with PP matrix were shown to exhibit low notch sensitivities, i.e. high toughnesses, at room temperature and in cold conditions [1,2]. A number of common failure criteria were applied in order to compare the predictions of notch sensitivity with the experimental results. Micromechanisms for toughening were also investigated and differences in toughness between different reinforcement structures were explained [3]. It was shown how the damage zone size of miniature specimens depends on fibre architecture.
The common failure criteria, such as the Whitney-Nuismer average and point stress criteria, are phenomenological and have the disadvantage that they are unable to explain the physical causes of the observed behavior. Also, the toughness-controlling material pa-

rameter can not be independently determined. A model to predict toughness based only on microstructural parameters and fiber and matrix properties would be a useful tool to understand toughening mechanisms and be able to design a material with a desired toughness. A possible route to achieve this goal, based on fiber bridging theory, was reviewed by Bao and Suo [4] and further discussed by Suo with coworkers [5]. A model is constructed where a damage zone is formed in the specimen close to the notch, see Figure 1. The damage zone is modeled as springs acting between surfaces of undamaged material. The interaction between the springs and the base material can then be treated with continuum mechanics.

The stress-displacement behaviour of the springs is a critical property in the model and is called the bridging law. An arbitrary bridging law is also depicted in Figure 1. From the bridging law and material properties a material parameter, a critical length d, can be calculated:

$$d = \frac{\Gamma E}{\sigma_0^2} \tag{1}$$

where Γ is the area under the bridging law stress-strain curve, E is the composite Young's modulus, and σ_0 is the maximum stress that the springs in the model can withstand. The maximum displacement of the springs is denoted by ∂_0. For some simple bridging laws such as when the stress increases linearly from zero at zero displacement to σ_0 at maximum displacement (linear bridging law) or when the stress is constant up to ∂_0, Γ is readily calculated and Eq. 1 simplifies to

$$d = k \frac{\partial_0 E}{\sigma_0} \tag{2}$$

where k is 1/2 or 1. In the present work a rectilinear bridging law is assumed, and so $k=1$.

The parameter d is a measure of the toughness of the material. When d is compared to the notch size a, brittle behaviour will be observed when $a>>d$ and ductile behaviour when $d>>a$. Therefore the notch sensitivity of any given material will depend on the notch size, and a material with high toughness will be able to withstand large notches without suffering severe strength reduction and vice versa. Normally the transition zone from notch insensitivity to notch sensitivity is expected to appear when $a\approx d$.

In order to understand the toughening mechanisms of a composite and to be able to optimise a composite for toughness, the dependence of d on microstructural parameters needs to be known. In order to calculate d, σ_0 is usually taken as the composite's strength (and in this work the composite unnotched strength will also be denoted in this way). Thus for a composite with a given stiffness and strength, the critical parameter controlling d is ∂_0. In the model, the maximum displacement ∂_0 of the springs is the displacement where any given element ceases to carry load. Marshall, Cox and Evans showed how ∂_o depends on constituent properties for an idealized unidirectional ceramic-matrix composite [6]. If the model is compared to a real material this displacement could be identified with the opening displacement of the damage zone formed close to the notch, at the moment when the zone ceases to carry load at the opening.

The objective of the present work is to evaluate whether ∂_o is a material parameter that can be used as an independent indicator of material toughness. The critical opening displacements of the damage zone or crack formed at the notch edge were measured for three different random glass mat reinforced polymeric composites. Furthermore, values of the proposed toughness parameter d were calculated from ∂_o and tensile properties and compared with the observed experimental notch sensitivities, measured with different notch sizes. A secondary objective was to compare tensile properties of wider specimens with those of more narrow specimens, since it has been shown that wider specimens should exhibit reduced scatter [7]. The use of different specimen widths also made possible the use of a wider range of notch sizes.

2. Experimental

Notch sensitivities in tensile specimens of three different materials were investigated. The materials were random glass mat reinforced composites: 1) a sheet moulding compound (SMC), which has a high filler content and a brittle thermosetting polyester matrix; 2) a glass mat thermoplastic (GMT) with a relatively stiff thermoplastic polyester (PET) matrix; and 3) a GMT with low modulus, tough polypropylene matrix.

Plates 700x1000 mm in size were prepared from the raw materials by press forming, a flow moulding process which induces flow patterns in the reinforcement. However, the flow patterns are difficult to predict or even to determine afterwards and therefore the materials are treated as having random fibre orientation. The plates were between 2.8 and 3.2 mm in thickness. Straight-sided tensile specimens of two sizes were cut from the plates: 40x300 mm and 80x600 mm. Circular center hole notches with diameters of 10 and 15 mm for the narrow specimens and 30 mm for the wide specimens were machined in the notched specimens. For the SMC and PP-GMT five specimens of each kind were tested for the narrow specimens and three of each kind for the wide specimens. For the PET-GMT the number of specimens was doubled. Tensile tests were performed in displacement-controlled mode in a servohydraulic tensile test rig. The stroke rate was 2 mm/min and the extensometer gage length was 50 mm.

Measurements by electronic speckle photography (ESP) were performed on the smaller specimens. An irregular pattern of drops of paint of contrasting colour was sprayed onto the specimen surfaces. During tensile testing the region close to the hole was filmed with an S-VHS video camera (25 pictures per second) equipped with special optics. Displacements at different locations in the region close to the hole could then be calculated from the recorded images using special code [8]. Of special interest in the present investigation was the critical opening displacement of the damage zone, which is formed at the notch edge prior to global specimen failure. As "critical" was chosen the moment of maximum global load carried by the specimen. On the assumption of a rectilinear or increasing bridging law this is likely to be correct; also from a practical viewpoint it is advantageous since it provides a well-defined point at which to make the evaluation. Fibre contents of the specimens were determined by burning off the matrix in an oven at 500°C for four hours.

3. Results and discussion

In Table 1 the tensile properties and notched strengths of the investigated materials are presented along with the measured values of ∂_0 and calculated values of d.

The SMC exhibited a high stiffness which is caused by its stiff thermosetting matrix and by the high total filler content, where glass fibers constitute only a smaller fraction. The stiffness difference between the two types of GMT is caused mainly by the difference in matrix stiffness. Both PP and PET are semicrystalline thermoplastics but PET has a stiffer molecular backbone, stronger secondary bonds and most importantly is in the glassy state at room temperature. It is also likely to have a higher fibre-matrix interface strength. The strain to failure was highest for the PP-GMT, followed by the PET-GMT and the SMC. This seems inuitively logical as a stiffer matrix is expected to be more brittle. However, there are several indications in the literature that the failure strain of glass mat reinforced composites is mainly controlled by the properties and architecture of the reinforcing fibers. Thus the high filler content of the SMC may play a role. The material (unnotched) strength can be regarded as the product of stiffness and strain to failure, and consequently the PET-GMT has the highest strength.

The notch sensitivity is the ratio of notched strength, calculated on the net section, to unnotched strength. A high relative notched strength equals a low notch sensitivity and reflects a high toughness. High toughnesses were observed for the three materials. For notch diameters 10 and 15 mm all three materials showed to be practically notch insensitive, with some reservation for effects caused by statistical deviation. With wide specimens and 30 mm notch diameter some reduction in notched strength appears. The notch sensitivities exhibited with the 30 mm holes were of the same order for the three materials.

The scatter was lower for the wider specimens, although fewer were tested (see Experimental). Some of the observations in Table 1, such as the difference in stiffness between narrow and wide specimens of SMC, the difference in unnotched strength for diffferent specimen sizes of PP-GMT, and the "negative notch sensitivity" of SMC 40/15, are likely to be statistical artefacts due to the large variance in tensile properties of the present type of materials.

In Figure 2 two plots of damage zone displacements are shown. They are from the same specimen of PET-GMT at two different stages of damage. The plots are of the zone at the notch edge, the notch is to the lower right in the plots. The discontinuities in the plot are caused by the apperance of a crack, which causes loss of correlation. Displacements in the specimen length direction of each point on the specimen, relative to the unloaded state, are plotted on the z-axis. The z-axis scale is in pixels (1 pixel = 0.016 mm). As can be seen in Table 1, the observed values of ∂_0 showed a high scatter, but nevertheless ∂_0 appears to be a material parameter. The dependence of ∂_0 on microstructure and constituent properties remains to be determined but is likely to be complex.

The measured values of ∂_0 were used to calculate values of the proposed toughness parameter d. These values turned out not to correlate with the observed toughnesses although it must be pointed out that the toughness differences in this investigation are too small to be statistically significant. However, the values of d were of the same order as the notch size a for the wide specimens (note that a equals half the notch diameter, i.e. $a = 15$ mm for the wide specimens). Since a transition from ductile to brittle behaviour is expected when the ratio $a/d \approx 1$ (see Introduction), this serves to explain the increased notch sensitivity for the larger notch diameter.

4. Conclusions

Notch sensitivities of tensile specimens with center hole notches were investigated for three different types of glass mat reinforced plastics. The ratio of notched to unnotched strength was high for all three materials, which indicates high toughnesses. Almost no notch sensitivity was exhibited for notch diameters of 10 and 15 mm, whereas 30 mm notches caused some strength reduction. Critical opening displacements of the damage zones at the notch edge were measured using electronic speckle photography. They were shown to be material dependent. A proposed toughness parameter d was calculated according to fibre bridging theory for the three materials. The values of d correctly predicted the appearance of a notch ductile-to-brittle transition in the region of 20-70 mm notch diameter.

References

[1] J. Lindhagen, L.A. Berglund, NOTCH SENSITIVITY AND DAMAGE MECHANISMS OF GLASS MAT REINFORCED POLYPROPYLENE - Polymer Composites, February 1997, vol. 18, no. 1, pp. 40-47.
[2] J. Lindhagen, L.A. Berglund, LOW TEMPERATURE STRENGTH AND NOTCH SENSITIVITY OF GLASS MAT POLYPROPYLENE - Journal of Cold Regions Engineering, September 1997, vol. 11, no. 3, pp. 181-197.
[3] J. Lindhagen, L.A. Berglund, MICROSCOPICAL DAMAGE MECHANISMS IN GLASS FIBRE REINFORCED POLYPROPYLENE - accepted for publication in Journal of Applied Polymer Science.
[4] G. Bao, Z. Suo, REMARKS ON CRACK-BRIDGING CONCEPTS - Applied Mechanics Review, August 1992, vol. 45, no. 8, pp. 355-366.
[5] Z. Suo, S. Ho, X. Gong, NOTCH DUCTILE-TO-BRITTLE TRANSITION DUE TO LOCALIZED INELASTIC BAND - Journal of Engineering Materials and Technology, July 1993, vol. 115, pp. 319-326.

[6] D.B. Marshall, B.N. Cox, A.G. Evans, THE MECHANICS OF MATRIX CRACKING IN BRITTLE-MATRIX FIBER COMPOSITES - Acta Metallurgica, 1985, vol. 33, no. 11, pp. 2013-2021.

[7] J. Varna, M.L. Ericson, L.A. Berglund, SPECIMEN SIZE EFFECTS ON MODULUS OF GMT AND OTHER INHOMOGENEOUS COMPOSITES - Journal of Thermoplastic Composite Materials, 1992, vol. 5, p. 105.

[8] M. Sjödahl, ELECTRONIC SPECKLE PHOTOGRAPHY: INCREASED ACCURACY BY NONINTEGRAL PIXEL SHIFTING - Applied Optics, Oct. 1994, vol. 33, no. 28, pp 6667-6673.

Table 1. Mechanical properties, notch sensitivities, measured critical displacements and the toughness parameter d. Mean values ± standard deviations [1].

Property	Specimen width [mm]	Notch diameter [mm]	Material SMC	PET-GMT	PP-GMT
Vf [%]			49.7 ± 0.93	16.6 ± 1.63	18.9 ± 0.63
E [GPa]	40		10.6 ± 1.1	10.1 ± 0.9	5.76 ± 0.79
	80		13.4 ± 0.7	10.5 ± 1.2	5.18 ± 0.44
ε^{*} [%]	40		1.23 ± 0.33	1.32 ± 0.24	1.84 ± 0.33
	80		1.03 ± 0.32	1.35 ± 0.21	1.76 ± 0.19
σ_0 [MPa]	40		70.4 ± 8.0	88.3 ± 6.3	75.7 ± 6.1
	80		75.8 ± 6.7	88.6 ± 3.8	65.9 ± 1.4
$\sigma_{n,}$ [MPa]	40	10	63.0 ± 16.2	85.11 ± 8.3	73.8 ± 6.0
	40	15	72.7 ± 10.9	82.1 ± 11.6	75.6 ± 1.8
	80	30	59.8 ± 7.4	72.4 ± 6.1	55.2 ± 5.8
σ_n/σ_0	40	10	0.90	0.96	0.98
	40	15	1.03	0.93	1.00
	80	30	0.79	0.82	0.84
∂_0 [mm]	40	10	0.182±0.096	0.157±0.088	0.149±0.058
	40	15	0.180±0.080	0.153±0.065	0.135±0.065
	overall mean		0.181±0.079	0.155±0.074	0.142±0.058
d [mm] $(=\partial_0 E/\sigma_0)$	40^2		27.1	17.9	10.8
	80^2		32.1	18.3	11.1

[1] The number of specimens tested varies from 3 to 10.
[2] Calculated using the overall mean value of ∂_0, and the tensile properties for each specimen width.

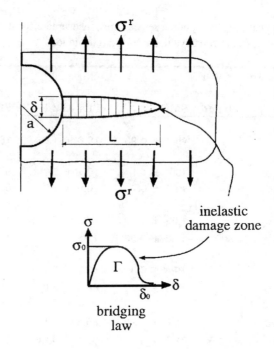

Figure 1. A model of a damage zone close to a center hole notch, and an example of an arbitrary bridging law.

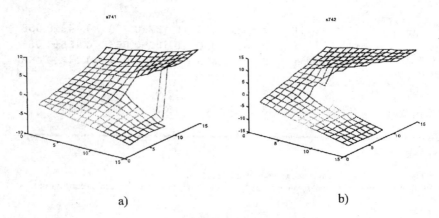

Figure 2. Two plots of the displacements in the damage zone at the notch edge in a specimen of PET-GMT. a) At maximum global load, b) just before final failure.

Predictive design and modelling failure problems in fibre composites

Peter WR Beaumont* and S. Mark Spearing**

*Cambridge University Engineering Depart., Trumpington St, Cambridge CB2 1PQ, UK
**Depart. of Aeronautics and Astronautics, M. I. T., Cambridge, MA 02139, USA.

Abstract

There exists a gap in our knowledge of the differences in behaviour between material and structure that can be explained in terms of a size of scale, microscopic vs macroscopic vs macroshape, and the changing nature of problems of failure as this sizescale moves from laminate to test coupon to component. This sizescale which spans several orders of magnitude provides a framework for understanding the differences in failure behaviour of material on the one hand and largescale structure on the other. A proposal is made for designing and certifying aerospace structures of composite laminates based on sound fundamental understanding of the mechanisms of damage accumulation in the material and structure across these orders of magnitude of size. Furthermore, it is important for the "design authority" to acquire in a timely manner the necessary "know-how" and procedure at an early stage of the development process for dealing with the identification of potential failure sites and failure mechanisms from the micron to the metre level of size.

1. Introduction

Frequently, the composites industry experiences hardware development problems due to design oversights that lead to matrix-dominated load paths and to the formation of a multitude of matrix cracks, splits and delaminations. Consequently, by far the largest number of development and in-service problems with composite hardware is associated with interlaminar shear and out-of-plane tension. This is of major concern. One approach is to assess in sequence the effect of cyclic stress on the residual strength and life-time of a test coupon, substructural element, component and finally largescale structure by a progression of experiments. This process of development testing has been termed the *"building block"* approach which is followed during the design and analysis phase wherein critical areas of the structure are selected for test verification. Any critical strength features are subsequently identified and isolated in the form of small test articles of progressively increasing complexity and size.

The "building block" methodology is based on transferring empirical information and relating experimental data from one point on the sizescale to another. Empirical methods, however, do have serious limitations and shortcomings. While a simple empirical law like Hooke's law may *describe* an elastic material's response to a tensile stress at room temperature, it provides no guidance in *predicting*, for example, the material's behaviour under cyclic stress at elevated temperature simply because empirical information contains no knowledge about mechanisms of microstructural change.

An alternative innovative way foward is to begin the design process at the constituent level with physical model-informed selection of fiber and it's surface treatment, matrix and it's processing conditions, and architecture (geometry) of laminate construction [1-3]. As the shift of information bothways along the sizescale proceeds, the design process at each level of size has to include the identification of the principal failure mechanisms. Making links or connections between our understanding of material and component behaviour at the various scale of sizes relies critically on a knowledge of microscopic and macroscopic structural changes over the entire span of size.

Differences in the nature of behaviour as the "size" of the problem varies, has resulted in the opening of a "gap" in our understanding on the sizescale where "design" normally changes hands from being a materials design issue to being a structural design issue. In this regime no one has really taken responsibility for "bridging" this void. If we take the materials side as going as far as establishing and understanding all sorts of material behavioural characteristics at the micro level of size, to do with fibres, matrices and their interaction, we might say that we characterise the properties of the material at that particular scale by reference to the fiber direction, and there is no real consideration of the effects of geometry or "architecture" of the laminate. Initial material tests are usually of unidirectional specimens, and a notch is considered as a geometrical abberation. Conversely, at the structural design size, we tend to look at the overall geometry of a component but we tend to look on the material properties as being set (in a geometric sense) at a global level. "Bridging" this "gap" appears to be a big a step, and is a key source of difficulty.

What we fail to look at is the geometry *within* the component, and it is by looking at this we gain an important link or bridge between both sides. If we say that a component carries various loads from loading points to attachment points, we can also say that this could be expressed as a stress field, and that this stress field would have some geometric pattern to it, generally a curving geometric pattern. Now if we look at this pattern and lay our fibers to follow accordingly by using the directional properties of the fibers to offer a vectorial solution to a vectorial problem, then at each micro point throughout the component, we have something approximating to the proper directional reference frame we gave ourselves at material level, i.e., the stresses are well referred to the fiber directions, and we can use our sequence of sizescales well. But what is normally done is to leave out this piece and define a global set of fiber orientations, which are then in general never in the right micro directions when we look at a particular point within the structure. This then wrecks the nice logic of the sizescale sequence. We know this because it is easy to think in terms of big sheets of fiber but we also know that people are looking at individual tow placement by embroidery techniques, and so on, to answer exactly the internal architecture question (Jim Platt, private communication, 1997).

2. A Brief Overview of Certification Methodologies

A basic element of any certification or qualification procedure is design development and verification testing which assesses and validates the design of critical hardware features [4]. All to frequently, the effect of local details are identified very late in the design cycle and unfortunate experiences of this kind contribute in large part to early over optimism in the material's initial performance figures and design rating. Furthermore, any structural analysis performed at this stage has rarely been made in sufficient detail to adequately evaluate the effect of these interlaminar stresses on potential failure modes and margins-of-safety.

Broadly speaking, three certification approaches have been considered at one time or another and all need to be underpinned by research in order to enhance the status of composite structure design, development and certification of flight hardware: (1) Damage Tolerance; (2) Safe Life/Reliability; (3) Wearout Model. To implement either one or more of these approaches demands that analysis procedures, material characterisation and component testing should be addressed in detail in order to provide an assessment and awareness of their applicability to the development and certification of composite flight hardware. For the purposes of this paper, the first approach will be briefly summarised only.

Damage Tolerance Methodology

The damage tolerance philosophy assumes that the largest flaw exists at the most critical location in the structure and that structural integrity can be maintained even though the flaw may grow until detected by periodic inspection. In this approach, "damage tolerance capability" which covers both flaw growth potential and residual strength of the damaged part is verified by analysis and test. The assessment of each component would include issues of high strain, strain concentrations, minimum margin of safety details, major load paths, damage prone areas, and special inspection areas. That part of the structure selected as critical would be included in the experimental and test validation of the substantiation procedures for damage tolerance. Those structural areas identified as critical after the analytical and experimental screening would then form the basis for the subcomponent and full-scale component level of evaluation, whichever is applicable, and would be developed or be available.

Potential damage initiation sites are free edges, bolt holes, ply terminations, and, with respect to repair of structures, at a composite patch interface. Hence, a technique for the evaluation of the conditions for growth/no growth of delaminations is an essential tool for the determination of the damage tolerance of composite structures. Numerical methods are generally available through the use of finite element analysis and the crack closure integral technique from fracture mechanics.

Prerequisites for an evaluation include: a structural analysis to indicate locations where critical interlaminar streses exist; empirical assessment of critical interlaminar strain energy release rates for failure modes I and II, combined with a subcritical fatigue crack growth law for both modes of failure. Whilst a fracture mechanics approach is not considered sufficiently mature at present to warrant a recommendation for general application to the certification of developmental composite hardware, nevertheless, we will consider the merits of fracture and damage mechanics principles.

3. A Brief Consideration of Fracture and Damage Mechanics

In most advanced composites, the earliest signs of damage comprises an array of matrix cracks in the individual transverse plies, often referred to as first ply failure. Following transverse ply cracking, delamination cracks accompanied by splits frequently form, including fiber breaks, precipitated by these transverse ply cracks [5-8]. Individual cracks of these kinds can be treated using the principles of fracture mechanics; collectively, their effects on the properties of the material can be quantified using relationships based on damage mechanics ideas.

Damage and fracture mechanics principles are based on quantitative relationships between microscopic and macrosopic parameters that attempt to "bridge" the gap that has opened up on the sizescale between classical materials science that focuses, for example, on single fiber reinforcement models at the microstructural level, and structural engineering and hardware development at the one metre level and above. It is precisely at this gap that "design" normally changes hands from being a materials design issue to being a structural one for which, as we said earlier, neither group has taken responsibility. At the micron level, basic research seeks a detailed understanding of a problem through elegant analysis or experimention with conspicuous absence of immediate need for solution or time constraints. Solutions to applied problems at the other end of the sizescale need not necessarily be complete and in fact a complete understanding of the problem is rarely required. The solutions require synthesis, optimisation, approximation and "feel", and generally have a time constraint.

In the stictest sense, "damage mechanics" refers to the *physical mechanisms* of multiple cracking by which damage forms and propagates in the material. The concept of continuum damage mechanics is a method of treating a "whole population of microscopic defects". Very few systematic attempts have been made to apply damage mechanics to advanced composites. One example goes back fifteen years to the pioneering work of Poursartip, Beaumont and Ashby [5]. In this and other reported work on fatigue damage accumulation [6, 8], a fatigue damage mechanics approach is developed which is based on the assumption of a relationship between the extent of damage accumulation and changes in stiffness (or strength) of a family of carbon fiber-epoxy laminates. It is noteworthy that in these studies, delamination induced by in-plane tension loading is the principal damage mechanism, translaminar cracking, and decohesion that occurs behind the advancing delamination crack front.

Turning briefly at this particular point to a related and also key aspect of design is the manner in which a material under stress reacts to the presence of stress concentrators such as holes or notches. At General Dynamics, more than twentyfive years ago, a paper was published on the macroscopic fracture mechanics and the fracture stress of advanced composites containing small holes or slits [9]. Numerous adaptations and variations embracing similar concepts have appeared since then [examples include 10-12].

Whilst these macroscopic empirical models offer the potential of working models they are considerably restricted in their application; for example, they do not provide insight into the mechanisms or causes of damage or the mechanics of ultimate failure. Neither do they predict a fracture stress for different intrinsic material variables or extrinsic (environmental and thermal) conditions other than those of the test; or provide insight into the connection between data at the laminate and microstructural levels; or even the reasoning behind the selection of fiber and matrix and process route in the first place. Such a design approach imposes a heavy burden of testing and data collection which is time-consuming and uneconomic in this current cost-conscious climate.

4. The Basis of a Model

A model is a gross simplification of the mechanism but containing the essential elements [13]. At the heart of the model lies the physical mechanism(s) which can best be identified by direct observation using, for instance, optical or in-situ scanning electron microscopy. Other indirect techniques include C-scan and X-radiography of damaged material.

In constucting a model, use is made of the tools of engineering and materials science: the equations and principles of mechanics, thermodynamics, kinetics, and so on. Many of the models have an empirical component which requires a more subtle approach or "fine tuning". An element of empiricism introduces unknown constants into the equation. Often, they simply multiply to produce only one, perhaps two, unknown constants. A calibration procedure is then required to set values on them; one experimental measurement for each constant. The design of critical experiments is a vital element to this process.

When the model is combined with experimental data of (say) fatigue strength, this approach leads to a design methodology having the power of prediction which comes from an understanding of the relevant (meaning dominant) mechanims. At best, the model encapsulates the physical basis of the problem in mathematical form; it summaries what has been observed and predicts behaviour under conditions that have not been investigated. The need to model the dominant mechanism(s) and to incorporate the model in the design process itself need only be pertinent in the first instance at the "microscopic" sizelevel.

5. The Needs and Requirements of a Predictive Design Methodology

Current design processes for aerospace composite structures are empirical. "Allowable stresses" in design and other material parameters are determined through extensive test programs which are then transferred to progressively higher levels of structural complexity. This "building block" approach is favoured because the overall design can be built up carefully and systemmatically from the results of individual tests conducted at various levels of size to meet the specific needs of the complete structural design. Such methodology lacks flexibility and is becoming uneconomic in today's cost-conscious operating climate [1-3]. It would be desireable to reduce the dependence on large test programs to generate design data or to justify the introduction of a new or alternative material system. This need is particularly acute at the early stages of the design process, when irreversible decisions can be made on the choice of material system and composite architecture. On the other hand, the physical modelling approach enables us to gain understanding of the origin of failure processes; to capture the material's response in an equation or code of design; to predict material behavior under conditions not easily simulated in the laboratory (for example, extremely long duration fatigue tests); and to guide the optimisation of micro(macro)structure of the material for longevity and durability [1-3].

6. Developing a Physical Model of Composite Fracture

At the beginning of the modelling process, start by asking what the model is for. It might, for example, require a predictive capability, as far as possible *a priori*, of the fracture and fatigue behavior of new composite material systems with a minimum of calibration or "fine tuning" of the model. To begin with, this necessitates the identification of the elements essential to such a model [3].

6.1 Desired Inputs and Outputs of a Damage Model

To be useful in structural design, in material system optimisation and selection, and in performance prediction, a damage-based model of fracture stress or fatigue strength would utilize material property data and information that would be routinely available as part of the overall design process. In the case of the fracture stress of cross-ply panels this would equate to the following inputs:

> *Geometry, (including hole size and shape); Lay-up (of the $(0_i/90_j)_{ns}$ family)*
> *Matrix and fiber properties; Laminate/ply elastic properties*
> *Delamination resistance (interlaminar toughness)*
> *Environment, temperature; Stress-state, cyclic load spectra, rate (or time).*

The desired output would be a prediction of fracture stress, fatigue life-time or post-fatigue (residual) strength (or stiffness) as a function of the inputs listed above.

6.2 Physical Mechanisms of Microstructural Change

Our observations show that failure involves similar components of damage in a variety of systems: carbon fibers in thermosetting and in thermoplastic matrices; Kevlar fibers in epoxy resin; and glass fibers in epoxy resin. Primarily, the common features of interacting matrix cracks within the notch tip damage zone include:

> *Splitting in the 0° plies*
> *Transverse ply cracking in the 90° plies*
> *Delamination cracking (of approximately triangular shape) at the 0/90 interfaces.*

In the carbon fiber-epoxy family of laminates, under increasing stress the damage zone simply grows in size without changing it's characteristic shape. Consequently, the extent of damage growth can, therefore, be completely quantified in a straightfoward manner. Under monotonic loading, the extent of damage simply increases in size until catastrophic (unstable) failure of the composite occurs by the simultaneous tensile fracture of the localised load bearing fibres adjacent to the notch tip.

From experimental evidence we postulate that two interrelated phenomena govern the failure of notched cross-ply laminates: firstly, that localised interlaminar (delamination) and intralaminar (splitting and transverse) cracks grow by either a quasi-static fracture mechanism or a fatigue (or a time-dependent) mechanism; and secondly, that these cracks modify the notch tip stress distribution that leads to notch tip blunting. Catastrophic failure of the composite occurs, therefore, when the local tensile strength of the $0°$ ply at the notch tip is exceeded by the local tensile stress.

It is important to consider, however, that the local tensile strength of the $0°$ ply may be affected by the presence of matrix-dominated transverse ply cracks in the damage zone. There is convincing evidence widely documented that composite strength is a stochastic quantity which depends on the volume of material under stress.

7. An Example: A Damage and Fracture Stress Model

Following identification of the dominant failure mechanisms, set the model up in two parts:

        1) a quantitative assessment of sub-critical damage growth-rate
        2) determination of the localised stress and $0°$ ply fracture stress at the notch.

The former would be determined by direct observation; the latter by finite element analysis and in-situ tensile strength measurement of the $0°$ ply. Details have been described extensively elsewhere[7, 8]. A brief synopsis follows here.

7.1 The Damage Growth

Initially, the global strain energy release-rate can be calculated solely on the applied tensile loading. These calculations utilize a crude finite element representation of the notched specimen combined with the idealized damage zone. Subsequently, the analysis could be extended to include the effects of elevated temperature and residual (thermal) stress. In the example of quasi-static loading, the global strain energy release-rate can be equated, via a Griffith-type energy balance, to the appropriate fracture energies of the splitting and delamination mechanisms. In the case of fatigue loading, an empirical cyclic crack growth-rate power law (or "Paris" law) is compared to the observed damage (split) growth-rate per cycle and to the calculated cyclic global strain energy release-rate.

7.2 The Damage-Modified Fracture Stress

The second part of the model would be to relate the fracture stress of the laminate to the actual damage-state at the notch tip. It would be reasonable to assume that catastrophic failure occurs when the localized tensile stress in the $0°$ plies exceed its local strength. Given that the strength of the $0°$ ply is dependent on the size of the damage zone and also on the equivalent volume of ply* under stress at the notch tip, then, not surprisingly, the localized tensile strength of the $0°$ ply follows a Weibull weakest link statistics model. We observe that the net effect of competition between the damage-growth mechanisms that can simultaneously blunt notches and reduce the $0°$ ply strength is to increase the residual fracture stress of the laminate.

---

*The equivalent volume is the volume of material which would have the same probability of failure as the material in the (non-uniform) stress distribution of interest if it were loaded uniformly to the peak stress (or other characteristic stress) in the stress distribution.

7.3 Model Integration

In monotonic loading, the two parts of the model would act in synergy. As the applied stress increases, the damage zone grows, reducing the stress concentration at the notch tip, while simultaneously increasing the effective volume of the 0° ply under stress. The latter would result in a reduction in the 0° ply strength. To predict failure, then, would require an evaluation of the current state of damage, a knowledge of the local notch tip stress concentration factor and the current volume of 0° ply under stress at the tip. This would establish whether the specimen survives or fails at that particular applied stress.
Such coupling between damage growth and the volumetric dependence of strength of the 0° ply would appear to be the origin of the so-called "hole size effect" frequently observed in composite materials and reported by Waddoups [9] and Whitney and Nuismer [10].

7.4 Interrogation of the Model and calibration procedure

1)   Perform at least one quasi-static tensile test to failure to obtain X-radiographs of the damage zone with increasing stress. Check to see if the notch tip damage grows in a self-similar manner (or not). Certain carbon fiber-epoxy laminates under increasing stress and temperature exhibit changes in shape, then further calibration steps are required.
2)   Construct a finite element representation of the damage zone. (Re-call that the finite element model is used to obtain a global strain energy release-rate for the combined growth of a split and delamination crack).
3)   Essentially, split energy, G_s , and delamination energy, G_d , (units of J/m²) are found by comparing the model to the experimental data of split length vs stress. G_s is obtained from the stress to initiate splitting, whilst G_d is determined from the rate of split growth with increasing stress.
4)   The *in situ* strength of the 0° ply within the laminate is determined by performing a series of tensile tests on waisted specimens. This data is fitted by a two parameter Weibull weakest link statistics model in order to determine the Weibull modulus.
5)   The finite element model of 2) is used to obtain the notch tip stress-volume integral (i.e., the effective volume of the 0° ply under localised stress at the notch tip) as a function of damage zone size which is characterised by split length.
6)   The model is now in a form that can be used to predict correctly the fracture stress as a function of hole size, ply thickness, and laminate stacking sequence (for the particular family of laminates) and it's dependence on the extent of damage growth.
For cyclic loading, there is an intermediate step, however, (1a), in which damage is monitored for a given loading waveform up to 10⁶ cycles. Such observations and measurements are necessary in order to determine the empirical parameters of the fatigue crack growth power law. Given this calibration sequence, then the fatigue version of the model now predicts correctly fatigue damage growth and post-fatigue residual strength, for a range of hole sizes, ply thickness, stacking sequences, and fatigue history.

8. Application of the Model

The economic advantage of reducing the high cost of vast experimental programs in assorted environments and stress-states having durations of many thousands of hours is potentially huge. In addition, physical modelling provides the means to assess the relative severity of different loading regimes, e.g., constant amplitude vs. variable amplitude spectra, frequency effects and R-ratio effects, as well as load/environment interactions. This capability is important in the design of experimental test programs that ensure critical loading regimes are examined. There is scope to integrate this modelling approach with the large experimental programs that are currently employed to design fracture critical components and structures. The key steps that underpin the modelling described here are:

mechanical testing, characterization of mechanisms by observation and modelling of these physical mechanisms to predict structural response. If these steps were routinely followed as part of the design process it would potentially have a profound effect on the overall cost-effectiveness of that process.

9. Final Remarks and Implications

Modelling the mechanisms of damage and relating them to experimental data and empirical correlations provides a first attempt at achieving a predictive capability in design, particularly with regard to issues of strength, durability and longevity [1-3]. It has the potential to provide a cost-effective alternative to the present largely empirical approach which takes an excessive amount of time and is expensive. Current mechanistic models can be used to bound problems, identify critical material properties, and map likely failure modes. This will lead to improved test programs which concentrate on critical issues at the important sizescale(s). In the longer term, improved modelling will allow the design process to begin with an intelligent choice of constituent materials and interfaces and to proceed with the prediction of performance at the various sizescale [1-3]. Existing design methodologies at the higher structural sizescales will be supported and justified by a fundamental understanding at lower sizescales. Added benefits of this approach include more options to the designer, a reduced need for extensive and costly testing and more efficient and shorter design iteration cycles.

Acknowledgements

We acknowledge the support of NATO in the form of a travel grant (CRG 950082) for the two authors.

References

[1] S. M. Spearing, P. A. Lagace and H. L. N. McManus (1995): "On the Role of Lengthscale in the Prediction of Failure of Composite Structures: Assessment and Needs", 10th International Conference on Composite Materials, Whistler, BC, Canada, IV-49. Published by Woodhead Publishers.
[2] S. M. Spearing, P. A. Lagace and H. L. N. McManus (1997): "On the Role of Lengthscale in the Prediction of Failure of Composite Structures: Assessment and Needs". Accepted for publication in *Applied Composite Materials*.
[3] S. M. Spearing and P. W. R. Beaumont (1997): "Towards a Predictive Design Methodology of Fibre Composite Materials". Tech. Report: CUED/C-MATS/TR 236. Accepted for publication in *Applied Composite Materials*.
[4] K.T. Kedward and P.W.R. Beaumont ((1992): "The Treatment of Fatigue and Damage Accumulation in Composite Design", *Intl. J. Fatigue* , 14 (5) 283-294.
[5] A. Poursartip, M. F. Ashby and P. W. R. Beaumont (1982) : "Damage Accumulation during Fatigue of Composites", *Scripta Metallurgica*, 16 601-606.
[6] A. Poursartip, M. F. Ashby, P. W. R. Beaumont (1986): "Fatigue Damage Mechanics of a Carbon Fibre Composite Laminate: Parts 1,2", *Composites Science and Technology*, 25, 193-218; 25, 283-299.
[7] M. T. Kortschot and P. W. R. Beaumont (1990): "Damage Mechanics of Composite Materials 1-4: *Composite Sci. and Tech.* 39, 289-302; 39, 303-326; 40, 147-166; 40, 167-180.
[8] S. M. Spearing, P. W. R. Beaumont (1992): "Fatigue Damage Mechanics of Composite Materials 1-4:*Comp. Sci. Tech.*, 44, 159-168; 44, 169-177; 44, 299-307; 44, 309-317.
[9] M. E. Waddoups, J. R. Eisenmann and B. E. Kaminski (1971): "Macroscopic Fracture Mechanics of Advanced Composite Materials", *J. Composite Materials*, 5, 446-454.
[10] J. M. Whitney and R. J. Nuismer (1974): "Stress Fracture Criteria for Laminated Composites Containing Stress Concentrations", *J. Comp. Materials*, 8, 253-265.
[11] J. Awerbuch and M. S. Madhukar (1985): "Notched Strength of Composite Laminates: Predictions and Experiments - A Review", *J. Reinf. Plas. Comp.* 4, 3-159.
[12] J. K. Wells and P. W. R. Beaumont (1982): "Correlations for the Fracture of Composite Materials", *Scripta Metallurgica*, 16 99-103.
[13] M. F. Ashby (1992): "Physical Modelling of Material Problems", *Mat. Sci. & Tech.*, 8, 102-111.

Tensile Damage in Ternary Melamine-Formaldehyde Composites

P.-O. Hagstrand and R. W. Rychwalski

Department of Polymeric Materials, Chalmers University of Technology
S-412 96 Göteborg, Sweden

Abstract

A new ternary planar random fibre composite called Reinforced Melamine Compound, abbreviated as RMC, is studied. Damage mechanisms are analyzed at the microlevel and related to macroscopic mechanical behaviour. For this purpose, *in-situ* scanning electron microscopy and cyclic tension tests are used. Damage in the material is quantified with a damage parameter determined completely from the unloading behaviour. Damage onset and evolution is characterized using the damage parameter and compared for different RMC grades. The comparative analysis is extended towards some other PRFC materials. SEM micrographs of samples loaded in a tensile stage, showing damage at different strain levels, are discussed.

1. Introduction

Planar random fibre composites (PRFC) is an interesting class of composite materials offering several advantages for high volume production. However, these materials are complex and, in spite of a research effort in the field, there remain unresolved problems reflecting on PRFC performance. As a consequence (and unfortunately), the full potential of these materials has not been achieved. For example, the notoriously high propensity to microcracking, manifested by the so called "knee" on the stress-strain plot, seriously limits the material in load-carrying applications, long-term performance, applications in certain environments etc. Generally, PRFC have a very low design strain value of about 0.2% [1]. In fact, even lower values around 0.1% can be observed to cause material damage. In this paper we study a recently developed new ternary type of PFRC. The material is called Reinforced Melamine Compound (RMC). It is a melamine-formaldehyde (MF) matrix filled with aluminum trihydrate (ATH) and reinforced with discontinuous chopped glass fibre bundles (50 mm in length). The first commercial RMC grade is launched onto the market under the tradename REMEL™ and is supplied by Perstorp AB (Sweden). MF itself is one of the hardest and stiffest isotropic polymeric material that exists. It has outstanding scratch resistance and surface gloss. Also advantageous, are temperature characteristics, flammability and environmental characteristics. It is therefore interesting to use this polymer as a matrix for composites. Damage mechanisms in RMC materials are studied in tension by means of *in-situ* scanning electron microscopic observations. Macroscopic damage is evaluated by means of stiffness reduction during cyclic straining.

2. Experimental

Moulding compounds were manufactured both, manually and using a commercial manufacturing plant. The moulding compounds were preheated at 110°C for 2 min before they were stacked up in a 25 x 25 cm mould, and were compression moulded at 150°C for 170 s at a pressure of 10 MPa to form flat plates of thicknesses 1 to 1.6 mm. The compositions of the investigated RMC grades are presented in *Tab. 1*. Test specimens with dimensions of 25 x 250 mm were produced by cutting out samples from the moulded plates with a saw equipped with a diamond wafering blade, followed by careful sanding using a 600 paper grade. Tests were conducted using an INSTRON tensile machine. The elongation was measured with a 50 mm INSTRON clip-on extensometer. Damage was evaluated by conducting cyclic tensile tests in which the maximum strain in the cycle was increased systematically until failure occurred. Between 10 and 13 strain cycles were performed on each specimen. Loading and unloading were undertaken at a crosshead speed of 0.5 mm/min. The unloading modulus in each cycle was used as a measure of the damage level rather than the modulus on loading occasionally adopted by some authors [1]. This was used since the effective damaged elastic modulus, E_D , is more accurately measured during unloading [2]. Typically, at the beginning and at the end of the unloading paths, non-linearities were found. We suggest that these non-linearities are caused by viscoelastic behaviour of the filled matrix and, particularly the near zero-stress level, are caused by deformed crack surfaces making contact, folding of fibres inside cracks etc. The presence of the viscoelastic mechanism was confirmed from stress-relaxation tests (not included here). Non-linearities at low load are also due to experimental devices (see *Figs. 1* and *2*). E_D was therefore measured from the central part of the unloading path. The measured values of E_D were then used to define a damage parameter D. The damage in the n'th cycle, D_n is related to the corresponding "damaged" unloading modulus $E_{D,n}$ and the "undamaged" unloading modulus E_0 (measured in the 1st cycle) as follows:

$$D_n = 1 - \frac{E_{D,n}}{E_0} \qquad\qquad (1)$$

Next, D_n was correlated with the corresponding maximum strain in each cycle, ϵ_{mc} (see *Fig. 3*).

Preparation of samples for *in-situ* SEM observations and the adopted procedure is described in the following. A part (50×100 mm) from a central section of a grade B plate was sanded using mainly 400 and 600 paper grades, followed by polishing with 7, 3 and 1 μm diamond pastes. A smaller part was cut from the polished plate and a waisted shape was achieved by using a fine file. Finally, two 4 mm in diameter holes were drilled near the ends. For this sample damage mechanisms were examined in the thickness direction. A second (grade B) sample was manufactured in a slightly different way. A rectangular shape was reinforced at the ends with 0.5 mm thick steel plates which were attached to each side with an epoxy adhesive. Finally, two 4 mm holes were drilled near the ends. This type of sample made it possible to study damage mechanisms in the width direction. Samples were then mounted in a Raith heating tensile stage which was fitted in a Zeiss scanning electron microscope (DSM 940 A). The crosshead speed was set to the lowest possible level of 2 μm/s. For both samples, the strain was measured from observed displacement with an estimated accuracy of ±0.7 μm, instead of using displacement measured with the stage. In the case of the waisted sample, the straining was stopped at elongations corresponding to strain levels of 0.086, 0.196, 0.306, 0.565, 0.761 and 0.879%. The straining of the rectangular sample was stopped at a strain of 0.246%.

3. Results and Discussion

Cyclic Stress-Strain Behaviour

Typical stress-straining behaviour following a procedure described in part 2, is shown in *Fig. 1*. The envelope of the loading paths reveals the presence of a "knee". After unloading, an irreversible strain is present; this strain increasing with progressive cyclic straining. As can be seen in *Fig. 2*, non-linear effects are present. However, on unloading the modulus is more constant. The damage development graphs constructed from data shown in *Fig. 2*, using *Eq. 1*, are presented in *Fig. 3*, where a damage threshold strain, ϵ_D, i.e. the strain below which no damage occurs, can be noted. There are some practical difficulties in measuring ϵ_D, and backward extrapolation was used here. This strain is lower than 0.15% for the four investigated grades. The values of ϵ_D were comparable with the strain levels in monotonic tensile tests where $d\sigma/d\epsilon$ rapidly decreases (not included here). The influence of fibre and filler contents on the damage threshold strain, ϵ_D, seems to be quite weak, however, by comparing the damage development of the B and D grades, it can be found that ϵ_D increases as the volume fraction of ATH, V_{ATH}, is reduced. For example, Bourban et al. [1] measured ϵ_D equal to 0.32 - 0.49% for two chopped strand mat (CSM) epoxy-based vinyl ester materials, and equal to 0.75% for a similar material with a rubber modified epoxy-based vinyl ester matrix with increased toughness. Their results are not strictly comparable with the present value since they used the initial loading elastic modulus as a measure of the damaged modulus, E_D. Their values are higher than for the present grades. The RMC composition affects the damage level. As can be seen grade G with less than 11 v/o of ATH exhibits a considerably lower damage level than the highly filled grades. The slope of the damage graph, $dD/d\epsilon$, yields information concerning the rate of damage with increasing strain. The damage rate in the B and D grades seems to evolve at a similar level around 150%/% in the beginning, and less than 20%/% at the end. The corresponding damage rate for grade G, with a relatively low amount of ATH and ~19 v/o of glass, is much lower being around 75%/% in the beginning, and less than 10%/% at the end. Consequently, the damage level at $\epsilon > \epsilon_D$ is considerably lower for this composition. The result suggests that the damage evolution is suppressed by reducing the stiffness of the matrix. The damage rate of grade I is also very low. However, this material was damaged already before the test due to high internal stresses from manufacturing. For comparison, the obtained value for the damage rate in a vinyl ester based PFRC measured by Bourban et al. [1] is approximately an order of magnitude lower, that is around 10%/% during the whole damage process. Failure at the mesoscale, that is when a crack propagates through the whole sample, in principle takes place at $D=1$. However, usually failure occurs at $D<1$ due to instability which suddenly induces decohesion in the remaining resisting area. This corresponds to a critical value of the damage parameter, D_c. For the investigated RMC materials D_c is in the range 0.36 to 0.67, as can be seen from *Fig. 3*. Bourban and co-authors measured D_c to be less than 0.2 for chopped strand mat epoxy-based vinyl ester materials. Thus, in RMC materials the instability does not occur until a relatively high damage level is reached. This is an advantageous feature of the composite in certain applications. The influence of the ATH content on ϵ_D, $dD/d\epsilon$ and D_c is discussed in the following. The elastic modulus of MF is around 9 GPa [3, 4], which is about two times higher than the stiffness of a typical epoxy resin. At ~48 v/o ATH (as for the B grade), the elastic modulus of the filled matrix increases to around 16 GPa [5]. The load carried by the matrix will therefore be considerably higher in a MF based composite than in a comparable composite based on an epoxy or polyester resin, as an example. It can be expected therefore, that RMC materials will have a relatively high damage propensity. Clearly, this effect will be more pronounced for highly filled RMC compositions. In addition, the filled matrix in such compositions has (as well known) a lower strength and strain to failure. It is difficult to draw conclusions about the influence of fibre content on ϵ_D, $dD/d\epsilon$ and D_c. From results published in the literature, a possible effect of fibre content is discussed in the following. It is well known that a planar random fibre composite can be approximated with a quasi isotropic laminate [6]. The strength and the strain to failure of the transversely loaded laminae in such a laminate are, typically, reduced with increasing fibre volume fraction. This is largely due to the inherent tendency for high local stresses and strains to develop in the matrix [7]. It is therefore likely, that a high glass content could have a negative effect on the damage initiation and evolution. This has been confirmed by Bourban *et al.* [1] for vinyl ester based PFRC. They found that an increasing glass content decreases ϵ_D and increases $dD/d\epsilon$ and D_c. We think this behaviour can also be representative in RMC materials.

In-situ SEM Observations

After loading to a strain level of 0.086%, the *in-situ* SEM study revealed debonding of large ATH particles (~100 μm) from the MF matrix, as shown in *Fig. 4*. The mean particle size of the ATH grade is around 6 μm, and particles ~100 μm are quite rare. The microstructure of RMC samples which had been mechanically loaded prior to SEM observations reveal, however, debonding not only at very large ATH particles but also at narrow elongated straight edged medium sized particles. This is shown in *Fig. 5*. As the strain was increased, no further debonding was observed. Furthermore, the discussed debonding was not observed to constitute crack precursors. *Fig. 6* shows fibre-matrix adhesive debonding at an overall strain level of about 0.2%, in a fibre bundle nearly transversely oriented to the applied load. As can be seen, some cracks have started to propagate from fibre tips of previously debonded fibres. These cracks propagated mainly perpendicular to the applied load. Debonding of nearly transversely oriented fibres can therefore be regarded as a crack initiating mechanism. At a strain slightly above 0.56%, a large number of damage mechanisms was possible to observe. The crack density had increased and some cracks reached a length of >1000 μm. When possible, the cracks propagated mainly along the fibre-matrix interface. Some cracks were bridged by fibres where they crossed nearly longitudinally oriented fibre bundles. As can be seen in *Fig. 7*, the stress in some of the crack bridging fibres exceeded the tensile strength of the glass fibres resulting in fibre failure. The fibre fracture enabled fibre debonding and fibre pull-out to some extent, also shown in *Fig. 7*. At a strain level around 0.88% several cracks had propagated through the whole sample. Finally, at a strain level of approximately 0.95% total failure occurred. *Fig. 8* shows a small part of the fractured surface. The pulled-out fibres are clean and not covered by resin. This observation indicates a weak fibre-matrix interface. The *in-situ* SEM tensile test of the rectangular sample, studied in the width direction, revealed the following. At an overall strain of about 0.25% relatively long transverse cracks appeared. These had started to propagate mainly from fibre bundles located very close to the sample surface. Some cracks were also initiated by fibre-matrix debonding in nearly transversely oriented fibre bundles located in the bulk. However, less than one out of ten bundles in the bulk had started to crack at this strain level. The observed cracks propagated, as expected, mainly perpendicular to the applied load. Fibre-matrix debonding inside bundles follows the fibre-matrix interface. This is shown in *Fig. 9* for a grade B which had been loaded prior to SEM observations. This observation indicates again a weak interface. Fibre-matrix debonding in nearly transversely oriented bundles, as well known, is an early damage mechanism in sheet moulding compounds (SMC) [8, 9].

4. Conclusions and future work

Tensile damage in MF based ternary composites containing ATH filler and random glass fibre bundles has been analyzed in the work using mechanical cycling and *in-situ* SEM. Occasionally, the carried out analysis is supported by measurements not included in the present report. Mainly two deformation mechanisms, microcracking and viscoelastic mechanism, are present in the composite. Separation of viscoelastic and plastic deformation was not carried out. Macroscopic damage was quantified using a damage parameter constructed completely from the unloading behaviour. Damage onset and development are compared for different MF/ATH/glass compositions. Type and source of damage leading to failure and that which does not dangerously propagate, have been determined. Based on the results from the present study, the future work will mainly focus on increasing the ductility of the composite.

Acknowledgement

We gratefully acknowledge Perstorp AB for the financial support. The authors are grateful to Mr. Göran Karlsson for initiating this work and to Dr. Stefan Lundmark, Mr. Hans Persson and Mr. Lennart Svensson for valuable discussions. We also thank DSM Resins (The Netherlands) for compression moulding.

Table 1. Volume Fractions of Glass and ATH for the Analyzed RMC grades.

RMC	V_{glass} [%]	V_{ATH} [%]
B	15.4	47.8
D	15.4	38.0
G	18.7	10.9
I	23.6	0.0

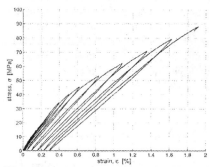

Fig. 1 Cyclic stress-strain behaviour of grade G.

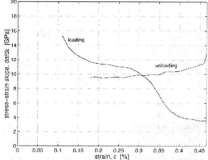

Fig. 2 Typical stress-strain loading and unloading behaviour.

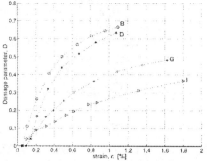

Fig. 3 Damage onset and evolution for different RMC grades (indicated by letters).

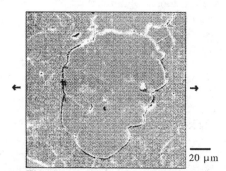

Fig. 4 Debonding at large ATH particle (ϵ=0.09%). Loading direction is shown with arrows, similarly in Figs. 6-8.

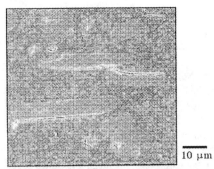

Fig. 5 Debonding at narrow elongated straight edged medium sized ATH particles (unloaded sample).

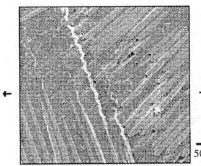

Fig. 6 Fibre-matrix debonding and crack propagation (ϵ=0.2%).

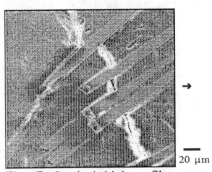

Fig. 7 Crack bridging, fibre fracture, fibre-matrix debonding and fibre pull-out (ϵ=0.56%).

Fig. 8 Fractured surface. Clean fibres indicate weak fibre-matrix interface.

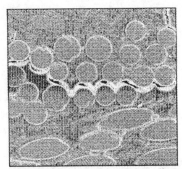

Fig. 9 Fibre-matrix debonding inside a fibre bundle (unloaded sample).

References

1. *P.-E. Bourban, W.J. Cantwell, H.H. Kausch and S.J. Youd,*
Proceedings of ICCM-9, Madrid, Spain (1993).
2. *J. Lemaitre*, A Course on Damage Mechanics, Springer Verlag, Berlin (1992).
3. *C. Ozinga*, DSM Research, Personal communication, (1995).
4. *H.H Landolt*, Zahlenwerte und Funktionen aus Physik, Chemie, Geophysik und Technik, Landolt-Börnstein, Berlin (1950).
5. *P.-O. Hagstrand, R.W. Rychwalski and C. Klason*, Polym. Eng. Sci., in press.
6. *J.C. Halpin and N.J. Pagano*, J. Comp. Mater. **3**, 720 (1969).
7. *D. Hull and T.W. Clyne*, An Introduction to Composite Materials, Cambridge University Press, (1996).
8. *D. Hull*, An Introduction to Composite Materials, Cambridge University Press, (1981).
9. *F. Meraghni and M.L. Benzeggagh*, Proceedings of ICCM-9, Madrid, Spain (1993).

STIFFNESS DEGRADATION IN CRACKED CROSS-PLY LAMINATES

J.Varna* and A.Krasnikovs**
*Div. of Polymer Engineering, Lulea University of Technology, S 971 87 Lulea, Sweden
**Institute of Mechanics, Riga Technical University, LV-1058 Riga, Latvia

ABSTRACT

A model for stiffness reduction in cross-ply laminates based on generalized plane strain assumptions is developed. Simple analytical expressions are obtained for longitudinal modulus and Poisson's ratio reduction as a function of transverse crack density. Additionally to the crack density, these expressions contain only elastic and geometrical properties of constituent laminae and the average crack opening displacement (COD) normalized with respect to the far-field strain.

Calculations of COD are performed using FEM and several analytical models based on shear lag assumptions and variational analysis. Their predictive capabilities are verified using comparison with experimental data and FEM results.

A simple procedure is suggested for COD determination, requiring for a given material only one FEM calculation using "an average laminate" with "an average crack spacing". The obtained COD is used with good results for stiffness prediction for a wide variety of laminates of this material.

INTRODUCTION

The stiffness reduction in cross-ply laminates containing transverse cracks has been in the focus of the composite community for a long time. Starting with the first systematic observations and shear lag analysis performed by Garett et.al [1], longitudinal modulus reduction experiments in GF/EP laminates by Highsmith et.al [2] and studies by Flaggs et.al [3] on 90-layer thickness effect on the first transverse cracking strain, numerous reports have been aimed to theoretical and experimental aspects of this subject. Modeling efforts to predict the laminate stiffness reduction recently have been often related to the crack opening displacement (COD) [4,5].

In this paper exact analytical expressions are obtained to describe stiffness reduction in cross-ply laminate as the crack density increases. The analysis is based on generalized plane strain assumptions and final expressions contain only geometrical and elastic parameters of the laminae, crack density and average crack opening displacement (COD) normalized with respect to the applied far field strain. Then the existing analytical models of different accuracy (shear lag[6], variational models of Hashin's [7], Varna et.al [8,9], and FEM are used to calculate the required COD. FEM calculations are based on plane stress assumptions.

Finally, based on the performed analysis, a simple calculation procedure is proposed that requires only one rather rough FEM calculation for the given material laminate to predict the stiffness reduction curves for the variety of laminates made of this material.

AVERAGED CONSTITUTIVE EQUATIONS AND STIFFNESS REDUCTION

Mechanically loaded $(0_n,90_m)_s$ cross-ply laminate with cracks in the $90°$ plies is shown in Fig.1. Each laminae in material symmetry axes has elastic constants E_1, E_2 v_{12}, v_{23}, G_{12}, G_{23}, where index 1 corresponds to the fiber direction, ($h=d+b$).

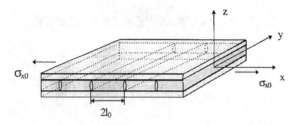

a)

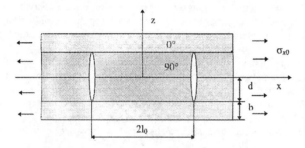

b)

Fig.1 The geometry of the cross-ply laminate.

Stresses in the undamaged laminate as well as its stiffness matrix have subscript $_0$. We assume that generalized plane strain state is realized

$$\varepsilon_y = \overline{\varepsilon_y^{90^\circ}} = \overline{\varepsilon_y^{0^\circ}} \tag{1}$$

As an example we model the reduction of laminate longitudinal modulus E_x, major Poisson's ratio v_{xy} and the average COD. These constants are defined by

$$E_x = \frac{\sigma_{x0}}{\overline{\varepsilon_x^{0^\circ}}} \qquad v_{xy} = -\frac{\overline{\varepsilon_y}}{\overline{\varepsilon_x^{0^\circ}}} \qquad u_a = l_0\left(\overline{\varepsilon_x^{0^\circ}} - \overline{\varepsilon_x^{90^\circ}}\right) \tag{2}$$

where average strains are

$$\overline{\varepsilon_i^{0^\circ}} = \frac{1}{2bl_0} \int\limits_{-l_0 d}^{+l_0 h} \varepsilon_i^{0^\circ}\,dz dx, \qquad \overline{\varepsilon_i^{90^\circ}} = \frac{1}{2dl_0} \int\limits_{-l_0 0}^{+l_0 d} \varepsilon_i^{90^\circ}\,dz dx \tag{3}$$

In the following analysis we will use the force balance in the y and z direction

$$\int\limits_0^d \sigma_y^{90^\circ}\,dz + \int\limits_d^h \sigma_y^{0^\circ}\,dz = 0 \qquad \int\limits_{-l_0}^{+l_0} \sigma_z^i\,dx = 0, \qquad i = 0^\circ, 90^\circ, \tag{4}$$

From Hooke's law

$$\varepsilon_y = -\frac{v_{12}}{E_1}\sigma_x^{0^\circ} + \frac{1}{E_2}\sigma_y^{0^\circ} - \frac{v_{23}}{E_2}\sigma_z^{0^\circ} \tag{5}$$

$$\varepsilon_y = -\frac{v_{12}}{E_1}\sigma_x^{90°} + \frac{1}{E_1}\sigma_y^{90°} - \frac{v_{12}}{E_1}\sigma_z^{90°} \tag{6}$$

we can express $\sigma_y^{0°}$ and $\sigma_y^{90°}$ and substitute them in the first of (4). Integrating the result with respect to x in the region between two cracks we obtain ε_y that depends on the average x - axis stress in both layers only.

$$\varepsilon_y = -\frac{v_{12}}{E_1(E_1 d + E_2 b)}\left(E_1 d\overline{\sigma_x^{90°}} + E_2 b\overline{\sigma_x^{0°}}\right) \tag{7}$$

Then from averaged (5) and (6) we can express also $\overline{\sigma_y^{90°}}$ and $\overline{\sigma_y^{0°}}$ through $\overline{\sigma_x^{90°}}$ and $\overline{\sigma_x^{0°}}$. Finally, averaging constitutive relationships for $\varepsilon_x^{0°}$ and $\varepsilon_x^{90°}$ may be also expressed through $\overline{\sigma_x^{90°}}$ and $\overline{\sigma_x^{0°}}$ only

$$\overline{\varepsilon_x^{0°}} = \overline{\sigma_x^{0°}}\frac{1}{E_1}\left(1 - \frac{v_{21}^2 E_2 d}{E_2 b + E_1 d}\right) + \frac{v_{12}v_{21}d}{E_2 b + E_1 d}\overline{\sigma_x^{90°}} \tag{8}$$

$$\overline{\varepsilon_x^{90°}} = \overline{\sigma_x^{90°}}\frac{1}{E_2}\left(1 - \frac{v_{12}v_{21}E_2 b}{E_2 b + E_1 d}\right) + \frac{v_{12}v_{21}b}{E_2 b + E_1 d}\overline{\sigma_x^{0°}} \tag{9}$$

Stresses in the laminate with cracks may be always written in the form

$$\sigma_x^{90°} = \sigma_{x0}^{90°} - \sigma_{x0}^{90°} f_1(\overline{x},\overline{z}) \qquad \sigma_x^{0°} = \sigma_{x0}^{0°} + \sigma_{x0}^{90°} f_2(\overline{x},\overline{z}) \tag{10}$$

where $\overline{x} = x/d$, $\overline{z} = z/d$, the first terms characterize stresses in undamaged laminate, but the second ones represent the stress perturbation caused by the presence of crack.
Averaging (10) and using the force equilibrium in the x-direction:

$$\overline{\sigma_x^{90°}} = \sigma_{x0}^{90°} - \sigma_{x0}^{90°} R(\overline{l_0})\frac{1}{2\overline{l_0}} \qquad \overline{\sigma_x^{0°}} = \sigma_{x0}^{0°} + \sigma_{x0}^{90°} R(\overline{l_0})\frac{1}{2\overline{l_0}b} \tag{11}$$

where
$$R(\overline{l_0}) = \int_{-\overline{l_0}}^{+\overline{l_0}\overline{h}}\int_{1} f_2(\overline{x},\overline{z})d\overline{x}d\overline{z} \tag{12}$$

Function $R(\overline{l_0})$ depends on $\overline{l_0}$ and represents in an average sense the stress perturbation caused by transverse cracks. Substituting (11) in (7), (8) and (9) we express all average strains needed in (2) through the perturbation function $R(\overline{l_0})$. Finally from (2)

$$\overline{u_a} = \frac{u_a}{\varepsilon_{x0}d} = \frac{d+b}{2b}\frac{E_{xv}}{E_1}\left(1 - v_{12}V_{xyo}\right)R(\overline{l_0}) \tag{13}$$

$$\frac{E_x}{E_{x0}} = \frac{1}{1 + \frac{E_2}{E_{x0}}\left(\frac{1 - V_{12}V_{xy0}}{1 - v_{12}v_{21}}\right)\frac{d}{(d+b)}\frac{1}{\overline{l_0}}\overline{u_a}} \tag{14}$$

$$\frac{v_{xy}}{v_{xy}^0} = \frac{1 - \frac{1}{(1 - v_{12}v_{21})}\left(\frac{E_1 - E_2}{E_{x0}}\right)\frac{db}{(d+b)^2}\frac{1}{\overline{l_0}}\overline{u_a}}{1 + \frac{E_2}{E_{x0}}\left(\frac{1 - V_{12}V_{xy0}}{1 - v_{12}v_{21}}\right)\frac{d}{(d+b)}\frac{1}{\overline{l_0}}\overline{u_a}} \tag{15}$$

Obtained expressions contain elastic properties and geometrical parameters of constituents, normalized crack density $1/\overline{l_0} = d/l_0$ and normalized average COD $\overline{u_a}$, corresponding to the unit of the applied far field strain.

MODELING THE AVERAGE COD

In order to use (14) and (15) we need to calculate the $\overline{u}_a$ for changing crack spacing $\overline{l}_0$ (using (13) or the COD profile (FEM)). $R(\overline{l}_0)$ from analytical models is (see (10), (12)) determined by $\sigma_x^{90°}$ distribution that is well known in each model.

Shear lag models [6]. There is a large variety of different modifications of this model but the main assumptions are similar. Integral force equilibrium conditions are used that leads to a second order differential equation with constant coefficients. The x-axis stress distribution in 0-layer can be expressed in the following form

$$\sigma_x^{0°} = \sigma_{x0}^{0°} + \sigma_{x0}^{90°}\frac{d}{b}\frac{\cosh\alpha\overline{x}}{\cosh\alpha\overline{l}_0}, \tag{16}$$

where

$$\alpha = \sqrt{k\frac{d}{E_2}\left(1+\frac{E_2 d}{E_1 b}\right)} \tag{17}$$

The differences between models appear in the interpretation of k in (17). One group of models [6] assume the shape of the longitudinal displacement across the 90-layer. Then $k = \gamma\frac{G_{23}}{d}$, where $\gamma = 1$ for uniform distribution and $\gamma = 3$ for parabolic distribution. From (16) (12), (13)

$$\overline{u}_a = \frac{d+b}{b}\frac{E_{xo}}{E_1}\left(1-\nu_{12}\nu_{xyo}\right)\frac{1}{\alpha}\tanh\alpha\overline{l}_0 \tag{18}$$

Hashin's model [7] was historically the first model that used variational approach to determine the stress state in the cross-ply laminate with cracks. The following assumptions are used: the x-axis normal stress in 90°- as well as in 0°-layer is independent on z co-ordinate

$$\sigma_x^{90°} = \sigma_{x0}^{90°} - \sigma_{x0}^{90°}\psi(\overline{x}) \tag{19}$$

$$\sigma_x^{0°} = \sigma_{x0}^{0°} + \sigma_{x0}^{90°}\psi(\overline{x})\frac{d}{b}. \tag{20}$$

The minimization procedure of the complementary energy lead to fourth order linear differential equation with constant coefficients for determination of $\psi(\overline{x})$. Finally

$$\overline{u}_a = \frac{d+b}{2b}\frac{E_{xo}}{E_1}\left(1-\nu_{12}\nu_{xyo}\right)\int_{-\overline{l}_0}^{+\overline{l}_0}\psi(\overline{x})d\overline{x} \tag{21}$$

where

$$\psi(\overline{x}) = A_1\cosh\alpha\overline{x}\cos\beta\overline{x} + A_2\sinh\alpha\overline{x}\sin\beta\overline{x} \tag{22}$$

Constants A_1 and A_2 follow from boundary conditions, parameters α and β are calculated using elastic properties and layer thickness.

2-D 0°-model [8] is a further development of Hashin's approach. The difference is in proposed stress profiles: a) In the 90°-layer the x-axis stress distribution in z-direction is uniform, shear stress is linear and normal z-axis stress is parabolic; b) In the 0°-layer the non-uniformity is included using $\varphi(\overline{z})$ that is described by an exponent with unknown shape parameter

$$\varphi(\overline{z}) = \frac{1-\cosh\Delta(\overline{z}-\overline{h})}{\Delta\sinh\Delta\overline{b}} \tag{23}$$

Stress assumptions are

$$\sigma_x^{90°} = \sigma_{xo}^{90°}\left(1-\psi(\overline{x})\right) \tag{24}$$

$$\sigma_x^{0°} = \sigma_{x0}^{0°} - \sigma_{rn}^{90°}\left(\psi(\overline{x})\varphi''(\overline{z})\right). \tag{25}$$

Function $\psi(\bar{x})$ is a solution of a fourth order differential equation with constant coefficients. The solution contains the unknown shape parameter Δ of the z-distribution. It is calculated by further numerical minimization of complementary energy.

Formally the expression for $\overline{u_a}$ is the same as in Hashin's model, see (21). Function $\psi(\bar{x})$ is still in the form (22) and constants A_1 and A_2 follow from boundary conditions. Expressions for parameters α and β are different than in Hashin's model and depend on the shape parameter Δ.

2-D 0°/90° model [9] includes a nonuniform distribution of x-axis stress in both 0- and 90-layer. It is the more advanced in the rank of variational models. We assume the following form of x-axis stress distribution in layers

$$\sigma_x^{90^o} = \sigma_{x0}^{90^o}\left[1 - \psi(\bar{x}) + \psi_1(\bar{x})\varphi_2''(\bar{z})\right] \qquad (26)$$

$$\sigma_x^{0^o} = \sigma_{x0}^{0^o} - \sigma_{x0}^{90^o}\left[\psi(\bar{x})\varphi_1'(\bar{z}) - \psi_1(\bar{x})\varphi_3'(\bar{z})\right] \qquad (27)$$

The arbitrary function $\varphi_2(\bar{z})$ is responsible for the non-uniform tensile x-axis stress distribution in the 90°-layer close to the crack. The x-dependence of this new stress term is characterized by $\psi_1(\bar{x})$. This function is expected to decrease very fast with distance away from the crack surface. The function $\varphi_3(\bar{z})$ in combination with $\psi_1(\bar{x})$ reflect the stress redistribution in the 0-layer when the non-uniform stress in the 90°-layer is present.

In the present paper we assume the following shape of the last three functions

$$\varphi_1(\bar{z}) = \frac{1 - \cosh\Delta_1(\bar{z} - \bar{h})}{\Delta_1\sinh\Delta_1\bar{b}} \qquad \varphi_3(\bar{z}) = \frac{1 - \cosh\Delta_3(\bar{z} - \bar{h})}{\Delta_3\sinh\Delta_3\bar{b}} \qquad \varphi_2(\bar{z}) = A + \frac{\bar{z}^{2n}}{2n} \qquad (28)$$

with arbitrary Δ_1, Δ_3 and n, where n is an integer shape parameters.

The minimization routine described in [9] leads to a pair of fourth order differential equations with constant coefficients and the solution can be expressed as a linear combination of trigonometric and hyperbolic functions. The numerical values of the constants depend on Δ_1, Δ_3 and n. Therefore the minimum value of the complementary energy also has such a dependence

$$V = V(n, \Delta_1, \Delta_3, \bar{l}_0) \qquad (29)$$

Minimization routine is continued in order to find the most accurate stress distribution across the thickness of each layer (constants Δ_1, Δ_3 and n).

Finite element analysis was performed using plane stress state simulation in cracked laminate by the FE code NISA. Only a quarter of the representative element was used. We have symmetry conditions on sides $x = -l_0$, $z \in [d,h]$ and $x \in [-l_0, 0], z = 0$, traction free conditions on $z = h$ and on the crack surface $x = -l_0$, $z \in [0,d]$, applied constant displacement in x-direction on $x = 0$, $z \in [0,h]$ The 2D four- node quadratic plane elements, uniform and non-uniform meshes were tested to achieve a better approximation in the crack tip vicinity. Contrary to analytical models FEM renders COD profiles in a direct way. The normalized average COD $\overline{u_a}$, needed in stiffness predictions may be obtained as follows

$$\overline{u_a} = \frac{1}{\varepsilon_{x0}d}\int_0^1 u(\bar{z})d\bar{z}. \qquad (30)$$

PREDICTIONS AND EXPERIMENTAL DATA

In this section the stiffness reduction in glass fiber/epoxy (Std GF/EP) cross-ply laminates is analyzed (E_L=45.5 GPa, E_T =12.7 GPa, ν_{LT} =0.28, ν_{TT} =0.42, G_{LT}=3.45 GPa, G_{TT} =4.5 GPa) and predictions are compared with test data. FEM and analytical modeling results are shown in Table 1. According to FEM (it will be used as the "correct solution" and other models compared with it) the crack spacing $\overline{l_0}$ has a very small effect on $\overline{u_a}$. Some effect is noticeable only for $\overline{l_0} = 1.5$ that corresponds to very high crack density. This observation allows to reduce significantly the number of FEM calculations needed for predictions. Increasing the thickness ratio b/d the COD's are reduced.

Table 1. Average COD $\overline{u_a}$ of Std GF/EP laminates

a) $(0,90_2)s$

lo/d	FEM	2D-0/90	2D-0	Hashin	Sh.lag $\gamma = 1$
8	1.399	1.524	1.853	1.965	2.050
2	1.396	1.528	1.843	1.938	1.850
1.5	1.330	1.456	1.746	1.807	1.652

b) $(0_2,90_2)s$

lo/d	FEM	2D-0/90	2D-0	Hashin	Sh.lag $\gamma = 1$
8	1.248	1.352	1.663	2.051	1.848
2	1.238	1.407	1.640	1.908	1.614
1.5	1.143	1.330	1.516	1.657	1.415

$(0_4,90_2)s$

lo/d	FEM	2D-0/90	2D-0	Hashin	Sh.lag $\gamma = 1$
8	1.180	1.196	1.455	2.412	1.729
2	1.143	1.271	1.440	1.953	1.477
1.5	1.050	1.277	1.335	1.575	1.282

The $\overline{u_a}$ data presented in Table 1 are used to predict stiffness reduction. The predictive capabilities of variational models are demonstrated in Fig.2 a). Hashin's model renders for the whole range of thickness ratio b/d the lowest values. It is consistent with the fact that this is the simplest of variational models. The 2D-0 model is the next in the order of complexity and the predictions are closer to FEM results. They are not very good (and are close to Hashin's model) for relatively thick 90-layers. The reason is the used stress assumptions: the stress state analysis is improved only in 0-layer. The 2D-0/90 model is the best among analytical models and the deviations from the FEM solution are very small, especially for thin 90-layers. For many practical cases, considering the existing scatter in experimental data, the predictive capabilities of FEM, 2D-0/90 and2D-0 model may be consider as equal.

The predictions of the two modifications of the shear lag model are presented in Fig. 2 b). The FEM solution is always between predictions of both these models.

The results in Tale 1 showed that the normalized ACOD $\overline{u_a}$ is insensitive to the changes in crack spacing $\overline{l_0}$ and layer thickness ratio b/d (less than 20%). When the COD's of FEM are compared with, for example, 2D-0 model the differences may reach 25%. Yet, we are claiming, based on Fig. 2, that this model is good. We further conclude, that without loosing much in accuracy we can take a rough COD approximate using only one FEM calculation for a given material cross-ply laminate.

Then the result ($\overline{u_a}$) could be used to predict elastic properties at different crack densities and for a large variety of layer thickness ratio. We suggest to use $(0_2,90_2)s$ laminate with a crack

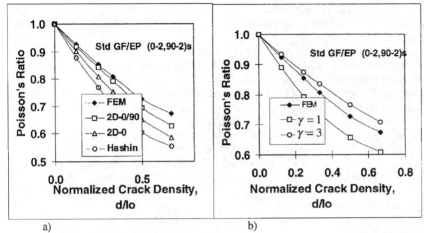

a)   b)

Fig.2 Predictions based on FEM and analytical models

spacing $\overline{l_0} = 4$ corresponding to an average crack density in order to calculate (using FEM) the average COD. After that the calculated $\overline{u_a}$ is used for predictions in laminates with different b/d. In this way for Std GF/EP laminates value $\overline{u_a} = 1.25$ was chosen. Predictions based on this values are presented in Fig. 3 together with FEM predictions based on accurate values of COD and test data. We see very small differences in prediction only at high crack densities and/or thick 90-layers. Agreement with test data is excellent using both approaches.

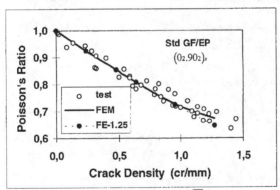

Fig. 3. Predictions compared with test data using $\overline{u_a}$ from FE analysis (FEM) and an approximate value 1.25 (FE-1.05).

CONCLUSIONS

Simple analytical expressions for stiffness reduction in the cracked cross-ply laminate are obtained using generalized plane strain assumptions in the specimen width direction. They contain only crack density, elastic and geometrical parameters of the laminae and the average crack opening displacement (ACOD) normalized with respect to the far field strain.

Expressions for ACOD according to two modifications of the shear lag model and three variational models of different accuracy are derived. The calculated values are analyzed and compared with FEM results.

Normalized ACOD is insensitive to changes in crack spacing and layer thickness ratio. This is used to develop a simple prediction procedure for stiffness reduction. FEM calculation for a given material is required only once for some "average" crack spacing and "average" 0/90-layer stiffness ratio. The normalized ACOD value obtained from FEM is used to predict stiffness reduction as a function of crack density in other laminates made of this material. Parametric analysis and comparison with test data demonstrates high accuracy of this approach.

Predictions of the used analytical models have different degree of accuracy. The most sophisticated 2D 0/90 model gives good predictions, 2D-0 model has problems for laminates with thick 90-layer, Hashin's model can not properly describe the 0-layer thickness effect and gives the most inaccurate predictions among variational models. Shear lag models can give in some cases good predictions, but the predictions are not consistent and the accuracy of models changes significantly with material change.

REFERENCES

1. Garett K.W. and Baily J.E.,". Multiple Transverse Fracture in 90^0 Cross-Ply Laminates of a Glass Fibre-Reinforced Polyester,"J. of Mater. Sci, 1977, Vol. 12, 157-168.
2. Highsmith A.L. and Reifsnider K.L.," Stiffness-Reduction Mechanisms in Composite Laminates," Damage in Composite Materials, ASTM STP 775, American Society for Testing and Materials, 1982, 103-117.
3. Flaggs D.L. and Kural M.H., "Experimental Determination of the In Situ Transverse Lamina Strength in Graphite/Epoxy Laminates,"J. of Composite Materials, 1982, Vol. 16, 103-115.
4. Zang W. and Gudmundson P.,."Damage Evolution and Thermoplastic Properties of Composite Laminates," International Journal of Damage Mechanics, 1993, Vol.2, No.3, July, 290-308.
5. Varna J., Berglund L.A., Krasnikovs A. and Chihalenko A.," Crack Opening Geometry in Cracked Composite Laminates," International Journal of Damage Mechanics, 1997, Vol.6, No.1, January, 96-118.
6. Fukunaga H., Chou T.W., Peters P.W.M., Schulte K." Probabilistic Failure Strength Analyses of Graphite/Epoxy Cross-Ply Laminates," J. of Composite Materials, 1984, Vol. 18, 339-356.
7. Hashin Z." Analysis of Cracked Laminates: A Variational Approach," Mechanics of Materials, 1985, Vol. 4, 121-136.
8. Varna J. and Berglund L.A." Multiple Transverse Cracking and Stiffness Reduction in Cross-Ply Laminates," Journal of Composites Technology & Research, JCT RER, 1991, Vol. 13, No. 2, 97-106.
9. Varna J. and Berglund L.A." Thermo-Elastic Properties of Composite Laminates With Transverse Cracks," Journal of Composite Technology & Research, January 1994, Vol. 16, No.1, 77-87.

Experimental Study of Stiffness Degradation and Crack Growth in Composite Materials

S.R. Abdussalam* and M.L. Ayari*

*Department of Mechanical & Industrial Engineering - University of Manitoba Winnipeg, Manitoba, Canada, R3T 2N2

Abstract

An experimental investigation to study stiffness degradation and crack growth in composite materials, when subjected to quasistatic load, is presented. The effect of various geometric parameters such as specimen thickness and fiber orientation on the material degradation is addressed. The load versus crack opening displacement presents a well defined post-peak regime which can be described by a simple bilinear model. Based on the experimental results, concepts of fracture mechanics are applied to evaluate stiffness degradation and fracture toughness in composite materials.

1. Introduction

Fiber reinforced composite materials consist of fibers of high strength and modulus embedded in, or bonded to a matrix material with distinct interfaces (boundaries) between them. Both fibers and matrix retain their physical and chemical identities, yet produce a combination of properties that cannot be achieved with any of the constituents acting alone. Fiber reinforced composites are becoming increasingly important in industrial application, particularly in aircraft industry, where the properties of high strength and low weight are desirable. As composites are being used to fabricate critical structural components, such as horizontal and vertical stabilizer skins and wing skins, their mechanical behavior must be clearly understood [1,2]. Like metals, these materials are notch sensitive and loose much of their structural integrity when damaged. however, the properties of these materials are not as well characterized as the conventional materials (primarily metals) which they are designed to replace. Some investigators (e.g. refs. [3,4,5]) studied the fracture behaviour and fatigue crack propagation of chopped-stand, glass-fiber reinforced composites. Experimental work were conducted in their investigations on composite specimens subjected to cyclic tensile fatigue loading. This work is part of an on-going effort to extend the results established for metals to the newer composite materials.

Because the mechanical response of composites is brittle in nature, it is important to study the applicability of fracture mechanics to predict failure. Despite the importance of this concept, there has been only limited studies concerning direct tensile tests to measure stiffness degradation in tension and relevant softening behavior of composite materials. In the present work, an experimental investigation is conducted to study the behavior of crack growth in composite materials. In particular, the effect of various geometrical parameters such as thickness and fiber orientation are investigated.

2. Experimental Program

A series of 50 uniaxial tension tests were conducted to evaluate the initial modulus of the material. In addition, in the fracture test, the vertical load and the crack opening displacement were monitored and recorded using a standard data acquisition system.

All specimens are prepared according to the following procedure. First, panels of fiberglass are manufactured using a vacuum bagging technique, with the chosen number of fiberglass cloth layers and fiber orientation as shown in Figure (1). Vacuum bagging (also referred to as vacuum bag laminating) is a clamping method which uses atmospheric pressure to hold the adhesive coated components of a lamination in place until the adhesive cures. The laminates are sealed within an airtight envelope. The envelope may be an air-tight mold on one side and an air-tight bag on the other. When the bag is sealed to the mold, pressure on the inside and outside of the mold is equal to atmospheric pressure, approximately 736.6 mm (29 inches) of Mercury (Hg), or 101.35 KPa (14.7 psi). A vacuum pump is then used to evacuate air from the inside of the envelope [6]. The pressure differential between the inside and outside of the envelope determines the amount of clamping force on the laminate. These are then allowed to cure for a minimum of 24 hours at room temperature. Next, the specimen size and orientation is recorded on the surface of each panel. The specimen have a length of 250 mm and a width of 25 mm as shown in Figure (2). The total gage length between grips is 140 mm. The specimens are cut using a bandsaw. Since the cutting procedure may cause local delamination, the specimens are cut 0.7 mm larger than required. Each specimen is then sanded to the final shape with successively finer grits of sandpaper. Dimensions are verified with a digital Vernier calipers. Each specimen must be within 0.03 mm in order to maintain decent accuracy within the experiment. For specimens requiring an initial crack, the location of the crack is marked on the specimen. The crack is then cut with a thin bandsaw blade at low speed in order to ensure minimal damage of the cut surface. The length of the crack is then verified to be 6.25 mm (see Figure 2).

After the testing apparatus is connected, the specimen is placed between the grips and locked into place. A crack opening displacement (COD) gage is fixed to the specimen between two knife edges which are either glued or screwed on either side of the crack mouth. Using the control keys of the Instron Cervo-controlled testing machine, we remove the slack prior to loading of the specimen. The Instron's self-calibration routine is initiated using a conventional controller. A computer program, written exclusively for this tests initiates loading and unloading of the specimen under displacement control. In order to have stable crack growth, the loading and unloading rates are chosen to be as slow as possible. The choice of a rate between 0.06 and 0.1 mm/min during loading and a rate between 0.09 and 0.15 mm/min in the case of unloading are recommended. Because of the slow rates each test takes from 7 to 9 hours to complete.

The load magnitude, total elongation and the COD voltage for a number of loading and unloading cycles are continuously recorded.

2.1 Test Results

The experimental work consisted of first testing several specimens with different thicknesses (number of layers) to determine the initial elastic modulus. The initial elastic modulus of the composite is found to be equal to 14.37 *GPa*. When the load transfer between the matrix and fibers takes place for the first time, the response remains linear with a secondary modulus of 9.87 *GPa*. corresponding to a drop of about 30%. The tensile strength is evaluated to be 312.2 MPa at the peak.

The load versus displacement measured at a gage length of 140 mm and the crack opening displacement are presented. As an example Table 1 shows the number and type of specimens used in the investigation. The results were repeatable with an accuracy of 3% in the evaluation of the peak load and compliances. The post peak response of the COD reading indicates that the specimen looses energy as the number of loading cycles increases. This decrease is caused by a localized energy dissipation in the vicinity of the crack tip.

From the results shown in Figures 3 and 4, it can be seen that following a linear regime, non-linear prepeak response is first observed. This is associated with slow crack growth and thus the formation of a fracture tip inelastic zone turning the notch into a crack. At around the peak load, the COD decreases, thus effectively reducing the load. The load is then reapplied and the response then extends into the post-peak regime. Following the peak load is the softening or the crack propagation stage where strong nonlinearity occurs. This nonlinearity is caused by the formation of micro cracks ahead of the notch as the matrix begins to fail. This is characterized by an increase in displacement and corresponding decrease in load see Figures 3 and 4.

According to the results obtained, the peak load increases with specimen thickness. For example, the peak load for a twelve layer cracked specimen is 3.4 kN, while for similar sixteen and twenty layers specimens peaks of 4.1 kN and 6.9 kN respectively are obtained. The strength of the specimen can also be affected by the fiber orientation. Other test results showed that the peak load carried by (0/45/90) fiber orientation is larger than the one carried by a specimen with the same number of layers but with fiber orientation (0/90/0).

Clearly, we have found that there is stiffness degradation and loss of carrying capacity throughout the test. The stiffness decreases as the number of cycles beyond the peak load increases. The values of the stiffness are determined for a number of cycles. It has been found that The degradation in stiffness is due to matrix failure a head of the crack. As a result the load is carried predominantly by the fibers. The compliance values were calculated from the load versus COD plot.

TABLE 1. Test group #1

Number of Layers	Fiber Orientation	Thickness (mm)	Peak Load (N)	Number of Cycles
12	0/90/0	2.05	3308	50
16	0/90/0	2.50	3886	50
16	0/45/90	3.00	5372	50
20	0/90/0	3.25	6908	50

TABLE 2. Typical Results for 16 Layers Specimen of (0/90/0) Fiber Orientation

Cycle Number	Applied Load in Newtons	Total Elongation (mm)	COD Reading (mm)	Crack Length (mm)	Fracture Toughness $MPa\sqrt{m}$
1	4678	1.55	0.25	6.25	15.8
5	4161	1.58	0.28	7.85	17.4
10	3490	1.50	0.33	9.80	20.2
15	2988	1.43	0.38	11.25	21.7
20	2326	1.36	0.39	11.95	18.9
25	1661	1.30	0.45	13.65	18.1
30	1266	1.30	0.52	15.15	18.2
35	1145	1.32	0.58	16.45	21.5
40	835	1.34	0.65	19.85	38.5

3. Determination of Fracture Toughness

Under the assumption of linear elastic fracture mechanics (LEFM), fracture toughness can be evaluated using the concept of effective crack length.The critical load and a calibration function for a particular specimen depends on the geometry of the specimen. The geometries used herein have the following calibration function:

$$f\left(\frac{a}{w}\right) = 0.265\left(1 - \frac{a}{w}\right)^4 + \frac{0.857 + 0.265\left(\frac{a}{w}\right)}{\left(1 - \frac{a}{w}\right)^{3/2}} \tag{1}$$

Where a is the crack length and w is the width of the specimen. The specimen used in this investigation have a constant notch to depth ratio equal to 0.25.

The crack advance da is measured using a microscope with an accuracy of 0.001mm which is mounted to the Instron machine. The effective crack length a_{eff} is defined as the sum of the initial notch length and the effective crack advance or extension. When the effective crack length is updated at each cycle the fracture toughness can be calculated using:

$$K_{IC} = \frac{P_C}{w \cdot t} \cdot \sqrt{\pi \cdot a_{eff}} \cdot f\left(\frac{a}{w}\right) \tag{2}$$

Where

K_{IC} is the fracture toughness of the specimen,

P_C is the corresponding critical load,

w is the width of the specimen,

t   is the thickness of the specimen,

a_{eff} is the effective length of the crack and,

a_0 is the length of the crack.

Table (2) show typical results of fracture toughness K_{IC} obtained for 16 layers specimen for different loading cycles. The fracture toughness versus crack length is plotted in Figure (5). It is clear the fracture toughness is affected by the increase in crack length as well as the number of loading cycles.

4. Conclusions

The effect of thickness and fiber orientation on the fracture of composite materials is studied. The stiffness degradation, compliances and crack length are measured. Based on the experimental measurements, the following conclusions are made:

1- The initial linear response is followed by a secondary response which signals the load transfer between the matrix and the fibers. Prior to the peak load, non linearity occurs due to the formation of micro-cracks in the vicinity of the crack tip.

2- The post-peak response is marked with a gradual decrease in stress accompanying an increase in displacement.

3- The strain energy accumulated in the specimen is released at the peak-load and a stable crack propagates with increase in displacement and reduction in load.

4- The strength of the specimen is clearly affected by the fiber orientation i.e. the load carried by (0/45/90) fiber orientation specimen is larger than the one carried by a specimen with the same number of layers but with fiber orientation (0/90/0). The reason is that the crack approaches inclined fibers, it changes direction so as to run parallel to them for a considerable distance, sometimes to the end of the fiber, but sometimes eventually turning again to cross the fiber and resume forward progress.

5- The results showed a gradual degradation in stiffness and a concurrent loss of carrying capacity, i.e. beyond the peak load, the value of stiffness decreases as the number of cycles increases.

References

[1] *R. A. Kline & F. H. Chang*, COMPOSITE FAILURE SURFACE ANALYSIS - Journal of Composite materials, vol. 14, pp. 315-324, (1980).

[2] *A. A. Rubinstein*, STRENGTH AND TOUGHNESS OF FIBER REINFORCED BRITTLE MATRIX COMPOSITES - Proceeding of the U.S. - Europe Workshop on Fracture and Damage in Quasibrittle Structures. Edited by Z.P. Bazant, Z. Bittnar, M. Jirasek and J. Mazars, Prague, Czech Republic, September (1994).

[3] *S. S. Wang, E. S.-M. Chim, T. P. Yu and D. P. Goetz*, FRACTURE OF RANDOM SHORT-FIBER SMC COMPOSITE - Journal of Composite Materials, Vol. 17, pp. 299-315, July 1983.

[4] *S. S. Wang, E. S. M. Chim and N. M. Zahlan*, FATIGUE CRACK PROPAGATION IN RANDOM SHORT-FIBER SMC COMPOSITE - Journal of Composite Materials, vol.17, pp. 250-266, May 1983.

[5] Sun Guofang, FRACTURE OF FIBERGLASS REINFORCED COMPOSITES - Journal of Composite Materials, Vol.15, pp. 521-530, (1981).

[6] A. C. Marshall, COMPOSITE BASICS - Marshall consulting, California, USA, 1985.

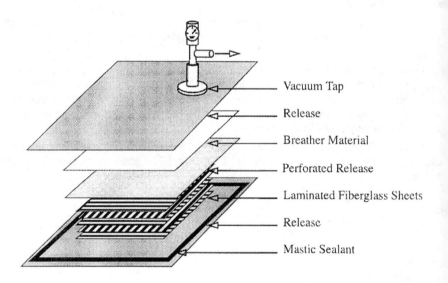

Figure.1. Specimen Preparation

Vacuum Tap

Release

Breather Material

Perforated Release

Laminated Fiberglass Sheets

Release

Mastic Sealant

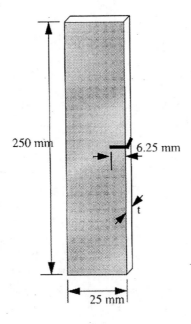

Figure. 2. Specimen Dimension

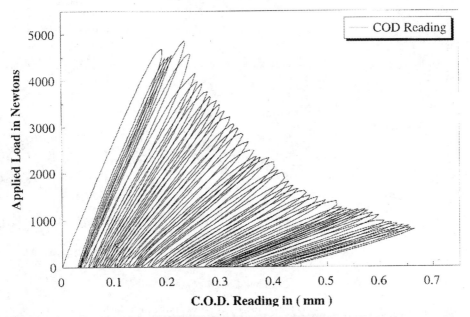

Figure. 3. Response of a 16 Layers Specimen of (0/90/0) Fiber Orientation

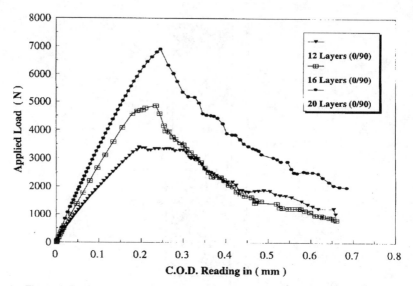

Figure. 4. Comparison of the Behaviour of Three Specimen of Different Thickness

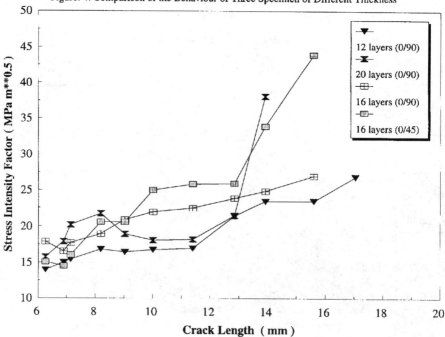

Figure. 5. Fracture Toughness Versus Crack Length

Failure Mechanisms of Carbon/Epoxy Laminated Composite Under Off-Axis Compression Loading

Tian Xiang Mao * and Vijay Gupta

Mechanical and Aerospace Engineering Department
University of California at Los Angeles

Abstract

An experimental investigation of the failure of selected compression-loaded $[\pm 10°]_{12s}$, $[\pm 45°]_{12s}$, $[\pm 80°]_{12s}$ Graphite/Epoxy composite laminates is described. A simple compressive test technique is used to obtain the experimental data. The dominant compression failure modes for the laminates in this study were found to be interlaminar shearing, fiber scissoring, in-plane matrix shearing and matrix compression.

1. Introduction

Long fiber-polymer matrix composites, such as carbon fiber in an epoxy matrix, possess excellent tensile properties resulting from the high tensile strength of the fibers. However, they fail in compression by plastic microbuckling at stresses of only 60% of their tensile strength. In many applications compressive strength is a design limiting feature. Over the past two decades   significant improvements have been made to the tensile strength, impact resistance and toughness of these composites. Unfortunately, compressive strength has shown little improvement.

In the past decade a lot of research work has been concentrated to the compressive failure of unidirectional composites and significant progress has been achieved. Budiansky and Fleck [1] carried out the elastic and plastic kinking analyses of microbuckling and it is able to account for some of the experimental observations for the compressive failure of unidirectional composites. They also pointed out that more detailed modeling of damage development in the off-axis plies is required, since the measured toughness G associated with delamination and splitting is more then two orders of magnitude greater than the dissipation due to microbuckling per unit area advance of the microbuckle.

Recently, extensive research work has been conducted in Professor Gupta's group on the compressive failure mechanism for various materials system under both unidirectional and bi-axial compression [2,3]. Gupta and Anand and Gupta and Grape described the failure mechanisms of the laminated carbon-carbon composites and two dimensional woven carbon/ polyimide laminates under unidirectional and bi-axial compression. They also

* On leave from the Institute of Mechanics, Chinese Academy of Sciences, Beijing, China

conducted experimental and analytical investigation of ±45° off-axis compression.

In this paper we investigate the failure mechanism of selected compression - loaded multidirectional composite laminates. Experimental results are presented for graphite-epoxy specimens. The specimens were tested using a simple compression test technique. The experimental results include compressive strength data and descriptions of laminate failure modes and failure mechanisms. The dominant compression failure modes for the laminates in this study were found to be interlaminar shearing, in-plane shearing and fiber scissoring and matrix compression.

2. Specimens, Apparatus and Tests

The graphite-epoxy composite specimens tested in this investigation were fabricated from commercially available unidirectional Hercules AS4 graphite fiber tapes preimpregnated with 450-K cure Hercules 3502 thermosetting epoxy resin. The tapes were laid to form 48-ply laminates approximately 6 mm thick. The laminates were cured with a computer controlled hot pressing machine in UCLA's composite manufacturing lab using the manufacturer's recommended procedure. All specimens were 25 mm long and 13 mm wide. The loaded ends of each specimen were machined flat and parallel to permit uniform compressive loading. The laminate stacking sequences are listed in Table 1.

The specimens were loaded to failure in axial compression using an Instron 8501, a 100 KN capacity hydraulic testing machine. The specimens were tested to failure by slowly applying a compressive load to simulate a static loading condition.

The test specimens and fixtures used in this study are simple and compact. The present compressive test method uses a specimen that is sized to be thick enough to avoid global buckling, and is adequate for providing a good distribution of the flows inherent in the microstructure. The limited specimen size also allowed generation of a reasonable set of data from a single composite panel, without the added panel-to-panel variability resulting from manufacturing. A schematic of the specimen showing the orientation of fibers and loading direction is shown in Fig. 1. All tests were performed at room temperature.

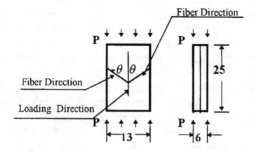

Fig. 1 Specimen (all dimension in mm)

3. Results and Discussion

The experimental results for this study are described in this section. The results include stress - strain curves and tabulated data. The laminates stacking sequence and failure stress data are given in Table 1. Failure strain is calculated using laminate end- shortening. Failure data for $[\pm10°]_{12s}$ and $[\pm80°]_{12s}$ laminates correspond to a catastrophic event that terminates the load-carrying capability of the laminates. Failure data for $[\pm45°]_{12s}$ laminates correspond to the near-zero slope of stress-strain curve. The failure modes are described and the experimental results are compared with previous results.

Table 1   Failure data for compression-loaded laminates

Stacking Sequence	Failure Stress MPa
$[\pm10°]_{12s}$	912
$[\pm45°]_{12s}$	181
$[\pm80°]_{12s}$	204
$[10°]_{48}$.	525
$[45°]_{48}$.	240

Typical stress-strain data for $[\pm10°]_{12s}$ specimens are shown in Fig. 2. These data are nearly linear to failure, and these specimens have the highest strength of all specimens tested. The dominant failure mode for this group is the interlaminar shearing. The interlaminar shearing failure mode is characterized by brooming and interlaminar cracking. Upon a closer look at the samples, the $[\pm10°]_{12s}$ samples fail by interlaminar shearing and some cases include the combination of interlaminar shearing and in-plane shearing.

Typical stress-strain data for $[\pm80°]_{12s}$ specimens are shown in Fig. 3. The data are very nonlinear, and the specimen also have very high failure strains. The dominant failure mode for this specimens is matrix compression. The matrix compression failure mode is characterized by a failure surface that extends through the laminate thickness and is oriented at 45 degrees to the middle surface.

Typical stress-strain data for $[\pm45°]_{12s}$ specimens are shown in fig. 4.   The degree of nonlinearity for $[\pm45°]_{12s}$ specimens is much higher and the failure strain for $[\pm45°]_{12s}$ specimens is higher than other two kinds of specimens tested in this investigation. The failure mode for $[\pm45°]_{12s}$ specimens is the combination of in-plane matrix shearing and fiber scissoring.

The specimens showed a linear stress-strain curve with no damage whatsoever on the composite surface prior to 150 MPa. Then there is a strong nonlinearity on stress-strain curve afterwards. At the peak load about 180 MPa , two sets of transverse cracks ,

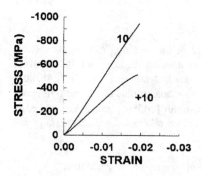

Fig. 2 Stress --Strain Curve for
±10° and 10° Laminates

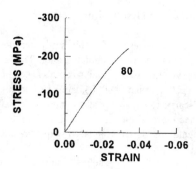

Fig. 3 Stress-- Strain Curve for
±80° Laminates

appearing as "shadow lines" to the naked eye, were observed on the specimens lateral face ABCD and EFGH (fig. 5). Upon closer examination, it is found that each shadow line on the upper half of the face ABCD has its counterpart on the lower half of the face EFGH, with each representing the two ends of a "scissoring fiber" emerging at the lateral surfaces. A SEM picture of one of the lateral surface is shown with cracks seen as white lines. During the compression, it was found that the specimen swelled in the middle part and the cracks were not found in the middle section of IJKL. On the other hand, on the outer layer of the $[\pm 45°]_{12s}$ specimen, in-plane matrix shearing was found with a crack along the fiber direction.

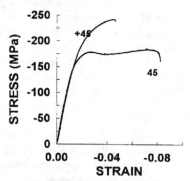

Fig. 4 Stress Strain Curve for ±45° and 45° Laminates

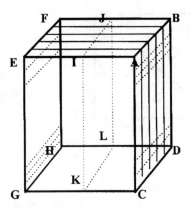

Fig. 5

For comparison with $[\pm 9°]_{12s}$ specimens, the compression test for specimens of $[10°]_{48}$ and $[45°]_{48}$ were also carried out.

The typical stress-strain curves for specimens $[10°]_{48}$ and $[\pm 10°]_{12s}$ are shown in Fig. 2. The failure load for $[10°]_{48}$ is about half the failure load for $[\pm 10°]_{12s}$, but the failure strains are almost the same. The failure mode for these two specimens are quite different. The failure mode for $[\pm 10°]_{12s}$ specimens is a combination of interlaminar shearing and in-plane matrix shearing and the failure mode for $[10°]_{48}$ specimen is in-plane matrix shearing.

The stress-strain curves for specimen $[45°]_{48}$ and $[\pm 45°]_{12s}$ are shown in Fig. 4. It is interesting to notice that both of the specimens behaved identically until the elastic limit of 150 MPa. After that they behaved quite different. The failure mode for $[45°]_{48}$ specimen is the combination of in-plane matrix shearing and the matrix compression failure. The fracture surface is parallel to the fiber direction, but not flat in the through-thickness direction just like the failure of matrix compression mode making a 45° angle with the middle plane. The failure mode for $[\pm 45°]_{12s}$ specimens is the combination of fiber-scissoring and in-plane matrix shearing. Fig. 6 shows the schematic view of the fiber scissoring. Under compression loading, the +45° fiber bundles in the $[\pm 45°]_{12s}$ specimens tend to scissor clockwise and the -45° fiber bundles tend to scissor counter-clockwise. On the left hand side lateral surface of the specimen, the lower parts of the +45° fiber bundles moved out from the surface and parts of the -45° fiber bundles moved out from the surface. This can be seen on the SEM picture. Both ends of the specimen attached with the loading fixture can't move because of friction, so the middle parts of the specimen swell. The most important evidence for fiber scissoring is that no cracks developed on the middle section of the specimen face IJKL. (Fig. 5)

Fig. 6

4. Conclusion

This paper describes the failure of selected compression loaded multidirectional Graphite/Epoxy composite laminates. Experimental results for $[\pm10°]_{12s}$, $[\pm45°]_{12s}$ and $[\pm80°]_{12s}$ laminates are presented and include failure stress, failure strain and laminate failure modes. For comparison, selected compression results for $[9°]_{48}$ specimens are also presented. Interlaminar shearing and in-plane matrix shearing are dominate failure modes for $[\pm10°]_{12s}$ specimens , in-plane matrix shearing and fiber scissoring are the failure modes for $[\pm45°]_{12s}$ specimens , and matrix compression is the failure mode for $[\pm80°]_{12s}$ specimens.

Acknowledge

The authors would like to thank Professor Tomas Hahn from the MAE department at UCLA for providing the materials and the manufacturing facilities in the Composite Manufacturing Lab. T. X. Mao would also like to thank the NSF and NSFC for its support.

Reference

[1] *B. Budiansky, N. A. Fleck,* COMPRESSION FAILURE OF FIBRE COMPOSITES - J. of Mech. Phys. Solids, 1993. Vol. 41, No. 1
[2] *K. Anand, V. Gupta,* A NUMERICAL STUDY OF THE COMPRESSION AND SHEAR FAILURE OF WOVEN CARBON-CARBON LAMINATES - J. of Composite Materials, 1995. Vol. 29, No. 18
[3] *J. A. Grape, V. Guppy,* FAILURE IN CARBON/POLYMIDE LAMINATES UNDER BIAXIAL COMPRESSION - J. of Composite Materials 1995. Vol. 29, No. 14
[4] *M. J. Shuart,* FAILURE OF COMPRESSION-LOADED MULTIDIRECTIONAL COMPOSITE LAMINATES - AIAA Journal, 1989, Vol. 27, No. 9
[5] *M. J. Shuart, J. G. Williams,* COMPRESSION BEHAVIOR OF ±45°-DOMINATED LAMINATES WITH A CIRCULAR HOLE OR IMPACT DAMAGE - AIAA Journal, 1986, Vol. 24, No. 1

Damage accumulation in non-crimp fabric based composites under tensile loading

Samantha Sandford[1], Lynn Boniface[1], Stephen Ogin[1], Subhash Anand[2], David Bray[3] and Clive Messenger[3].

[1]School of Mechanical and Materials Engineering, University of Surrey, Guildford, Surrey, GU2 5XH, UK; [2]Faculty of Technology (Textiles), Bolton Institute, Deane Road, Bolton, BL3 5AB, UK; [3]Structural Materials Centre, Defence Evaluation and Research Agency, Farnborough, Hampshire, UK.

Abstract

Damage development in a non-crimp fabric glass/epoxy cross-ply composite has been studied under tensile loading conditions. The main form of damage, transverse matrix cracking, has been characterised using microscopy and quantified as a function of applied strain. The form of cracking developed is very similar macroscopically to that seen in non-stitched cross-ply composites, although the stitching loops influence the crack location. The effect of matrix cracking on the residual laminate modulus has been measured and the experimental results have been compared with a prediction based on a shear-lag model which takes into account a feature of the laminate geometry resulting from the stitching. The shear-lag model based gives reasonable agreement with the experimental data.

Introduction

Textile structural composites have the potential for extending the engineering applications of composite materials significantly. Textile processes such as weaving, however, introduce deformation of the straight fibre tows with a consequent reduction in properties, an effect which has been modelled extensively by many workers including Chou, Naik and co-workers (e.g. Chou and Ko, 1990; Naik and Shembekar, 1992). A more recent addition to the class of textiles for composites is non-crimp fabric (NCF) which allows layers of continuous aligned fibre tows to be stitched together to produce multi-axial reinforcement with little or no crimp of the fibre (see for example Hogg and Woolstencroft, 1991).

Due to the lack of fibre crimp, damage accumulation during tensile loading in NCF-based composites might be expected to be very similar to damage accumulation in non-stitched composite laminates produced using conventional methods. However, the complication introduced by the stitching yarn leads to local variations in fibre volume fraction, resin-rich pockets and possible fibre misalignment (Godbehere et al, 1994), all of which can modify the laminate response. The present paper is concerned with characterising, quantifying and modelling the damage observed in a bi-axial NCF composite with a tricot knit stitching pattern.

Materials and experimental method

The reinforcing fabric used was Cotech® ELT566, a bi-axial 0/90 non-crimp, glass fibre fabric with polyester tricot knit stitching and balanced weights of fibres (283 gm^{-2}) in the warp and weft direction. A two-layer symmetric fabric (0/90)$_s$ laminate was manufactured in-house using a wet lay-up technique and an epoxy resin matrix (Astor Stag) to produce a transparent laminate. Figure 1 shows low and high magnification views of the fabric with figure 1(a) showing that the fibres in the warp direction are slightly misaligned during fabric manufacture, whereas the weft fibres are well aligned. Figure 1(b) shows the stitching loop structure and the resulting separation of the fibre bundles. Specimens (20 mm wide by 230 mm long, with 50 mm aluminium alloy end tabs) were cut with the length parallel to the warp (0°) direction and loaded in quasi-static tension using an Instron 1196 tensile testing machine. Strain was measured using a 50 mm gauge length extensometer and the stress-strain response was recorded using a computer data logger. Specimens were loaded to 0.4 % initially (which is below the onset of damage) and then incrementally at 0.1 % strain increments between 0.4 % and 1.4 % strain, recording the change in the Young's modulus as a function of applied strain and crack density. Young's modulus measurements were taken over the range 0.1 % to 0.4 % strain, which is below the onset of matrix cracking. In-situ plan view photographs were used to quantify damage development as a function of strain. In addition, photomicrographs were taken of the plan view and polished edge sections after failure in order to observe the damage morphology and laminate microstructure in detail.

Results and discussion

The total fibre volume fraction of the composite was measured from a burn-off test and found to be 46.2 %. However, edge sections (figure 2) showed that there was a resin-rich layer at the surface of about 0.06 mm thickness, which is probably due to the presence of the polyester stitching at the specimen surface. The overall laminate thickness is around 1.11 mm and the thicknesses of the 0° ply (b) and the half-thickness of the 90° ply (d) is approximately 0.3 mm and 0.2 mm, respectively.

Under tensile loading, crack initiation occurred at an applied strain, ε, of about 0.5 % (see figure 3, which shows experimental data from three specimens for crack density, 1/2s, as a function of applied strain, where 2s is the average crack spacing) and increases with applied strain in a very similar manner to non-stitched cross-ply laminates (e.g. Garrett and Bailey, 1977). Figure 4 shows plan view photographs after damage has initiated and just prior to failure. The cracks generally extend across the full width of the specimen at high strains. Edge sections (figure 2) show that the majority of the cracks extend across the full thickness of the 90° ply, in common with non-stitched cross-ply laminates and woven glass fabric composites made with untwisted tows (Boniface et al, 1995). Detailed observations indicate that cracks which extend across the full specimen width grow within the resin-rich regions formed as a consequence of the stitching and hence that they pass very close to, or through, the loops (figure 5).

Measurements of the Young's modulus during incremental loading showed that the modulus increased initially (prior to crack initiation) by about 1.4 % and then decreased once cracks began to form. Figure 6 shows Young's modulus (E) as a function of applied strain for three specimens. The origin of the initial modulus increase is not entirely clear but may be related to straightening of the misaligned warp fibre tows. The subsequent modulus reduction, which is related to crack development, is of the order of 12 % prior to failure. Figure 7 shows experimental data for the modulus normalised by the peak modulus prior to crack initiation (E_0) as a function of crack density. The initial increase in normalised modulus (E/E_0) reflects the small rise in the measured modulus prior to crack initiation.

Figure 7 also includes the results of a shear-lag model (Steif, 1984) relating to this laminate geometry, modified to take into account the resin-rich regions at the laminate surface. Figure 8 is a schematic diagram of the idealised laminate lay-up used for the modelling. Due to the surface resin-rich region, the fibre volume fraction in the $[0/90]_s$ sub-laminate is higher than the burn-off volume fraction and is calculated as 51.5 %. The values of the transverse ply modulus (E_2 =9.5 GPa) and the Poissons ratio, v_{23} ($v_{23} = 0.3$) were obtained from Sih (1979) for this fibre volume fraction.

For the modelling, it is necessary also to have a value for E_1', which is the effective modulus of the warp plies which have the wavy fibre tows (figure 1a). The effective 0° ply modulus (E_1') can be found from the rule of mixtures expression applied to the overall laminate fitted to the experimentally determined average peak undamaged modulus (E_0) of 20.3 GPa, giving a value for E_1' of 30.6 GPa. The rule of mixtures expression can again be used to find the effective undamaged modulus (E_0') of the $[0/90]_s$ sub-laminate, which is 22.2 GPa.

The reduced modulus (E') of the $[0/90]_s$ sub-laminate as a consequence of cracking is found by applying the shear-lag expression, to give:

$$\frac{E'}{E_0'} = \frac{1}{1 + \left[\frac{E_0'}{E_1'}\right]\left[\frac{b+d}{b} - \frac{E_1'}{E_0'}\right]\left[\frac{\tanh(\lambda s)}{\lambda s}\right]} \qquad \textbf{eqn. 1.}$$

where: $$\lambda^2 = \frac{\alpha\, G_{23}\,(b+d)E_0'}{d^2 b E_2 E_1'} \quad \text{and} \quad G_{23} = \frac{E_2}{2(1+v_{23})} \qquad \textbf{eqn 2.}$$

Here G_{23} is the out-of-plane shear modulus and s is half the crack spacing.

Having obtained a value for the reduced modulus, E', of the sub-laminate, the rule of mixtures can again be used to find the overall reduced laminate modulus, and hence E/E_0 as a function of crack density can be predicted. Predictions using $\alpha = 1$ and $\alpha = 3$ are shown in figure 7; $\alpha = 1$ assumes a linear variation of the longitudinal displacements in the transverse ply and $\alpha = 3$ assumes a parabolic variation. The prediction for $\alpha = 1$ is in reasonable agreement with the experimental data. It is interesting to note that in non-stitched cross-ply laminates, the assumption of a parabolic variation of the longitudinal displacements (i.e. $\alpha = 3$) usually gives better agreement with experimental data (e.g. Boniface et al, 1991). The much closer agreement here of the predictions based on $\alpha = 1$ may be related to differences in laminate geometry and response to cracking caused by the stitching, but further work is required to clarify this observation, including, possibly, the use of more sophisticated analyses (e.g. Nairn, 1989; McCartney, 1990).

Conclusions

Damage accumulation in a bi-axial non-crimp glass fabric based cross-ply laminate has been studied under tensile loading. Observations indicate a marked similarity in crack accumulation and damage morphology between this laminate and non-stitched cross-ply laminates. However, it is clear that due to the more complex fibre architecture of the NCF material there are clearly identifiable sites for crack growth formed by the resin-rich regions between the stitched fibre bundles. The experimental data for modulus reduction as a function of crack density are described reasonably well by a model based on a shear-lag analysis.

Acknowledgements

The authors would like to thank the EPSRC and DERA (Farnborough) for financial support during the course of this work, Tech Textiles International Ltd. for providing the fabric and Mr. R. Whattingham for technical assistance.

© *British Crown Copyright 1998/DERA. Published with permission of the Controller of her Britannic Majesty's Stationary Office.*

References

Boniface L, Ogin S.L. and Smith P.A. (1991), "Strain energy release rates and the fatigue growth of matrix cracks in model arrays in composite materials". *Proc. R. Soc. Lond. A. 432, p 427.*

Boniface L, Gao F, Marsden W.M, Ogin S.L, Smith P.A. and Greaves R.P. (1995), "Matrix cracking phenomena in glass and carbon woven fabric composites". *Proc. 3rd Int. Conf. 'Deformation and Fracture of Composites', University of Surrey, UK, p 317.*

Chou T.W. and Ko F.K. (1990), "Textile structural composites". *Elsevier Science publishers BV, Amsterdam.*

Garrett K.W. and Bailey J.E. (1977), "Multiple transverse fracture in 90° cross-ply laminates of a glass fibre-reinforced polyester". *J. Mats. Sci., Vol. 12, p 157.*

Godbehere A.P, Mills A.R. and Irving P. (1994), "Non crimped fabrics versus prepreg CFRP composites - a comparison of mechanical performance". *Proc. 6th Int. Conf. 'Fibre Reinforced Composites', Newcastle University, UK, p 6/1.*

Hogg P.J. and Woolstencroft D.H. (1991), "Non-crimp thermoplastic composite fabrics: Aerospace solutions to automotive problems". *Proc. 'Advanced Composite Materials: New Developments and Applications', Detroit, Michigan, USA, p 339.*

McCartney N. (1990), "Theories of stress transfer in a cross-ply laminate containing a parallel array of transverse cracks". *NPL Report DMA(A)189, National Physical Laboratory, Teddington, England.*

Naik N.K. and Shembekar P.S. (1992), "Elastic behaviour of woven fabric composites: 1 - Lamina analysis". *J. Comp. Mats., Vol. 26, p 2196.*

Nairn J.A. (1989), " The strain energy release rate of composite microcracking: a variational approach". *J. Comp. Mats, 23, p. 1106.*

Sih G.C. (1979), "Fracture mechanics of composite material", *Proc. 1ˢᵗ USA-USSR Symp. 'Fracture of Composite Materials', p 111.*

Steif P.S. (1984), Appendix to Ogin S.L, Smith P.A, Beaumont P.W.R. Technical report CUED/C/MATS/TR105, *Cambridge University Eng. Dept.*

Figures

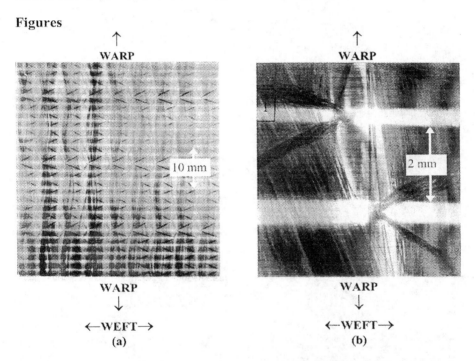

↑
WARP

10 mm

↑
WARP

2 mm

WARP
↓
←WEFT→
(a)

WARP
↓
←WEFT→
(b)

Figure 1   (a) Low and (b) high magnification views of one surface of the fabric.

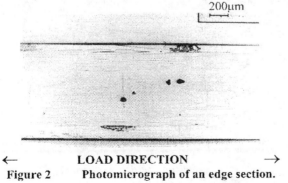

200μm

←   **LOAD DIRECTION**   →
Figure 2   Photomicrograph of an edge section.

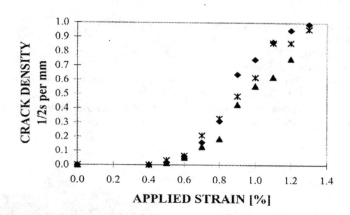

Figure 3   Crack density as a function of applied strain.

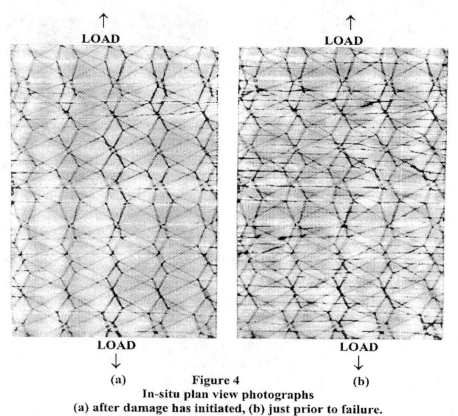

(a)   Figure 4   (b)
In-situ plan view photographs
(a) after damage has initiated, (b) just prior to failure.

(specimen width = 20 mm)

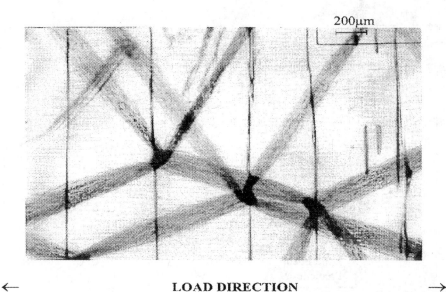

Figure 5
Photomicrograph of cracks intersecting stitch knots.

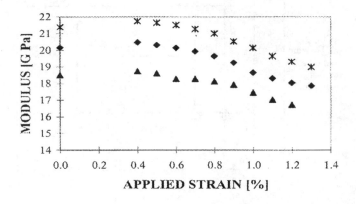

Figure 6
Young's modulus as a function of applied strain.

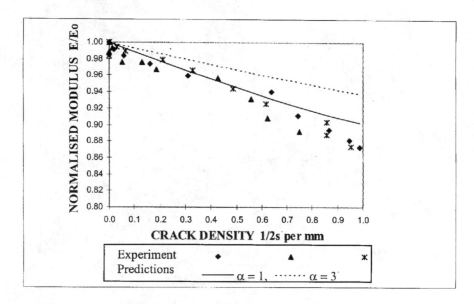

Figure 7
Normalised modulus as a function of crack density.
(experimental data and model for $\alpha = 1$ and $\alpha = 3$)

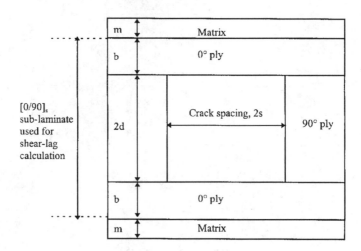

Figure 8
Schematic diagram of the idealised laminate lay-up used for modelling.

Analytic Characterization of
3-D Angle-Interlock Woven Composites

Joon-Hyung Byun, Sang-Kwan Lee, and Moon-Kwang Um

Composite Materials Lab, Korea Institute of Machinery and Materials
66 Sangnam-Dong, Changwon, Kyungnam, South Korea

Abstract

The engineering constants of 3-D (three-dimension) angle-interlock woven composites have been predicted based upon the geometric model and the elastic model. The geometric model defines the thickness of yarns, the cross-sectional aspect ratio of yarns, and the distance between two yarns in the machine direction and in the thickness direction, respectively. The elastic model utilizes the coordinate transformation and the averaging of stiffness and compliance constants. Since two types of unit cells are identified in the thickness direction of the preform, analytic characterization is based upon the macro-cell, which essentially occupies the repeating length in the machine direction, the unit width in the fill yarn direction, and the whole thickness. Relatively good agreement has been observed between the model prediction and the test results of carbon/epoxy composite samples.

1. Introduction

The development of innovative fiber architecture and textile manufacturing technology has significantly expanded the potential of fiber reinforced composites. The integrated fiber network of 3-D preforms provides stiffness and strength in the thickness direction, thus reducing the potential for interlaminar failure, which often occurs in laminated composites. Among several forms of 3-D textiles, angle-interlock weaves, which have two sets of fibers in the in-plane direction with the third fiber interconnecting the layers, can have better application in structural parts due to the possibility of wider panels or beams for various cross sections.

When textile preforms are designed, its microstructure, dimension and fiber volume fraction should be carefully controlled to meet the design requirements of the final composites. Thus, the objective of this study is to establish the analytic model for predicting the geometric characteristics and elastic constants of 3-D angle-interlock woven composites.

Several approaches have been applied to determine the elastic constants of 3-D fiber-reinforced composites: a 3-D lamination analogy [1], the fiber inclination model [2], and the averaging technique [3,4,5]. Although the lamination theory has been applied to 3-D textile composites [1,2], only the in-plane properties were obtained due to the assumption of the theory. Since 3-D textile composites frequently have a thick cross-section by introducing the through-the-thickness reinforcements, it is necessary to determine the elastic properties in all directions for a tailored design. The stiffness averaging method utilized by the above researchers [3,4,5] has some limitations in application: (1) the averaging was based on only stiffness, and (2) the unit cell does not represent the whole structure in the case of 3-D textile composites.

In this paper, the averaging technique based on stiffness and compliance is applied to predict the elastic constants of angle-interlock woven composites. Since there is no repeatable unit in the thickness direction of the preform, a representative volume of the composites is a macro-cell, which comprises of different types of unit cells.

2. 3-D Woven Structure

In 3-D angle-interlock textile preforms, yarns are interlaced in a manner similar to that of 2-D woven fabrics, except that warp yarns interlace several fill yarns in the thickness direction. The pattern of angle-interlock weaves can be represented in the x-y-z coordinate system. The x-axis is along the warp yarn direction or the length-wise axis of the textile forming direction. The y-axis and z-axis is the fill direction and thickness direction, respectively. Figure 1 shows the schematic 3-D woven architecture considered in this study. Since a warp yarn interlaces consecutive layers of fill yarns, the type of this preform is termed as *layer-to-layer* structure. Dots and lines indicate fill yarns and warp yarns, respectively. The unit cell, which is defined as the smallest repeating unit, is shown as dotted lines in Fig. 1. Unit cell 1 fails to represent the whole structure because it does not repeat on the top and bottom surfaces of the preform. Thus, the macro-cell is defined in the figure, which comprises unit cell 1 and 2, and occupies the repeating length in the machine direction, the unit width in the fill yarn direction, and the whole thickness. The number of fill yarns per unit length (N_f) is five in Fig. 1. The unit width is the same as the dent size of the reed, in which the same pattern of fill yarns repeats in the width direction.

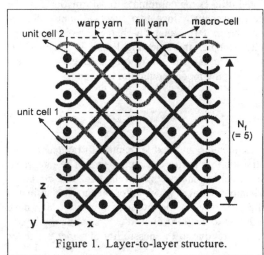

Figure 1. Layer-to-layer structure.

3. Geometric Relation

The properties of textile structural composites depend on the geometric pattern of the textile preform, the properties of its constituent materials, and their volume fractions. In order to link the microstructure of the preform to the performance of the composite, a geometric model for layer-to-layer structure is developed. The "lenticular" shape [6] of yarn cross section was adopted in the analysis due to the simple description for the wavy geometry of the crossing yarn. Since the interlacing of warp yarn and fill yarns is different in the 3-D interlock-woven structure, the geometry of yarns is discussed separately in the warp and fill yarn directions.

In order to understand the yarn geometry, the microstructures of preforms were observed. Figures 2 (a) and (b) show the yarn sections along the warp yarn and fill yarn directions.

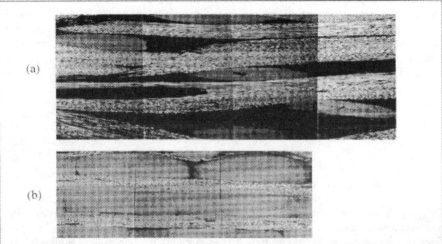

(a)

(b)

Figure 2 Microscopic yarn sections (a) along the warp yarn and (b)fill yarn directions.

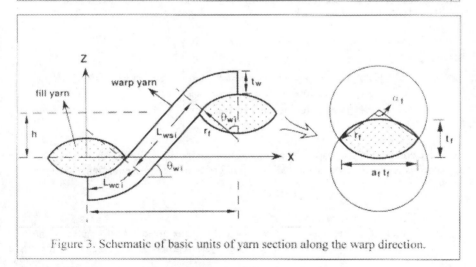

Figure 3. Schematic of basic units of yarn section along the warp direction.

Now, the basic unit of yarn is defined to identify the yarn geometry. They consist of a set of one or two warp and fill yarns, which can give basic geometric relations. Figures 3 shows the schematic of basic units of yarn sections along the warp direction. Following parameters are defined: t = the thickness of yarns; a = the cross-sectional aspect ratio of yarns; S, H = the distance between two yarns in the machine direction and in the thickness direction, respectively. The subscripts w and f denote the warp yarn and fill

4. Elastic Constants

The mechanical properties of 3-D woven composites have been predicted based upon the fiber and matrix properties and the three-dimensional fiber architectures resulted from the geometric model. The basic assumption in the analysis is that the yarns are considered unidirectional composite cylinders after resin impregnation. The composite cylinder is transversely isotropic in elastic properties, and its compliance matrix is expressed as

$$[S] = \begin{bmatrix} 1/E_{11} & -v_{21}/E_{22} & -v_{21}/E_{22} & 0 & 0 & 0 \\ -v_{12}/E_{11} & 1/E_{22} & -v_{32}/E_{22} & 0 & 0 & 0 \\ -v_{12}/E_{11} & -v_{23}/E_{22} & 1/E_{22} & 0 & 0 & 0 \\ 0 & 0 & 0 & 1/G_{23} & 0 & 0 \\ 0 & 0 & 0 & 0 & 1/G_{12} & 0 \\ 0 & 0 & 0 & 0 & 0 & 1/G_{12} \end{bmatrix} \tag{10}$$

The Young's and shear moduli are obtained from the fiber and matrix properties using micro-mechanics analysis.

Since a yarn composite locates spatially in the unit cell, its deformation properties are transformed to the orthogonal (reference) coordinate system. The direction cosines between the reference coordinate system, x-y-z, and the 1-2-3 coordinate system associated with the unidirectional composite can be established.

$$[T] = \begin{pmatrix} m^2 & 0 & n^2 & 0 & 2mn & 0 \\ 0 & 1 & 0 & 0 & 0 & 0 \\ n^2 & 0 & m^2 & 0 & -2mn & 0 \\ 0 & 0 & 0 & m & 0 & -n \\ -mn & 0 & mn & 0 & m^2-n^2 & 0 \\ 0 & 0 & 0 & n & 0 & m \end{pmatrix} \tag{11}$$

Thus, the compliance matrix of the unidirectional composite cylinder, referring to the 1-2-3 coordinate system, is transformed to [S'], referring to the x-y-z coordinate system.

$$[S'] = [T]^t [S] [T] \tag{12}$$

where, $[T]^t$ is a transpose matrix of [T].

By transforming the stresses and strains via the generalized transformation matrix defined in Eqn. (11), the compliance of a straight yarn segment is expressed as follows:

$$S_{11}^2 = S_{11} \cos^4\theta_w + S_{33} \sin^4\theta_w + (2S_{13} + S_{55})\cos^2\theta_w \sin^2\theta_w$$

or

$$S_{11}^x = U_1 + U_2 \cos 2\theta_w + U_3 \cos 4\theta_w \tag{13}$$

Here, $U_1 = (3S_{11} + 3S_{33} + 2S_{13} + S_{55})/8$; $U_2 = (S_{11} - S_{33})/2$; $U_3 = (S_{11} + S_{33} - 2S_{13} - S_{55})/8$.

For a crimp yarn segment, the average compliance is obtained by integrating the transformed compliance from the angle 0 to $2\theta_w$.

4. Elastic Constants

The mechanical properties of 3-D woven composites have been predicted based upon the fiber and matrix properties and the three-dimensional fiber architectures resulted from the geometric model. The basic assumption in the analysis is that the yarns are considered unidirectional composite cylinders after resin impregnation. The composite cylinder is transversely isotropic in elastic properties, and its compliance matrix is expressed as

$$[S] = \begin{bmatrix} 1/E_{11} & -v_{21}/E_{22} & -v_{21}/E_{22} & 0 & 0 & 0 \\ -v_{12}/E_{11} & 1/E_{22} & -v_{32}/E_{22} & 0 & 0 & 0 \\ -v_{12}/E_{11} & -v_{23}/E_{22} & 1/E_{22} & 0 & 0 & 0 \\ 0 & 0 & 0 & 1/G_{23} & 0 & 0 \\ 0 & 0 & 0 & 0 & 1/G_{12} & 0 \\ 0 & 0 & 0 & 0 & 0 & 1/G_{12} \end{bmatrix} \qquad (10)$$

The Young's and shear moduli are obtained from the fiber and matrix properties using micro-mechanics analysis.

Since a yarn composite locates spatially in the unit cell, its deformation properties are transformed to the orthogonal (reference) coordinate system. The direction cosines between the reference coordinate system, x-y-z, and the 1-2-3 coordinate system associated with the unidirectional composite can be established.

$$[T] = \begin{pmatrix} m^2 & 0 & n^2 & 0 & 2mn & 0 \\ 0 & 1 & 0 & 0 & 0 & 0 \\ n^2 & 0 & m^2 & 0 & -2mn & 0 \\ 0 & 0 & 0 & m & 0 & -n \\ -mn & 0 & mn & 0 & m^2 - n^2 & 0 \\ 0 & 0 & 0 & n & 0 & m \end{pmatrix} \qquad (11)$$

Thus, the compliance matrix of the unidirectional composite cylinder, referring to the 1-2-3 coordinate system, is transformed to [S'], referring to the x-y-z coordinate system.

$$[S'] = [T]^t [S] [T] \qquad (12)$$

where, $[T]^t$ is a transpose matrix of [T].

By transforming the stresses and strains via the generalized transformation matrix defined in Eqn. (11), the compliance of a straight yarn segment is expressed as follows:

$$S_{11}^2 = S_{11} \cos^4 \theta_w + S_{33} \sin^4 \theta_w + (2S_{13} + S_{55}) \cos^2 \theta_w \sin^2 \theta_w$$

or
$$S_{11}^x = U_1 + U_2 \cos 2\theta_w + U_3 \cos 4\theta_w \qquad (13)$$

Here, $U_1 = (3S_{11} + 3S_{33} + 2S_{13} + S_{55})/8$; $U_2 = (S_{11} - S_{33})/2$; $U_3 = (S_{11} + S_{33} - 2S_{13} - S_{55})/8$.

For a crimp yarn segment, the average compliance is obtained by integrating the transformed compliance from the angle 0 to $2\theta_w$.

$$S_{ij}^c = \frac{1}{2\theta_w} \int_0^{2\theta_w} S_{ij} d\phi \quad (i, j = 1 - 6) \tag{14}$$

Here, S'_{ij} are the transformed compliance of the infinitesimal yarn segment inclined with respect to the x-axis. Thus, the compliance of a crimp yarn segment is

$$S_{11}^c = U_1 + \frac{U_2}{2\theta_w} \sin(2\theta_w) + \frac{U_3}{4\theta_w} \sin(4\theta_w) \tag{15}$$

The compliance of a warp yarn are obtained by averaging the compliance of a straight yarn segment and those of a crimp yarn segment. Denoting λ_1 and λ_2 as the length fractions of a straight yarn segment and a crimp yarn segment, respectively, the average compliance of a warp becomes:

$$S_{ij}^w = S_{ij}^s \lambda_1 + S_{ij}^c \lambda_2 \quad (i, j = 1 - 6) \tag{16}$$

where, $\lambda_1 = L_{ws} / (L_{ws} + 2L_{wc})$ and $\lambda_2 = 2L_{wc} / (L_{ws} + 2L_{wc})$. Using trigonometric identities, all the compliance constants of a warp and fill yarn can be written as follows:

$$
\begin{aligned}
S_{11}^w &= U_1 + U_2\xi_1 + U_3\xi_2 \quad ; \quad S_{12}^w = U_6 + U_7\xi_1 \\
S_{13}^w &= U_4 - U_3\xi_2 \quad ; \quad S_{15}^w = U_2\lambda_1 \sin 2\theta_w + 2U_3\lambda_1 \sin 4\theta_w \\
S_{23}^w &= U_6 - U_7\xi_1 \quad ; \quad S_{25}^w = 2U_7\lambda_1 \sin 2\theta_w \\
S_{33}^w &= U_1 - U_2\xi_1 + U_3\xi_2 \quad ; \quad S_{35}^w = U_2\lambda_1 \sin 2\theta_w - 2U_3\lambda_1 \sin 4\theta_w \\
S_{44}^w &= U_8 + U_9\xi_1 \quad ; \quad S_{46}^w = -U_9\lambda_1 \sin 2\theta_w \quad ; \\
S_{55}^w &= U_5 - 4U_3\xi_2 \quad ; \quad S_{66}^w = U_8 - U_9\xi_1
\end{aligned} \tag{17}
$$

where, $U_4 = (S_{11} + S_{33} + 6S_{13} - S_{55})/8$; $U_5 = (S_{11} + S_{33} - 2S_{13} + S_{55})/2$

$U_6 = (S_{12} + S_{32})/2$; $U_7 = (S_{12} - S_{32})/2$; $U_8 = (S_{44} + S_{66})/2$; $U_9 = (S_{44} - S_{66})/2$

$\xi_1 = \lambda_1 \cos 2\theta_w + \lambda_2 \sin 2\theta_w$; $\xi_2 = \lambda_1 \cos 4\theta_w + \lambda_2 \sin 4\theta_w / 4$

It should be noted that θ_{w1} and θ_{w2} should be substituted for the unit cell 1 and 2, respectively. In order to determine the effective stiffness of warp yarns, warp yarn stiffness of unit cell 1 and unit cell 2 are averaged over the macro-cell volume.

$$\bar{C}_{ij}^w = C_{ij}^1 (\frac{N_f - 1}{N_f}) + C_{ij}^2 \frac{1}{N_f} \quad (i, j = 1 - 6) \tag{18}$$

Here, 1 and 2 denote unit cell 1 and 2, respectively.

Next, the fill direction is considered. Since fill yarns are assumed to be straight as shown in Fig. 2(b), the effective compliance is the same in Eqn. (10). For the determination of the effective stiffness of the composites, the compliance of fill yarns are inverted to stiffness ($\bar{C}_{ij}^f$), and then they are averaged over the macro-cell volume. The average should include the warp yarns, fill yarns, and matrix material.

$$C_{ij}^c = \bar{C}_{ij}^w \frac{V_w}{V_t} + \bar{C}_{ij}^f \frac{V_f}{V_t} + C^m (1 - v_y) \quad (i, j = 1 - 6) \tag{19}$$

where v_y is defined in Eqn. (9), and m denotes the matrix material. Finally, the stiffness are inverted to the compliance (S^c_{ij}), resulting in the elastic constants of the composite:

$$E_{xx} = 1/S^c_{11} ; \quad E_{yy} = 1/S^c_{22} ; \quad E_{zz} = 1/S^c_{33};$$

$$G_{yz} = 1/S^c_{44} ; \quad G_{xz} = 1/S^c_{55} ; \quad G_{xy} = 1/S^c_{66}; \quad\quad (20)$$

$$v_{xy} = -S^c_{12}/S^c_{22} ; \quad v_{xz} = -S^c_{31}/S^c_{11} ; \quad v_{yz} = -S^c_{32}/S^c_{22}$$

5. Corelation

In order to verify the analytic model, tensile tests have been conducted for layer-to-layer structure of composite samples. The preform shown in Fig. 4 was fabricated using Toray 12K T700 carbon fibers, and it was consolidated with epoxy resin by the resin transfer molding technique. The samples were cured at 120°C for two hours. The dimension of specimens was 25.4mm x 3.5mm x 230mm, and they were instrumented with a biaxial gage of 10mm in gage length. Table 1 summarizes the input, and the results of model prediction and experiments. The geometric data were obtained from the microscopic picture of the sample section. The measurement of the yarn packing fraction, κ, was not made directly.

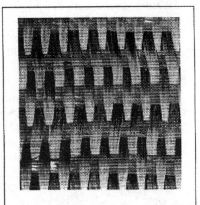

Figure 4. Pattern of the preform.

Instead, the fiber volume fraction of the sample was measured by the acid digestion method, then it was checked with the predictions in Eqn. (9) for a trial value of κ until these two fractions coincides. The fiber packing fraction of 0.86 gave the fiber volume fraction of the measurement and the prediction gave 0.462.

Table 1: Summary of input data, the model predictions and experimental results.

Input Data		Model Prediction	Test Results
Mechanical Data	Geometric and Physical Data	Engineering Constants	Engineering Constants
T700 Carbon Fiber	$t_w = 0.35$ mm	$E_{xx} = 76.5$ GPa	$E_{xx} = 74.0$ GPa
E_{1f}=250 GPa	$t_f = 0.35$ mm	$E_{yy} = 23.2$ GPa	$E_{yy} = $ - GPa
E_{2f}=20 GPa	$a_w = 7.2$	$E_{zz} = 13.7$ GPa	$E_{zz} = $ - GPa
G_{12f}=26 GPa	$a_f = 9.5$	$G_{yz} = 3.95$ GPa	$G_{yz} = $ - GPa
G_{23f}=6 GPa	$s_f = 9.0$	$G_{zx} = 16.4$ GPa	$G_{zx} = $ - GPa
v_{12f}=0.23	$h_f = 0.63$	$G_{xy} = 6.35$ GPa	$G_{xy} = $ - GPa
Epoxy	$n_f = 5$	v_{yz}= 0.54	v_{yz}= -
E_m=6 GPa	$\kappa = 0.86$	v_{zx}= 0.39	v_{zx}= -
v_m=0.38		v_{xy}= 0.05	v_{xy}= 0.156

Relatively good agreement between the model prediction and the test results can be observed. Although the experimental data was not enough to support the predictions due to the lack of established test methods and size limitation of the current samples for this study, the methodology proposed in this paper can be effectively utilized in obtaining the three-dimensional elastic constants of 3-D textile composites.

6. Conclusion

(1) An analytic model has been proposed to predict the engineering constants of the layer-to-layer structure of 3-D angle-interlock woven composites. Two types of unit cells were identified from the sample cross-section. The analytic approach was based upon the macro-cell, which basically occupies the repeating length in the machine direction, the unit width in the fill yarn direction, and the whole thickness of the composites.

(2) The coordinate transformation and the averaging of stiffness and compliance constants are utilized in the prediction of elastic constants. Depending on the assemblage of the unit cells in serial or parallel connections, the averaging was based upon compliance or stiffness. In order to verify the analytic model, tensile tests have been conducted for the carbon/epoxy samples. The experimental results compared favorably with the model predictions.

References

[1] T. J. Whitney, T-W. Chou, MODELING OF 3-D ANGLE-INTERLOCK TEXTILE STRUCTURAL COMPOSITES – J. Composite Materials, Vol. 23, 1989, pp. 890 – 911.

[2] J-M. Yang, C. L. Ma, T-W. Chou, FIBER INCLINATION MODEL OF THREE-DIMENSIONAL TEXTILE STRUCTURAL COMPOSITES, J. of Composite Materials, Vol. 20, 1986, pp. 472 – 484.

[3] A. F. Kregers, G. A. Teters, USE OF AVERAGING METHODS TO DETERMINE THE VISCOELASTIC PROPERTIES OF SPACIALLY REINFORCED COMPOSITES, Mechanics of Composite Materials, Vol. 4, 1979, pp. 617 – 624.

[4] A. F. Kregers, G. A. Teters, STRUCTURAL MODEL OF DEFORMATION OF ANISOTROPIC THREE-DIMENSIONALLY REINFORCED COMPOSITES, Mechanics of Composite Materials, Vol. 18, No. 1, 1982, pp. 10 – 17.

[5] F. K. Ko, THREE-DIMENSIONAL FABRICS FOR COMPOSITES, Chapter 5 in Textile Structural Composites, Composite Materials Series 3, edited by T-W. Chou and F. R. Ko, Elsevier, Amsterdam, 1989.

[6] W. J. Shanahan, J. W. S. Hearle, AN ENERGY METHOD FOR CALCULATIONS IN FABRIC MECHANICS, PART II: EXAMPLES OF APPLICATION OF THE METHOD TO WOVEN FABRICS, Journal of Textile Institute, Vol. 69, 1978, pp. 92 – 100.

ANALYSIS OF SATIN AND TWILL WEAVE COMPOSITE DAMAGE UNDER TENSILE LOAD: ANALYTICAL MODELLING OF FAILURE BEHAVIOUR

D. SCIDA *, Z. ABOURA *, M.L. BENZEGGAGH * & E. BOCHERENS **

* Université de Technologie de Compiègne LG2mS - « Polymères & Composites »
UPRESA 6066–CNRS BP 20529 60205 COMPIEGNE cedex France
E-mail: Daniel.Scida@utc.fr

** Délégation Générale pour l'Armement
Atelier Industriel de l'Aéronautique de Cuers Pierrefeu-Division « Radômes et composites »
BP 888 83800 TOULON France

ABSTRACT

An analytical model called MESOTEX (MEchanical Simulation Of TEXtile) is proposed in this paper with the aim to predict 3D elastic and failure properties of satin and twill woven composite materials. Using the Classical Laminate Theory (CLT), applied to these woven structure, this analytical model takes into account the strands undulation in the two directions and integrate also geometrical and mechanical parameters of each constituent (resin, fill and warp strands).
Different stages of failure, such as failure of matrix and strand in the longitudinal and transverse tension direction, have been considered to predict the stress-strain behaviour.
The study of these damage mechanics have been also identified by the use of acoustic emission (AE) and scanning electron microscopic (SEM) observations. A good agreement was observed between these mechanisms viewed on the SEM photographs, the AE analysis and the different stages of failure predicted with the model.

KEY WORDS : Woven composite materials, Laminate theory, Elastic and failure properties, Failure behaviour, Acoustic emission, SEM photograhs.

1. INTRODUCTION

A growing interest in textile composites has been observed in recent years. Their use in the fabrication of high mechanical performance structures is more and more frequent in the field of aerospace, aeronautic, naval construction or automobile.
Woven composite material represents a type of textile composites where strands are formed by the process of weaving; these strands are interlaced in the two mutually orthogonal (warp and fill) directions to one another and impregnated with a resin material. Composite materials reinforced with woven fabric have better out-of-plane stiffness, strength and toughness properties than laminate composites. However, the geometry of this composite class is complex and the possible architectures as well as constituents choice are unlimited. Many parameters of woven composite materials can

be changed, as microstructure geometry, weaving type, hybridisation or constituents choice (geometrical and mechanical parameters of strands and resin).

Thus, in order to select woven composite materials with the best possible combination for weight, cost, stiffness and strength properties, their mechanical performance must be predicted by validated analytical models. In the 1980's, Ishikawa and Chou [1-3] proposed models based on the classical laminate theory to predict the elastic stiffness of woven composite materials. They presented three analytical models for the stiffness and strength investigation of woven composites (plain and satin weaves). Many other researchers have attempted to define mathematically the geometry of 2D orthogonal plain weave fabric [4-9]. Naik and Ganesh [4-6] considered the strands continuity along both fill and warp directions and the inter-strand gap presence and also simulated in detail strand cross section and strand undulation. Vandeurzen and al. [7-9] proposed a 3D geometry description of several woven composite architecture. The full geometry of woven architecture is built from rectangular macro-cells assemblage. An analytical model, called combi-cell model, is developed; it's based on modelling each strand system with a matrix and a strand layer. The stiffness values predicted by applying the complementary principe are compared with finite element models and show a good correlation.

Aboura and al. [10] and Chouchaoui and al. [11] presented a similar model to Naik and Ganesh [4-6]; it takes into consideration the strands undulation according to the x and y-directions and the possibility to superpose several layers in the right and wrong side with or without a relative translatory motion. Scida and al. [12-15] proposed a similar model extended to other woven composites materials and to failure behaviour prediction. More, this software is adapted to the hybridisation principle which is an advantageous solution to satisfy specific cost and performance requirements.

The present paper is concerned with an analytical analysis of two woven composite materials. Called MESOTEX (MEchanical Simulation Of TEXtile) and based on the use of the classical laminate theory to the woven structure, this modelling is presented in order to predict 3D elastic properties, damage initiation and progression and strength in satin and twill woven composite materials.

The calculated stiffness properties is first compared with test data and followed by the prediction of failure strengths under tensile load in these woven composite materials. Using a point-wise stiffness reduction technique, the calculated ultimate failure strength, stresses at different stages of failure and the stress-strain history are correlated well by available test data, acoustic emission measurements and scanning electron microscopic observations.

2. ANALYTICAL MODELLING

The technique proposed for the mechanical behaviour prediction is a point-wise lamination approach using CLT in which undulated fibres are considered as an assemblage of many infinitely small pieces of unidirectional lamina oriented at an off-axis angle. So, from CLT, the resultant forces and moments, N and M respectively, can be expressed in terms of strains and curvature, ε and κ respectively, as

$$\begin{Bmatrix} N \\ M \end{Bmatrix} = \begin{bmatrix} A(x,y) & B(x,y) \\ B(x,y) & D(x,y) \end{bmatrix} \begin{Bmatrix} \varepsilon^0 \\ \kappa \end{Bmatrix} \qquad (1)$$

in which $A(x,y)$, $B(x,y)$ and $D(x,y)$ are the stiffness coefficients for each infinitesimal element defined by,

$$A_{ij}(x,y), B_{ij}(x,y), D_{ij}(x,y) = \int_{-\frac{h}{2}}^{\frac{h}{2}} (1, z, z^2) \overline{Q}_{ij}^I dz \qquad (i,j = 1 \text{ to } 6) \quad (2)$$

in which the « I » superscript refers to either fill strand (F), warp strand (W) or matrix (M) and the $\overline{Q}_{ij}^I$ transformed stiffness matrix (with respect to a global coordinate system) is evaluated for the « I » element (warp or fill strand and matrix) by the Q_{ij}^I stiffness matrix, as follows

$$[\overline{Q}_{ij}^I] = [T_{ij}^I]^{-1} [Q_{ij}^I][R_{ij}][T_{ij}^I][R_{ij}]^{-1} \qquad (3)$$

in which R and T are respectively the Reuter matrix and the stress transformation matrix.

Once the $\overline{Q}_{ij}^I$ terms are calculated, the A, B and D 3D stiffness matrices can be evaluated for each element in different regions using equation 2 [12-15].

Moreover, this analytical model is extended in order to predict failure strength in this woven fabric composites with a point-wise stiffness reduction technique, whereby the global stiffness is reduced as local failures occur.

Actually, in order to understand the failure mechanism in woven composites, the local stresses and strain in the unit cell must be computed, as follows, for a N_i given in-plane load,

$$\begin{Bmatrix} \varepsilon_x \\ \varepsilon_y \\ \gamma_{xy} \end{Bmatrix} = \begin{bmatrix} a_{11} & a_{12} & a_{16} \\ a_{12} & a_{22} & a_{26} \\ a_{16} & a_{26} & a_{66} \end{bmatrix} \begin{Bmatrix} N_x \\ N_y \\ N_{xy} \end{Bmatrix} \qquad (4)$$

To determine when failure occurs in the unit cell, a failure criterion is applied to the calculated stresses after a transformation from the global coordinate system to their principal direction.

The criterion used for matrix is based on the maximum stress failure criterion and a Tsai-Wu failure criterion is applied in the fill and warp strands, as follows,

$$F_i \sigma_i + F_{ij} \sigma_i \sigma_j = 1 \qquad (i,j = 1 \text{ to } 6) \qquad (5)$$

Due to the difficulty in obtaining the F_{ij} out-of-plane parameters [17], the Tsai-Wu failure criterion is applied in the in-plane direction and the relation 5 can be written as,

$$F_1 \sigma_1 + F_2 \sigma_2 + F_{11} \sigma_1^2 + F_{22} \sigma_2^2 + F_{66} \sigma_{66}^2 + 2 F_{12} \sigma_1 \sigma_2 = 1 \qquad (6)$$

in which $\quad F_1 = \dfrac{1}{X^+} - \dfrac{1}{X^-} \qquad F_{11} = \dfrac{1}{X^+ X^-} \qquad F_{66} = \dfrac{1}{S^2}$

$$F_2 = \frac{1}{Y^+} - \frac{1}{Y^-} \qquad F_{22} = \frac{1}{Y^+ Y^-}$$

and the F_{12} term is estimated by Von Mises criterion as $F_{12} = -0.5 \sqrt{F_{11} F_{22}}$

3. MATERIALS AND EXPERIMENTAL PROCEDURE

Materials

Two woven composite materials have been considered in this experimental study, an eight-harness satin weave composite and a 2/2 twill weave composite with glass reinforcement and epoxy resin. The (V_f) overall fibre volume fraction of the composites was determined by loss of fire method. Because woven composite materials contain regions of pure matrix, the (V_f^s) fibre volume fraction in the strand – as estimated with photomicrographs observations – exceeds the overall fibre volume fraction. The photomicrographs observations have been used also to determine weave geometry parameters of each woven composite material such as the strand width, the strand thickness, and the unit-cell dimensions as schematically depicted in figure 1 and 2. The experimental data used to validate analytical and numerical results was obtained from two tests. The (E_x and E_y) Young's moduli and the (υ_{xy}) Poisson's ratio were measured from tensile test on an INSTRON 1186 universal static testing machine with a 2 mm/min moving rate. Through the Lekhniskii procedure, a shear machine marketed by PRODEMAT company (Villeurbanne, France) was used to measure the (G_{xy}) shear modulus. Due to the difficulty in obtaining experimental specimens with high thickness, the (E_z, υ_{xz} and G_{xz}) mechanical properties in the third direction could not be estimated experimentally. Each specimen was instrumented with a biaxial strain gage and an acoustic emission captor.

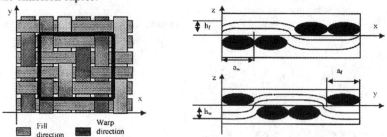

Figure 1 : 2/2 twill woven fabric
Unit cell (a) and fibre undulation in fill and warp direction (b)

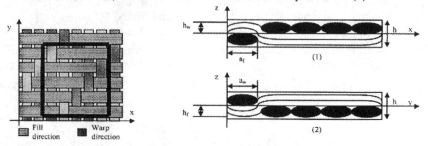

Figure 2 : Five-harness satin weave
Unit cell (a) and fibre undulation in fill and warp direction (b)

Acoustic emission procedure

AE monitoring is carried out by using Dunegan Endvco-3000 instrumentation. The acoustic waveforms are detected by a piezo electric transducer (Pac micro 80) with a wide band frequency (200kHz - 1 MHz), instrumented on the specimens with silicon oil [18]. A high pass filter with 40 dB fixed gain preamplifier is used to obtain a total of 94 dB. In order to have maximum sensivity during tests, a predetermined threshold is set providing a ring down count of 15-25 counts/h simply from background noise but higher than the level of electronical and mechanical noises. The total ring down cumulative counts and the amplitude distribution are recorded during the test.

4. RESULTS

Elasticity

The first table summarises the mechanical properties of each strand used in this study. Tables 2 and 3 compare the predicted elastic properties using the analytical and numerical model [15] discussed in this paper with test data. The tables show a good agreement between measured and predicted Young's moduli, Poisson's ratios and shear moduli.

Table 1. Mechanical properties of strands

E-glass/epoxy Strand	E_1 GPa	E_2 GPa	G_{12} GPa	G_{23} GPa	υ_{12}	υ_{23}	σ_{1t} MPa	σ_{1c} MPa	σ_{2t} MPa	σ_{2c} MPa	τ_{12} MPa
75 %	55,7	18,5	6,89	6,04	0,22	0,34	1551	721	46	141	85
80 %	59,3	23,2	8,68	7,60	0,21	0,32	1655	769	50	150	89

In table 2, the E_z Young's modulus and the G_{xz} shear modulus predicted with the analytical modelling are much higher than the numerical model values. Experimental tests with high thickness specimens will allow to estimate the E_z, υ_{xz} and G_{xz} mechanical properties and to compare predicted and measured values.

Table 2. Comparison Model / Experience for a E-glass eight-harness satin weave/epoxy

	$E_X=E_Y$ [GPa]	E_Z [GPa]	G_{XY} [GPa]	$G_{XZ}=G_{YZ}$ [GPa]	ν_{XY}	$\nu_{XZ}=\nu_{YZ}$	CPU (s)
Experience	25,6 ± 0,2	n/a	5,7 ± 0,3	n/a	0,13 ± 0,005	n/a	
Analytical Model	26,03	15,65	5,67	5,42	0,125	0,283	8,7
Numerical Model	25,84	12,50	5,47	3,60	0,152	0,325	462

n/a: Not available

Table 3. Comparison Model / Experience for a 2/2 twill E-glass woven fabric /epoxy

	$E_X=E_Y$ [GPa]	E_Z [GPa]	G_{XY} [GPa]	$G_{XZ}=G_{YZ}$ [GPa]	ν_{XY}	$\nu_{XZ}=\nu_{YZ}$	CPU (s)
Experience	19,2 ± 0,2	n/a	3,6 ± 0,1	n/a	0,13 ± 0,005	n/a	
Analytical Model	19,54	10,92	3,92	3,78	0,122	0,305	3,5
Numerical Model	17,97	8,05	3,31	2,20	0,14	0,38	250

n/a: Not available

Failure

Under on-axis uniaxial static tensile loading, the E-glass/epoxy satin and twill woven composite behaviour show different stages of failures; the figure 4 presents two analytical modelling examples of the stress-strain plots. The experimental result of tensile loading and of total count in acoustic emission measurement are also given. The different stages of failure are indicated by *a-c* on the stress-strain plots (fig.4) and by the SEM photographs (fig.5).

The *a* point indicates the initiation and the complete failure in the warp strand. The stress-strain curve between *0-a* is the region when no failure has taken place. The *b* point indicates that failure criterion was right in the matrix region.

The fill strand failure under longitudinal tension is shown by the *c* point; this failure is assumed to be the final failure of the unit-cell and actually of the woven composite materials. the different stages of failure in these two woven composites (fig.4) are:

> - warp strand failure *(a)*
> - matrix failure *(b)*
> - fill strand failure *(c)* corresponding of the ultimate failure of the woven

composite.

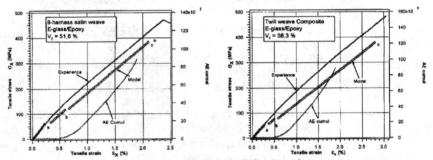

Figure 4 : Stress-strain behaviour for
(a) E-glass/Epoxy satin weave composite (V$_f$ = 51,6 %)
(b) E-glass/Epoxy twill weave composite (V$_f$ = 38,3 %)

The predicted stress-strain curve correlated well with experimental curve, in particular between *0-a*, when no damage and accordingly no AE events have occurred (fig.4). AE events are associated with an increased level of damage in material. Fig. 4 shows that the AE accumulation begins at 0,5 % strain level. This value attests the woven composite materials are damaged and have to tolerate damage (D>0). It's important to note that the total count initiation in AE measurement, characteristic of composite materials damage initiation, happens when warp and matrix failure criterion are right. After the first failure criterion occurs, the predicted stress-strain curve between *a-b* presents a discontinuity in *a* and a small downward movement compared with experimental curve. This discontinuity and downward movement are the effect of stiffness reduction procedure: when failure initiation in the warp strand happens, the global stiffness is reduced because the local stiffness for the warp strand is equal to zero. The same procedure was used for the two others stages of failure, indicated by *b* and *c*.

This simplest and penalising stiffness reduction technique shows good agreement between predicted and experimental results : the predicted strengths are between 17-21% much lower of the measured strengths.

Load direction

Load direction

Figure 5 : Failure in transverse strand of an E-glass/epoxy satin weave composite

5. CONCLUSION

A 3D mechanical modelling of two woven composite materials has been examined in this work. This approach is an analytical modelling based on the use of the classical laminate theory to the woven structure.

This method allows predicting the 3D elastic properties of several woven composite materials. The effectiveness of this analysis is validated with experimental data because a good correlation is obtained between analytical prediction and experimental results.

With this analytical technique easy to use, the prediction of the ultimate failure strength, stresses at different stages of failure and the stress-strain histories of several woven composite materials under tensile loading were determined.

The present analytical modelling provides a useful software package for the prediction of the 3D elastic mechanical properties, damage and strength for a wide range of woven composite materials, hybrid and non-hybrid, with different geometry architecture. It also provides a help decision to select the material in respect with constituents (fibres and resin), weaving type, hybridisation and microstructure geometry. This help decision using the analytical technique presented in this work is all the more effective because it needs a very low calculation time.

ACKNOWLEDGEMENT

The authors would like to acknowledge gratefully the « Délégation Générale pour l'Armement » and actually the « Radômes et composites » department of « Atelier Industriel de l'Aéronautique » for the financial support of this study, a supply of materials examined in this work and for providing necessary assistance.

REFERENCES

[1]   Ishikawa, T. & Chou, T.W., Stiffness and strength behaviour of woven fabric composites, *J. Mater.Sci.*, **17** (1982) pp 3211-3220.

[2]   Ishikawa, T. & Chou, T.W., One-dimensional micromechanical analysis of woven fabric composites, *A.I.A.A. J.*, **21** (1983) n° 12 pp 1714-1721.

[3]   Ishikawa, T. and al., Experimental confirmation of the theory of elastic moduli of fabric composites, *J.Composite Mater.*, **19** (1985) pp 443-458.

[4]   Ganesh, V.K. & Naik, N.K., Failure behavior of plain weave fabric laminates under on-axis uniaxial tensile loading : I-Laminate geometry, *J. Composite Mater.*, **30** (1996) n° 16 pp 1748-1778.

[5]   Ganesh, V.K. & Naik, N.K., Failure behavior of plain weave fabric laminates under on-axis uniaxial tensile loading : II-Analytical predictions », *J. Composite Mater.*, **30** (1996) n° 16 pp 1779-1822.

[6]   Ganesh, V.K. & Naik, N.K., Failure behavior of plain weave fabric laminates under on-axis uniaxial tensile loading : III-Effect of fabric geometry, *J. Composite Mater.*, **30** (1996) n° 16 pp 1823-1856.

[7]   Vandeurzen, Ph., Ivens, J. & Verpoest, I., Structure-performance analysis of two dimensional woven fabric composites, ICCM 10, vol. IV, ed. K. Street & A. Poursartip, Canada, 1995, pp 261-270.

[8]   Vandeurzen, Ph., Ivens, J. & Verpoest, I., A three-dimensional micromechanical analysis of woven fabric composites: I. Geometric analysis, *Comp. Sci. Technol.*, **56** (1996) pp 1303-1315.

[9]   Vandeurzen, Ph., Ivens, J. & Verpoest, I., A three-dimensional micromechanical analysis of woven fabric composites: II. Elastic analysis, *Comp. Sci. Technol.*, **56** (1996) pp 1317-1327.

[10]   Aboura, Z., Chouchaoui, C.S. & Benzeggagh, M.L., Analytical model of woven composite laminate superposition effect of two plies », ECCM 6, Bordeaux (1993).

[11]   Chouchaoui, C.S., Aboura, Z. & Benzeggagh, M.L., Une comparaison entre un modèle analytique et numérique pour l'analyse élastique d'un composite à renfort tissu, Compte rendu des neuvièmes journées nationales sur les composites, Saint-Etienne (1994) pp 245-254.

[12]   Scida, D., Aboura, Z. & Benzeggagh, M.L., Analytical modelling of elastic behaviour in textile composite, 18th International SAMPE Europe, April 1997 pp 393-396.

[13]   Scida, D., Aboura, Z., Benzeggagh, M.L. & Bocherens, E., Elastic behaviour prediction of hybrid and non-hybrid woven composite , *Comp. Sci. Technol (in press)*.

[14]   Scida, D., Aboura, Z., Benzeggagh, M.L. & Bocherens, E., An analytical analysis of woven composite materials: 3D elasticity and failure , submitted to *Comp. Sci. Technol.*

[15]   Scida, D., Aboura, Z., Benzeggagh, M.L. & Bocherens, E., 3D elastic properties modelling of woven composite materials: an analytical and numerical approach, submitted to *Composite part A*.

[16]   Leischner, U. & Jonhson, A.F., Micromechanics analysis of hybrid woven fabric composites under tensile and compression load, *Composite Materials Technology*, **IV** pp 397-405.

[17]   Benzeggagh, M.L., Khellil, K. & Chotard, T.; Experimental determination of Tsai failure tensorial terms F_{ij} for unidirectionnal composite materials, *Comp. Sci. Technol.*, **55** (1995) pp 145-156.

[18]   Bouden, A., Aboura, Z., Benzeggagh & Lambertin, M., Signal processing of acoustic emission for damage identification in woven composite materials, *in Proceedings of ICCM-10*, Whistler, B.C., Canada, (1995) pp V-453 - V-459.

Micromechanical Analysis of a Woven Composite

Zheng-Nong Feng, Howard G. Allen and Stuart S. Moy

Department of Civil & Environmental Engineering
University of Southampton, Southampton, SO17 1BJ, UK

Abstract

The present paper provides a detailed micromechanical analysis of a plain woven composite. Detailed modelling allows one to study the stress distributions and concentrations inside the composite. Plotted results clearly show the stress concentrations in the yarns and in the matrix due to the interlacing and waviness (undulation) of the fabric. The term *stress variation factor* is used to study the effects of various characteristics of the composite on the stress concentrations. The results help to identify failure initiation and facilitate the study of the failure mechanism of the composite.

1. Introduction

Woven composites are used for structural applications in ship and aircraft construction because of their ease of handling, low fabrication costs and high structural performance. The interlacing of the woven fibre bundles can provide many obstacles to the growth of damage. However, there are also negative effects due to the interlacing and the undulation of the yarns. The undulation of the yarn may induce many local stress concentrations which can result in early initiation of diffuse damage, particularly in the matrix. This in turn, reduces the stiffness and strength of the composite material.

The complicated geometry of the woven composite makes a detailed analysis quite challenging. Early stress analyses were based on laminate theory and on a simplified model of the woven composite, such as the mosaic model and the bridging model in references [1]. In recent years there have been a few attempts to model the architecture in detail, to predict internal stress states, and hence to model progressive failure [2-3]. Glaessgen at el. [2] made finite element models of a plain woven composite at microstructural level. The distribution of strain energy density inside the composite was predicted. However, little information was provided on the stress concentrations and the effects of various parameters on them.

In this paper the stress distributions and concentrations inside the woven composite are studied by modelling the architecture in detail by finite element method. The effects of various characteristics of the composite, such as waviness ratio, yarn fibre content, the yarn gap, etc., on the stress variation factor are studied. Several important conclusions are drawn from the study.

2. Geometry of a Plain Woven Composite

The commonest weave is plain woven composite. The periodicity of the repeating pattern in a woven fabric can be used to isolate a small *repeating unit element* (RUE) which is sufficient to describe the fabric architecture. Figure 1 demonstrates the geometry of the RUE of a plain woven composite.

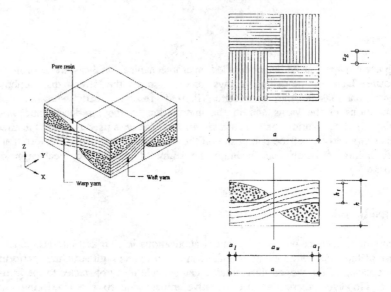

Figure 1 Geometry of the RUE of a plain woven composite.

The main geometrical parameters used to describe the architecture of the woven composite are also shown in the Figure 1. They are overall thickness of the RUE h ; overall width of the RUE a ; yarn maximum thickness h_t ; undulating length a_u ; length of the straight portion of the yarn a_l ; and yarn gap a_g .

To determine the volume of the pure resin and /or volume of the yarns, the shape function of the yarn needs to be defined beforehand. Based on photomicrographs of the woven composite cross-section, the basic shape function of the yarn is assumed as a cosine function. Because of the presence of the pure resin regions, the fibre content by volume of the yarn (referred as yarn fibre content) is different from the *overall* fibre content by volume.

3. Finite Element Modelling

ABAQUS finite element package was used to model the plain woven composite and to construct the finite element meshes.

(1) Solid model for the finite element analysis

To model a woven composite, firstly, the cross-section of the yarn was input as a series of discrete points calculated from a shape function. These control points were fitted to Loft Spline curves in ABAQUS commands to create the outlines of the cross-section desired. Secondly, the yarn centre line was created in the same manner. Area Group commands were used to form the area of the cross-section. Then, the cross-section was slid along the centre line to develop the solid object of a yarn by using Solid/Glide commands. The procedures for creating additional yarns were the same.

The outer boundaries of the RUE are pure resin regions and a pure resin pocket penetrated through the gap between the yarns. Boolean commands were used to subtract the yarn solids from the outer block. The remaining objects represent the surrounding resin contained within the RUE.

(2) Finite element meshes

Once the solid model of the RUE was formed, the finite element meshes were set up. The final element mesh was composed of about 600 elements in the RUE with C3D6 and C3D8 elements.

(3) Boundary conditions

The actual woven composite may consist of only one layer (woven lamina) or several layers stacked together. Each of these layers is subjected to different boundary conditions. Therefore the response of a RUE to a given load depends on its position within the overall structure. Two extreme cases were considered in this paper. One of the boundary conditions was suitable for the RUE embedded in the inner layer of the composite. For in-plane extensional loading in the x-direction, these boundary conditions were described as:

$$u(0, y, z) = 0 \qquad u(a, y, z) = \text{specified constant value}$$
$$v(x, 0, z) = v(x, a, z) = 0 \qquad\qquad\qquad (1)$$
$$w(x, y, -h_t) = w(x, y, h_t) = 0$$

Other boundary conditions may be suitable for the RUE in woven lamina or outer layer of the composite with top and bottom surfaces free. For in-plane extensional loading in the x-direction, these boundary conditions were described as:

$$u(0, y, z) = 0 \qquad u(a, y, z) = \text{specified constant value}$$
$$v(x, 0, z) = v(x, a, z) = 0 \qquad\qquad\qquad (2)$$

4. Stress Distributions and Concentrations

The finite element analysis of the RUE was carried out in ABAQUS IBM SP2 parallel processing computer. The stress distributions and concentrations inside the composite were investigated. In the *basic model* the RUE was subjected to an axial displacement in the " 1 " direction (warp direction) as illustrated in the following figures, with the boundary conditions of expression (1). The material prototype was a woven composite used in our panel research in which the fibre is E-glass, and the matrix is polyester resin. The material properties and geometrical parameters used in the *basic model* are given in Table 1.

Table 1. Material properties and geometrical parameters used in the *basic model*

Property		Yarn	Pure resin
E_{xx}	*(GPa)*	60.6	3.5
$E_{yy} = E_{zz}$	*(GPa)*	20.2	3.5
$G_{xy} = G_{xz}$	*(GPa)*	7.38	2.1
G_{yz}	*(GPa)*	7.76	2.1
$v_{xy} = v_{xz}$		0.3	0.3
v_{yz}		0.3	0.3

Geometrical parameters	
yarn gap a_g	0
length of the straight portion of yarn a_l	0
waviness ratio (h_t / a_u)	0.125 (1/8)
yarn fibre content	0.858

To study concentration of stress, the term *stress variation factor* (SVF) is defined as:

SVF = Maximum stress / Nominal average stress

in which *Nominal average Stress = (Maximum Stress + Minimum Stress) / 2*

Figure 2a provides an overall view of a RUE and a typical stress distribution diagram. Figure 2b shows the distribution of the stress *along* fibre in the *warp* yarns. Stress concentrations occur at the interlacing areas. The SVF in the warp yarns is about 1.23. The results indicate that (1) the failure of the warp yarns in longitudinal tension would start at the interlacing areas; (2) due to the stress concentrations, fibres in a woven composite may fail earlier than those in a unidirectional composite.

Figure 2c shows the distribution of the stress *along* fibre in the *weft* yarns. The maximum value of the longitudinal stress in the weft yarns is just about 14% of the longitudinal stress in the warp yarns. From this one may conclude that failure is unlikely to happen in the fibre direction of the weft yarn if in-plane extension is applied in the warp direction, although there are the stress concentrations in the weft yarns.

Figure 2d demonstrates the distribution of the stress *across* fibre in the *weft* yarns. Stress concentrations are at the tip of the weft yarn in the interlacing areas. The SVF in the weft yarns is about 1.27. As the failure strength of composite in transverse direction is smaller than that in longitudinal direction in tension, the transverse cracks may occur in the tip of the weft yarns because of intensive transverse stress.

Figure 2e shows the distribution of the stress *across* fibre in the *warp* yarns. The maximum value of the transverse stress in the warp yarns is just about 13% of the transverse stress in the warp yarns. Therefore, failure may happen later in the transverse direction of the warp yarns than it does in the weft yarns.

Figure 2f shows the distribution of the stress in the pure resin regions (isotropic body). The stress is concentrated at the interlacing areas. The peak transverse stress is about 60 percent higher than lowest stress. The SVF in the pure resin regions is about 1.22. This confirms that the undulation induces the local stress concentrations, and that the pure resin cracks around the interlacing areas first.

5. Parametric Studies

The effect of various parameters on the SVF are studied. These parameters include the waviness ratio, the yarn gap, different material, boundary conditions.

(1) Effect of waviness ratio

Three waviness ratios were studied. One is zero. This represents a laminate which consists of four unidirectional layers: top and bottom pure resin sheets, two orthogonal unidirectional flat yarns sandwiched between pure resin sheets. Other two waviness ratios were 0.125 (*basic model*) and 0.3. The material properties for three cases are the same the *basic model* in Table 1. Table 2 lists the effect of the waviness ratio on the SVF. The effect of waviness ratio on the SVF is significant. The bigger waviness ratio, the bigger the SVF. If the strengths of the fabric yarn were the same in the three cases, under the same loading condition the strength loss for the case of waviness ratio = 0.3 would be as high as 50% compared with the case of waviness ratio = 0. This results indicates that the waviness ratio of fabric yarn is so critical that must be controlled in the woven composite structures.

Table 2 The effect of waviness ratio on the SVF

Waviness ratio	SVF		
	Longitudinal stress in warp	Transverse stress in weft	Principal stress in pure resin
0	1	1	1
0.125 (*basic model*)	1.23	1.27	1.22
0.3	1.55	1.38	1.5

(2) Effect of yarn gap

A woven composite with the yarn gap $a_g/a = 0.1$ was analysed. The material properties and waviness ratio were the same as the *basic model* in Table 1. The results were compared with these from the *basic model* without the yarn gap. Table 3 lists the effect of the yarn gap on the SVF. The comparison shows that the yarn gap has no significant effect on the SVF.

Table 3 The effect of yarn gap on the SVF

Yarn gap a_g/a	SVF		
	Longitudinal stress in warp	Transverse stress in weft	Principal stress in pure resin
0 (*basic model*)	1.23	1.27	1.22
0.1	1.26	1.28	1.29

(3)Effect of different materials

A woven composite with Carbon/Epoxy material (yarn fibre content 0.7 and waviness ratio 0.125) was analysed. The stiffness of Carbon/Epoxy in fibre direction in the fabric yarn is much stronger than transverse direction. Table 4 shows that there is a significant increase of the SVF in fibre direction, comparing with E-glass/Polyester material under the same geometric conditions. This indicates that the stress concentration is more important in Carbon/Epoxy composite than E-glass/Polyester composite.

Table 4 The effect of different materials on the SVF

Material	SVF		
	Longitudinal stress in warp	Transverse stress in weft	Principal stress in pure resin
E-glass/Polyester	1.22	1.20	1.26
Carbon/Epoxy	1.49	1.24	1.36

(4)Effect of boundary conditions

The *basic model* was also calculated under the boundary conditions in the expression (2). In the expression (2) the top and bottom surfaces are free. The stress variation factors for the two boundary conditions are listed in Table 5. The SVF in the pure resin for the boundary conditions (2) is increased significantly. This may indicate that the pure resin near the top and bottom free surfaces of the composite cracks earlier than that inside the composite.

Table 5 The effect of different boundary conditions on the SVF

Boundary Conditions	SVF		
	Longitudinal stress in warp	Transverse stress in weft	Principal stress in pure resin
Expression (1)	1.23	1.27	1.22
Expression (2)	1.30	1.29	1.57

6. Conclusions

The architecture of a plain woven composite was modelled in detail by using the finite element method in ABAQUS code. The stress distributions and concentrations were investigated. From the analyses some important conclusions can be drawn.

(1) Due to interlacing and waviness of the fabric yarn, there are significant stress concentrations inside the yarns and the pure resin regions. The stress concentrations will result in resin cracking, transverse damage and fibre breakage in the woven composite earlier than in unidirectional laminate made of the same materials as the woven composite. In other words, a woven composite will be weaker than an equivalent non-woven composite.

(2) The waviness ratio has a significant effect on the stress concentrations. Therefore the waviness should be controlled in the manufacture of woven composites.

(3) A carbon fibre composite is strong in the fibre direction and weak in the transverse direction. The undulation of a woven composite made of carbon fabrics causes higher stress concentrations in fibre direction than a geometrically similar composite made of E-glass fabrics. Special attention needs to be paid to the stress concentration in a carbon fabric composite.

References

1. Ishikawa, T. and Chou, T.W., "Stiffness and strength behaviour of woven fabric composites", *Journal of Materials Science*, 1982, 17:3211-20 .

2. Glaessgen, E.H. et al., "Geometrical and finite element modelling of textile composite", *Composites: Part B*, 27B, 1996, pp43- 50.

3. Blackketter, D.M., et al., "Modelling damage in a plain weave fabric-reinforced composite material", *Journal of Composite Technology and Research*, 1993, 15:136- .

4. Whitcomb, J., "Three-dimensional stress analysis of plain weave composite", *Composite Materials: Fatigue and Fracture*, ASTM STP 1110, ed. T. K. O'Brien, American Society for Testing and Materials, 1991, pp. 417- 438.

5. Whitcomb, J., et al., "Effect of various approximations on predicted progressive failure in plain weave composites", *Composite Structures*, 1996, 34:13-20 .

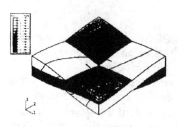

Figure 2a. Contour plotted stress
distribution for woven fabrics

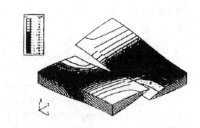

Figure 2b. Longitudinal stress (s_{11})
in warp yarns

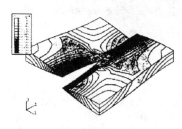

Figure 2c. Longitudinal stress (s_{22})
in weft yarns

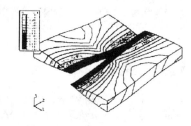

Figure 2d. Transverse stress (s_{11})
in weft yarns

Figure 2e. Transverse stress (s_{22})
in warp yarns

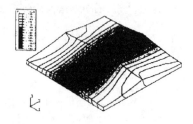

Figure 2f. Stress (s_{11}) in pure
resin region

An Analytical Model of the Failure Process in Woven Fabric Composites

Ph. Vandeurzen*, J. Ivens*, I. Verpoest*

* Department of Metallurgy and Materials Engineering, Katholieke Universiteit
Leuven, de Croylaan 2, B-3001 Leuven, Belgium

Abstract

With the emerging ability to engineer textile composite materials comes the need to develop computationally efficient thermo-mechanical models. In this paper, a simple substructuring concept for solving the stress analysis problem of a general woven fabric composite is presented. For this development, the complementary variational principle was chosen to calculate the micro-stress fields. With the application of failure criteria, the model allows the simulation of the failure process of woven fabric composites under complex stress states. Parametric studies showed a wide range of variability in the onset of first-cell-failure. A stiffness degradation scheme was adopted to predict the complete stress-strain curve and the final failure of the composite. The model accounts for fabric geometry, interacting effects between intersecting yarns and matrix distribution. Good correlation with experimental data was observed.

1. Introduction

Current applications of textile composite materials require an in-depth knowledge of the composite thermo-mechanical properties. Therefore, it is important to develop micromechanics approaches that can predict, with sufficient accuracy, the effect of the microstructural details on the internal and macroscopic behaviour of these materials. In literature, two classes of micromechanics approaches for woven fabric composites were reported. The first class are the iso-strain or iso-stress methods [1-6] that obtain a rough estimate of stresses only. Second, the *'finite element method'* can be used to analyse the elastic behaviour, the internal stress state and the damage propagation [7-8]. The studies on the failure behavior are however limited [6-8].

In this paper, we present the Complementary Energy Model (CEM) (Figure 1). This model uses an automatic decomposition scheme for the geometric description while the *complementary variational principle* [9-10] is employed for the stress analysis. From the stress fields, the three-dimensional compliance and thermal expansion tensors are computed. Using appropriate failure criteria and stiffness degradation tensors, the stress-strain curves and the ultimate failure can be predicted.

2. Geometric model

The geometric model deals with a perfect, regular, one-layer fabric composite, described by the unit cell. This unit cell is decomposed into different material levels as shown in Figure 1 and is described in detail elsewhere [11]. Only the weave construction pattern from the weaving company, and a small number of measured geometric parameters (the yarn aspect ratios, the filling crimp and the composite thickness) are sufficient to decompose the unit-cell automatically. The method can be applied to any kind of weave. Therefore, this five-level substructuring scheme could be considered as an 'intelligent mesh generator' for 2D woven fabric composites. The *geometric meshing of the unit-cell* is essential for the computation of the composite mechanical properties.

3. Homogenisation procedure

3.1. CALCULATION OF THE ELASTIC CONSTANTS

The knowledge of the average stress (strain) in each fibre or matrix phase is sufficient to calculate the 3D effective elastic moduli. The tensors relating the average stress (strain) σ_i (or ε_i) in each phase to the uniform stress (strain) σ (or ε) on the boundary of the unit-cell, are called 'concentration tensors':

$$\{\sigma_i\} = [A_i]\{\sigma\} \quad or \quad \{\varepsilon_i\} = [B_i]\{\varepsilon\} \tag{1}$$

with A and B the fourth-order concentration tensors. The proposed *multistep homogenisation technique*, proceeding bottom-up from level 5 to level 1 (Figure 1), is based on computing these 'stress concentration tensors' and is described in detail elsewhere [12]. For this development, the complementary variational principle is chosen. It is assumed that from all the admissible stress fields, the true field minimises the total complementary energy. The function to be minimised is the complementary energy U:

$$U = \sum_i \frac{k_i}{2} \{\sigma_i\}^T [S_i]\{\sigma_i\} \tag{2}$$

where S_i are the 3D compliance matrices of the subcells at each level and k_i are the fractional volumes of these subcells. Taking into account the constraints and by using the method of the Lagrangian multipliers, the stress concentration tensor A_i of each layer is obtained. From this, the overall compliance matrix S can be calculated:

$$[S]_{CC} = \sum_{i=Y,M} k_i [A_i]^T [S_i][A_i] \tag{3}$$

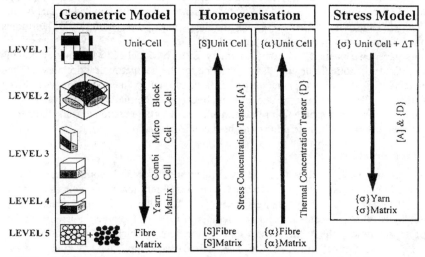

Figure 1. Modelling of woven fabric composites by variational principles: overview.

This procedure is repeated at each decomposition level, resulting in the computation of the overall symmetric 3D compliance matrix of the woven fabric composite *unit cell*.

3.2. CALCULATION OF THE THERMAL EXPANSION CONSTANTS

For textile composites, a knowledge of the thermal expansion constants is indispensable in estimating the mechanical consequences of temperature variation under service conditions, or the consequences of processing conditions. The multistep homogenisation technique is therefore extended to include the computation of several important thermal parameters [13]. The expression for the complementary energy U becomes:

$$U = \sum_i \frac{k_i}{2} \{\sigma_i\}^T [S_i]\{\sigma_i\} + \sum_i k_i \Delta T \{\sigma_i\}^T \{\alpha_i\} \tag{4}$$

where ΔT is the temperature difference, and α_i is the second-order thermal expansion tensor for each subcell.. Solving the new quadratic optimisation with the external mechanical stress tensor equal to zero, will yield the 'thermal concentration tensor' D.

$$\{\sigma_i\} = \{D_i\} \Delta T \tag{5}$$

The computed second-order thermal concentration tensors D_i are crucial in the calculation of the effective thermal expansion constants:

$$\{\sigma\}_{cell} = \sum_i k_i \left([S_i]\{D_i\} + \{\alpha_i\}\right) \tag{6}$$

4. Local damage and onset of damage

Owing to a great difference between the thermal expansion constants of matrix and fibres great residual stresses can occur in the fabric composite structure after it is cooled down from its processing temperature. In the stress analysis of such components, externally induced stresses must be superimposed on these pre-existing residual stresses. Through the calculation of the concentration tensors, a direct link is established between the external forces on the fabric composite and the cell stress at each geometric level:

$$\{\sigma_i\} = [A_i]\{\sigma_j\} + \{D_i\} \Delta T \tag{7}$$

To evaluate the first-cell-failure of woven fabric composites, the stress model is employed to calculate the yarn and matrix stresses. For the isotropic matrix material, the parabolic failure locus applied on the principal stresses within the matrix cell is used in this paper. Failure of the yarns is predicted by using a modified Tsai-Hill failure criterion.

5. Strength Analysis

In the *progressive failure analysis*, the effects of matrix and yarn failure are taken into account in an average sense. It is based on the assumption that the damaged material could be replaced with an equivalent material of degraded properties. The properties of the damaged material are adjusted as the loading and progression of damage continues. In the present study, the stiffness reduction method as proposed by Blackketter [14] will be used: (1) First, the method accounts for the damage mode when modelling degradation of yarn materials. If failure is detected, the appropriate moduli are reduced. (2) Second, the matrix failure is introduced by reducing the Young's modulus and the shear modulus. After failure, the matrix is no longer isotropic.

Finally, to clarify the present approach, some important remarks are provided in the following:

- The woven fabric composite is assumed to be *initially free of damage*
- The *non-linearity of the matrix* is not taken into account for two reasons. First, the non-linear nature of the different matrix zones in the composite is different form that of the bulk material. This is mainly caused by the presence of local multiaxial stress states and thermal stresses. This information is usually not available. Second, the non-linear stress-strain behaviour of woven fabric composites was shown to be mainly influenced by damage propagation[25] and not by the non-linearity of the matrix.
- The model does not calculate the *fabric geometric deformation* at each load step. It is expected therefore, that the model will predict less accurate results for off-axis tensile tests.
- The proposed model is *deterministic*.

6. Parametric study

6.1. STRESS DISTRIBUTION IN A HYBRID TWILL WEAVE

A 2x2 hybrid twill weave fabric composite is considered in this study (Figure 2). The the elastic properties of the constituents are shown in Table 1. The calculated overall fibre volume fraction for this configuration is 40.2%. A uniaxial load σ_x is applied on the unit cell, and the stress distribution at different decomposition levels is calculated.

The stress fields at the *micro-cell* level are discussed, because they illustrate in a clear way the effect of yarn interactions. Figure 3 depicts the σ_{zMC} component of the micro-cell stress tensor. The xy plane corresponds to the surface of one unit-cell. At each yarn cross-over point a compressive stress is predicted. For this geometric configuration, the predicted compressive stress can reach 10% of the applied longitudinal stress. Of course, the other regions of the woven fabric composite will display tensile stresses

Material	E_1 (GPa)	E_2 (GPa)	G_{12} (GPa)	G_{23} (GPa)	ν_{12}
Carbon	220	13.79	8.96	4.83	0.2
Dyneema	87	87	36.25	36.25	0.2
Epoxy	3.448	3.448	1.27	1.27	0.35

Table 1. Properties for the carbon fibre, the dyneema fibre and the epoxy matrix.

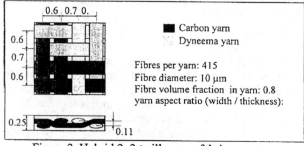

Figure 2. Hybrid 2x2 twill weave fabric structure.

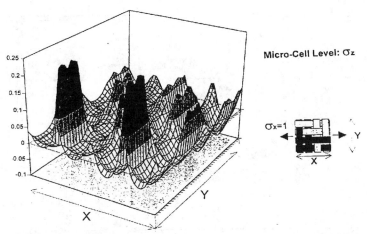

Figure 3. Prediction of the out-of-plane σ_{zMC} micro-cell stresses: 3D view.

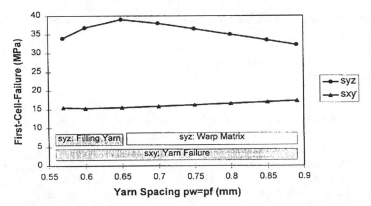

Figure 4. First-cell-failure as a function of the yarn spacing p (V_f=50%,ΔT=0): shear.

because the resultant stress in the z direction should be equal to zero. Figure 3 shows two specific regions were the micro-cell tensile out-of-plane stress gets to 25% of the applied stress. These regions correspond to micro-cells with highly undulated warp yarns.

6.2. FIRST DAMAGE: PARAMETER STUDY OF A PLAIN WEAVE

For this parametric study, a glass-epoxy balanced plain woven fabric composite is loaded in shear. The influence of the yarn spacing on the first cell failure is shown in Figure 4 (at the same time, the layer thickness was adjusted to keep the fibre volume fraction constant at 50%). Different failure modes can be distinguished in the woven fabric composite: off-axis failure of the warp and/or filling yarn (due to transverse tension or shear), and matrix cell failure. The maximum or minimum in the curves can be explained by considering the failure mode and position of the cell that fails first.

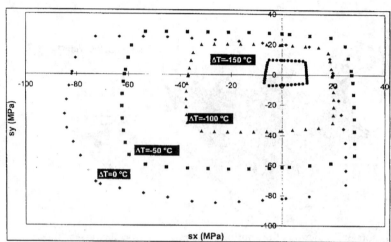

Figure 5. In-plane failure locus as a function of the temperature difference ΔT.

The residual thermal stresses have an important influence on the onset of damage. Figure 5 plots the in-plane first-cell-failure locus as a function of the temperature difference ΔT. Clearly, the presence of thermal stresses in the plain weave fabric composite is shown to influence the position of the failure locus, especially the compression/compression quadrant.

6.3. STRENGTH ANALYSIS

In this parametric study, the glass/epoxy plain weave fabric composite of section 6.2 is considered. The influence of the yarn aspect ratio on the stress-strain curve of the composite, loaded in the warp direction, is examined (Figures 6). A very strong influence of the yarn aspect ratio f on the failure behaviour is observed. If the yarn aspect ratio f equals 3, the ultimate strength is only 60 MPa and the first cell failure is due to matrix cell failure. However, if the ratio equals 12, the strength reaches 240 MPa and the first cell failure is due to transverse weft yarn failures at points of maximum yarn curvature. These differences should be attributed to the different geometric architectures. The maximum yarn orientation plays a key role. Basically, the predicted strength decreases considerably with increased yarn undulation. For all four woven fabric composites considered, the ultimate failure is due to warp fibre breakage. The non-linearity of the curves is a result of progressive damage development.

7. Experimental validation

Tensile tests were conducted for a glass/epoxy fabric composite. The glass fabric RE280, a basket weave, was supplied by Syncoglass, Belgium. Epoxy resin films F533 from Hexcel were selected as matrix. Composite plates were produced in the autoclave by curing stacked systems of resin films and woven fabric layers during 2 hours at 125°C and 3 bar pressure. The fibre volume fraction was measured experimentally by a matrix burning method and was 33.7 %. Test specimens were cut in warp, weft and bias direction.

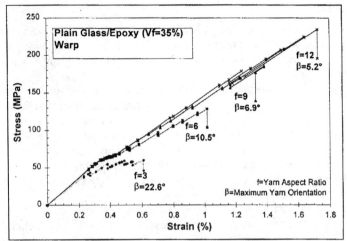

Figure 6. Predicted warp stress-strain curves as a function of the yarn aspect ratio f.

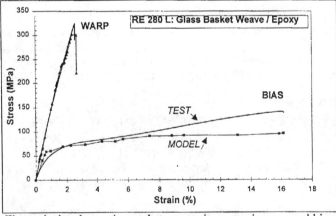

Figure 7. Theoretical and experimental stress-strain curves in warp and bias direction.

Figure 7 compares the experimental and theoretical stress-strain curves for the RE280 glass/epoxy composite for the warp and the bias loading direction A good correlation is found between the theoretical and experimental stress-strain curves in the *warp direction*. The characteristic knee-behaviour is observed for woven fabric composites. The knee is the result of transverse weft yarn failures. The position of the knee is predicted very accurately by the Complementary Energy Model.

In the *bias specimen*. a discrepancy is found between the theoretical and experimental stress-strain curves at high strain levels. This is due to the fact that the model does not account for the fabric yarn reorientation which in reality results in a further increase of the stresses. Therefore, the ultimate bias strength is not predicted very accurately.

8. Conclusions

The proposed Complementary Energy Model for woven fabric composites is an addition to the known micromechanical approaches in the mechanics of composites. We believe, it can serve as a useful tool for solving the stress analysis problem of various textile composites.

A practical drawback of the method is of course the absolute need for a geometric model of the textile architecture. That is, the substructuring technique requires a detailed description of the orientation and the position of the yarns in the fabric composite. Advantages of the proposed method are the automated substructuring of the unit cell, the accurate prediction of the 3-dimensional stiffness matrix and a detailed estimation of the stresses in yarn and matrix phases as a function of the position in the composite unit cell, allowing the prediction of the onset of damage, and the prediction of the complete stress-strain curve of the composite.

This analysis showed that damage critical parameters, such as the transverse stress in the yarns, are highly sensitive to the fabric morphology. It should be concluded, that the *positions of the yarns and matrix phases* play a key role in the understanding of the thermo-elastic and damage behaviour of woven fabric composites.

9. Acknowledgements

This text presents research results of the Belgian programme on Interuniversity Poles of Attraction, funded by the Belgian state, Prime Minister's Office, Science Policy Programming. Ph Vandeurzen and J Ivens are financed through grants of the Flemish Institute for the Promotion of the Scientific-Technological Research in Industry (IWT).

10. References

1. *F.K. Ko, T.W. Chou.*, COMPOSITE MATERIALS SERIES 3 -TEXTILE STRUCTURAL COMPOSITES, Elsevier Science Publishers Amsterdam, 1989.
2. *T. Ishikawa, T.W. Chou.*, J. Mater. Sci., 1982, Vol. 17, 3211-3220.
3. *N.K. Naik, P.S. Shembekar.*, J. Compos. Mater., 1992, Vol. 26, 2196-2225.
4. *H.T. Hahn, R. Pandy.*, Journal of engineering materials and technology, 1994, Vol. 116, 517-523.
5. *Ph. Vandeurzen, J. Ivens, I. Verpoest.*, Comp. Sci. Tech., 1996, Vol. 56, 1317-1327.
6. *R.A. Naik*, J. Comp. Mater., 1995, Vol. 29, 2334-2363.
7. *J. Whitcomb, W. Kyeongsik*, Composite Structures, 1994, Vol. 28, 385-390.
8. *A. Paumelle, A. Hassim, F. Léné*, La recherche aérospatiale, 1990, Vol. 6, 47-62.
9. *C.M. Leech, J. Kettlewell*, International journal of mechanical engineering education, 1981, Vol. 9, 157-180.
10. *D. Chen, S. Cheng*, J. Reinf. Plas. Comp., 1993, Vol. 12, 1323
11. *Ph. Vandeurzen, J. Ivens, I. Verpoest*, Comp. Sci. Tech., 1996, Vol. 56, 1303-1315.
12. *Ph. Vandeurzen, J. Ivens, I. Verpoest*, J. Compos. Mater., 1998, Vol. 32, (in print)
13. *Ph. Vandeurzen, J. Ivens, I. Verpoest*, Proc. Conf. Deformation and Fracture of Composite Materials, Manchester (UK), J. Hodgkinson Ed., 1997, 319-328.
14. *Blackketter, D.M., D.E. Walrath and A.C. Hansen*, Journal of Composites Technology and Research, 1993, Vol. 15, 136-142.

Targeting cost reduction by FEA-designed reinforcement textiles

K. Hörsting, M. Huster

LIBA Maschinenfabrik GmbH
Postfach 11 20, 95112 Naila
Oberklingensporn, 95119 Naila
Germany
Tel.: +49 92 82 67-0, Fax: +49 92 82 57 37, email: liba.nt@t-online.de

Summary

An ideal approach for design and manufacture defines the way towards cost-effective long-fiber reinforced parts. Thus the link between component design and reinforcement textile design is significant as shown by the example of a positioning unit of a tie sewing machine. Using the FEA the fiber orientation as well as the component geometry result in an optimum load considering design.

The described PARAMAX multiaxial layer reinforcement fabric offers excellent pre-requisites for the important transfer of calculation results to the reinforcing textile due to variable fiber orientation and selectable drapeability. Further advantages contributing to cost-effectiveness are displayed.

By further integrating a suitable impregnation process (e.g. RTM process), stochastic variables as well as costs in composite part manufacture could be cut down noticeably. The implementation of this complete manufacturing approach will continue because of the rising constraints of using lightweight materials and cutting down manufacturing costs especially in the automotive industry.

1. Introduction

The manufacture of long-fiber reinforced composite components has been marked by the efforts to eliminate labor-intensive manufacturing steps such as manual cutting, impregnation and hand lay-up of the reinforcement semi-finished part by increasing automation. This is the only way to make full use of the advantages of this material technology for commonplace applications, apart from high-tech applications such as aircrafts, spacecrafts and racing cars. In the course of these efforts new impregnation processes such as RTM (Resin Transfer Molding), vacuum resin injection and SCRIMP (Seemann Composites Resin Infusion Manufacturing Process) have been developed in the field of plastics processing to supersede wet lay-up of sheet composite parts /1/.

This progress in plastics technology on its own could not bring the manufacturing costs so far down to introduce composite materials economically into other fields of application. That is why the current trend is towards complex preconfectioned reinforcement textiles to increase the degree of automation in the manufacture of long-fiber reinforced composite components /2/. This would allow e.g. stitched fiber preforms, PARAMAX multiaxial layer reinforcement fabrics and three-dimensional braids to be sewn together into a three-dimensional construction, to make a very complex reinforcement textile such as a floor panel structure of a car body (cf. Fig. 1).

This cost-efficient approach to component manufacture which is essentially marked by the developments in the field of reinforcement textiles can be further optimized if both the design of parts and reinforcement textiles and the impregnation process are well-matched. The component design by means of the finite-element method (FEM) helps to find a fiber orientation of the reinforcement textile that is appropriate for the load. Based on flexible textile manufacturing processes, it is possible to design the complex reinforcement preform. The use of a suitable impregnation technology allows the implementation of a rational, reproducible manufacturing approach for composite components.

2. Layout of composite part using FEM

This manufacturing approach shall be illustrated by a manufacturer component, the positioning unit of a LIBA tie sewing machine (cf. Fig. 2). This machine is part of a completely automated production line for ties. Its task is to sew the tie fabric and the lining together.

For this purpose the tie fabric placed on a movable table (1) is moved under the positioning unit (2) (cf. Fig. 2). The positioning unit then takes the lining out of the magazine (3) located above it. Crossed needles prick into the lining and fix it to the positioning unit. The positioning unit then moves down and rotates 180° around its vertical axis to place the lining onto the tie fabric. Both are then sewn together by a sewing yarn device (4).

The processing time of sewing the tie fabric to the lining takes 7 seconds. Only 1.4 seconds are available to move the positioning unit down and back to its initial position /3/. The cycle times mentioned and their associated accelerations which the functional elements of the positioning unit are subject to result in great torsion and bending moments. Due to the stiffness of the aluminum made part is completely exploited it makes sense to replace the aluminum U-shaped beam by a carbon-fiber reinforced composite U-shaped beam to improve its performance.

The process of designing the component by means of FEM is in three steps: Preprocessing, calculation and postprocessing. These steps are performed with the calculation software program Systus+/Mosaic of Framasoft, a French firm, which has optimized its software by adding the Composic module for the designing of composite components. The goal of preprocessing is the provision of all information needed by the calculation software for a realistic simulation.

In a first step the geometry of the component is generated. To cut down calculation times, the FE model consists only of one component half or component segments, component symmetry permitting. The missing segments are replaced by appropriate boundary conditions. This option is used for the generation of the U-shaped beam of the positioning unit (cf. Fig. 3). Complex component geometries are preferably generated with 3D-CAD software. The geometry data are then transferred to the FE-software via a suitable interface.

The basis for the calculation of tensions and deformations in a component by means of FEM is the finite element mesh which consists of small elements. The number of elements and the element properties determine not only the quality of the simulation but also the calculation time. The experienced calculation engineer must therefore assess the accuracy of the simulation versus the calculation effort to get reasonable results. The knots of the mesh serve to link both the loads the component is subject to, and the boundary conditions which represent the bedding of the component. Individual elements are assigned material data in order to determine deformation and material failure.

The Composic module of the calculation program Systus+/Mosaic provides special elements for the calculation of long-fiber reinforced composite parts. In such a calculation the orientation of the element coordinate system of individual elements is indispensable due to the anisotropy of the composite material. The material parameters of different materials are then defined for the individual layer. Based on that, all the fibers of the individual layer or each layer coordinate system is oriented in relation to the element coordinate system and each layer thickness is defined (cf. Fig. 4, left). In this way laminates can be generated from different materials, with individual layers in any possible orientation. The material properties of the laminate that depend on direction are represented in the x/y plane of the element coordinate system (cf. Fig. 4, right). The transfer of various calculation parameters to the system concludes preprocessing. The system now knows all required data to start calculating.

The calculation is followed by the interpretation of the results, the so-called postprocessing. The layout of the U-shaped beam of the positioning unit with a view to component rigidity is done along the z-axis /4/. The measure for that are the displacements of the geometry in this direction. The results can then be represented in isocolors on the geometry. The different colors are assigned to the respective range of resulting values in the legend (cf. Fig. 5). The maximum deformation of the U-shaped beam of the positioning unit is 0.3 mm along the z-axis and is therefore uncritical.

The strengths of the U-shaped beam are also represented by means of isocolors. In the design of the U-shaped beam for rigidity the tensions occurring in the component are uncritical, i.e. the resulting values of the strength criterion according to Hashin are noticeably smaller than one (cf. Fig. 6). Using this criterion, it is possible to identify the individual layer of the laminate where the maximum tensions occur at any place of the component geometry. This option allows an efficient way to optimize a laminate.

3. Transfer of the layout results to the design of the reinforcement textile

The quality of long-fiber reinforced composite component is largely determined by fiber orientation, besides the quality of the laminate. That is why the link between the component design by means of FEM and the design of the reinforcement textile plays a central role. The semi-finished reinforcement part used is essential for the calculation results to be utilized effectively. A good solution is the PARAMAX multiaxial layer reinforcement fabric (cf. Fig. 7). Such a multiaxial layer fabric is built up with up to 8 individual layers having different fiber orientations, fixed by a system of stitches. Within a layer the fiber rovings are parallel and straight /5/.

Besides its improved mechanical properties, the PARAMAX multiaxial layer reinforcement fabric also offers economic advantages over woven fabrics such as better handling of the layer fabric, drapeability that can be set by the selection of suitable stitches and the strongly reduced cutting expenses. Furthermore, the use of the PARAMAX multiaxial layer reinforcement fabric can cut down manual labor in component manufacture and increase reproducibility.

The increased rigidity of the PARAMAX multiaxial layer reinforcement fabrics as compared with woven fabrics, which is achieved by the straightness of the fiber roving, is the technical advantage of these reinforcement semi-finished part. The improved specific rigidity values lead to a reduction in weight that depends solely on the reinforcement semi-finished part.

Preconfectioned reinforcement textiles such as the PARAMAX multiaxial layer reinforcement fabric only make sense if a suitable impregnation process such as the RTM-process is available /6/. This largely automated process can yield good laminate qualities and can further increase reproducibility in component manufacture.

5. Literature

/1/   N. N.:
       Reinforced Plastics, January 1997
       Verlag: Elsevier Advanced Technology, Oxford
/2/   Hörsting, K.:
       Rationalisation of the manufacturing process of the composite motorcycle wheel
       for series production methods
       Intern report, CSIR, Aerotek Division, Pretoria, RSA, November 1994
/3/   N. N.:
       Information of the engineering department of the LIBA company, Naila,
       November 1996
/4/   Michaeli, W., Huybrechts, D., Wegener, M.:
       Dimensionieren mit Faserverbundkunststoffen
       published: Carl Hanser Verlag, München 1995
/5/   Hörsting, K.:
       Rationalisierung der Fertigung langfaserverstärkter Verbundwerkstoffe durch
       den Einsatz multiaxialer Gelege
       published: Shaker Verlag, Aachen 1994, ISBN 3-8265-0236-1
/6/   Michaeli, W., Wegener, M.:
       Einführung in die Technologie der Faserverbundwerkstoffe
       published: Carl Hanser Verlag, München 1989

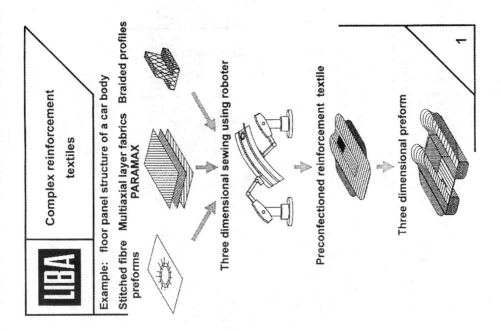

LIBA

Complex reinforcement textiles

Example: floor panel structure of a car body

Stitched fibre preforms   Multiaxial layer fabrics PARAMAX   Braided profiles

Three dimensional sewing using roboter

Preconfectioned reinforcement textile

Three dimensional preform

1

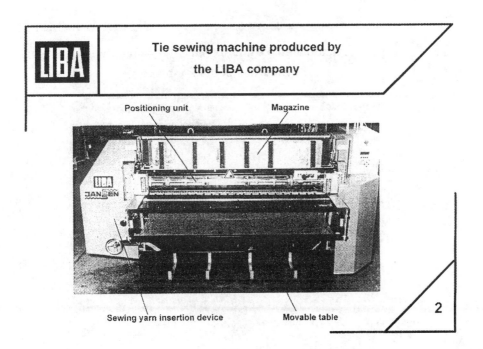

LIBA

Tie sewing machine produced by the LIBA company

Positioning unit

Magazine

Sewing yarn insertion device

Movable table

2

Component geometry with mesh, loads and boundary conditions

3

Material editor and display of the material properties in the x-y-plane

Element coordinate system

Material properties refering to the element coordinate system

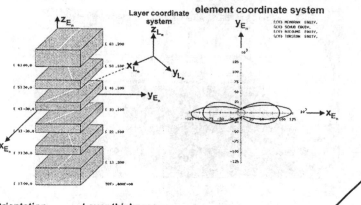

Orientation   Layer thickness

4

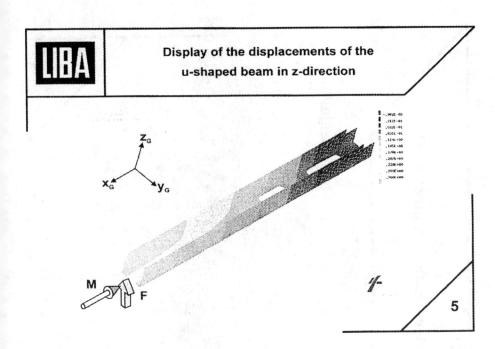

Display of the displacements of the u-shaped beam in z-direction

5

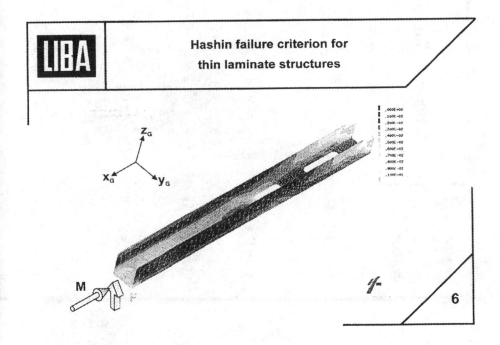

Hashin failure criterion for thin laminate structures

6

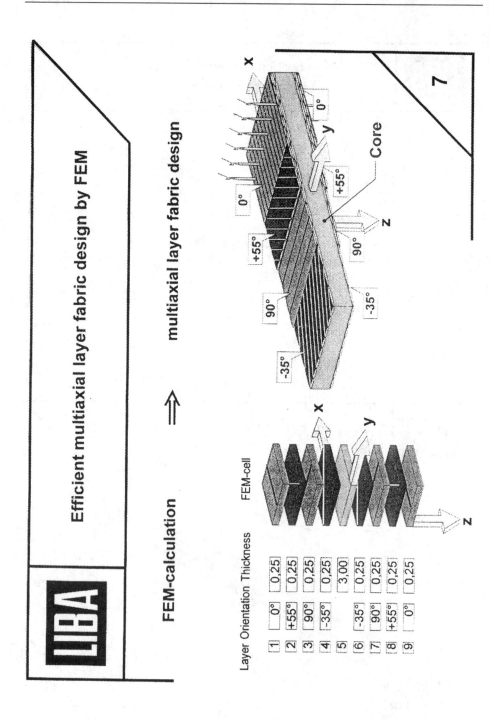

LIBA

Efficient multiaxial layer fabric design by FEM

FEM-calculation $\Longrightarrow$ multiaxial layer fabric design

FEM-cell

Layer	Orientation	Thickness
1	0°	0,25
2	+55°	0,25
3	90°	0,25
4	-35°	0,25
5		3,00
6	-35°	0,25
7	90°	0,25
8	+55°	0,25
9	0°	0,25

Core

7

New Nonisocyanate Polyurethane Matrix

Shapovalov Leonid, Chemonol Ltd., Nesher, Israel
Blank Nelly, Chemonol Ltd., Nesher, Israel
Figovsky Oleg, Eurotech Ltd., La Jolla, USA

Composite polymeric materials based on polyurethanes are widely used in industry. Exiting polyurethanes have good mechanical properties, but have a lot of pores, isocyanates are very toxic and sensible to moisture.

Nonisocyanate polyurethane compounds (NPC) are formed from olygomers of cyclocarbonates and primary amines. Because of forming an intramolecular bond and blockage of carbonyl oxygen it is considerably lowers the susceptibility of the whole urethane group to hydrolyses. The chemical resistance of materials is 1.5-2 times greater than that of materials without such bonds. The cyclocarbonates are synthesed from epoxy compounds by reaction with CO_2.

For the fist time in the world NPC were created by authors under guidance of Prof. Figovsky and based on 3-5 functional olygomeric cyclocarbonates and 2-6 functional primary amines.

The normal temperature carries out the hardening of NPC. Also there are created the special composites with high benzene and fire resistance and minimum absorption of the toxic and radioactive contamination.

Design and modeling of short fiber-polypropylene matrix composites

G.L. Gigli**, S. Puzziello*, M. Baroni*, S.E. Barbosa, J.M. Kenny**

*ISRIM - Loc. Pentima Bassa, 21 - 05100 TERNI
** Institute. of Chemical Technologies, University of Perugia - Loc. Pentima Bassa, 21 - 05100 TERNI

Abstract

Short fiber reinforced thermoplastics are already very important commercial materials. Whereas good processability properties are typical of short fiber composites, the mechanical properties of the final products mainly depend on the residual fiber length, orientation and distribution which are determined by the processing conditions. In last years new efforts have been dedicated to fundamental research [1] in the field of processing of a short fiber composites as a consequence of their massive use for automotive applications and in other important industrial sectors.
Among short fiber composites, polypropylene (PP) reinforced with glass fiber is one of the most attractive materials for a wide number of industrial products processed by Injection Molding; but some aspects of its behavior are still uncertain.
Most of these uncertainties are related to the behavior of injected glass reinforced polypropylene (GFPP) and in particular to the effect of the process parameters on the properties of the final part.
In this work, in order to evaluate the effect of injection pressure and temperature on the morphological and mechanical properties an experimental analysis and a computer modeling of the injection process on small scale PP and GFPP parts are presented.

1. Introduction

Polypropylene matrix composites reinforced with short glass fibers (GFPP) are used in the industrial fields from at least 10 years, and these applications are due to their very good mechanical properties connected with a not elevated density. In any case, the use of GFPP for these applications requires high assurance and reproducibility to conform to high quality standards required for industrial parts. So the use of modeling for the prevision of the molding behavior and of the final mechanical properties is becoming more important.
In order to meet these requirements the aim of this work is to analyze the relationship between processing, structure and final properties in the Injection molding process of pure and reinforced Polypropylene. First of all, to characterize the production process a numerical simulation of the filling of a simple mold has been performed applying a commercial software such as I-DEAS. Afterwards, the results obtained in terms of filling time, injection pressure, fiber orientation, mechanical properties and morphological anisotropy have been compared with experimental data obtained from thermal analysis (DSC), electron microscopy and mechanical testing of samples produced in a laboratory

scale Injection Molding apparatus.

The results of this work lead to the possibility to predict the properties of polypropylene based composites produced by Injection Molding.

Although the study has been conducted on a very simple mold, the application of appropriate scale factors could permit the application of these analytical and simulation techniques to real scale process and components.

2. Experimental

Several plates of pure PP and glass reinforced (40% of short glass fibers) parts in a SANDRETTO 60-T Injection machine. The produced parts are rectangular 200x100x3 mm plates obtained from a mold with single injection point. The production condition are reported in Table 1.

For the numerical simulation a Silicon Graphics "Indigo 2" workstation have been used.

For the thermal characterization a Perkin Elmer Pyris 1 DSC, able to work between -160 °C and 750 °C with heating or cooling rate up to 200 °C/min, has been used. The electron scanning microscopy have been performed using a CAMBRIDGE Stereoscan 360.

The mechanical properties have been evaluated using an Instron 4500 and applying the ASTM standard D638 for the tensile test, while the Izod impact resistance have been measured with a CEAST impact pendulum with maximum energy of 7.5 J.

The specimen for all the tests have been obtained from several plates as reported in Figure 1 and 2 where also the injection gate is showed.

Figure 1: Tensile Test

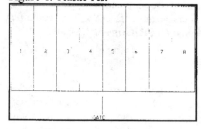

Figure 2: DSC & Izod Test

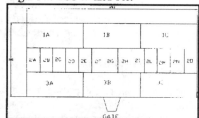

The polypropylene used is a commercial Moplen 20 - Himont® (now Montell) in powder form, in the composite the reinforcement is made of Vetrotex® EC5168 short glass fibers. The initial length of the fibers was 4.5 mm while the initial diameter is 13 µm.

To obtain a mononodal distribution of the initial length of the fibers in the composite the Injection molding equipment the premiscelated material has been inserted.

Table 1 - Processing parameters

Variabili di processo nello stampaggio del PP e del GFRP		
Screw Speed	Hold Pressure	Barrel Temperature Profile
200 rpm	5 MPa	180-200-220-220
Back Pressure	Transfer Point	Injection Speed
1 MPa	2% screw run	50% max inj. speed
Hold Time		
5 sec		

3. Results and Discussions

Filling stage simulation

The filling stage begins when the mold is closed and ends when the material have completely filled the cavity. The duration of this process is the filling time. During all this phase the screw do not rotate but act as a plunger; the advancing velocity of the plunger, generally constant, is chosen by the operator determining a constant volumetric flow rate during all this stage.

The performed work has been focalized on the filling procedure and on the parameters of industrial interest such as pressure, time and temperatures. The program used has been the I-DEAS Plastics module of the SDRC, this program permit the complete numerical simulation of the Injection molding process and is an hybrid finite element - finite difference program for the flow analysis in the injection channels and in the mold cavity.

The polymer flow, inside the mold, has been approximated to an Hele-Shaw flow of a non-Newtonian fluid [2]. The Finite Element Model (FEM) is shown in Figure 3 while in Figure 4 is shown the diagram of the clamping force. The filling time analyzed have been of 2 and 5 seconds.

It is very important to notice that due to the simplicity and the symmetry of the model, the simulation of the rectangular plate is a very interesting case because every small error is greatly enhanced.

The pressure difference between different points is the training force of the process, obviously the polymer flow in direction where the pressure gradient is negative; the maximum pressure it always at the injection gate while the filling front is at P_a.

The simulation results obtained with a filling time of 5 seconds show that there is a great temperature variation inside the mold cavity and that an high percentage of polymer (more than 30% of the plate thickness) solidifies before the total filling of the mold creating a "skin" that causes such a great difference of pressure between the gate and the other part of the front.

Therefore, the simulation filling time has been reduced at 2 seconds. The trend of all properties is similar so the time reduction does not influence the flow stability; but the pressure diagram changes significantly. In fact the pressure at the gate decreases of about 1 MPa. This reduction can be justified with the difference in the "skin" thickness.

In fact the difference between the hotter and the cooler point with the 5 seconds filling time is of 40 °C while with 2 seconds the difference is only of 10 °C maintaining the polymer always at a temperature near 210 °C. This fact reduces the viscosity of the system so facilitating the mold filling, reducing the pressure and reducing the skin thickness of about 12%.

Analyzing the shear force that even if its value increase at the gate they generally decrease inside the mold cavity; this is very important because, as can be seen from the experimental mechanical test results, the shear forces give an indirect indication of the macromolecule and fiber orientation. To have a component with an high level of isotropy is necessary to minimize the shear forces.

Figure 3: FEM Model   *Figure 4:Clamping Force vs. Time*

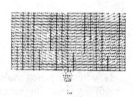

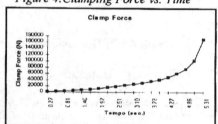

Thermal analysis

Experimental results made examining the plates' skin in some of the point shown in Figure 2 are reported in Figure 5.

Figure 5: DSC Dimanic Test

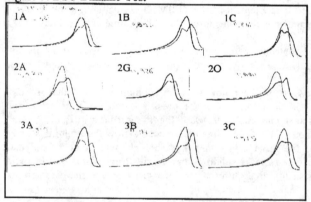

These tests have been performed in a range of temperature between 50 and 220 °C with an heating rate of 20 °C/min; for every sample two consecutive dynamic scans have been performed as shown in Figure 5.

The form of the DSC melting curve shows two peaks; this phenomenon could be related to structural reasons or could be a direct consequence of to the injection process [3,4].

In this case the presence of two peaks is certainly due to recristallization processes that can be seen only on the skins due to the strong orientation of the macromolecules caused, near the mold surface, by the shear stresses. Another phenomenon which leads to recristallization is the reduction of the thickness of the lamellae due to the strong orientation. The DSC scans relative to the middle part of the plate do not show any double peak confirming the previous assumption.

In conclusion some considerations can be made analyzing the results showed in Figure 5.

The symmetry of the mold is respected, in fact the scan of the portion 2A and 2O are very similar. This indicates that, even if a little diversity in the cristallinity grade is present, the crystalline structures and their spatial disposition are almost identical. This is a very important fact from the industrial point of view because it assures a structural, morphological and tensional equilibrium inside the component.

Mechanical experiments

The results obtained in tensile tests are summarized in Figure 6, which reports the elastic modulus as a function of the position in the plate as described in Figure 1 for 5 different plates. For each plate the elastic modulus behaves in a similar way confirming the different orientation of the fiber inside the plate. The scattering between the data is certainly due to the mold temperature difference experienced during the production of the samples that influences the mechanical properties. In fact at higher mold temperatures the cooling rate inside the mold is lower with the development of more stable crystalline structures.

Figure 6: Tensile Test Results

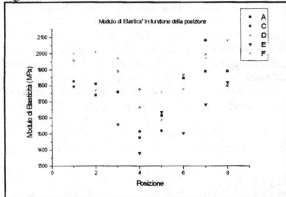

On the side of the plates the macromolecules are essentially oriented in the direction of the force applied while in the middle part the orientation is perpendicular to the force. This is the reason for the difference of some hundreds of MPa between the side and the center of the plates. So, the polymer flow in the mold is very relevant for the properties of the final product. Therefore, it is possible to enhance the mechanical properties changing some process variables. In fact, an high injection pressure (90 MPa), an high injection rate (18 cm/sec), a low polymer melt temperature (160-190 °C) and an high temperature of the mold (50-80 °C) permit the creation of a *selfreinforced* PP with a molecular structure oriented in the flow direction. Moreover, it has been experimentally proven that an increase in the mold temperature and an increase in the volumetric flow cause a strong reduction of the crystallinity gradient lowering the anisotropy in the mechanical properties [5].

Impact tests

The mean value and the related plot of the Izod impact resistance as a function of the position on the plate are reported in Table 2 and in Figure 7. As can be seen the glass reinforced plates have an impact resistance greater of a factor 10-20 then neat PP.

Table 2: Izod Test Results

Provino	Energia per unitá di superficie (kJ/m^2)	
1	1.28	12.34
2	1.59	21.01
3	1.32	11.31
4	1.26	23.14
5	1.19	26.49
6	1.25	24.39
7	1.25	26.3
8	1.51	22.05
9	1.51	22.4
10	1.51	20.07
11	1.49	18.72
12	1.32	14.17
13	1.21	12.26
14	1.19	14.99
15	1.23	17.39
16	1.15	17.85
17	1.27	20.36
18	1.5	22.63
19	1.49	27.25

Figure 7: Impact Strength vs. Position

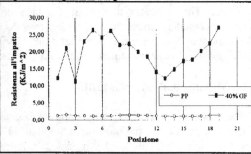

A so bigt increase of this property, very important for a lot of industrial applications, merges with the fact that the technology for the production does not change too much compared to that of the pure polymer, making the reinforced materials very interesting from an industrial point of view.

In any case, it is very interesting how the impact resistance change as a function of the position in the plate. Remembering that the middle of the plate is represented by the specimens from 7 to 19, a very similar behavior in this zone is observed between impact resistance and tensile resistance of the pure PP. This fact confirms the validity of the assumption of macromolecule orientation also for the short fibers. In fact if fiber-fiber. interaction is neglected, is the flow which determines the orientation. In practical cases other variables have to be considered and an acccurate prevision of the final orientation as a function of the process variables is, at the moment, not possible.

Moreover, if the impact data for the pure PP are analyzed more carefully, a similar but very small variation in the properties is found similar to that of the reinforced PP.

Fiber orientation

Properties of composite parts made with short fibers are highly influenced by the fiber content, by the fiber length distribution and by their orientation [6,7]. This last parameter is the most difficult to determine because is a function of the local mass flow inside the mold.

The analysis of the SEM micrographs of the position 2A and 2O leads to the following qualitative consideration about the fiber orientation in the composite part:

- A symmetrical behavior of the fiber orientation in the plate has been observed.
- This observation is in accordance with the results of the mechanical properties.
- The fibers are bent as along as transverse to the mean axis
- A concentration difference between the two zones is present but density measurements have shown that there is no density gradient inside the plate.

Figura 8: Skin Orientation

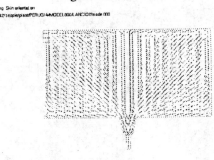

In Figure 8 the representation of the "skin" fiber orientation obtained from the numerical simulation is reported. Comparing the results of the SEM analysis, see Figure 9 and 10, the orientation is very similar. In any case, to obtain more precise results a tridimensional analysis is required due to the change of orientation through the thickness.

4. Conclusions

The numerical simulation of the injection process is able to predict, with very good accuracy, the behavior of pure and reinforced polypropylene and gives useful information on the effect of the most important process variables. Using this approach it will be possible to optimize the process looking for the better combination between injection channels and gates, mold temperature, polymer injection temperature and flow rate of the injected material. Thermal analysis has shown the presence of a little gradient in crystallinity of the polymer and the existence of a skin-core structure. Mechanical properties have shown the existence of anisotropy essentially related to macromolecules orientation and to crystallinity differences. Moreover, a variation of the elastic modulus as a function of the mold temperature is observed. The analysis of the reinforced plates has shown that the impact resistance depends on the distribution of the short fibers.
SEM analysis has individuated the existence of preferential directions in fiber orientation in the "skin" of the composite and this observation is in accordance with the numerical model results.
Results obtained lead to the conclusion that this methodology could be applied with success to real systems so reducing production time and costs.

Figure 9: Sample 2A

Figure 10: Sample 2O

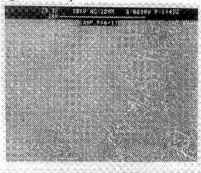

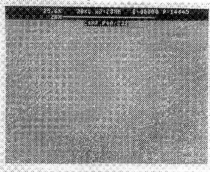

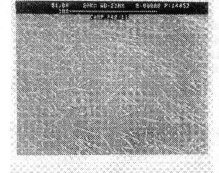

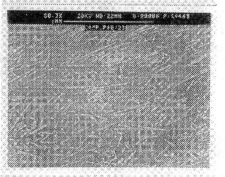

References

[1] *C.L. Tucker III*, COMPUTER MODELING FOR POLYMER PROCESSING, Ed. Hanser, 1989

[2] *S Middleman*, FUNDAMENTAL OF POLYMER PROCESSING, Ed. McGraw-Hill, 1977

[3] *J. Karger Kocsis*,, POLYPROPYLENE Vol. 1,2,3, Ed. Chapman & Hall, 1995

[4] *W. Sifflet*, Polymer Engineering and Science, 1973, Vol. 13, p. 10

[5] *S.E. Barbosa, J.M. Kenny*, Proc. Antec 96, 1996

[6] *R.S. Bay, C.L: Tucker III*, Polymer Composites, 1992, Vol. 13, p. 317

[7] *Y.C. Kim, W. Han, C.Y. Kim*, Polymer Engineering and Science, 1997, Vol. 37, p. 1003

Mechanical Properties of Nonwoven Glass Fiber Composite

Tae Jin Kang*, Sung Ho Lee*

* Department of Fiber and Polymer Science, College of Engineering, Seoul National University, San 56-1, Shinlim-Dong, Kwanak-Ku, Seoul 151-742, Korea

ABSTRACT

Mechanical and impact resistant properties of needle-punched nonwoven glass fiber composite have been studied. Fourteen plies of carded webs of E Glass fiber were stacked and needle punched with different punching densities of 15, 30, 60, 90 penetrations per square centimeter(ppsc). The needle punched webs were formed into composite with unsaturated polyester resin. The structural changes in nonwoven composite at different punching densities were studied by means of the fiber orientation as well as the length changes in the fibers using image analysis system. The effect of changes in fiber structure on the mechanical properties of nonwoven composite was investigated through the testing of tensile, bending, impact, compression after impact(CAI), fatigue, friction/wear and mode-I interlaminar fracture properties.

The nonwoven composites showed an improvement in mechanical properties with increasing fiber entanglements caused by needle punching. At the same time the fiber damage during the needle punching process caused an impairment of performance of the nonwoven composite. In the range of adequate punching densities, the nonwoven composite showed substantially improved mechanical and impact properties compared with woven laminate composite. The nonwoven composite showed improved properties of flexural and compression strengths as well as mode-I interlaminar fracture toughness.

1. INTRODUCTION

The needle punched nonwoven fabric has been used for long time ever since the felted structure was used as textile substrate. The new evolution of punching needles as well as punching equipment has made it possible to use the needle punched nonwoven fabric as a major products in many applications. The experimental and analytical studies on nonwoven fabrics have been performed since 1960's. Hearl etc. investigated parameters that influencing the mechanical properties of needle punched nonwoven fabric[1]. They reported that punching density is the most important parameter which influences the mechanical properties of the nonwoven fabric. The orientation, fiber length and the entanglement of fibers in nonwoven composites are affected by the needle punching process due to the through-the-thickness reinforcement and breakage of constituent fibers caused by the mechanical punching during the process[2-5]. The increase in punching density generally increases the mechanical properties of nonwoven fabric due to the increase of entanglement and interlocking of fibers. However excessive needle punching decreases the mechanical properties of the fabric because the needle punching process results in damage and breakage of the fibers as well as the redistribution of the fiber orientation. Thus the needle punched nonwoven fabric has the optimum punching density at which the reinforcing effect becomes maximum in the composite.

In this study, the mechanical and impact properties of needle punched nonwoven E glass fiber composites with different punching densities of 15, 30, 60, 90 ppsc were investigated and compared to those of comparable woven laminate composite. In order to verify the structural changes induced by the needle punching process, the fiber length and the orientation of fibers were measured using image analysis system. The tensile, bending, impact, compression after impact, fatigue, friction/wear and mode-I interlaminar fracture tests were performed to investigate the effect of changes in fiber structure on the mechanical properties of nonwoven composite. From these tests and examinations the governing parameters of the needle punched nonwoven composites in the determination of mechanical properties have been identified and their relationships have been determined.

2. EXPERIMENTAL

E-glass fibers of $13 \mu m$ diameter and 7cm length were carded to form parallel webs. The 14 plies of parallel-laid webs were needle punched with different punching densities of 15, 30, 60, 90 ppsc(penetrations/cm^2). The nonwoven fabrics were impregnated with unsaturated polyester resin and then cured using hot press. The resin content of nonwoven composite was kept about 45wt% and thickness was 2.8mm. The comparable E-glass woven laminate composite was also prepared to compare the properties to the nonwoven composites. The resin content of woven laminate composite was 40 wt% with the thickness of 2.7mm.

The length and the orientation of fibers were measured using image analysis system to verify the structural changes in nonwoven composites. The tensile, flexural and drop weight impact test were performed to investigate the effect of structural changes on the mechanical and impact properties of nonwoven composites. The mode-I interlaminar fracture energy, G_I, was determined by the compliance method using double cantilever beam(DCB) specimen[6,7]. The damage tolerant properties of woven and nonwoven composites were also studied from the impact fatigue life as well as the compression-after-impact(CAI) strength. Fatigue life[8] and fracture behavior of nonwoven and woven laminate composites in load-control tension-tension fatigue tests have been examined at 10 Hz sinusoidal loading condition. The friction and wear characteristics[9] of nonwoven composite were examined in the pin-on-ring type wear tester at the sliding speed of 2.64 m/sec.

3. RESULTS AND DISCUSSION

Characterization of the reinforcing web structures by image analysis

Figure 1 shows the schematic diagram of image capture and analysis system. The fiber orientations were measured both in the surface and in the cross-section of the composites using image analysis. The fiber length was also measured at four different punching densities of nonwoven fabric to access the extent of fiber breakage during the punching process.

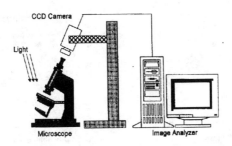

Figure 1. Schematic diagram of image processing

The mean fiber orientation angle and the mean fiber length of woven laminate composite are presented in Figure 2. The mean fiber orientation angle was slightly increased with punching density. The mean fiber length in nonwoven fabric with punching density of 90 ppsc showed 30% reduction compared with that of 15 ppsc nonwoven fabric. And this causes the reduction in mechanical properties of nonwoven composite with high punching densities.

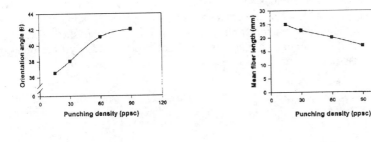

(a)   (b)

Figure 2. Mean fiber orientation angle(a) in loading direction and mean fiber length(b) measured by image analysis

Mechanical properties

The tensile strength and modulus of needle punched nonwoven and woven laminate composites are presented in Figure 3. As the punching density increases, fiber entanglements were increased and the tensile strength of nonwoven composite was improved due to the efficient load transfer between fibers. However, the reinforced fibers in machine direction were bent against the axial direction by needle punching process, so that the tensile modulus was decreased with punching density. The tensile strength and modulus in transverse direction were increased with punching density due to the fiber entanglement and the transverse fiber orientation which were increased by needle-punching.

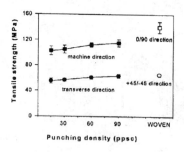 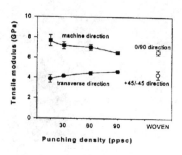

(a)   (b)

Figure 3. Tensile strength(a) and modulus(b) of needle punched nonwoven composites at different punching densities compared to woven laminate composite.

The flexural strength and flexural modulus of nonwoven composites were decreased with increasing punching density as shown in Figure 4. The fiber damage and structural changes of reinforced fibers, especially in fiber orientation angle, caused by needle-punching resulted in decreased flexural properties. However, the flexural properties of needle punched nonwoven composites showed much higher value than those of woven laminate composite. The nonwoven composites resisted against bending force as one phase material without delamination, while the woven laminate composite showed little resistance against the delamination and thus it did not experienced any fiber breakage during the flexural deformation.

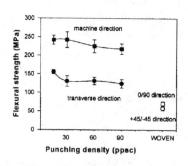 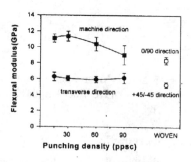

(a)   (b)

Figure 4. Flexural strength(a) and modulus(b) of needle punched nonwoven composites at different punching densities compared to woven laminate composite.

Impact properties and damage tolerance

The energy absorption capability of nonwoven composites showed its maximum at the punching density of 30 ppsc. The energy absorption capability of woven laminate composite was similar to that of nonwoven composite with the punching density of 60 ppsc. The woven laminate composite suffered with large deformation area and bulging when the impact load was applied, while the nonwoven composite did show little bulging.

The nonwoven composite at the punching density of 30ppsc effectively transferred impact energy through the three dimensionally entangled reinforcing fibers, so it showed high resistance against the repeated impact. However, the excessive needle-punching resulted in easy initiation of cracks and confined energy propagation due to the fiber damage as well as the brittle fracture phenomena of the composite.

The compression strength of nonwoven composite was higher at the low range of punching density due to the high orientation of the fibers. The reduction in compression strength after impact was decreased with increasing punching density, because the increase in punching density resulted in poorer energy propagation and the damaged area confined in a relatively small area.

Mode-I interlaminar fracture toughness

The adhesion strength of nonwoven composite was 1.8 times higher than that of woven laminate composite. The crack propagation velocity of nonwoven composite was decreased with punching density. It is obvious that the crack propagation was retarded with the needle-punching. As shown in Figure 5, the mode-I fracture energy of needle punched nonwoven composite showed 3 times higher than that of woven laminate composite due to the presence of the three dimensionally reinforced fibers. The fracture energy of nonwoven composite was increased with increased punching density due to the increased number of z-axis fibers that resist the delamination in nonwoven composite.

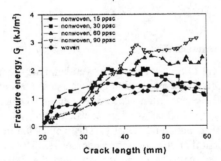

Figure 5. Mode I fracture energy(G_I) of nonwoven and woven laminate composites

Fatigue properties of nonwoven and woven laminate composites

The fatigue test results of nonwoven and woven laminate composites are shown in Figure 6. The fatigue life of woven laminate composite was much longer than that of needle punched nonwoven composites. The fatigue life of nonwoven composite was decreased with increasing punching densities. The fatigue life of nonwoven composite was determined by the length of the constituent staple fibers. The fiber length of nonwoven composite with higher punching density has been reduced compared to the one with low level of punching densities, due to the higher fiber breakage during the needle punching process. The woven laminate composite which is consisted with continuous fibers showed much longer fatigue life than that of the staple fiber reinforced nonwoven composites.

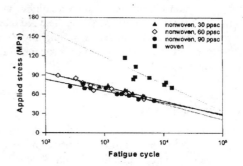

Figure 6. S/N curves for nonwoven and woven laminate composites with stress level in the tension-tension fatigue test.

The modulus of nonwoven composite showed little decrease during the low range fatigue cycle and abrupt decline of modulus just before the fracture of the composite. While the woven laminate composite showed gradual degradation of modulus as the fatigue cycle increased.

Friction and wear characteristics

The friction coefficient of nonwoven and woven laminate composites were measured for 5 hours using pin on ring type friction and wear test machine. The mean value of friction coefficient and wear resistance of nonwoven and woven laminate composites were presented in Figure 7. The friction coefficient of nonwoven composite was increased with increasing punching density. The woven laminate composite showed higher friction coefficient than the nonwoven composites. The weight loss of nonwoven composite was increased with punching density, and the woven laminate composite showed less weight loss than the nonwoven composite due to the length of the constituent fibers.

As the punching density increased, the fibers in the nonwoven composite getting shorter, and the fiber separation can become easier when the frictional force was applied. On the other hand, continuous fibers as in woven laminate composite reduced wear rate by restricting the separation of reinforcing fibers. The amount of wear-out particles of nonwoven composite was doubled with the 30% reduction of fiber length.

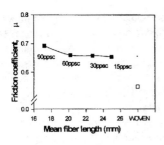
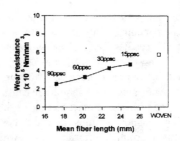

(a)   (b)

Figure 7. Mean friction coefficient(a) and Wear resistance(b) of nonwoven and woven laminate composites at the sliding speed of 2.64 m/sec.

4. CONCLUSIONS

Mechanical and impact resistant properties of needle-punched nonwoven E Glass fiber/Polyester composite with different punching densities of 15, 30, 60, 90 penetrations per square centimeter have been studied. The structural changes in nonwoven composite at different punching densities were studied by means of the fiber orientation as well as the length changes of the constituent fibers using image analysis system. The mean fiber length was decreased up to 30% by the needle-punching process and the fibers were more randomly distributed with increasing punching densities. The effect of changes in fiber structure on the mechanical properties of nonwoven composite was investigated through the testing of tensile, bending, impact, compression after impact, fatigue, friction/wear, and mode-I interlaminar fracture properties.

The tensile strength and mode I interlaminar fracture toughness of the nonwoven composite was increased with increasing punching densities due to the increased fiber entanglement. The tensile strength of nonwoven composite was increased with increasing fiber entanglements through the increased load transfer efficiency although the fiber length was decreased with increasing punching densities. The woven laminate composite showed higher tensile strength than the nonwoven composites due to the continuous nature of reinforcing fibers. The mode-I interlaminar fracture toughness(G_I) of nonwoven composite was increased with punching densities due to the contribution of the through-the-thickness reinforcing fibers and it showed three times higher value than that of woven laminate composite.

The flexural, impact, fatigue and wear properties were decreased with punching densities due to the reduction of fiber length caused by the damage of fibers during the punching process. The flexural strength and modulus of nonwoven composites were decreased with increasing punching densities due to the fiber damages during the needle-punching process. The flexural properties of woven laminate composite were lower than those of nonwoven composites due to the easy delamination. The impact energy absorption capability and damage tolerance of nonwoven composites was higher for the punching density of 30 penetrations/cm^2. The reduction in compression strength after impact was decreased with punching densities due to the confined damage area of highly needle-punched nonwoven composite. The fatigue life of the nonwoven composite was reduced with the shorter fiber length caused by the high number of needle punchings. The fatigue life of woven laminate composite was much longer than that of needle-punched nonwoven composites due to the continuous nature of reinforcing fibers. The friction coefficient of nonwoven composite was increased with increasing punching densities. The wear rate of the nonwoven composite was increased with decreasing fiber length, because the short fiber was more easily pulled out or removed from the contact surface of the nonwoven composite.

The tensile modulus and the compression strength were determined from the orientation of the reinforcing fibers which was affected from the punching densities. The tensile modulus was predominantly determined by the orientation of the reinforced fibers, so that the increased orientation angle with increased punching densities resulted in reduced tensile modulus. The increase in orientation angle of reinforcing fibers with respect to loading direction resulted in reduced compression strength as well. The flexural properties were also influenced by the orientation angle of the fibers same as in the case of the compression strength. The mechanical properties of nonwoven composite showed rather isotropic properties with increased random distribution of the fibers at high punching densities. In general, the nonwoven composite offered some advantage in flexural properties, impact resistant properties, and especially in mode-I interlaminar fracture toughness.

REFERENCES

1. J. W. S. Hearl and M. A. I. Sultan, "A Study of needled fabrics Part 1 : Experimental Method and Properties", Journal of the Textile Institute, 1967.
2. R. M. Christensen and F. M. Waals, "Effective stiffness of randomly oriented fiber composites", Journal of Composite Materials, Vol. 6, 1972.
3. C.-T. D. Wu and R. L. McCullough, "Constiutive relationships for heterogeneous materials", Developments in Composite Materials - I, G. S. Hollister, ed., Applied Science Publishers, Ltd., London, 1977.
4. J. C. Halpin and N. J. Pagano, "The laminate approximation for randomly oriented fibrous composites", Journal of Composite Materials, Vol. 3, 1969.
5. J. C. Halpin and J. L. Kardos, "Strength of discontinuous reinforced composites : I. Fiber reinforced composites", Polymer Engineering and Science, Vol. 18, 1978.
6. Lin Ye, "Evaluation of mode-I interlaminar fracture toughness of fiber-reinforced composite materials", Composite Science and Technology, Vol. 43, 1992.
7. M. A. Solar and F. J. Belzunce, "Fracture toughness and R-curves of glass fiber reinforced polyester", Composite, Vol. 20, No. 2, 1989.
8. Z. Hashin and A. rotem, "A cumulative damage theory of fatigue failure", Material Science and Engineering, Vol. 34, 1978.
9. N.S.El-tayeb and R.M.Gadelrab, "Friction and wear properties of E-glass fiber reinforced epoxy composites under different sliding contact conditions", Wear, 1996.

Eigenstrain models for complex textile composites

G. Huysmans*, I. Verpoest*, P. Van Houtte*

* Katholieke Universiteit Leuven, Dept. Metal. & Mat. Eng.
  De Croylaan 2, 3001 Leuven - Belgium

Abstract

This paper presents a novel modelling approach based upon eigenstrain concepts for the prediction of the elastic behaviour of textile reinforced composites. The emphasis in the development of the current models was put on composites with a very complex (and in some cases partially random) geometry. Nevertheless, the concept can be generalised to other fibre architectures. The heterogeneous composite is substituted by a homogenous medium by introducing equivalent eigenstrains. The relationship between the perturbation strains in the yarns and the eigenstrains, given in the form of an integral expression, is solved in an approximate way using two simplifying assumptions. A mean field approximation was used to avoid the explicitation of the exact spatial distribution of the yarns. Secondly, the autocorrelating Eshelby tensors for curved yarn segments were approximated by a short fibre equivalent. The model was subsequently used to predict the elastic constants of several types of E-glass/epoxy warp knitted fabric composites, using both a Mori-Tanaka and a singly embedded self-consistent averaging scheme. Predicted engineering constants agree well with experiments.

1. Introduction

During the last decade, an emerging range of - existing - textile technologies is being adapted for application in composite materials. Textile based composites offer specific advantages over traditional laminates such as improved formability and damage tolerance, higher production rates at lower manufacturing costs etc. Knitted fabrics in particular, which are characterised by an interlaced loop structure, have excellent drapability properties coupled with coherence of the fabric. Elastic properties of knitted fabric composites were shown to lie in between those of short fibre mats and woven fabric composites. However, knitted fabric composites were shown to have an excellent interlaminar fracture toughness coupled with improved impact resistance [1]. Knitted fabrics can also be combined with short fibre mats to form hybrids which are more stable that pure mats. These hybrids also offer some advantages in RTM-processing of composites.
The mechanical behaviour of knitted fabric composites has mainly been investigated from an experimental point of view, including stiffness and strength properties, impact resistance and fatigue, see e.g. [2,3,4]. In addition, the triclinity of most knitted fabric composites needs special attention regarding the measurement of the elastic constants [5].

For a better understanding of the mechanical behaviour of knitted fabric composites, predictive models are desired. Attempts to describe the mechanical behaviour currently found in literature are mainly based on analytical approaches using iso-stress and/or iso-strain trial fields for the yarn segments [6]. Their major advantage is that only the orientation distribution of the fibres or yarns is required, and that they provide an estimate of the full stiffness tensor. However, only bounds can be obtained, and these models cannot be used for damage or strength prediction. Numerical, finite element based models on the other hand lack simplicity and efficiency [7]. Moreover, knowledge about the spatial distribution of the yarns is required, and geometrical simplifications are needed in order to be able to create the finite element model.

In the following, a more efficient analytical model is developed which apart from predicting the elastic constants, additionally provides estimates for the internal stresses and strains, so it can be used as a basis for damage modelling. The model is not only applicable to knitted fabric composites, but under certain conditions as well to other textile reinforced composites, such as 2D wovens and 2D braids.

2. Geometrical description of knitted fabrics

Knitted fabrics are produced by the subsequent interlacing of yarn loops. According to the manufacturing technique or production direction, one distinguishes weft knitting, in which the loops are formed sequentially in the course direction of the fabric, and warp knitting with simultaneous loop formation over the whole width of the fabric (Figure 1).

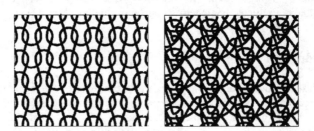

Figure 1. Schematic representations of knitted fabrics (a) Plain weft knit - (b) Warp knit produced using two guide bars acting in a 2-1 and 1-1 lapping movement.

The plain weft knitted fabric as shown in Figure 1a has a relative simple loop structure with some symmetry elements. Some authors used a combination of straight line sections and circular arcs to describe the fabric geometry. The parameters controlling the shape of these functions are determined from the machine settings such as needle spacing and loop density. However, such an approach is impossible for complex warp knits with skewed loop geometries and a total lack of symmetry.

Figure 1b schematically shows a warp knit used in this study, which was produced using two guide bars. The large diversity in possible loop arrangements does not allow to use special shape functions to describe the fabric geometry. Instead, cubic interpolation functions $\mathbf{p}(t)$ with continuous second derivatives were used to reconstruct the yarn path from a set of measured keypoints. The in-plane keypoint co-ordinates were measured on the dry textile, with the implicit assumption that these positions do not change significantly during composite production. For the out-of-plane co-ordinates, levels were assigned to each keypoint based upon microscopic observation of subsequent yarn cross-overs. In a second step, the levels are scaled to the thickness of a single composite layer.

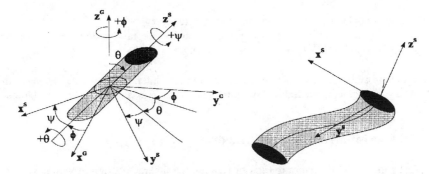

Figure 2. Definition of yarn orientation angles (ϕ,θ,ψ) and relationship of the local segment co-ordinate system (x^S,y^S,z^S) to the yarn centreline path

The spline functions are used to derive the local yarn orientations (Figure 2) and to obtain the yarn curvature radius $R(t)$ at every point $\mathbf{p}(t)$:

$$R(t) = \frac{\|\mathbf{p}'(t)\|^3}{\|\mathbf{p}'(t) \otimes \mathbf{p}''(t)\|} \tag{1}$$

Using a separate keypoint set for the yarn loops produced by every single guide bar, the splines are interpolated to obtain a mathematical representation of the yarn centreline paths within the fabric unit cell. The splines are subsequently split up into a number of yarn sections. Following assumptions are made regarding the yarn sections:

1. All yarn segments S have a circular cross-section (with radius r^S). In reality, the aspect ratio can vary between 1:2 and 1:5. It is however very difficult to obtain the aspect ratio distribution and orientation from geometrical principles; hence the assumption of a circular cross-section is the simplest and most straightforward.
2. Yarn twist is ignored. Yarn twist is the superposition of yarn centreline twist, which can be derived from the interpolation functions, and the fibre twist within the yarn. The twist effect is only of secondary importance for the elastic behaviour of multi-oriented preforms, so it will be assumed that all fibres within the yarn run parallel to the yarn centreline.

This geometrical treatment of the knitted structure leads to a set $\left\{f^S, R^S, r^S, a^S_{ij}\right\}$ for each yarn section S, where f^S is the total volume fraction of the segment and a^S_{ij} the direction cosines of the local co-ordinate system $(\mathbf{x}^s, \mathbf{y}^s, \mathbf{z}^s)$ associated with the segment. This information of the knit structure will form the input of the elastic model. Before describing the model, it is noted here that numerical models in most cases (for simplicity) do not consider the mechanical interaction of different fabric layers. This is an important effect in knitted fabric composites, which is clearly shown by their extremely high mode I fracture toughness. The geometrical arrangement of the yarn loops is however difficult to assess, due to:

1. the unknown phase-shift between different fabric layers
2. small local deformations occurring during composite lay-up and manufacturing, causing this phase shift to be position dependent.

Therefore, it seems reasonable to consider *average* interaction effects between the different yarn segments, including both intra-ply interactions and inter-ply interactions.

3. Description of the elastic model

Let $\mathbf{A}^S$ be the strain concentration tensor relating the (average) strain in yarn segment S to the composite strain. If $\mathbf{A}^S$ is known for each yarn segment, the composite stiffness $\overline{\mathbf{C}}$ is obtained as [8]:

$$\overline{\mathbf{C}} = \mathbf{C}^M + f\left\langle \left(\mathbf{C}^S - \mathbf{C}^M\right)\mathbf{A}^s\right\rangle \tag{2}$$

where f is the total yarn volume fraction, $\mathbf{C}^M$ is the stiffness of the matrix and $\mathbf{C}^S$ is the stiffness of the segment under consideration. The notation $< . >$ stands for a configurational average. The computation of the strain concentration tensors is performed using Eshelby's concepts of eigenstrains. To achieve this, the heterogeneous composite is replaced by a homogenous domain with non-vanishing eigenstrains $\mathbf{e}^{*S}$ distributed over the segment domains. The eigenstrains are defined by:

$$\sigma^S = \mathbf{C}^S \mathbf{e}^S = \mathbf{C}^M\left(\mathbf{e}^S - \mathbf{e}^{*S}\right) \tag{3}$$

It is now possible to set up an elasticity problem in terms of the unknown eigenstrains. The strains in each segment S can be related to the eigenstrains in the segment domain Ω^S itself, leading to an (average) Eshelby tensor $\mathbf{S}^S(\Omega^S; \mathbf{C}^M)$, and to the eigenstrains in the other segment domains Ω^R, through the interaction Eshelby tensors $\mathbf{S}^{*S}(\Omega^S; \Omega^R; \mathbf{C}^M)$. The evaluation of the Eshelby tensors $\mathbf{S}^*$ of the second type (inter-inclusion interaction) needs to be carried out for each segment pair (R, S), and depends upon the shapes Ω and the relative orientations of both segments. Because of the large complexity and the problems mentioned in §2, averaging methods where used to estimate an *average* inter-segment interaction. Two schemes which are well-known from short-fibre composite modelling and the description of poly-crystalline materials are the Mori-Tanaka method and the self-consistent method respectively. With $\mathbf{Q}^S$ a localisation tensor defined as:

$$Q^s(\Omega^s;C^*;C^s) = -\left[S^s(\Omega^s;C^*)+(C^s-C^*)^{-1}C^*\right]^{-1} \qquad (4)$$

the strain concentration tensors A^S are obtained as:

$$A_{MT}^s = \left[I+S^s(\Omega^s;C^M)Q^s(\Omega^s;C^M;C^s)\right]\left[I+f\langle S^R(\Omega^R;C^M)Q^s(\Omega^R;C^M;C^R)\rangle\right]^{-1} \qquad (5a)$$

for the Mori-Tanaka method, and

$$A_{SC}^s = I+S^s(\Omega^s;\overline{C})Q^s(\Omega^s;\overline{C};C^s) \qquad (5b)$$

for the self-consistent method, see e.g. [8].

For solution of the model, an expression is still needed for the Eshelby tensors $S^s(\Omega^s,C^*)$ for the (curved) yarn segments. Solutions are already available (either in closed form or as a series expansion) for ellipsoids, cuboids and general polygonal shaped inclusions. In order to avoid the tedious calculations related to the evaluation of S^S for curved shaped segments, use was made of the attractive solutions for the ellipsoid by introducing a short fibre equivalent for each yarn section. The equivalent aspect ratio is related to the local yarn curvature by (see also [9] for a more in-depth discussion):

$$l^s = \beta R^s \qquad (6)$$

with l^s being the length of the equivalent ellipsoid. The determination of the empirical factor β in (6) for a given material combination can be performed by calibrating the model for a single fabric type.

For isotropic matrices, the Eshelby tensors can now be approximated either analytically or by very fast numerical calculations when using the Mori-Tanaka scheme. For the self-consistent method, the Eshelby tensor is a function of the (anisotropic) composite stiffness, and the evaluation has to be done using numerical quadrature [10].

4. Results and discussion

The model was applied to six different types of warp-knitted E-glass/epoxy fabric composites. Material properties are listed in table I. The tensile properties of the composites were experimentally determined by tensile tests according to ASTM D3039-93. The shear modulus was measured from multiple tensile test data using the regression scheme outlined in [5].

Material	E (GPa)	ν (-)	Packing (%)
Epoxy F533 (Hexcel)	3.0	0.35	
E-Glass	72.5	0.235	60

Table I. Material properties for the knitted fabric composites

In the model, every spline was subdivided into approximately 50 segments, resulting in about 100 segments for each fabric type. The yarn properties were calculated from the matrix and fibre properties and the packing degree using a local uni-directional Mori-Tanaka submodel. The in-plane elastic constants for each fabric type were computed using both the Mori-Tanaka and the self-consistent method.

From (5b), it is seen that both the Eshelby tensors and the localisation tensors are a function of the (unknown) composite properties. Therefore, the model has to be solved in an iterative way. As an initial guess, one can use an isostrain (Voigt) estimate for the composite stiffness, as for many fabric types the E-modulus lies close to the Voigt-bound.

Figure 3 compares the experimental and predicted E-moduli and shear moduli for all fabrics. The agreement for both approaches is very good. It is also remarkable that the shear modulus is predicted quite accurately. This was indeed one of the weak points in the models developed previously.

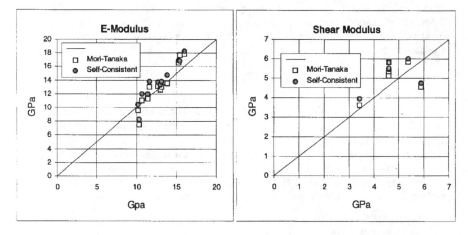

Figure 3. Comparison between experimental and predicted in-plane E-moduli (a) and shear moduli (b) for six types of E-glass/epoxy warp knitted fabric composites (wale and course), using the Mori-Tanaka and self-consistent method

The self-consistent predictions are slightly higher than the Mori-Tanaka predictions. Nevertheless, both predictions are lying close to each other. This indicates that the iterative self-consistent method can be speed up by starting from an initial Mori-Tanaka estimate rather than a Voigt-estimate. It reduces the number of iterations from 10-12 till typically 7-8.

5. Conclusions

An efficient analytical micro-mechanical model has been proposed for the prediction of the elastic constants of complex textile reinforced composites. The model was applied to knitted fabric composites, but can also be used for other (solid) composite materials, including short fibre composites, 2D wovens and 2D braids.
Although the model must be solved in an approximate way, the application to knitted fabric composites showed that they have a large potential. Currently, the model is extended to include thermal effects, non-linear matrix behaviour and progressive damage development in the yarns and the yarn/matrix interface.
In the future, the model will be applied to hybrid composites and knitted fabric composites with other matrix and/or yarn systems to validate the generality of the approach.

Acknowledgements

The authors gratefully acknowledge the Flemish Government for their support of this research in the IWT-project "The Use and the Optimisation of Knitted Fabrics in Composite Structures" (INM/93145). This paper presents research results of the Belgian program on Interuniversity Poles of Attraction initiated by the Belgian State, Prime Minister's Office, Science Policy Programming. Part of this research also fits in the framework of GOA 98/005 on "Development of unified models for the mechanical behaviour of textile composites".

References

[1] I. Verpoest, B. Gommers, G. Huysmans, J. Ivens, Y. Luo, S. Pandita, D. Philips, THE POTENTIAL OF KNITTED FABRICS AS A REINFORCEMENT FOR COMPOSITES - in proc. ICCM-11, Goldcoast (AU), 14-18 July 1997, Vol. I, pp. 108-133.
[2] S. Ramakrishna, D. Hull, TENSILE BEHAVIOUR OF KNITTED CARBON-FIBRE FABRIC/EPOXY LAMINATES - PART I & II - Comp. Sci. & Techn., 1994, Vol. 50, pp. 237-258.
[3] S. Ramakrishna, H. Hamada, N.K. Cuong, Z. Maekawa, MECHANICAL PROPERTIES OF KNITTED FABRIC REINFORCED THERMOPLASTIC COMPOSITES - in proc. ICCM-10, Whistler (Canada), 1995, Vol. IV, pp. 245-252.
[4] S. Chou, H.-C. Chen, C.-C. Chai, THE FATIGUE PROPERTIES OF WEFT-KNIT FABRIC REINFORCED EPOXY RESIN COMPOSITES - Comp. Sci. & Techn., 1992, Vol. 45, pp. 283-291.
[5] B. Gommers, I. Verpoest, P. Van Houtte, DETERMINATION OF THE MECHANICAL PROPERTIES OF COMPOSITE MATERIALS BY TENSILE TESTING I & II - accepted in J. Comp. Mat., 1997.
[6] B. Gommers, I. Verpoest, P. Van Houtte, MODELLING THE ELASTIC PROPERTIES OF KNITTED FABRIC COMPOSITES - Comp. Sci. & Techn., 1996, Vol. 56, pp. 685-694.

[7] *G. Huysmans, B. Gommers, I. Verpoest*, A BINARY FINITE ELEMENT MODEL FOR THE EFFECTIVE STIFFNESS PREDICTION OF 2D WARP KNITTED FABRIC COMPOSITES - in proc. 4th Int. Conf. on Deformation and Fracture of Composites, Manchester (UK), 24-26 March 1997, pp. 309-318.

[8] *M. Hori, S. Nemat-Nasser*, DOUBLE-INCLUSION MODEL AND OVERALL MODULI OF MULTI-PHASE COMPOSITES - J. Eng. Mat. & Techn., July 1994, Vol. 116, pp. 305-309.

[9] *G. Huysmans, I. Verpoest, P. Van Houtte*, A POLY INCLUSION APPROACH TO THE MODELLING OF KNITTED FABRIC COMPOSITES - accepted in Acta Mater., 1998.

[10] *A. Molinari, G. R. Canova, S. Ahzi*, A SELF-CONSISTENT APPROACH OF THE LARGE DEFORMATION POLYCRYSTAL VISCOPLASTICITY - Acta Metall., 1987, Vol. 35, No. 12, pp. 2983-2994.

Microtribological Performance of a Short Carbon Fiber Reinforced PEEK/PTFE-Composite Blend

Y. C. Han, S. Schmitt, P. Klein and K. Friedrich

Institute for Composite Materials Ltd. (IVW), University of Kaiserslautern, Erwin-Schrödinger-Str. 58, D-67663 Kaiserslautern, Germany

Abstract

Atomic force microscopy (AFM) and lateral force microscopy (LFM) are used to conduct microtribological studies on a short carbon fiber reinforced PEEK/PTFE-composite blend. The relative microfriction coefficients of different filler particles have been compared with the PEEK matrix. The effects of the anisotropic fiber structure as a function of local fiber orientation (in-plane parallel or antiparallel vs. normal) can be identified with this technique. Microscratching and nanoindentation on different phases were made. In the microscratch test, grooves consisting of a central trough with pile-ups on each side can be seen clearly. Plastic deformation is the main deformation process in the microscratch and nanoindentation tests of the two polymeric phases.

1. Introduction

Tribology is a key-technology for mechanical systems and is an interdisciplinary area that covers a wide range, from basic physics and chemistry to machine development [1]. One of the goals of tribology is to explain the friction and wear phenomena occurring when two material surfaces are sliding against each other in intimate contact. The advent of new techniques [2] has led to the development of a new field referred to as microtribology or nanotribology; it helps to explore the origin of friction and wear on an atomic or molecular level [3].

Short carbon fiber reinforced PEEK/PTFE composite blends are being increasingly used as engineering structural materials for technical applications in which tribological properties are of considerable importance. A fundamental understanding of the nano- and microtribological performance can further enhance their potential applications.

2. Experimental

A commercial AFM/LFM, as produced by Digital Instruments Ltd. (Nanoscope III), was used to conduct studies of microfriction, microscratching and nanoindentation on short carbon fiber reinforced PEEK/PTFE-composite blends. Microfriction was carried out in AFM/LFM contact mode, whereas microscratching and nanoindentation were performed by lithography in AFM Tapping Mode.

The test sample used in this study was an injection-molded plate consisting of a polyetheretherketone (PEEK) matrix filled with 10 weight % polytetrafluoroethylene (PTFE) particles, 10 weight % short PAN based carbon fibers, and 10 weight % graphite flakes. The composite material was supplied by Kureha Chemicals GmbH, Düsseldorf, Germany. Before testing, the sample was well polished and cleaned.

3. Results and Discussion
3.1 Microfriction

The AFM/FFM contact mode was used to conduct studies on microfriction of different parts of a short carbon fiber reinforced PEEK/PTFE composite blend. Figure 1 (a) illustrates the topography of a PTFE particle in a surrounding PEEK matrix. The bright color corresponds to high points in topography, whereas the dark color refers to the corresponding valleys. As the PTFE particle appears darker than the PEEK matrix, it is clearly below the rest of the surface. From the corresponding Figure 2 (a) one can see that there exists, in fact, a height difference of about 90 nm. It indicates, that the much softer PTFE (relative microhardness of 11, as measured by the use of a Shimadzu DUH 202 microhardness testing device) gets much easier worn during polishing than the harder PEEK matrix (relative microhardness of 31).

If the data result from friction between tip and sample, the relative intensity of the signal will invert as the scan direction is reversed. For most color tables, the image produced from the trace is darker in the low-friction areas than it is in the high-friction areas. Conversely, for the same color table, the image produced from the retrace would be brighter in the low-friction areas. On the friction pictures (Figures 1 (b) and (c)), one can see that in the trace direction the PTFE particle is darker than the PEEK matrix, whereas PTFE is brighter than PEEK in the retrace image. This means that the relative friction coefficient of PTFE is lower than that of PEEK. This can also be verified by the results of Figure 2 (b). The difference between the mean height level of trace and retrace is smaller in the area of the PTFE particle than for the surrounding PEEK matrix. It, therefore, can be concluded that the trace and retrace signals move closer together when regions of low friction are encountered, since the frictional forces are reported as the average difference in the lateral-force-derived voltages measured at the photodetector during left-to-right versus right-to-left scanning. The larger amplitude of the signal in case of the PTFE particle is an indication of a more pronounced scratch pattern due to higher wear during polishing of the softer PTFE (mainly by microploughing and microcutting effects [4]). This can also be seen by carefully looking, for instance, at Figure 1 (b).

Figure 3 demonstrates the topographic and friction mode images of graphite relative to the surrounding PEEK matrix. The torsional deflections of the cantilever in both modes are shown in Figure 4. Both figures clearly indicate that the surface roughness of the graphite filler is remarkably higher than that of the PEEK around it. But the distance of the mean value of the graphite's trace and retrace curves in the friction mode seems to be slightly shorter (about 3.5 divisions) than that of PEEK (about 4 divisions). This means a slightly lower coefficient of friction can be considered for the graphite filler. In addition, this filler got slightly more worn during polishing than the surrounding PEEK environment, which becomes obvious from the topographic trace lines of Figure 4 (a). Opposite to the PTFE inclusions, however, wear did not primarily occur due to intensive microploughing effects (almost no remarkable scratches are visible on the graphite), but most probably by microcracking of very small graphite chips from the bulk filler. These chips may help in a real sliding wear of the composite material against a metallic counterpart to reduce the coefficient of friction as long as they maintain in the contact region, e. g., as graphitized transfer film elements on either surface of the two partners in sliding contact.

Figure 5 shows the topographic and friction mode images of two carbon fibers having different orientation. In the trace mode friction picture, the color of the nearly in plane oriented carbon fiber is darker (Figure 5(b)) than that of the fiber under normal orientation to the polished surface; this means that the normal direction result in a higher coefficient friction than the almost in plane fiber when being scratched in its transverse (or anti-parallel) direction. This can also be seen clearly from Figure 6. In the topographic mode, the in-plane oriented carbon fiber is higher than the carbon fiber under normal orientation, which in turn is slightly higher than the PEEK matrix. On the other hand, the trace line of the carbon fiber under normal orientation is a little bit higher than that of the PEEK, and the latter is higher than that of the in-plane oriented carbon fiber. Therefore the order of the coefficient of friction is: carbon fiber in normal orientation > PEEK matrix > carbon fiber with in-plane orientation. With increasing the tip deflection voltage, this difference can even be seen more clearly. In addition one can conclude, that the carbon fibers are more wear resistant than the PEEK matrix under the same polishing procedure .

In general one can conclude here, that in some cases particles in the composite can not easily be recognized in the surface topography. However, frictional force distribution is a useful method to distinguish them, because in the trace image, dark parts correspond to low-friction regions and bright parts correspond to high friction regions.

A final plot of the half differences between the mean values of the frictional trace and retrace signals (ΔZ in [V]), as a measure for the active frictional force, against the deflection voltage (applied normal load Z_N in [V]) gives an idea about the coefficient of friction of the various phases relative to each other (Figure 7). It reflects for all the components, i. e. the PEEK matrix, the PTFE particles, the Graphite flakes, and the CF under normal and in-plane orientation that the macroscopically derived Leonardo da Vinci/Amontons-relationship (linear proportionality between normal load and frictional force) is not simply transferable to the atomic level [5]. An extrapolation of the data points, achieved for the various phases at different deflection voltages, to a deflection voltage of zero yields to a positive shear force ΔZ_0, necessary to keep the sliding movement going on. This intrinsic value, in turn, was different for the various phases in the reinforced blend (i. e. PEEK, PTFE, Graphite, CF). The origin of it may be based the energy losses due to vibration of the crystal lattices of the two materials in sliding contact [5]. The slopes of the ΔZ vs Z_N plots can be considered as the nanoscopic "average coefficients of friction" (μ_{avg}), that result from the adhesive and ratchet mechanisms [6]. Their absolute values are listed in Figure 8, and they illustrate even more clearly the differences in friction resistance between the various blend phases. These numbers are, however, much lower than those known for these phases from macroscopic measurements against smooth metallic surfaces (e. g. PTFE vs. steel: $\mu \approx 0.05 \div 0.12$; PEEK vs. steel: $\mu \approx 0.25 \div 0.4$) [7-9]. An explanation for these differences cannot really given by us at the moment.

3.2 Microscratch

Figure 9 reflects the image of scratch marks generated on PEEK and graphite. The scratch was started in the PEEK matrix towards a graphite particle by drawing the tip over the surface under constant applied load at constant velocity. On the surface of the PEEK matrix, a clear groove consisting of a central trough with pile-ups on both sides becomes visible. The graphite, on the other hand, does not reflect a clear evidence for a scratch. This means that graphite is much more scratch resistant than the PEEK matrix. Using the section view analysis method (Figure 9 (b)), the height of the pile-up and the depth of the groove can be determined. By measuring both, the pile-up height from the level of the non-scratched surface and the depth of the groove, it is possible to obtain more information about the deformation of the sample (which will be discussed in detail later). For PEEK, at a tip deflection voltage of 18V, the depth of the groove was 64.4 nm, and the height of the pile-up 47.2 nm. In case of the scratch measurement on graphite, the depth was of the same order as the variation in the surface roughness. Therefore it is difficult to measure any accurate value for it.

Figure 10 illustrates the image of a scratch mark generated on PEEK and the carbon fiber. Scratching was carried out from the matrix PEEK towards the carbon fiber. Again a clear scratch can be seen only on the surface of the PEEK matrix, whereas no scratch is visible on the carbon fiber. This means that a carbon fiber is harder and more scratch resistant than PEEK. Consequently, the order of scratch resistance is: PEEK matrix ‹ graphite ‹ carbon fiber.

From Figures 9 and 10, the formation of grooves and pile-ups can be clearly recognized. The grooves consist of a central trough with different pile-ups on each side. It can be assumed that these grooves are caused by a wear mechanism called microploughing [4].

In the case of PEEK (Figure 9 (b)), it is obvious that (A_1+A_2) ((A_1+A_2) is the cross-sectional area of the material displaced as pile-ups at edges of the groove) is greater than A_v (A_v is the cross-sectional area of the wear groove). As this is impossible according to the volume conservation principle (there can not be more material piled up than being removed from the groove), one possible explanation that remains is that the whole area below the groove and the pile-ups was elastically pushed due to the build up of internal stresses during the scratching procedure. Shifting down of the groove and the pile-ups to the previous level of the surface leads to the situation that A_v increases and (A_1+A_2) decreases. Even though this is an assumption, it seems to be realistic and yields to the conclusion that (A_1+A_2) and A_v become very equal. This means, in turn, that microploughing is the most dominant wear mechanism in the nanoscale scratch experiment of PEEK. Another explanation for $(A_1+A_2) > A_v$ is, that the pile-ups do not consist of compacted material but are instead loose material tongues that were pushed out sidewise during the scratching procedure. As the topography mode of the AFM cannot detect any underlying cavities, however, it reflects the total cross sections A_1 and A_2 as being consistent of piled up material.

3.3 Nanoindentation

Figure 11 shows the gray scale image of an indentation made on PTFE with a tip deflection voltage of 11 V. The residual impression resembles the shape of the tip. The very smooth side-walls of the residual impression and the absence of any surface or subsurface cracks indicate that plastic flow was the principal deformation process in this microindentation experiment. The depth of indentation into PTFE is 596 nm, as analyzed by section view analysis.

The same experimental condition was used to conduct an indentation into PEEK. Figure 12 shows the gray scale image of such an indentation (tip deflection voltage of 11 V). The slightly raised material at the indentation edges, in conjunction with the absence of cracks, is characteristic for the plastic flow behaviour of an indented ductile material. From the section view analysis, the obtained depth of indentation in PEEK is 148 nm, i. e. much less than that observed for PTFE (similar as found for the ratio of microhardness, mentioned in the first section of chapter 3.1).

Measuring the volumes of indentation and pile-ups is possible through the *Bearing* inverted image after indentation of PTFE and PEEK, respectively. The volume of pile-ups and indentation of PEEK are $4.99 \times 10^6 nm^3$ and $2.21 \times 10^6 nm^3$, respectively. As expected from the scratching studies, the volume of the pile-ups above the starting surface is larger than the volume of the indentation. This is opposite to the case of pile-ups around indentation points on metals, where the volume of the groove is almost equal to the volume displaced by the indentation [11]. The reason why the volume of the piles-ups exceeds the volume of the indentation in the polymeric material may be due to the formation of internal stresses underneath the indented area, as a result of plastic deformation and temporary compaction of the volume per molecular unit. After release of the external load, these stresses may be released and push the whole area upwards relative to the surface level of the material away from the indentation site.

For the PTFE phase, the volumes of pile-ups and indentation are $1.44 \times 10^8 nm^3$ and $2.92 \times 10^8 nm^3$, respectively. This means, that the much softer PTFE does not react in the same way as the stiffer and stronger PEEK. This is probably due to the fact that PTFE deforms much easier than PEEK, and that it forms very easily a transfer film on the counter-part surface due to easy shear failure events, e. g. during sliding wear [12].

4. Conclusions

The microtribological properties of short carbon fiber reinforced PEEK/PTFE composite blends were investigated by atomic force microscopy (AFM) and lateral force microscopy (LFM). The following conclusions were obtained:

(1) As expected, PTFE showed a lower coefficient of friction than PEEK, when moving the tip of the AFM needle over their surface.

(2) Normally oriented carbon fiber ends had a slightly higher coefficient of friction than PEEK, but in-plane oriented carbon fibers, measured in their length direction, had a clearly lower frictional coefficient.

(3) The classical friction law, i. e. linearly between frictional force and normal load, is not valid on this atomic level, especially in the lower load range.

(4) The average values of the coefficient of friction measured under LFM-conditions are by an order of magnitude lower than comparable values known from macroscopic experiments.

(5) Carbon fibers are harder and more scratch resistant than graphite, PEEK, and PTFE. In the microscratch test, grooves consisting of a central trough with pile-ups on each side can be seen clearly. Plastic deformation (microploughing) is the main deformation process in the microscratch and nanoindentation tests of the two polymeric phases.

Acknowledgments

Dr. Y. C. Han wishes to thank the Alexander von Humboldt-Foundation (AvH) for the research fellowship at the University of Kaiserslautern. P. Klein is grateful for his personal "Forschungspraktikum" by the Stiftung Industrieforschung, Düsseldorf. Further thanks are due to the Deutsche Forschungsgemeinschaft (DFG FR 675/19-1) for their support of research in the general field of polymer composites tribology.

References

1. K. Friedrich (ed.), *Advances in Composite Tribology*, Composite Materials Series, 8, Elsevier, Amsterdam, 1993.

2. R. Kaneko, K. Nonaka, K. Yasuda, Scanning Tunneling Microscopy and Atomic Force Microscopy for Microtribology, *J. Vac. Sci. Technol.*, A6(2), 291-292 (1988).
3. B. Bhushan, J. N. Israelachvili, U. Landman, Nanotribology: Friction, Wear and Lubrication at the Atomic Scale, *Nature*, 374(13), 607-616 (1995).
4. H. K. Zum Gahr, *Microstructure and Wear of Materials*, Elsevier, Amsterdam, 1987.
5. J. Krim, Friction at the Atomic Scale. *Scientific American*, October, 48-56 (1996).
6. Y. C. Han, S. Schmitt and K. Friedrich, Microfriction of Short Carbon Fiber Reinforced PEEK/PTFE-Composite Blends, *Tribology International*, (1997), submitted.
7. Z. P. Lu and K. Friedrich, On Sliding Friction and Wear of PEEK and its Composites. *Wear*, 181-183, 624-631 (1995).
8. B. J. Briscoe, Interfacial Friction of Polymer Composites: General Fundamental Principles, in: K. Friedrich (ed.) Friction and Wear of Polymer Composites, Elsevier, Amsterdam, pp. 25-59 (1986).
9. H. Czichos, and K.-H. Habig, Tribologie Handbuch Reibung und Verschleiss. Vieweg, Braunschweig / Wiesbaden, 1992.
10. Y. C. Han, S. Schmitt and K. Friedrich, Nanoscale Indentation and Scratch of Short Carbon Fiber Reinforced PEEK/PTFE-Composite Blends by Atomic Force Microscope Lithography, *Applied Composite Materials*, (1998), in press.
11. K. L. Johnson, *Contact mechanics*, Cambridge University Press, Cambridge, 1985.
12. K. Friedrich, Z. Lu, A. M. Häger, Recent advances in polymer composites' tribology. *Wear*, 190, 139-144 (1995).

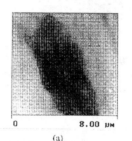 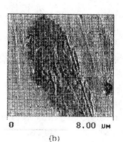 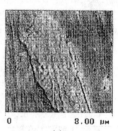

Figure 1 Atomic force micrographs of PEEK and PTFE: (a) topographic mode; (b) trace in friction mode and (c) retrace in friction mode at the tip deflection voltage of 7 V.

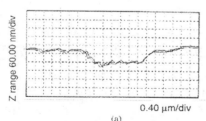 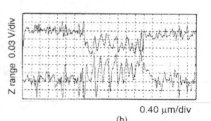

Figure 2 A plot of (a) the AFM signal of PEEK and PTFE for a trace and retrace in topographic mode; (b) trace and retrace in friction mode at the tip deflection voltage of 7 V. All of these plots were taken across the middle of the image. Trace: upper line; retrace: lower line.

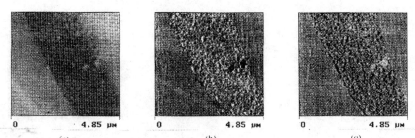

(a)   (b)   (c)

Figure 3 Atomic force micrographs of PEEK and graphite: (a) topographic mode; (b) trace in friction mode and (c) retrace in friction mode at the tip deflection voltage of 4 V.

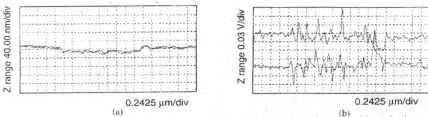

(a)   (b)

Figure 4 A plot of (a) the AFM signal of PEEK and graphite for a trace and retrace in topographic mode; (b) trace and retrace in friction mode at the tip deflection voltage of 4 V. All of these plots were taken across the middle of the image. Trace: upper line: retrace: lower line.

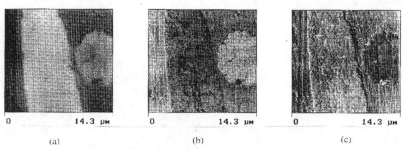

(a)   (b)   (c)

Figure 5 Atomic force micrographs of PEEK and carbon fiber in normal and anti-plane orientation: (a) topographic mode; (b) trace in friction mode and (c) retrace in friction mode at the tip deflection voltage of 6 V.

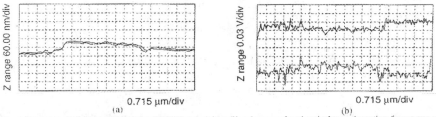

(a)   (b)

Figure 6 A plot of (a) the AFM signal of PEEK and carbon fiber in normal and anti-plane orientation for a trace and retrace in topographic mode; (b) trace and retrace in friction mode at the tip deflection voltage of 6 V. All of these plots were taken across the middle of the image. Trace: upper line; retrace: lower line.

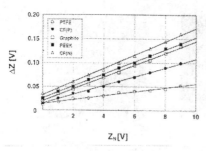

Figure 7 ΔZ vs ZN for PEEK, PTFE, Graphite, CF(P) and CF(N), respectively.

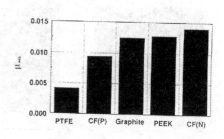

Figure 8 Relative coefficient of friction of PEEK, PTFE, graphite, CF(P) and CF(N).

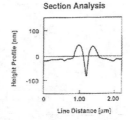

(a)   (b)

Figure 9 (a) Image of a scratch generated from the PEEK matrix towards the graphite particle; (b) section view of the same image showing the depth of the scratch and height of the pile up along the black line indicated on the corresponding photo.

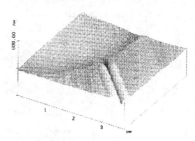

Figure 10 Image of scratch generated from the PEEK matrix towards the edge of the carbon fiber end.

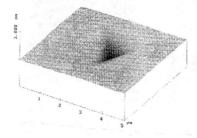

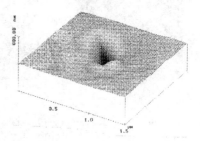

Figure 11 Gray scale image of an indentation mark on PTFE.   Figure 12 Gray scale image of an indentation mark on PEEK.

Table 1 Volume of pile-ups and indentation of PEEK and PTFE

Materials	Volume of Pile-ups [nm^3]	Volume of Indentation [nm^3]
PEEK	4.99×10^9	2.21×10^5
PTFE	1.44×10^8	2.92×10^8

Composites with various kinds of interfacial adhesion: Compression along layers

Igor A. Guz

Timoshenko Institute of Mechanics
Nesterov str. 3, 252057 Kiev, Ukraine

Abstract

The problem of failure caused by the internal instability was solved in this paper within the scope of the exact statement based on the application of the model of piecewise-homogeneous medium and equations of the three-dimensional linearized theory of deformable bodies stability (TLTDBS) [1], which allows to eliminate the restrictions imposed on using the approximative theories and schemes as well as inaccuracies they involve. To estimate critical loads the upper and the lower bounds were suggested for laminated structures in compression along the interlayer defects utilizing the results for rigid-connected and sliding layers. Substantiation of the bounds is based on the one of the general principles of mechanics on the influence of liberation from a part of connections on value of critical loads for the mechanical system. Numerical investigations for composites with elastic-plastic matrix had shown that suggested bounds present reasonable fair results for particular cases of composites.

1. Introduction

The abundant structure of composites is a laminated one. Therewith, different cases of the interface adhesion breakdown in composites (such as cracks, separations onto layers, delaminations, exfoliations, slippage zones and similar imperfections) can arise due to the manufacturing technology or operating conditions. One of the most interesting, peculiar and inadequately investigated phenomena in mechanics of composites (mechanics of non-homogeneous media) is failure in compression, in which mechanisms of failure not specific for homogeneous media but for composites (non-homogeneous media) only are revealed. One of the mechanisms referred to is failure owing to the loss of stability in the composite structure – so called the internal instability following [2].

At the present time there is a large amount of studies devoted to the stability problems of composites and in the most of them it is assumed that structural elements of the material are rigidly attached. However, in real composites the usual concept of ideal contact between structural elements does not always correspond to reality. Different cases of reduced adhesion (interface contact) in composites such as cleavages cracks, non-adhesion, slippage zones and similar imperfections can arise due to the manufacturing technology or operating conditions.

Investigations of stability of laminated composites with intercomponent cracks are fulfilled mostly within the scope of various approximative applied theories (such as continual theory) or schemes (bar, shell and others). But the most accurate results can be obtained only within the scope of the three-dimensional linearized theory of deformable bodies stability (TLTDBS) [1]. Application of the model of piecewise-homogeneous medium and fundamental equations of TLTDBS, which are used in this paper, allows to eliminate the restrictions imposed on using the approximative theories and schemes as well as inaccuracies they involve.

2. On the problem statement

The general problem statement for a composite of an arbitrary layered structure, which may consist of an arbitrary combination of layers and half-spaces, was considered in [3]. It was supposed there that the composite has interfacial cracks and is situated in conditions of the plane strain state in compression along layers by "dead" loads applied at infinity in such a manner that equal deformations along all layers are provided in direction of loading. (This is the uniform precritical state.) Cracks were simulated by mathematical sections regardless of reasons of their occurrence. Layers of the investigated structure were assumed to have various mechanical characteristics and thicknesses. They were simulated by compressible or incompressible, elastic or elastic-plastic, isotropic or orthotropic (with elastically equivalent directions which are parallel and perpendicular to interfaces) bodies. In the case of elastic-plastic layers the generalized conception of the continuous loading, which allows to neglect the changes of loading and offloading zones during the stability loss, was utilized. All investigations were fulfilled within the Lagrangian coordinate system (which is Cartesian one in the non-deformed state) using the static method of investigation of static problems of TLTDBS [1].

Besides that, the classification of cracks was introduced in [3]. It was based on certain indiscrepant simplifications, connected not with the changing of equations, but only with geometrical characteristics of the composite and the cracks. Macrocracks, structural cracks and microcracks were discerned in this classification. Problems of internal instability for such structures had been solved for the case of absence of cracks (perfectly rigid-connected or sliding without friction layers) in [4-8] within the scope of the above-mentioned exact approach.

3. Bounds for critical parameters

The problem statement indicated above was formulated for the crack model which is ideal in a certain sense. Such model (the free of stresses crack surfaces) is used in classical fracture mechanics. However, other kinds of the interlayer adhesion breakdown can also occur in composite materials. For example, a change in the nature of contact of layers is possible, when interaction of layers is implemented by friction forces. In this case, there are no gaps between layers and the continuity of normal components of stresses and displacements at the interface is retained. If the friction force between the layers is sufficiently large, it can be assumed that rigid contact does not break down. If the friction force is very small, zones of such defects can be assumed to be slippage zones without friction with boundary conditions for perturbations stresses and displacements boundary where imperfections must be located, i.e.

$$f < f^+ \tag{2}$$

At the same time, if in a structure with imperfections all remaining connections (zones of the rigid connection) are released, structure of the same type with ideal slippage (sliding without friction) between all the layers and half-planes is obtained. Applying the same principle, it can be said that the critical loading parameter f^- for a structure with sliding layers must be smaller than the critical loading parameter f for a structure with imperfections (1)

$$f^- < f \tag{3}$$

And finally, from (2), (3)

$$f^- < f < f^+ \tag{4}$$

The parameters f^+ are determined from the solution of equations (2) from [3] with boundary conditions (3), (5), (6) from [3], where in this case

$$S = S_r \cup S_s, \quad S_c = 0 \tag{5}$$

The parameters f^- are determined from the solution of equations (2) of I.A. Guz (3) with boundary conditions (5), (6) from [3] and (1) from the present paper, where in this case

$$S = S_c \cup S_s, \quad S_r = 0 \tag{6}$$

When the f^+ and f^- are found, the bounds required follows from (4).

4. Ivestigation of the particular model of composite

Let us consider a composite consisting of alternating layers of a linear-elastic isotropic compressible filler (of thickness $2h^{(1)}$), which constitutive equation is as follows below

$$\sigma_{ij}^0 = \delta_{ij} \frac{E\nu}{(1+\nu)(1-2\nu)} \epsilon_{nn}^0 + \frac{E}{1+\nu} \epsilon_{ij}^0 \tag{7}$$

and an elastic-plastic incompressible matrix (of thickness $2h^{(2)}$) with power-mode dependence between equivalent (effective) stresses (σ_I^0) and strains (ϵ_I^0) in the form

$$\sigma_I^0 = A(\epsilon_I^0)^k \tag{8}$$

Instability problems for the composites with various kinds of layer models, including the above-mentioned one, were considered in [4-6] for the purpose of determining f^+ and in [7,8] for the purpose of determining f^-.
The bounds obtained similarly for the case of three-dimensional non-axisymmetric problem are shown in Fig. 1-3. The shaded region is located between f^+ and f^-. According

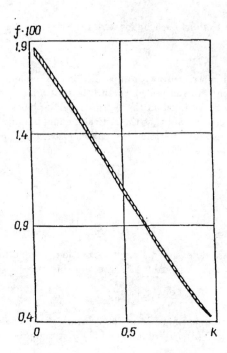

Figure 1: Bounds for critical loading parameters for metal-matrix composites with interface defects ($\nu = 0.2$, $A/E = 0.0003$, $h^{(1)}/h^{(2)} = 0.015$).

in the form

$$t_{21}^{(i)} \big|_{S_c} = 0, \quad t_{21}^{(i+1)} \big|_{S_c} = 0; \quad t_{22}^{(i)} \big|_{S_c} = t_{22}^{(i+1)} \big|_{S_c}, \quad u_2^{(i)} \big|_{S_c} = u_2^{(i+1)} \big|_{S_c} \tag{1}$$

The problem statement (2)-(6) from [3] remains in force also for the defects described above, which will be called below "defects with connected edges". It is only necessary to replace the boundary conditions for cracks (4) from [3] by the boundary conditions (1).

To estimate the critical loading parameters in composite material with various imperfections (1) between layers the following bounds can be proposed using the well-known principle of mechanics, namely, stating that release from some connections cannot increase the value of the critical load, under which stability loss of the system takes place.

According to the above-said, the critical parameter f^+, under which stability loss takes place in a structure without imperfections (i.e. with rigid contact of layers) must be larger than the critical loading parameter f for the same structure with imperfections (1). In other words, it can be said that a structure with imperfections is obtained from a structure without imperfections by releasing from connections those parts of the

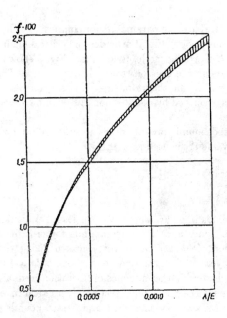

Figure 2: Bounds for critical loading parameters for metal-matrix composites with interface defects ($k = 0.34$, $\nu = 0.2$, $h^{(1)}/h^{(2)} = 0.015$).

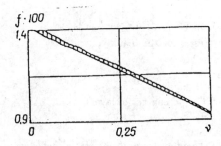

Figure 3: Bounds for critical loading parameters for metal-matrix composites with interface defects ($k = 0.34$, $A/E = 0.0003$, $h^{(1)}/h^{(2)} = 0.015$).

to the above-said, the critical loading parameters for composite with interlaminar cracks are located precisely in this region. Fig. 2 provides the critical loading parameters as a function of A/E where E is Young modulus for filler. Fig. 3 shows the dependence of critical loading parameters on the Poisson ratio ν, Fig. 1 – on k. The results presented in Fig. 1-3 are obtained within the theory of small initial deformations of TLTDBS [1], since the real composites described by adopted for this example model even break down under small deformations.

As one sees, the suggested bounds provide quite accurate result for f, particularly for engineering (the strain value ϵ_{11}^0 is used here as f).

5. Conclusions

The estimation of critical loading parameters was suggested and substantiated. This estimation establishes the upper and the lower bounds for critical loads for laminated structures in compression along the interlayer cracks utilizing the results for rigid-connected and sliding layers. Substantiation of the bounds is based on the one of the general principles of mechanics on the influence of liberation from a part of connections on value of critical loads for the mechanical system. Numerical investigations for composites with elastic-plastic matrix had shown that suggested bounds presents reasonable fair results for particular cases of composites and, therefore, may be considered as one of the first approximations on the way to exact solution of the problem of stability in compression along the interlaminar defects.

6. List of simbols

ϵ_{ij}^0 = strains in precritical state;
σ_{ij}^0 = strains in precritical state;
ϵ_c = critical strain;
$u_i^{(l)}$ = perturbations of displacements for the l-th layer;
$t_{ij}^{(l)}$ = perturbations of stresses for the l-th layer;
σ_I^0 = intensity of stresses;
ϵ_I^0 = intensity of strains;
$2h^{(i)}$ = thickness of the i-th layer;
S_r = set of interfaces with rigid contact of layers;
S_s = free surface;
S_c = set of the crack surfaces;
f = critical parameter for composites with interfacial defects;
f^+ = critical parameter for composites with rigid-connected layers;
f^- = critical parameter for composites with sliding layers.
E = Young modulus;
ν = Poisson ratio.

References

[1] *A.N Guz*, FUNDAMENTALS OF THE THREE-DIMENSIONAL THEORY OF DE-FORMABLE BODIES STABILITY - Vyshcha Shkola, Kiev, 1986. (In Russian)

[2] *M.A. Biot*, MECHANICS OF INCREMENTAL DEFORMATIONS - Wiley, New York, 1965.

[3] *I.A. Guz*, STABILITY AND FAILURE OF LAYERED COMPOSITES WITH IN-TERFACE CRACKS - "Computational Mechanics '95". Vol. 2. Theory and Applications (Eds.: S.N. Atluri, G. Yagawa, T.A. Cruse). - Springer-Verlag, New York, 1995. - pp.2317-2322.

[4] *I.A. Guz*, SPATIAL NONAXISYMMETRIC PROBLEMS OF THE THEORY OF STABILITY OF LAMINAR HIGHLY ELASTIC COMPOSITE MATERIALS - Soviet Applied Mechanics. 1989, Vol.25, No.11. - pp.1080-1085.

[5] *I.A. Guz*, THREE-DIMENSIONAL NONAXISYMMETRIC PROBLEMS OF THE THEORY OF STABILITY OF COMPOSITE MATERIALS WITH A METALLIC MA-TRIX - Soviet Applied Mechanics. 1989, Vol.25, No.12. - pp.1196-1201.

[6] *I.A. Guz*, INTERNAL INSTABILITY OF LAMINATED COMPOSITES WITH A METAL MATRIX - Mechanics of Composite Materials. 1990, Vol.26, No.6. - pp.762-767.

[7] *I.A. Guz*, PLANE PROBLEM OF THE STABILITY OF COMPOSITES WITH SLIPPING LAYERS - Mechanics of Composite Materials. 1991, Vol.27, No.5. - pp.547-551.

[8] *I.A. Guz*, ESTIMATION OF CRITICAL LOADING PARAMETERS FOR COM-POSITES WITH IMPERFECT LAYER CONTACT - International Applied Mechanics. 1992, Vol.28, No.5. - pp.291-296.

INVERSE SCATTERING FOR EVALUATING CHARACTERISTIC QUANTITIES OF PARTICULATE COMPOSITES.

M. HADJINICOLAOU AND G. KAMVYSSAS

ICEHT/FORTH

GR 265 00 PATRAS, GREECE

ABSTRACT

Powder particles in plastic materials offer advantages in terms of cost and easy handling and processing. Such particles are used in a wide variety of composite materials, either as cheap fillers (e.g. talc, clay, in polymers and rubbers), or as a mean of changing the physical and or the mechanical properties of the composite material, such as density, stiffness and strength. Particulate composites can be described naively as spherical particles, coated in some cases, distributed within a continuous medium.

The aim of the present work is to use the inverse scattering theory in order to identify geometrical and physical characteristics of the powder particles which in this case are considered to be coated. Precisely, we consider an acoustic wave field emanating from a point source, which propagates through the medium . The existence of the particles causes scattering of the incident field. We assume low concentration of particles (up to 25%) in order to avoid multiple scattering effects. When dealing with coated particles we consider the scatterer as consisting of two concentric penetrable spheres. The medium occupying the spherical shell can be either lossy or lossless. We further assume that the wavelength of the incident field is much larger than the radius of the spherical scatterer, so that low frequency approximation theory can be employed. From the solution of the direct scattering problem[2] we use the expressions obtained for the leading order approximation of the far field (scattering cross section) which is measurable quantity in inverse scattering. Making the necessary measurements, we develop algorithms that enable us to calculate the position of the particle as well as its geometrical and physical characteristics, such as the thickness of the shell and the mass density. This can be done as follows.

If the position and the external radius of the scatterer are unknown, then we consider an orthogonal system of coordinates and we measure the value of the scattering cross section σ. As it is shown in [1] using the least square method we can isolate the individual measurement m_i from the total measurement of σ. Let m_i, $i=0,1,2,3,4$ be the measurements of the second low frequency term of σ when the excitation point is located at different points each time, that are at the unknown distances r_0, r_1, r_2, r_3, r_4 respectively, apart from the center of the scatterer. Then the spherical scatterer is located at the intersection of the four spheres centered at the previous points, with radii r_0, r_1, r_2, r_3, r_4 respectively, given below

$$r_i = \sqrt{\frac{2/m_i}{1/m_0 - 2/m_3 + 1/m_4}}$$

It is wearthnoting that although we seek for four unknowns (the three coordinates of its center and the radius of the outer sphere) we need five equations due to the fact that the intersecting spheres are second degree surfaces. Furthermore, the thickness of the shell is evaluated assuming the physical parameters (two different densities of the two spheres) are known. Otherwise, when knowing the geometrical quantities (position, thickness of the shell), from one single measurement we can identify the density of either the kernel, provided the shell' s is known or the shell' s when the kernel' s density is known. This is obtained by solving accordingly the equation:

$$m_i r_i^2 = \frac{(1-B)(B_0+2) + (1-B_0)(2B+1)\ell^3}{(B+2)(B_0+2) + (B-1)(B_0-1)\ell^3}$$

where B and B_0 imply the density of the coat and of the inner sphere respectively.

REFERENCES

1. Dassios, G., and Kamvyssas, G., "Point Source Excitation in Direct and Inverse Scattering: The soft and the Hard Sphere", IMA J. Appl. Math., 55, pp.67-84,1995

2. Dassios, G., M. Hadjinicolaou and G. Kamvyssas, "Direct and Inverse Scattering for Point Source Fields. The Penetrable Small Sphere" submitted.

SYNTACTIC FOAMS REINFORCED WITH FIBERS: A NOVEL APPROACH

Michel Palumbo*, Ezio Tempesti*, Giorgio Donzella**

* Università di Brescia, Dip. di Chimica e Fisica per l'Ingegneria e per i Materiali,
Via D.Valotti n° 9 - 25133 Brescia.
palumbo@bsing.ing.unibs.it - tempesti@bsing.ing.unibs.it

** Università di Brescia, Dip. di Ingegneria Meccanica,
Via Branze n° 32 - 25133 Brescia.
donzella@bsing.ing.unibs.it

Abstract

The Young's moduli and the compressive strength of composites filled with (51.5% v/v; 25% wt/wt) hollow glass spheres in an epoxy resin matrix have been measured. The data have been compared with the elastic moduli obtained from syntactic foams charged with lower amounts of filler. The increase of elastic moduli and normalized compressive strength on lowering the density of the composite confirm that the syntactic foams are the most eligible candidates for high stiffness structural elements.

Introduction

One of the most prominent peculiarities of polymers is the possibility to change the inherent physical and mechanical properties by chemical or physical modifications[1]. The simplest way of physical modifications of common polymers is their compounding with particulate fillers. Organic fillers are mostly used to increase toughness while inorganic rigid fillers are mostly used for rigidity improvement, creep compliance and price/volume ratio reduction[2-4]. Interest has also been shown in the use of hollow microspheres as fillers (syntactic foams) for polymeric composites due to their unusual and useful characteristics, including low density, high stiffness compared to normal foams (especially in compression), low thermal conductivity and interesting electrical properties[5].

The selection of resin matrix, fillers (microspheres and fibers) and manufacturing technology are all important factors to be considered when reinforced syntactic foams are candidates for different specific structural uses (high-impact lightweight structural components, honeycomb cores to alluminium skins for the aerospace industry, structural flotation elements such as launch platforms or deep-submergence ocean vehicles, electronic components due to the excellent electrical and thermal insulation properties or low dielectric-constants)[6,7].

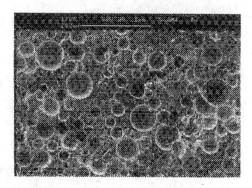

Figure 1 Scanning electron micrograph of a section of a typical moulding (filler fraction: 15% wt/wt)

Experimental

The epoxy prepolymer used in this study was a diglycidyl ether of bisphenol A (DER 332) from Dow Chemical hardened with 4,4'-diamino diphenyl methane (DDM). The formulation of the matrix was a stochiometric mixture of DER 332 and DDM. The glass microspheres used (3M Scotchlite type K37; microsphere density: $\rho_f = 0.37$ g/cm^3, glass density: $\rho_{f.bulk} = 2.4$ g/cm^3) had wall thickness between 1 and 3 µm, with particle diameters ranging from 50 µm to 70 µm. They were given no special surface treatment. The formulation was mixed and gently stirred at 60°C and measured quantities of hollow glass microspheres (respectively 5-15-25% by weight) were added. The mixtures were stirred at a slightly lower temperature (50°C) for a further period of time (1.5 hours) to allow air bubbles generated during mixing to escape, and then poured under vacuum into the mold. The composites were cured at 60°C for 24 hours.

Results and Discussion

If we except a few articles on the impact strength[8,9] of hollow glass filled fiber-glass composites, no structural observations have ever been made relative to the structure of syntactic foams reinforced with glass fibers.

Our results[10] show (see Figure 2) that in absence of glass fibers on increasing the microsphere fraction volume the composite rigidity rises while its density decreases. As a consequence the lowest density syntactic foams are the most eligible candidates for high stiffness structural elements.

It is known that the density (ρ) is determining in characterizing the stiffness of geometrical simple solid figures. Indeed the mass of a component of a given axial stiffness can be minimized by selecting a material with the maximum value of E/ρ. On the other hand, according to Ashby[11], the mass of a beam of a given geometry and bending stiffness can be minimized by selecting a material with the maximum value of E/ρ^2 while the mass of a plate of a given bending stiffness is minimized by selecting the material with the maximum value of E/ρ^3.

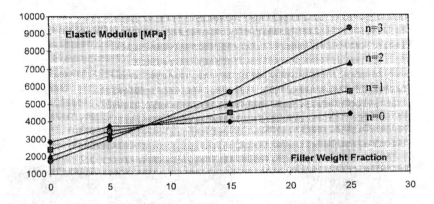

Figure 2 Absolute and normalized (by ρ^n n=1, 2, 3) Elastic modulus of syntactic foams vs filler weight fraction.

Whatever is the shape of the structural element considered, the optimum ratio E/ρ^n n=1,2,3 corresponds, in our case, to the 25% wt/wt filler content since the normalized compressive elastic modulus rises monotonically with increasing the filler fraction volume, as shown in Figure 2 and Table 1.

Filler Weight Fraction	0%	5%	15%	25%
E/ρ^n n=0 [Mpa]	2850	3727	3978	4409
E/ρ^n n=1 [Mpa]	2415	3451	4470	5652
E/ρ^n n=2 [Mpa]	2047	3195	5022	7246
E/ρ^n n=3 [Mpa]	1734	2959	5643	9290
σ/ρ^n n=0 [Mpa]	121	114	101	89
σ/ρ^n n=1 [Mpa]	102	105	114	114

Table 1 Absolute and normalized by ρ^n n=1, 2, 3 values of Elastic Compressive Modulus and Compressive Strength of syntactic foams.

From Table 1 it can also be seen that even if the absolute compressive strength values decrease with increasing the filler content, the normalized values of compressive strength rise with decreasing the density. This result confirms the suitability of choosing the lowest density syntactic foam as an eligible candidate for structural applications.

Before discussing the structural applications of syntactic foams we must first investigate both the low and high strain properties of such a composite. Particulate composites do have good compressive performances since at the microstructural scale loads are transmitted by simple contact. The predominant tough behaviour of epoxy matrixes in compression[12] (see Figure 3) is evident at low filler fraction volume, while the composite becomes brittle if the filler content rises (see Figure 4).

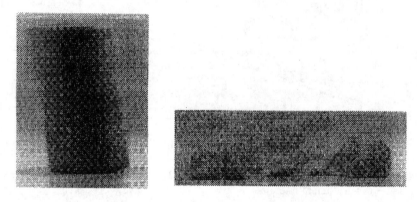

Figure 3   **Figure 4**
Photographs of syntactic foams tested in compression filled respectively with
5% wt/wt (figure 3) and 25% wt/wt (figure 4) hollow glass microspheres

On the other hand, if tested by uniaxial traction loads, sintactic foams show brittle behaviour, also at low filler densities and break at lower stress values12. The asimmetric behaviour of untreated hollow glass microspheres syntactic foams can be shown in Figure 5.

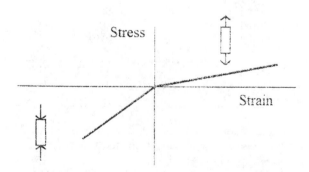

Figure 5 Asimmetrc mechanical behaviour of syntactic foams

Since bending moment implys the structural element being charged simultaneously with tensile and compressive loads, we need a composite which behaves similarly in both cases. According to our preliminary results[13] the values of the elastic constants of our composite are linked in accordance with the well known relationship of isotropic materials:

$$E = 2G(1 + \nu)$$

Conclusions

The need of a stiff lightweight material resistant to both axial and torsional loads induced us to consider two possibilities.

-       to lower the density of our composite to obtain the highest E/ρ^n ratio,
-       to increase the torsional and traction stiffness of syntactic foams by adding glass fibers.

The design of such a new structural composite is based on new methods of increasing normalized stiffness so as to allow for more reliable beams and plates when subjected to any state of stress. Since it is possible to increase the stiffness by simply adding a fiber reinforcement in the outer envelope of beams or plates designed for structural applications, we can reach very high values of specific rigidity without loosing the lighweight prerequisite. According to some preliminary results of ours obtained using syntactic foams reinforced with fibers, we have found that by duly shaping the outer envelope of the fibers it is possible to observe at least a three fold increase of breaking strength (see Fig. 6).

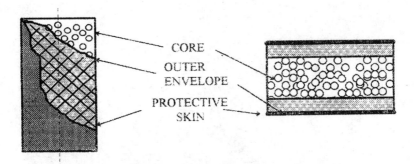

CORE
OUTER ENVELOPE
PROTECTIVE SKIN

Figure 6 Draft of syntactic foams reinforced with glass fibers

Based on our experience[10] we have also shown that in the absence of glass fibers the elastic moduli of hollow spheres increase with a power law on decreasing the diameter while increasing the shell thickness. Using syntactic foams reinforced with fibers, further work is now in progress to evaluate the elastic moduli as a function of different shapes of the outer envelope of fibers while varying the nature of the filler and/or the ratio of its internal to external radii.

References

1   D.M. Bigg, *Polym. Compos.*, **8**, 115 (1987)

2   B. Pukanszy, F. Tudos, and T. Kelen, *Polym. Compos.*, **7**, 106 (1986)

3   J. E. Stamhuis, *Polym. Compos.*, **5**, 202 (1984); J. Kolarik, G. L. Agrawal, Z. Krulis, and J. Kovalr, *Polym. Compos.*, **7**, 463 (1986)

4   Gy. Marosi et al., *Coll. Surf.*, **23**, 185 (1986)

5   F.A. Shutov, *Adv. in Polym. Sci.*, **73**, 63 (1986)

6   K. Young, *Modern Plastics*, **4**, 92 (1985)

7   A.R.Lyle and M.H.Collins, *Polym. in a Marine Environment*, paper 23 (1987)

8   G. Prithcard and Qiang Yang, *J. Mat. Sci.* 29 (1994) 5047-5053

9   C. Hiel, D. Dittman and O. Ishai, *Composites* 24 (1993) 447-450

10   M. Palumbo and E. Tempesti, *Polym. and Polym. Compos.* 5 (1997) 217-221 see also M. Palumbo, G. Donzella, E. Tempesti and P. Ferruti J. Appl. Polym. Sci. 60 (1996) 47-53

11   M.F. Ashby, *Acta Metall.* 37 (1989) 1273

12   A. Ferrante, "Materiali sintattici a matrice epossidica: effetti dell'interfase sulle caratteristiche meccaniche", thesis, Brescia, 1998

13   D. Rossi, "Comportamento meccanico dei materiali sintattici ottenuti da matrice epossidica rinforzata con microsfere di silice cave", thesis, Brescia, 1997

Effects of Concentration of the Stabilizer on the Mechanical, Thermal and Electrical Properties of Polyvinyl

H. DJIDJELLI*, T. SADOUN* AND D. BENACHOUR**

*Laboratory of Organic Materials, Institute of Industrial Chemistry , University of Bejaia
06000, Algeria.
**Intitute of Industrial Chemistry, University of Setif, 19000, Algeria

Abstract

A study has been done on the mechanical , thermal and electrical properties of several formulations of polyvinylchloride (PVC) stabilized with lead trisulfate and plasticized with diisodecyle phtalate (DIDP).We have investigated plastized and stabilized samples aged in artificial conditions (for 0 to 400 hours at 80, 100, 110°C), 80°C corresponds to the service température of electrical cables. In these conditions, we have not observed a change in the chemical composition and the structure of the material. It was found that a mass loss is attributed to the plasticizer migration and it is not influenced by the concentration of stabilizer. The plasticizer loss results in the decrease of elongation values and the increase in the glass transition température (Tg). Results show that a stabilizer improves the electrical properties by a decreasing the dielectric loss factor (tgδ). In general, published investigation relate to the study of the dehydrochlorination reactions at hight températures, leading to the formation of polyene sequences[1-2-3], oxydation of polymer with formation of carbonyl groups [4, 5, 6], formation of aromatics groups [7] and a kinetic models[8].

1. Introduction

PVC has a large sales volume, second only to polyethylene[9]. Its high chlorine content provides it with a very high level of combustion resistance for building products, eletrical enclosures, and wire & cable insulation. PVC has a unique ability to be compounded with a wide variety of additives, making it possible to produce materials in a range from flexible elastomers to rigid coumpounds. In service, cracking cable may results from thermal expansion and contraction or from insulating material shrinkage with aging. In the present study the effect of concentration of the stablizer on the properties of virgin and aged PVC were investigated by the determination of the: Plasticizer loss, elongation at breack , glass transition temperature,dielectric loss factor.

2- Experimental

2.1 Materials

Polyvinyl chloride,suspension grade (PVC 3000H, Product of Algeria ENPC, K= 42, Dp= 1260 à 1400, density= 0.42-0.47) was used in this investigation.
Calcium carbonate (CaCO3) was used as filler, tribasic lead sulfate (TBLS) as stabilizer, diisodecyle phtalate (DIDP) as plasticizer and lead stearate was used a processing aid (lubricant).

2.2 Preparation of formulations

Mixtures of PVC with 1 phr of lubrifiant, 40 phr of filler, 50 phr of plasticizer, 2, 4, 6 and 8 phr of stabilizers were prepared.
Several formulations at different ratio of stabilizer 2, 4, 6 and 8 phr were noted by M2, M4, M6 and M8 respectively. The dry- blends were obtained using a mixer at 50°C and at 3000 tr/min rotationnal speed. The mixed compounds were molded in the form of 2 mm thick sheets in a compression molding machine. The molding was carried out at 180°C and 300 bars for 15 min. After the molding, the samples were cut from the molded sheet for tensile strength, electrical and thermal studies.

2.3 Thermal degradation

The samples were thermally degraded at 80, 100 and 110°C in a forced air circulating oven, for up to 400 hours. Specimens were periodically removed and elongation at breack, mass loss, dielectric loss factor (tgδ) and glass transition temperature (Tg) values were determined. The elongation tests were determined using a dynamometre ZE400 machine at a crosshead speed of 21 cm/min.
The electrical properties were determined using a Schering bridge Tetex AG 2809a. The voltage and frequency applied were 100 Volts and 50Hz.

3. RESULTS AND DISCUSSION

3.1 Thermal stability

Figure 1 shows the dependance of the themal stability values of specimens of differents formulations on the ratio of stabilizer. The thermal stability increases linearely with the increase of the ratio of the stabilizer.

3.2 Mass loss

The mass loss of PVC specimens of the formulations stabilised at 2%, 8% and at 80, 100 and 110°C are plotted against aging time in figure 2. The result show that as the PVC specimens are aged the mass loss increase whatever the temperature is. The loss mass is slow at 80°C, moderate at 100°C and increaes charply at 110°C. IR and UV spectroscopy analysis confirmed that the loss mass enregistred in these conditions is attributed to the migration of the plasticizer (carbonyles groups and polyene sequences are not detected). Similar results have been reported by K. Anandakumaran and D.J. Stonkus[4]. Comparizon of mass loss of different formulations is repported in figure 2 which indicate that the mass loss is not influenced by the ratio of stabilizer .

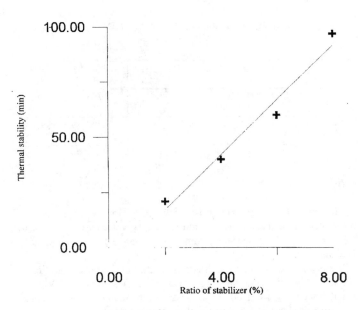

Figure 1.Effect of stabilizer on thermal stability

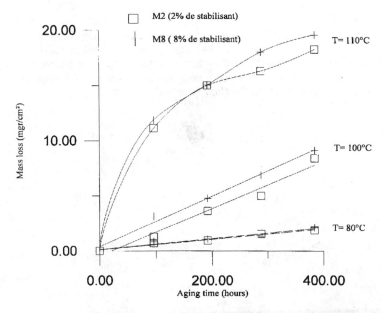

Figure 2. Effect of oven aging at 80, 100 and 110°C
on the mass loss for PVC samples (2and 8% stabilized)

Table 1. Dependance of changes of glass transition temperature on the mass during aging of plastized PVC (M6) at 110°C.

Aging time (hours)	0	96	192	288	384
Mass loss (mg/cm²)	0	10.40	15.34	17.76	18.90
Glass transition temperature (°C)	64	73	75	77	85

3.3 Thermal properties

Table 1. shows shifts in glass transition temperature (Tg) with aging. However, the extent of aging time on the Tg of the material is represented in figure 3. the Tg values are measured by an electrical method using a Schering bridge and a minimum of tan δ versus temperature corresponding to the glass transition temperature.

The results show that Tg increase with aging time, this can be attributed to the migration of the plasticizer.The rigid PVC Tg, of about 80°C, the plasticizer is soluble in PVC and separates the molecules, it lowers Tg from 80 °C to below room temperature, if enough plasticizer is added, thus making a flexible, elastomeric PVC at room temperature.

3.4 Elongation break

The results of tensile elongation against aging time for PVC specimens at 110°C are summurised in figure 4. The elogation values decrease with aging time (without showing an induction period) up to 200 hours and became stable after this time. In these conditions the decrease of elongation break is not influenced by the percentage of the stabiliser and it is caused by a migration of the plasticizer. There is a good agreement between the decrease of PVC elongation and loss of plasticizer and the glass transition temperature.

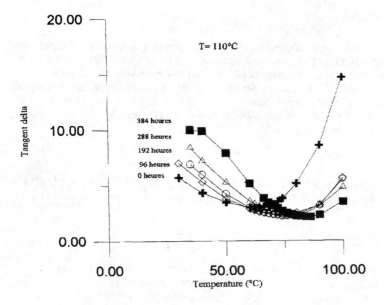

Figure3. Dielectric loss factor and Tg of aged PVC (M6).

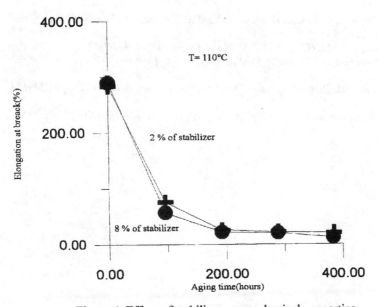

Figure 4. Effect of stabilizer on mechanical properties

4- CONCLUSIONS

In this study, we found that changes in the PVC properties (mechanical, thermal and electrical) are dependant on aging time and temperature. At 80, 100 and 110°C aging for 0 to 400 hours,we have not observed a change in the chemical composition and the structure of the material. It was found that a mass loss is attributed to the plasticizer migration and it is not influenced by the concentration of stabilizer. The plasticizer loss results in the decrease of elongation values and increase in the glass transition temperature. Results show that a stabilizer improves the electrical properties.

5- ACKNOWLEDGEMENT

The autors would like to thank Ms C. MIJANGOS, Director of the C.S.I.C Madrid, Mr. J. M. Barrales-Rienda for his technical assistance and ENICAB company for material and technical supports.

6- REFERENCES

[1] Gerardo Martinez , Carmen Majingos, José-Luis-Millan and Donald L. Gerrardo, William F. Maddams, macromol. Chem. Vol.180, pp. 2937-2945, 1979.
[2] G. Martinez, C. Mijangos, and J. Millan, Journal of Applied Polmer Science, Vol. 28, pp. 33-43, 1983.
[3] Jean-Luc Gardette and Jaques Lemaire, Journal of Vinyl Technology, Vol.15, No.2, June 1993.
[4] K. Anandakumaran and D. J. Stonkus, Journal Of Vinyl Technology, Vol. 14, No. 1, 1992.
[5] Radu Bacaloglu and Michael H. Fisch, Journal of Vinyl and AdditiveTechnology, Vol. 1, No.4, pp. 241-249, 1995.
[6] M. Beltran and A. Marcilla, Eur.Polym. J. Vol.33, pp. 1135-1142, 1997.
[7] Giorgio Montando and Concetto Puglisi, Polymer Degradation And Stability, Vol. 33, pp. 229-262, 1991.
[8] A. Marcilla and M. Beltran, Polymer Degradation And Stability, Vol.53, pp-251-260, 1996.
[9] James W. Summers, Journal of Vinyl Technology, Vol. 3, No. 2, pp. 130-139, juin 1997.

AGING BEHAVIOR OF SHORT FIBER REINFORCED THERMOPLASTIC POLYMERS

Marie DAVANT, Yves REMOND and Christiane WAGNER

Laboratoire des Procédés et Matériaux Polymères EP 647 CNRS
Université Louis Pasteur - ECPM
4, rue Boussingault - 67000 Strasbourg - FRANCE
Y.Remond@eahp-ulp.u-strasbg.fr   or   C.Wagner@eahp-ulp.u-strasbg.fr

Abstract

In this work, we test a classical method to predict the change of rigidity and strength performances as function of the time for thermoplastic polymers, in the case of glass fibers reinforced thermoplastic polymers. A medium density polyethylene reinforced with 20% short glass fibers is used. In order to evaluate the long term performance of this material, appropriate tests are developed. Samples were submitted to creep solicitation in tension in stainless steel boxes with thermoregulated water at 30°C, 60°C and 80°C. An extension to the time temperature equivalent method permits to identify a shift factor a_{T/T_0} by analysis of the strain evolution versus time under different test temperatures, and different stress levels. The creep master curves obtained appear to be a quite good method to obtain informations on the aging behaviour.

I – Introduction

Few papers have studied the aging behaviour of thermoplastic reinforced with short glass fibers. Simon [1] has shown the aging effects of water on the fiber matrix interface. Aging increases the critical length for rupture of the glass fibers, and increases interfacial damage. In that case, the reinforcement effect of the short fibers is reduced. Cardon [2] has reviewed succinctly the different aging models : time temperature equivalent method (WLF), numerical analysis of structure with critical element inducing rupture from Reifsnider [3], and analysis of the damage evolution of the structure versus time, Taljera in [4]. We use here an artificial aging made by creep under different temperatures and an analysis of the associated damage processes up to rupture. This double approach is used in this study and allows greater accuracy of the sample lifetime prediction.

Figure 1 : View of the three creep thermoregulated boxes with 180 samples

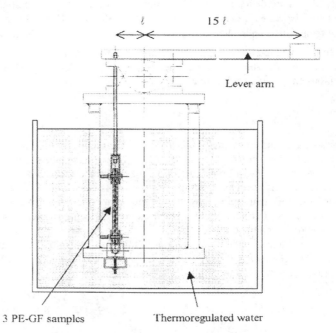

Figure 2 : Detail of the creep samples setup

II – Experimental setup

We designed and built an aging test stand for GF-PE tension samples. It consists of three stainless steel boxes with regulated water at 30°C, 60°C and 80°C. The samples were prepared under controlled processing conditions and submitted to creep solicitation in these boxes. Figure 1 presents the three experimental boxes in activity. 60 samples can be tested in each of them. Figure 2 presents the detail of the setup in each box. Three samples were tested which each same loading system : a mass applied with a lever arm amplifying the stress in the ratio of 15. The stress levels used here are : 1.5 MPa, 3 MPa, 5 MPa and 7.5 MPa. The vertical displacements of the lever arm extremity are measured with a displacement sensor and recorded each 60 minutes during the test. At the beginning of the test or during the relaxation phase displaying at the end of the test, the registration period was 1 per minute. At the same time, creep tension tests were carried out in a hot box, with a tension machine Instron, up to 12.5 MPa. In this case, the temperature used is 30°C, 40°C, 50°C, 60°C, 70°C and 85°C.

The time of experimentation was between 0.5 to 5 hours in a hot box, with a tension machine, and between 250 to 1000 hours in the creep thermoregulated boxes.

III – Results

The registration of the creep strain versus time under different temperatures is presented figure 3 in the case of a creep stress of 7.5 MPa. Figure 4 displays the creep strain versus time at the same temperature T=60°C under different creep stresses. The analysis of this results with the help of the time temperature equivalent method is founded on a large hypothesis : indeed, this method gives a good prediction for thermoplastic polymers behaviour in the range of the second order glass transition temperature, between T_g and T_g + 50°C. Here, we are not in this case, but the extension of this method in a larger temperature interval is frequently made.

The identification method consists in the measurement of the parameters a_{T/T_0} displaying in the equation (1).

$$J(T, t) = \frac{\rho_0 T_0}{\rho T} J(T_0, a_{T/T_0} * t)$$

In this equation, the material density considered is unchanged $T_0 = T$.
Then, we can obtain the values of a_{T/T_0} as follows :

T_0 (°C)	T (°C)	$\log a_{T/T_0}$
30	40	0.66
40	50	1.50
50	60	1.42
60	70	2.48
70	80	0.38

Table 1 : Identification of the shift factors $\log a_{T/T_0}$

Finally, the displacement of the curves with these shift factors allows to construct the creep master curves in the figures 5 and 6.

For the same tests, figure 5 describes the evolution of log J(t) versus log t, where : $J(t) = \varepsilon(t)/\sigma_0$, and figure 6 describes the evolution of log $\varepsilon(t)$ versus log t.

Figures 5 and 6 present a linear evolution up to 70°C and up to 10 MPa. After that, the behaviour is more complex and the evolution of these creep master curves is nonlinear.

With the same methodology, we can identify the values of the shift factors during the tests at the same temperature and under different creep stresses.

σ_0 (MPa)	σ (MPa)	$\log b_{\sigma/\sigma_0}$
30	40	0.66
40	50	1.50
50	60	1.42
60	70	2.48
70	80	0.38

Table 2 : Identification of the shift factors $\log b_{\sigma/\sigma_0}$

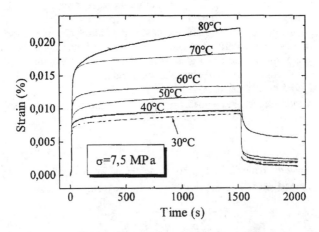

Figure 3 : Creep strain evolution versus time at different temperatures of a GF-PE

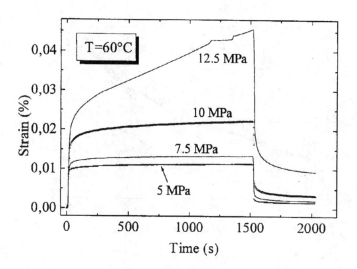

Figure 4 : Creep strain evolution versus time at different temperatures of a GF-PE

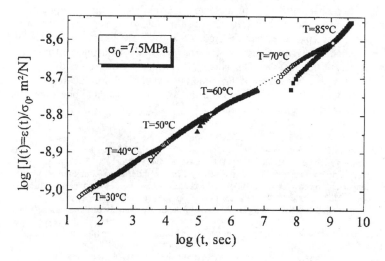

Figure 5 : Creep master curve σ =7,5 MPa

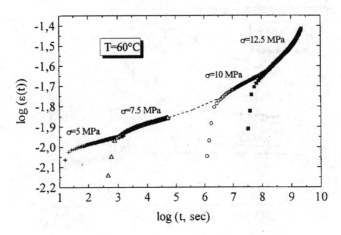

Figure 6 : Creep master curve T=60°C

IV – Discussion and conclusion

The use of the time equivalent method outside the good temperature interval can give an information on polymers and composites aging. However, it is necessary to complete these informations with a characterization of the damage evolution during the aging processes. Indeed, the variation of the shift factors between the steps of temperature or stress show that the use of the creep master curves (figure 5 and 6) are interesting as engineering construction method. They could not be associated directly to damage processes or to physical evolutions of the material. This is why, it is necessary to explore the evolution of the mechanical behaviour (evolution of the moduli and evolution of the strength) during the aging processes, to identify these evolutions in terms of microscopical phenomena and to modelize them in a macroscopical behaviour.

[1] - H. Simon ,Propriétés interfaciales dans les composites : énergie de surface et adhésion fibre de verre/matrice macromoléculaire; Thèse de 3ème cycle, Mulhouse 1984 (in french).

[2] - A.H. Cardon, Comportement à long terme des systèmes composites à composante polymèrique : objectifs, développement récents et orientations de recherche, Annales des composites, Durabilité des matériaux composites, JST AMAC 1996 (Moret sur Loing), pp. 5-13 (in french).

[3] – A.H. Cardon, H. Fuduka, K. Reifsnider, Progress in durability analysis of composite systems, E.A. Balkema Publ. 1996

[4] – A.H. Cardon, G. Verchery, Durability of polymer based composite systems for structural applications, Elsevier Applied Science Publ. 1991

[5] – P. Mele, A. Agbossou, N. Alberola, Relation entre vitesse de déformation, contrainte et endommagement d'un composite polypropylène chargé de fibres de verre, Actes des JNC8, Palaiseau, 1992, pp. 739-749

A model of formation and failure of the structure of short fibre composites

N.S.Sarkissyan, N.A.Prokopenko and S.T.Mileiko
Solid State Physics Institute of the Russian Academy of Sciences, Chernogolovka
Moscow district, 142432 RUSSIA

Systematic experimental observations of the failure behaviour of ceramic[1] and metal[2] matrix composites reinforced with short fibres randomly oriented in the matrix reveal both strength/fibre-volume-fraction and fracture-toughness/fibre-volume-fraction dependencies which cannot be certainly interpreted in terms of pure micromechanical models accounting normally for important microstructural features of one structural level. Perhaps a most important attribute of the composite behaviour is the occurrence of maximum on the dependencies mentioned at rather low values of the fibre volume fraction. This can be a result of the interaction of two processes happening on two structural scales, the first one is shortening the fibres down to the critical length which causes decreasing the composite strength, and the second one is homogenization of the structure at a many-fibres scale which yields increasing the strength.

In the present paper, an attempt to evaluate macroscopic (effective) failure and fracture characteristics of the composites by constructing two-level models is performed. The micro-level of the model relates to the fibre/matrix-interactions in a composite. The meso-level of the model describes both formation of a structure on the many-fibres agglomerate scale when obtaining a composite body and its fracturing on the stage of the loading of the body.

A model of the structure formation describes both homogenization of the structure in terms of a spatial distribution of the fibre volume fraction and fibre length shortening.

Analyzing the strength/fibre-volume-fraction dependence, $\sigma^*(v_f)$, we take into account changing the effective fibre strength with changing the fibre length and perform computer simulation of the failure process of a non-homogeneous rod loaded by tension assuming a simple stress redistribution among cells serviced after a condition of failure is fulfilled for some cells. In particular, this yields dependencies $\sigma^*(v_f)$ to be strongly influenced by mixing time in the ball mill. They can be either non-monotonic, or monotonic. The dependencies obtained can be a base for optimization a fabrication process of the composites.

The work was performed under financial support of Russian Foundation for Basic Research, Project # 96-01831, and International Science and Technology Center, Project # 507-97.

---

[1] Mileiko,S.T., Khvostunkov,A.A., Kiiko,V.M., and Gelachov,M.V., Graphite-fibre/carbide-matrix composites - III: Fracture behaviour of the composites with boron-carbide matrix, *Compos. Sci. and Technol,* in press.

[2] Mileiko,S.T., Gelachov,M.V., Khvostunkov,A.A., Kiiko,V.M., and Skvortsov,D.B., 'Short-fibre/titanium-matrix composites', in: *Proc. of ICCM-10,* Vancouver, August 1995, Eds. A.Poursartip and K.Street, Woodhead Publish. Ltd., Vol. 2, 131 - 138.

Nanocomposites in the Biomedical Field

G. Carotenuto, G. Spagnuolo[*], L. Nicolais

Department of Materials and Production Engineering, University of Naples. Piazzale Tecchio, 80 - 80125 Naples, Italy.

(*) Department of Dentistry, University of Naples, School of Medicine. Via S. Pansini, 5 - 80131 Napoli, Italy.

Abstract

The incorporation of nanosized inorganic particles into a polymeric matrix represents one of the most difficult problems in the fabrication of nanocomposites. The success in the manufacturing of such materials can be achieved only if the the particle aggregation is avoided, such a characteristic is essential when high trasparency is required. Here, a method is described for the preparation of a polymeric nanocomposite, containing a well-dispersed red pigment synthesized by adsorption of an organic dye onto nanosized silica particles. The resulting material was characterized by scanning electron microscopy (SEM) showing a homogeneous, low-defect microstructure.

1. Particulate composites: general aspects

The introduction of an inorganic phase in a polymeric material modifies the physico-mechanical properties of the polymer. This phenomenon is used in order to obtain materials with the desired characteristics. Generally, the principal reasons for using particles in a composite are: to increase the system rigidity, to increase the viscous flow, to change the permeability to gasses or liquids, to increase the resistance to abrasion, and to modify electrical and thermal properties.

In particular, the addition of a rigid filler causes an increase of the elastic modulus in proportion with the volume percentage of filler. If spherical particles are introduced in the polymeric matrix, the material behaviour is isotropic and the elastic properties can be easily predicted by using the Kerner's equation [1÷3]:

$$E_c = E_m \cdot \left(\frac{1 + A \cdot C \cdot V_f}{1 - C \cdot V_f} \right) \qquad (1)$$

where:

$$A = \frac{7 - 5 \cdot v_m}{8 - 10 \cdot v_m} \qquad\qquad C = \frac{E_f / E_m - 1}{E_f / E_m - A}$$

v_m is the Poisson's ratio of the matrix, V_f is the filler volume fraction, and E_f, E_m are the filler and matrix Young's moduli, respectively. Many other analytical equations are available for more complicate reinforcement shapes (for example, the Halpin-Tsai's equation that can be used for composites reinforced with short fiber [4]).

In addition to the elastic modulus, other tensile properties of the polymer are modified after filler introduction [5]. For example, the effective surface fracture energy is higher in a composite than in an unfilled polymer for three principal reasons: dispersed particles make the crack propagation path longer, absorb a portion of energy, and enhance the plastic deformation of the matrix. Consequently, the composite strength should increase with the volume percentage of filler, but this is not the case. When a deformation is applied to the composite the matrix can detach from the particle surface, and voids are produced inside the material. In addition, because the voids are produced at filler-matrix interface, the presence of a large quantity of filler decreases the distance between voids making easier the interaction of neighbour voids during the crack propagation.

The introduction of a filler modifies the mechanical properties of the polymer as a consequence of changes in its microstructure, and we have analysed the effect on the material elasticity and mechanical strength, in the case of finely dispersed fillers. However, many complicating factors must be considered [6]. For example, to process high-filled compositions, whose viscosity is accordingly high, one has to use elevated temperatures and shear stresses. These two factors promote degradation processes in the matrix. Such a variation of the polymer molecular weight distribution during the shaping process is frequently observed, and it must be taken into account in the comparison of filled and unfilled material characteristics.

In addition, high content of voids in filled systems is an inevitable side effect of their fabrication process. Since voids initiate cracks under deformation, they reduce the composite strength. The void content in a composite can result also very high in the case of poor filler wettability, and for the presence or release of water from the particle surfaces.

A special attention has to be given to adhesion at interface. Three different polymer-filler interactions are possible: (i) simple filler-nonpolar polymer mixture, (ii) wetting of the filler surface by the polymer with good physical contact between the two phases, and (iii) chemical bond across the interface. A strength reduction results in the first case, instead, a significative improvement of the strenght value characterizes the second one. The last case is the most convenient because it is the ideal situation, characterised by the highest strength improvement.

In the case of "good" adhesion at interface the maximum stress that can be transmitted from the matrix to the reinforcement phase is equal, for a plastic matrix, to its shear yield point, and for a brittle matrix to its shear strength. In the case of "poor" adhesion the maximum stress transmissible from the matrix to the reinforcement is smaller than the adhesion strength. From these considerations, it follows that in complete absence of adhesion even a very small stress applied to the matrix can cause the detachment of the matrix from the reinforcement surface with void formation. No stresses at all are transmitted to the reinforcement in this case. Actually, the mechanism of stress transfer at the interface in composites is more complex, and one must distinguish between the normal and shear stresses present at interface. Under tension, the normal stress is applied to each particle. The points at which such stress is at a maximum are the most vulnerable since the separation is most likely there. Voids are formed which destroy the integrity of the composite and, as they grow and interact with each other, initiate cracks and promote the material failure. Obviously, higher is the resistance to the interfacial separation, higher is the stress that can be applied to the specimen. In particular, void size, void content, and stress-causing-dewetting are the factors determining the composite resistance. The void size is determined by the size of the filler particles. Smaller is the particles size, lower results the dimension of voids produced for dewetting. But, one may expect a considerable statistical scattering of the mechanical behaviour since a single void, when critically large, may initiate the crack. The void content (total volume of voids) is determined by the volume fraction of filler in the composite. The stress required to cause dewetting depends on the adhesion strength between filler surface and matrix. At higher filling ratios, the voids formed near individual particles begin to interact with each other, consequently, it becomes likely that the voids will merge and cause a rapid specimen failure, as soon as the matrix begins to separate. If a very good adhesion between matrix and filler is assured (i.e., the adhesion strength is higher than the matrix adhesion force), the stress value required to break the specimen is controlled by the stress concentration present in the vicinity of the filler particles.

2. Nanocomposites

The particulate composites can be classified on the basis of the filler size. Conventional particle-dispersed type composites are usually processed by mechanical mixing of each component. So the controllable smallest limit of microstructure is in the order of 10^{-6} m. These composite materials are usually called macrocomposites when the filler dimension is higher than 10^{-6}m, and microcomposites when the filler size is in the order of some microns. Instead, the term "nanocomposite" is used for composite whose microstructure has inhomogeneity in the scale range between 10^{-7} m (submicron) and 10^{-9} m (nanometer) order [7].

Nanocomposites are a new class of particulate composites which exhibit ultra fine phase dimensions. Experimental works conduced on these materials show as all types and classes of nanocomposites lead to new and improved properties when compared with their macro-counterparts [8÷10], and the actual research results indicate a tremendous promise for this new class of engineering materials.

In particular, the nanophase composites exhibit characteristics rather different relative to those of conventional composite, because in these materials microstructure and microchemistry of interface control the properties. A nanophase material with 5nm average reinforcement particle size has about 60% of its atoms associated with interface. This percentage falls only to about 40% for a 10nm particle size, and is as low as 10% for a 100nm particle size. Since such a large fraction of their atoms reside in the interface, the interface structures play a significant role in determining the mechanical properties of nanocomposite.

Composites between polymers and inorganic particles in the micrometer range are often opaque. Light scattering, responsible for the opacity, can be suppressed either by using materials with nearly matching refractive indices or by decreasing the filler's dimensions to a range below ca. 50nm. Therefore, nanocomposites of polymers and inorganic colloids can act as optically homogeneous materials with modified optical properties [11÷16].

3. Nanocomposite preparation

Nano-scale polymer-base composites cannot be processed by the mechanical mixing used for conventional composite materials. To make the most uniform distribution of particles in the matrix, the particle surface must be organically modified. In these conditions the polymer coat the particles in the starting precursor solution, and large interface extension, improved interfacial bond, and, consequently, high mechanical strength are achieved in the resulting material. Such a processing avoid the presence of residual stresses, porosity, matrix degradation, agglomeration and other problems involved in the conventional composite fabrication. In this work we have developed a method for the fabrication of thermoplastic matrix composites containing some-micron or submicron-scale colloidal particles (See Figure

1), and we have used this method to prepare monolithic specimens of poly(methyl methacrylate) - monodisperse silica particles.

In particular, the composite material was obtained as follows. Dispersion of uniform spherical silica particles in water were prepared by the Ströber-process [17]. The particle dispersions in water were transferred to ethanol and then to methoxy propylacetate (PMA, Aldrich). The particle suspensions were mixed to the polymer solution (50wt.% of polymer in PMA) and then sonicated for 5min. The samples were centrifuged to remove bubbles, and then dispersed into rectangular poly(tetrafluoroethylene) molds. All samples were heated at 100°C for 15 min (in vacuum). The particle surface modification was obtained by treatment at room temperature of the silica particles with an organic colorant (Methyl Red), as described earlier [18].

The nano-size particles can be synthesized using precipitation process in homogeneous solutions. Such a synthetic methods is controlled to yield uniform particles of different chemical composition, structure, shape, and size. However, when used to make composite materials, the colloidal particles cannot be dried before the introduction in an hydrophobic polymeric matrix, because for the high value of surface free energy an agglomeration process would follow. Therefore, a "wet" preparation way becomes required.

The particle surface is hydrophilic and, to make a good dispersion with the hydrophobic PMMA-PMA system, it is required its modification by a coupling agent. If such reaction is performed in aqueous media the water molecules contained on the particle surface must be accurately removed. Here, it was possible to eliminate totally the water by washing the dyed particles with solvents of growing apolarity. Initially, a water-mixable solvent (ethanol) was used, and then it was removed by washing with an ethanol-mixable organic solvent. In the present case, the organic solvent was exactly the polymer solvent (PMA).

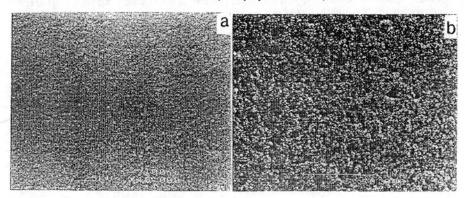

Figure 1 - Monodisperse silica particle microstructure before (a) and after (b) the compatibilisation treatment.

Such operations are required also because, to produce a material transparent and with good mechanical properties, the polymeric matrix must contact the full particle surface, and the interfacial bond must be increased as much as possible. If the unmodified inorganic surface of the particles contact the polymeric solution the only possible interactions at interface are by hydrogen bonds between the methyl-ester groups of the polymer and the hydroxyl groups contained on the particle. These interactions are not strong enought. But, if the surface of the particles is organically modified, operation that was here realized by reaction with a dye (a water-soluble organic molecule having both a hydrophylic and a hydrophobic side), the surface of each particle results hydrophobic and can interact with the large quantity of not polar segments contained in the polymeric molecules. This situation produces a remarkable improvement in the quality of dispersion. As shown in Figure 2, when a dispersion of organically modified particles is mixed with the polymer solution, the polymer wraps from the particles since the interaction with the modified surface is stronger than with PMA.

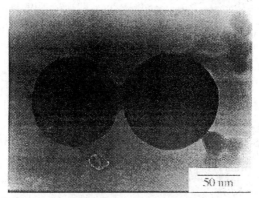

Figure 2 - TEM-micrograph showing the presence of a thin layer of polymer around the particles contained in the starting liquid precursor of organically modified particles.

Consequently, in addition to improved mechanical properties, an easy solvent removal and, for the low viscosity of the suspension, simple casting operations result. The T_g of the nanocomposite matrix (50wt.%SiO$_2$ particles) was about 96°C, i.e. 5°C up to the T_g value characteristic of the pure PMMA (91°C). Obviously, such a behavior can be explained on the bases of the interactions between the modified SiO$_2$ surface and the PMMA macromolecular segments.

4. Applications in the biomedical field

Nanocomposites have properties that make them really interesting as materials for the biomedical field. In particular, nanocomposites can find useful applications in the ophthalmic field. Ophthalmic lenses are lenses intended for use in spectacles. They consist of a thin-walled, shallow, spherical shell made of a rigid polymer. Ophthalmic lenses require a number of properties such as scratch resistance, rigidity and impact resistance, in addition to their optical characteristics. The improvement of these characteristics and particularly of the scratch resistance and rigidity, without a reduction of transparency and optical purity, can be easily performed by using a nanometer-in-size filler. Therefore, the PMMA-SiO$_2$ nanocomposites represent a new and inexpensive ophthalmic material.

References

[1] L. Nicolais, M. Narkis, E. Joseph. THE ELASTIC MODULUS OF PARTICULATE FILLED POLYMERS. J. Appl. Polym. Sci. 22(1978)2391-2394

[2] *L. Nicolais, M. Narkis,* STRESS-STRAIN BEHAVIOUR OF SAN/GLASS BEAD COMPOSITES ABOVE THE GLASS TRANSITION TEMPERATURE, J. Appl. Polym. Sci. 15(1971)469-476.

[3] *L. Nicolais, M. Narkis,* STRESS-STRAIN BEHAVIOUR OF SAN/GLASS BEAD COMPOSITES IN THE GLASSY REGION, Polym. Eng. and Sci. 11(1971)194-199.

[4] *J.M. Kenny, L. Nicolais,* Science and technology of polymer composites in: Comprehensive Polymer Science, Pergamon Press, pgg. 472-525, (1991).

[5] *L. Nicolais,* Composites - Materials and Process Engineering, CEI-EUROPE Elsevier, Courses in Advanced Technology, Castelvecchio (Pisa), Italy, October 15-19, 1990.

[6] *A.A. Berlin, S.A. Volfson, N.S. Enikolopian, S.S. Negmatov,* Principles of Polymer Composites, Springer-Verlag, Berling Heidelberg, 1986.

[7] *N. Koshizaki, K. Yasumoto, K. Suga,* FUNCTIONALITIES OF NANO-SCALE COMPOSITES, Materiaux & Technique, 4(1994)7-11.

[8] *H.B. Sunkara, J.M. Jethmalani, W.T. Ford,* COMPOSITE OF COLLOIDAL CRYSTALS OF SILICA IN POLY(METHYL METHACRYLATE), Chem. Mater. 6(1994)362-364.

[9] *J.P. Lemmon, M.M. Lerner,* PREPARATION AND CHARACTERIZATION OF NANOCOMPOSITES OF POLYETHERS AND MOLYBDENUM DISULFIDE, Chem. Mater. 6(1994)207-210.

[10] *G. Sergeev, V. Zagorsky, M. Petrukhina,* NANOSIZE METAL PARTICLES IN POLY(P-XYLYLENE) FILMS OBTAINED BY LOW-TEMPERATURE CODEPOSITION, J. Mater. Chem. 5(1995)31-34.

[11] *L. Zimmerman, M. Weibel, W. Caseri,* U.W. Suter, HIGH REFRACTIVE INDEX FILMS OF POLYMER NANOCOMPOSITES, J. Mater. Res. 8(1993)1742-1748.

[12] *L. Nicolais, G. Carotenuto, X. Kuang,* SYNTHESIS AND CHARACTERIZATION OF NEW POLIMER-CERAMIC NANOPHASE COMPOSITE MATERIALS, App. Comp. Mater. 3(1996)103-116.

[13] *G. Carotenuto, L. Nicolais,* CORRELATION BETWEEN FRAGILITY AND MICROSTRUCTURE IN ORGANIC-INORGANIC NANOCOMPOSITES, Sci. and Eng. of Comp. Mater. 5(1996)57-61.

[14] *G. Carotenuto, L. Nicolais,* Preparation and characterization of nanocomposite films, Adv. Comp. Lett., 5(1996)87-90

[15] *G. Carotenuto, Y.S. Her, E. Matijevic',* PREPARATION AND CHARACTERIZATION OF NANOCOMPOSITE THIN FILMS FOR OPTICAL DEVICES, Ind. & Eng. Chem. Res., 35(1996)2929-2932.

[16] *G. Carotenuto, L. Nicolais, X. Kuang, Z. Zhu,* A METHOD FOR THE PREPARATION OF PMMA-SIO$_2$ NANOCOMPOSITES WITH HIGH HOMOGENEITY, Appl. Comp. Mater., 2(1995)385-393.

[17] *H.Giesche, E. Matijevic,* J. Mater. Res., 9(1994)436.

[18] *W.P. Hsu, Y.Rongchi, E. Matijevic,* Dye and Pigments, 19(1992)179.

Advanced composites. New generation of high performance pre-pregs.

P. Grati, M. Parente, E. Vercelli

Seal S.p.A. - Via Quasimodo 33, 20025 Legnano (MI) - Italy

Introduction

In the last years, advanced composites have revealed big developments from the point of view of mechanical properties, behavior to fire and aging resistance.
These points have convinced the designers to look with more interest at these products and to exploit completely all the potentialities of composites, especially related to the particular properties of the carbon and aramid fibers and of the resin systems.

Abstract

In front of these new market demands, Seal SpA has developed a range of pre-pregs having specific properties, using carbon fibers with high modulus and very high modulus and renewing the resin systems.
Following this work, several kind of matrix, based on epoxy resin system, have been generated to be used for the impregnation of carbon, aramid, glass and hybrid fiber fabrics or carbon yarns with tensile modulus up to 800 GPa.
The matrix that Seal SpA has developed are:
- ET441
- EF431
- ES252

The matrix called ET441 has been studied for manufacturing visible parts which need an optimum finish. It has also an high Tg.
This resin system is already well-known in the market, so it won't be subject of the present work.
The other two matrix are of very late study and will be analyzed from the point of view of their properties and applications.

1. EF431 resin system

This system has been studied for the production of fire resistant pre-pregs with low emission of smokes and with cure temperature of 125°C.
It can be used for manufacturing internal and external parts of coaches and trains and has the following properties (evaluated on pre-preg based on a glass-fiber fabric of 380 g/sqm):
- reaction to fire, according to the standard NF P 92-501 = **M1**
- smokes density, according to the standard NF X 10-702: **VOF4 = 21**

• toxicity of the smokes, according to the standard NF X 70-100:

Tipo di prodotto		resina epossidica		
Modello		TEXIPREG TR 380 EF 431		
Colore		rosso		
GAS	C.C. (mg/m³)		t, (mg/m³)	t,/C.C
CO2	90.000		251,9	0,0028
CO	1.750		49,2	0,0281
HCL	150		0,015	0,0001
HBr	170		0,033	0,0002
HF	17		0,008	0,0005
HCN	55		2,209	0,0402
SO2	260		0,026	0,0001

Table 1: Analysis of smokes emission of ES252

2. ES252 resin system

This matrix is based on a epoxy resin suitable for carbon-tools construction.
This resin system is used for the impregnation of carbon-fiber fabrics of 200 and 700 g/sqm (CC206 and CC700), which are the most suited and economically valid for the construction of a carbon-resin tool.

2.1 Cure cycle of the tool
Here below is summarized an indicative cure cycle:

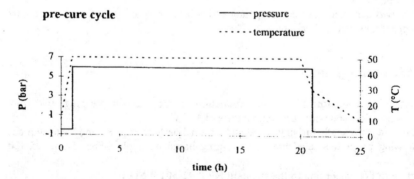

Fig. 1: Pre-cure cycle for ES252

Post-cure cycle:
- place the tool in a oven
- heat the tool from 20°C up to 180°C in 16 hours
- hold temperature for 6 hours
- cool to 150°C at a rate of 0,5°C/min
- cool to room temperature at a rate of 2-2.4°C/min

The tools so produced are ready to work in autoclave at a process temperature of 120-140°C for a considerable number of produced pieces.

The effect of matrix on the fatigue behaviour of short glass fibre reinforced PEK

G. Caprino and A. D'Amore

Department of Materials and Production Engineering - University of Naples "Federico II" - Piazzale Tecchio, 80 - 80125 Naples - Italy

Abstract

The classical S-N curve of polyphenylene ether ketone (PEK) was obtained by tension-tension fatigue tests. The results were treated in the light of a previous two-parameter model, proposed for composite materials, and very good agreement was found between theoretical predictions and experimental data. The two parameters appearing in the model were used to predict the S-N curve in the case of flexural fatigue of short glass fibre reinforced PEK. In this case, the matrix was slightly modified by adding 15% polyphenylene sulfide (PPS). An excellent correlation was verified between theory and experimental tests, performed in four-point bending, where the failure was precipitated on the tensile side. This suggests that the composite response in fatigue is mainly governed by matrix, whose behaviour seems to be negligibly affected by the presence of PPS.

1. Introduction

In the last years, a variety of random Glass Fibre Reinforced Thermoplastics (GFRT) has appeared on the market. Despite their relatively low mechanical properties, these materials offer the advantages of easy storage and processability, together with high production rates and recycling perspectives. Therefore, they can be considered as ideal candidates when components destined to support moderate loadings are concerned.

One of the most intricate tasks in designing with GFRT is the definition of reliable allowable strengths. This is partly due to the usually large scatter in mechanical properties, strongly affected by the fabrication methods utilised. However, another drawback comes from the difficulty to predict the long-term behaviour of the material after exposure to adverse conditions, e.g. aggressive environments or cyclic loadings.

Looking at the fatigue response, it has been long recognised [1, 2] that many damage types, i.e. matrix cracking or yielding, fibre/matrix debonding, fibre fracture can happen within the composite volume as a consequence of cycle evolution. When short fibres are used as a reinforcement, the role of matrix and interface in determining final failure seems to be predominant, in that the process of nucleation, propagation and coalescence of cracks rarely involves fibre fracture. Therefore, it can be supposed that the composite behaviour in fatigue is strictly correlated to the resin response, as also suggested from limited data published in the literature [3].

Due to the great number of matrices and fibre forms available nowadays, together with the intrinsic time-consuming nature of fatigue tests, models and procedures allowing for a quick fatigue characterisation of GFRT should be appreciated. Of course, this task could be more easily accomplished if: a) a relationship could be found between the component behaviour and the composite behaviour; b) a suitable fatigue model, possibly accounting for the features of fatigue cycles (frequency, stress ratio, etc.) could be assessed.

In [4], a two-parameter model was proposed, aimed at the prediction of the classical S-N curve of a composite material. The model, statistically developed in [5] and taking into account the effect of the stress ratio, was assessed for various types of continuous random glass fibre reinforced plastics subjected to flexural fatigue [4-6]. In this work, the same model is applied to a Polyphenylene Ether Ketone (PEK) resin, loaded in tension-tension fatigue using two different stress ratios. It is shown that also in this case the S-N curve is correctly predicted by theory. The parameters appearing in the model, calculated from the experimental results concerning the resin, are utilised to predict the flexural fatigue behaviour of a short glass fibre reinforced PEK, blended with 15% polyphenylene sulfide (PPS). From the tests, carried out in four-point loading, a very good correlation is found between theory and experiments. This suggests that the resin governs the fatigue behaviour of the composite, provided consistent failure modes are observed in tension and in flexure. In addition, the results presented indicate that the addition of PPS to PEK does not sensibly affect the fatigue performance of the material.

2. Background

In [4], the following closed-form formula was proposed to predict the fatigue life of Fibre Reinforced Plastics (FRPs) containing randomly oriented fibres:

$$N = \left[1 + \frac{1}{\alpha \cdot (1-R)} \cdot \left(\frac{\sigma_o}{\sigma_{max}} - 1 \right) \right]^{1/\beta} \tag{1}$$

In eq. (1), N is the critical number of cycles to failure, σ_{max} the maximum applied stress, R the stress ratio (i.e. the ratio of minimum to maximum stress), σ_0 the virgin material strength. In [4], the parameters α and β were supposed to be constant for a given material, but dependent on the load scheme adopted.

Eq. (1) was derived from the hypotheses that the material strength during cycle evolution suffers a progressive decay according to a power law, and that final failure is precipitated when the residual strength matches σ_{max}.

In order to assess the model, eq. (1) was rearranged in the form:

$$\left(\frac{\sigma_o}{\sigma_{max}} - 1 \right) \cdot \frac{1}{1-R} = \alpha \cdot \left(N^\beta - 1 \right) \tag{2}$$

from which all the fatigue data, irrespective of R, should group in a single curve when the quantity on the left side (indicated by the symbol K hereafter) is plotted against N.

In [4], eq. (2) was also usefully employed to calculate the constants α and β. In fact, reporting K against (N^β-1) should result in a straight line of slope α, passing through the origin. Therefore, different attempt values were adopted for β, until the best fit straight line fitting the experimental data actually passed through the origin.

In [5], the fatigue model in eq. (1) was statistically developed. It was postulated that the scatter in fatigue data strictly depends on the scatter in monotonic strength, so that a lower ultimate strength results in lower fatigue life, for given fatigue test conditions. A two-parameter Weibull distribution was assumed for σ_0. In order to assess the statistical model, in [5, 6] Eq. (1) was solved for σ_0, obtaining:

$$\sigma_{o=}\sigma_{oN} = \sigma_{max} \cdot \left[1 + \alpha \cdot (1-R) \cdot (N^\beta - 1) \right] \qquad (3)$$

By eq. (3), the virgin material strength (indicated by the symbol σ_{oN} to distinguish it from that, σ_0, directly measured in a monotonic characterisation test) was calculated from the fatigue data. Comparing the Weibull distributions of σ_0 and σ_{oN}, they were found to be in excellent agreement, supporting the hypotheses done.

3. Materials and experimental methods

The matrix examined in this work was PEK-C provided by Xuzhou Engineering Plastics Co., China. The specimens for tensile tests, 4.6 mm in thickness and 10 mm in width, were dumbbell-shaped according to ASTM D638 Type I, directly obtained by injection moulding.

The previous resin was mixed with 15% PPS, then charged with short glass fibres and injected moulded to fabricate rectangular specimens 4.25 mm in thickness, 16.35 mm in width and 80 mm in length. These samples were subjected to four-point loading adopting a support span to load span ratio 3 and a support span to specimen thickness ratio 16. The nominal fibre content by weight in the composite was 25%.

All the tests were carried out up to failure on an Instron 8501 servo-hydraulic machine. The monotonic tests were performed at a crosshead speed v=100 mm/min, to provide a load rate comparable with the fatigue tests. In fatigue, a sinusoidal waveform was adopted, with frequencies in the range 0.8 - 2 Hz. Two different stress ratios, namely R = 0.1 and R = 0.3, were used in the tensile tests, whereas R = 0.3 was set in bending.

4. Results and discussion

After final failure, the samples were visually examined, to assess failure modes. Apparently a single crack, fastly propagating across the entire specimen width, determined the collapse in tension tests on the resin, irrespective of the type of loading (monotonic or fatigue). Similarly, the failure in flexure tests on the composite was initiated by a distinct crack perpendicular to the specimen length, nucleating at the tension side and quickly propagating along the thickness direction. No visible failure phenomena were observed on the compression side of the beam. Therefore, from the previous failure mechanisms it can be concluded that the composite collapse in bending is governed by its tension, rather than by its compression behaviour.

In Fig. 1, the maximum applied stress, σ_{max}, is reported on a semi-log scale against the fatigue lifetime, N, for pure PEK. From the figure, the virgin tensile material strength, measured from three monotonic tests, is 90.4 N/mm^2.

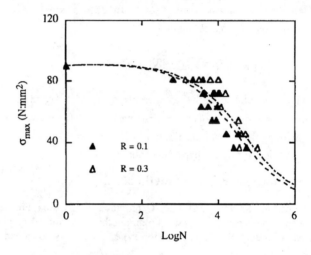

Fig. 1 - Tensile fatigue curves for PEK. Dashed lines: predictions based on eq. (1).

It is evident from Fig. 1 that, for a given σ_{max}, a higher stress ratio results in a shift of the experimental points through higher lifetimes. This finding qualitatively agrees with both the results presented in [4-6] for composites, and eq. (1). Interestingly, the fatigue curves are quite flat up to about 10^3 cycles, whereas a more and more evident strength decay is recorded beyond this limit.

As stated above, in previous papers [4-6] the effectiveness of eq. (1) was verified for random GFRPs subjected to flexure. However, its validity was never demonstrated in the case of tension fatigue of pure resin: this was the first scope of the present workAt this aim, all the tensile fatigue data were utilised to calculate the quantity K (term on the left of eq. (2)), which was plotted against N on a semi-log scale (Fig. 2). As predicted from eq. (2), the experimental points concerning R = 0.1 and R = 0.3 actually converge to a single master curve, demonstrating the efficiency of eq. (2) in modelling the effect of stress ratio.

To ascertain the validity of the model in describing the classical S-N curve of the material, it is necessary to demonstrate that, as results from eq. (2), a suitable value of the constant β can be found, such that the K values are well fitted by a straight line passing through the origin, when reported against (N$^\beta$-1). The results of such analysis are shown in Fig. 3.

From the figure, it is seen that the experimental data, although affected by a quite high scatter, actually follow a linear trend. The continuous straight line in the figure was drawn assuming $\beta = 0.625$, and represents the best fit to the data. The slope of the straight line provided the value $\alpha = 1.71 \times 10^{-3}$.

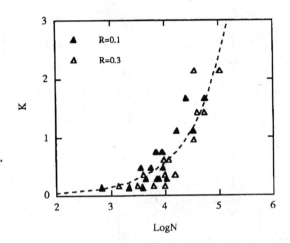

Fig. 2 - Semi-log plot of the term on the left side of eq. (2), K, against the critical number of cycles, N. Dashed line: theoretical prediction. Material: PEK.

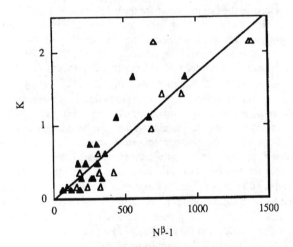

Fig. 3 - Diagram for the evaluation of the material constants α, β appearing in eq. (2). Material: PEK.

Substituting in eqs. (1) and (2) the α, β values found, the dashed lines in Figs. 1 and 2 were drawn, respectively: a favourable correlation can be appreciated between theory and experiments, demonstrating the effectiveness of the theoretical model not only in the case of composites, but also for resin alone.

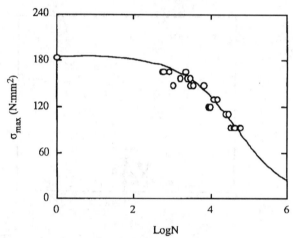

Fig. 4 - Flexural fatigue curves for glass fibre reinforced PEK. Continuous line: prediction based on eq. (1).

As discussed previously, the failure modes observed in the composite specimens in flexure were consistent with those resulting from the unreinforced resin loaded in tension, developing on the beam tension side. Therefore, the question arises whether a correlation exists between the fatigue curves of the materials under study. To verify this, the S-N curve of the composite in bending was calculated using in eq. (1) the flexural strength resulting from monotonic tests, together with the α and β values calculated from the tension fatigue data concerning the matrix. The prediction is represented by the full line in Fig. 4, from which an excellent agreement with the experimental points (symbols in the figure) can be appreciated. It can be concluded that the fatigue behaviour of the composite can be simply calculated, if the matrix response is known. This suggests that, for matrix dominated materials, the fibre effect on the fatigue sensitivity is negligible. Of course, the same does not hold for absolute strength. From Fig. 4, the virgin flexural strength of the GFRT considered is 183.7 N/mm^2; since the ratio of the flexural to the tensile strength of a composite is generally in the range 1.3-1.5 [7, 8], a value of 120-140 N/mm^2 is foreseen for the tensile strength. Therefore, the presence of short glass fibres contributes to increase absolute strength. Singularly, also the addition of PPS in the matrix seems to be ineffective in modifying the fatigue curve of the composite.

The limited number of monotonic tests carried out within the present work did not allow for a statistical analysis of the monotonic experimental data. Therefore, in order to gain some valuable information on the statistical distribution of virgin strength, eq. (3) was applied only to the data provided from tensile and bending fatigue tests.

In Fig. 5, the probability of failure of the calculated monotonic strength both in tension (unreinforced resin) and in flexure (composite) is reported. Referring to the tensile data, different symbols have been used to distinguish the points pertaining to the two different stress ratios adopted. The black and open triangles in Fig. 5 are quite uniformly interspersed: this suggests their belonging to the same population, and indirectly supports the validity of the fatigue model proposed.

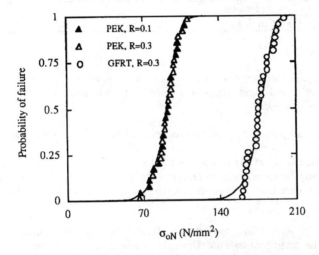

Fig. 5 - Probability of failure of the calculated monotonic strengths.

Each data set in Fig. 5 was fitted by a two-parameter Weibull curve, and the values of the parameters were calculated according to the best fit method (Table I). From them, the solid lines in the figure were drawn. It is seen that the probability of failure of the PEK tensile data closely follows a Weibull distribution. Also in the case of flexure of GFRT, the agreement is favourable.

From Table I, the δ value pertaining to the composite is about double than the one resulting from the tensile strength of the unreinforced resin, indicating that the scatter affecting the virgin strength is far higher for PEK than for GFRT. It must be recognised that a direct comparison of the δ values shown in Table I could be misleading: in fact, it has long been noticed by other researchers [7, 8] that the shape parameter is strongly affected by the type of loading, being lower in tension than in flexure for the same material. However, interestingly the shape parameters resulting in [7] from tensile tests were in the range 13 to 20, whereas flexure tests on the same composites yielded $\delta = 23 - 43$. Therefore, in that case also the shape parameter pertaining to flexure was about double than the corresponding values found in tension. If the same would happen for the present GFRT, it could be concluded that not only its fatigue sensitivity, but also its scatter in virgin tensile strength is correlated to the matrix behaviour.

Table I - Characteristic strength, γ, and shape parameter, δ, of the Weibull distribution of the calculated monotonic strength.

Material	γ (N/mm^2)	δ
PEK in tension	96.6	9.68
GFRT in flexure	182.3	18.87

5. Conclusions

From the data presented and discussed in this work, concerning fatigue tests carried out on unreinforced PEK and short glass fibre reinforced PEK blenden with 15% PPS, the conclusions are as follows:

- the fatigue model previously presented for composites can be reliably applied also to the resin alone;
- the fatigue sensitivity of the composite considered in this work strictly depends on the matrix sensitivity, being negligibly affected by the presence of both the fibres and PPS;
- perhaps, also the scatter in virgin tensile strength of the composite is correlated to the matrix behaviour.

It must noted that, at this time, the latter conclusion does not rely on a direct experimental evidence, so that additional tests are required to assess it on a sound basis.

References

[1] *S. S. Wang and E. S.-M. Chim*, J. Compos. Mater., Vol. 17, 1983.
[2] *J. F. Mandell*, Fatigue of Composite Materials, K. L. Reifsnider Ed., Elsevier Publ., Amsterdam, 1991.
[3] *J. F. Mandell, F. J. McGarry, D. D. Huang and C. G. Li*, Polym. Compos., Vol. 4, 1983.
[4] *A. D'Amore, G. Caprino, P.,Stupak, J. Zhou and L. Nicolais*, Sci. and Eng. of Compos. Mater., Vol. 5, No. 1, 1996.
[5] *G. Caprino and A. D'Amore*, in press on Compos. Sci. Technol., 1998.
[6] *G. Caprino, A. D'Amore, and F. Facciolo*, in press on J. Compos. Mater., 1998.
[7] *J. M. Whitney and M. Knight*, Exptl; Mech., Vol. 20, 1980.
[8] *M. R. Wisnom*, Composites, Vol. 22, No, 1, 1991.

ECCM-8

EUROPEAN CONFERENCE ON COMPOSITE MATERIALS
SCIENCE, TECHNOLOGIES AND APPLICATIONS

AUTHOR'S INDEX

Finito di stampare
nel mese di maggio 1998
nella Tipolitografia
R. ESPOSITO
Via Diocleziano, 154 - Napoli - Italia

Woodhead Publishing Ltd
Abington Hall
Abington
Cambridge CB1 6AH   ISBN 1 85573 410 9 (Vol 4)
England   ISBN 1 85573 377 3 (Four volume set)

WOODHEAD PUBLISHING LIMITED